POLARITONS IN PERIODIC

AND

QUASIPERIODIC STRUCTURES

POLARITONS IN PERIODIC

AND

QUASIPERIODIC STRUCTURES

Eudenilson L. Albuquerque
Departamento de Física, Universidade Federal do Rio, Grande do Norte
Natal-RN, 59078-970, Brazil

Michael G. Cottam
Department of Physics and Astronomy, University of Western Ontario
London, Ontario N6A 3K7, Canada

ELSEVIER B.V.
Sara Burgerhartstraat 25
P.O. Box 211, 1000
AE Amsterdam
The Netherlands

ELSEVIER Inc.
525 B Street,
Suite 1900 San Diego,
CA 92101-4495
USA

ELSEVIER Ltd
The Boulevard,
Langford Lane
Kidlington, Oxford
OX5 1GB UK

ELSEVIER Ltd
84 Theobalds Road
London WC1X 8RR
UK

© 2004 Elsevier B.V. All rights reserved.

This work is protected under copyright by Elsevier B.V., and the following terms and conditions apply to its use:

Photocopying
Single photocopies of single chapters may be made for personal use as allowed by national copyright laws. Permission of the Publisher and payment of a fee is required for all other photocopying, including multiple or systematic copying, copying for advertising or promotional purposes, resale, and all forms of document delivery. Special rates are available for educational institutions that wish to make photocopies for non-profit educational classroom use.

Permissions may be sought directly from Elsevier's Rights Department in Oxford, UK: phone: (+44) 1865 843830, fax: (+44) 1865 853333, e-mail: permissions@elsevier.com. Requests may also be completed on-line via the Elsevier homepage (http://www.elsevier.com/locate/permissions).

In the USA, users may clear permissions and make payments through the Copyright Clearance Center, Inc., 222 Rosewood Drive, Danvers, MA 01923, USA; phone: (+1) (978) 7508400, fax: (+1) (978) 7504744, and in the UK through the Copyright Licensing Agency Rapid Clearance Service (CLARCS), 90 Tottenham Court Road, London W1P 0LP, UK; phone: (+44) 20 7631 5555; fax: (+44) 20 7631 5500. Other countries may have a local reprographic rights agency for payments.

Derivative Works
Tables of contents may be reproduced for internal circulation, but permission of the Publisher is required for external resale or distribution of such material. Permission of the Publisher is required for all other derivative works, including compilations and translations.

Electronic Storage or Usage
Permission of the Publisher is required to store or use electronically any material contained in this work, including any chapter or part of a chapter.

Except as outlined above, no part of this work may be reproduced, stored in a retrieval system or transmitted in any form or by any means, electronic, mechanical, photocopying, recording or otherwise, without prior written permission of the Publisher.
Address permissions requests to: Elsevier's Rights Department, at the fax and e-mail addresses noted above.

Notice
No responsibility is assumed by the Publisher for any injury and/or damage to persons or property as a matter of products liability, negligence or otherwise, or from any use or operation of any methods, products, instructions or ideas contained in the material herein. Because of rapid advances in the medical sciences, in particular, independent verification of diagnoses and drug dosages should be made.

1st edition 2004

Library of Congress Cataloging in Publication Data
A catalog record is available from the Library of Congress.

British Library Cataloguing in Publication Data
A catalogue record is available from the British Library.

ISBN: 0 444 51627 1

∞ The paper used in this publication meets the requirements of ANSI/NISO Z39.48-1992 (Permanence of Paper).
Printed in The Netherlands.

To

Juliana M. L. Albuquerque

In Memoriam

Contents

Preface   xi

1 Basic Properties of Excitations in Solids   1
    1.1 Symmetry and Crystal Lattices 1
    1.2 Reciprocal Lattices and Brillouin Zones 5
    1.3 Bulk, Surface, and Superlattice Excitations 7
    1.4 Phonon: Quantum of the Lattice Vibrations 9
    1.5 Plasmon: Quantum of the Plasma Oscillations 12
    1.6 Exciton: Bound Electron–Hole Pair 14
    1.7 Magnon: Quantum of the Spin Wave 18
    References 22

2 Periodic and Quasiperiodic Structures   25
    2.1 Periodic Structures 26
    2.2 Quasiperiodic Structures 30
    2.3 Examples of Quasiperiodic Structures 32
       2.3.1 Cantor 33
       2.3.2 Fibonacci 34
       2.3.3 Thue–Morse 36
       2.3.4 Double Period 37
    References 38

3 Bulk Polaritons   41
    3.1 The Frequency-Dependent Dielectric Function 42
    3.2 Bulk Plasmon- and Phonon-Polaritons 45
    3.3 Bulk Exciton-Polaritons 48
    3.4 Magnetic Susceptibility 54
    3.5 Bulk Magnetic-Polaritons 58
    References 62

4 Surface Plasmon- and Phonon-Polaritons   65
    4.1 Single-Interface Modes: Isotropic Media 65
    4.2 Single-Interface Modes: Anisotropic Media 71
    4.3 Charge-Sheet Modes 72

	4.4	Thin Films	73
	4.5	Experimental Studies	82
	References	86

5 Plasmon-Polaritons in Periodic Structures — 89

5.1	Two-Component Superlattices	90
	5.1.1	Infinite Superlattices	90
	5.1.2	Semi-Infinite Superlattices	94
	5.1.3	Finite Superlattices	97
5.2	Superlattices with Charge Sheets	101
5.3	Doped Semiconductor Superlattices	105
5.4	Piezoelectric Superlattices	107
	5.4.1	Piezoelectric Layer	110
	5.4.2	Superlattice Structure	112
5.5	Magnetoplasmon-Polaritons in Finite and Infinite Superlattices	115
References	122

6 Plasmon-Polaritons in Quasiperiodic Structures — 125

6.1	Two-Component Quasiperiodic Structures	125
	6.1.1	Numerical Examples	128
6.2	Localization and Scaling Properties	131
6.3	Multifractal Analysis	134
6.4	Quasiperiodic $nipi$ Structures	139
6.5	Thermodynamic Properties	144
	6.5.1	Theoretical Model	145
	6.5.2	Specific Heat Profiles	147
References	155

7 Magnetic Polaritons — 157

7.1	Exchange Spin Waves in Thin Films	159
7.2	Magnetostatic Modes in Thin Films	161
	7.2.1	Magnetization Parallel to the Film Surfaces	163
	7.2.2	Magnetization Perpendicular to the Film Surfaces	167
7.3	Spin Waves in Magnetic Superlattices	168
	7.3.1	Exchange Region	168
	7.3.2	Magnetostatic Region	170
7.4	Rare-Earth Superlattices	171
7.5	Metamagnetic Thin Films	177
7.6	Quasiperiodic Structures	186
References	191

8 Magnetic Polaritons in Spin-Canted Systems — 195
- 8.1 The Magnetic Hamiltonian — 198
- 8.2 Magnetic Polaritons in Canted Antiferromagnets — 202
- 8.3 Magnetic Polaritons in Spin-Canted Thin Films — 209
- References — 213

9 Metallic Magnetic Multilayers — 215
- 9.1 Magnetoresistance Self-Similar Spectra — 217
- 9.2 Magnetization Profiles — 224
- 9.3 Ferromagnetic Resonance Curves — 231
- 9.4 Thermodynamic Properties — 239
- References — 246

10 Exciton-Polaritons — 249
- 10.1 Thin Films — 250
- 10.2 Superlattice Modes — 256
- 10.3 Superlattice Modes in the Presence of a Magnetic Field — 259
- References — 265

11 Experimental Techniques — 267
- 11.1 Raman Scattering in Periodic Structures — 267
  - 11.1.1 Two-Component Superlattices with 2D Charge Sheets — 271
  - 11.1.2 *nipi* Superlattices — 278
- 11.2 Raman Scattering in Quasiperiodic Structures — 281
- 11.3 Brillouin Light Scattering — 286
- 11.4 Resonant Brillouin Scattering — 289
  - 11.4.1 Reflection and Transmission Spectra — 289
  - 11.4.2 Light-Scattering Formalism — 291
  - 11.4.3 RBS Cross Section — 295
- 11.5 Far-Infrared Attenuated Total Reflection — 300
- 11.6 Other Techniques — 305
  - 11.6.1 Light-Emitting Tunnel Junction — 305
  - 11.6.2 Far-Infrared Fourier-Transform Spectroscopy — 306
  - 11.6.3 Magneto-Optical Kerr Effect — 306
  - 11.6.4 Ferromagnetic Resonance — 307
- References — 308

12 Concluding Topics — 311
- 12.1 Non-linear Dielectric Media — 311
- 12.2 Non-linear Excitations in Single-Interface Geometries — 314
- 12.3 Non-linear Excitations in Double-Interface Systems — 317
- 12.4 Non-linear Excitations in Multilayer Systems — 321
- 12.5 Conclusions and Future Directions — 323
- References — 325

A Some Theoretical Tools   327
  A.1 Perturbation Theory 327
  A.2 Second Quantization 330
  A.3 Basic Properties of Green Functions 332
  A.4 Diagrammatic Perturbation Theory 335
  References 338

Subject Index   339

Preface

Artificial (or fabricated) periodic systems have been of considerable interest in physics and materials science since 1970, when Esaki and Tsu proposed a synthesized semiconductor superlattice of a one-dimensional periodic structure of alternating ultrathin layers, with its period less than the electron mean free path and the de Broglie wavelength. They envisioned two types of synthesized superlattices, namely the doping and the compositional ones, where in either case a superlattice potential was introduced by a periodic variation of impurities or composition during their growth. It was shown theoretically that such synthesized structures would possess unusual physical properties, not seen in the constituent semiconductor materials, due to predetermined quantum states that are of a two-dimensional character. Because of the potential device applications of such systems, their achievement and understanding has been a mixture of strong motivation from basic interest and technical applications, and much work has been devoted to understanding their unique physical properties. Further stimulus arose when modern growth techniques, such as molecular-beam epitaxy and metal-organic chemical vapor deposition, made it possible to fabricate these periodic layered materials with sharp, high-quality interfaces. Nowadays the fabrication of material structures with dimensions of the order of micrometers and nanometers is feasible to a high degree of precision.

On the other hand, the subject of quasicrystals first achieved prominence in 1984, when measurements using high-resolution X-ray scattering techniques produced electron diffraction patterns consisting of sharp spots but showing specific symmetries, forbidden by the rules of crystallography for an infinite lattice. Theoretical studies explained these types of symmetry through the aperiodic two- and three-dimensional Penrose tilings and their diffraction patterns (tiling is the geometrical operation that results in filling space with an arrangement of regular polyhedra). One important feature of these quasicrystal structures is that in one dimension they behave like the quasiperiodic structures formed by the incommensurate arrangement of periodic unit cells following a given mathematical sequence (like the well-known Fibonacci one). In turn, such structures can be tailored using the modern layer-growth techniques mentioned earlier. An appealing extra motivation for studying these quasiperiodic structures is that they exhibit a highly fragmented energy spectrum displaying a self-similar pattern. From the mathematical perspective, it has been proved that their spectra are Cantor sets in the thermodynamic limit. Moreover, the localization of electronic states, which is one of the most active fields in condensed-matter physics, could thus occur not only in disordered systems but also in deterministic quasiperiodic systems. Another interesting feature of these structures is that they exhibit collective properties that are not shared by their constituent materials. They are due to the presence of long-range correlations, and are expected to be reflected somehow in their various spectra (light propagation, electronic transmission, density of states, polaritons, etc.), defining a novel description of disorder. One of the main reasons for this is the fact that they represent an accessible and intermediate case between a periodic crystal, with extended Bloch states, and random disordered solids, with exponentially localized states. Furthermore, it obviously opens the way to many theoretical approaches in attempts to understand and foresee their physical properties, without the degeneracy rules of periodic invariance. Theoretical treatments (based, for example, on the transfer-matrix method descrebed later) show that a common factor shared by all these excitations is a complex fractal or multifractal energy spectrum.

The purpose of this book is to present an overall account of the dynamical properties of these periodic and quasiperiodic structures, in terms of the polaritons (bulk and surface modes) that propagate in them. In general, the term polariton refers to a mixed excitation (or wave) made up from a dipole-active elementary excitation (such as a phonon, plasmon, magnon, exciton, etc.) coupled to a photon (a quantum of light). The basic properties of polaritons may be obtained using simple theories that are related to the frequency-dependent dielectric, optical, and magnetic characteristics of the media.

Motivated by the potential device applications of such systems, our intention here is to provide a text, at the graduate level, for students, researchers, and academic staff working in this field who have an interest in understanding the unique physical properties of polaritons in these artificial systems. This includes the methods of generating polaritons in laboratories at frequencies of interest to experimentalists and the physics that may be learned from them. The book addresses the fundamentals of the propagation process for polaritons in such artificial structures, keeping in mind that, since experimental reality is approaching theoretical models and assumptions, detailed analysis and precise predictions are being made possible.

The book is organized so that we start with the basic properties of excitations in solids, highlighting their main concepts that can be found in some solid-state physics textbooks (Chapter 1). Next we define the periodic and quasiperiodic structures of interest (in the sense that they either can be or already have been grown by experimentalists), and we give the mathematical properties of some of them, namely Cantor, Fibonacci, Thue–Morse, and Double-period (Chapter 2). A discussion then follows of bulk and surface polariton modes of various types (mainly plasmon, phonon, magnetic, and exciton), stressing the role played by the dielectric function as well as the magnetic susceptibility (Chapters 3 and 4). This serves as the main introduction to these excitations. From that point on, the book presents the wide-ranging and interesting physical concepts behind the properties of polaritons in periodic and quasiperiodic artificial structures, stressing their spectra, localization and scaling properties, and power laws, which are a guide to their *universality classes*, and defining a novel description of disorder. In particular, important questions are addressed such as the propagation of polariton modes in doped semiconductors, piezoelectric, metamagnetic and rare-earth materials, among others, as well as the behavior of the thermodynamic quantities (particularly the specific heat spectra) in these systems (Chapters 5–10). Experimental techniques to probe these spectra are described in Chapter 11, with emphasis given to the (currently) most powerful spectroscopic method of Raman and Brillouin scattering of light, as well as to the so-called attenuated total reflection spectroscopy. Finally, in Chapter 12, we present some additional topics (in particular, systems with non-linear dielectric properties) and we point to future directions for this research field. A few important theoretical tools are presented in Appendix A to help readers with the theoretical methods employed throughout this book.

Both of us have been engaged heavily for many years in research programs focused on the physical properties of these elementary excitations, with two review articles already published on this subject. We believe that our book devoted to this burgeoning area will be valuable in covering the many new developments that have occurred since the, now classical, books *Polaritons* (edited by E. Burstein and F. de Martini) in 1974 and *Surface Polaritons* (edited by V.M. Agranovich and D.L. Mills) in 1982. Furthermore, as this field is rapidly changing, a good comprehension of the fundamental concepts presented here should be important for readers interested in this subject and for researchers seeking to make further advances.

We are indebted to a large number of friends and collaborators who directly or indirectly have influenced this book and provided ideas. We gratefully acknowledge the award of a fellowship from the Brazilian Research Agency CAPES, and leave granted by the Universidade Federal do Rio Grande do Norte to one of us (E.L.A.) to spend the summer of 2003 at the University of Western Ontario, where the main ideas of this book were established. Last, but not least, we would like to thank our families for their invaluable support and unfailing encouragement in addition to sustaining us through many difficult and challenging moments.

June 2004

Eudenilson L. Albuquerque **Michael G. Cottam**
Natal-RN London, Ontario
Brazil Canada

Chapter 1

Basic Properties of Excitations in Solids

The dynamical properties of a crystalline solid in terms of its constituent particles are of great interest to solid-state physicists and materials scientists. In particular, the concept of excitations in solids, especially in bulk materials, forms a significant part of the standard textbooks (see e.g. Refs. [1–3]). Typically these books cover a wide range of matters related to the dynamical response of the crystal to various kinds of external stimuli (such as temperature, electric field, magnetic field, etc.). All the excitations have at least one feature in common: they are associated with the whole crystal collectively and not just with a particular atom. As such, they depend sensitively on the structure of a solid and the interactions within it.

In general, the topic of excitations in solids is very complex, and it is not our intention to give a detailed account in this chapter, since this is done elsewhere. Instead, it is our aim to present here those fundamentals of the theory at a level sufficient to provide readers with the necessary background to understand better the specific material to be covered in the following chapters related to periodic and quasiperiodic structures.

We start this chapter with general considerations about the periodic arrangements of atoms, leading to the definition and characterization of a crystalline solid. This brings us to symmetry-related restrictions on the excitations themselves (e.g. through the well-known Bloch's theorem). Then we introduce the basic concepts of the main excitations to be considered in this book, namely the phonons, plasmons, magnons, and excitons. Later we shall be considering the properties of these excitations in various artificially structured materials, both individually and especially as "mixed" excitations in which they couple with a photon (or light quantum) to form a *polariton*.

1.1 Symmetry and Crystal Lattices

A crystalline solid is essentially an ordered array of atoms, bound together by electrical forces that may be attractive (as for the Coulomb interaction between electrons and protons) and repulsive (as for the electron–electron and proton–proton Coulomb interactions) to form a very large system. The different strengths and types of bond are determined by the particular electronic structures of the atoms involved and may, in principle, be found from quantum mechanics [4,5].

Magnetic forces have only a weak effect on cohesion and the gravitational forces are negligible. In a typical solid there are as many as 10^{23} nuclei and 10^{24} electrons in a cubic centimeter, which, at first sight, implies that it is almost impossible to study effectively such a large number of interacting particles theoretically. Fortunately, this complication can be overcome due to the high symmetry of a solid.

Indeed, although a bulk crystalline solid may be arbitrarily large, it can be viewed effectively as an infinite three-dimensional (3D) regular repetition of much smaller identical building blocks (or *repeating units*), which can be set up following specific symmetry relationships among its various physical parameters. The arrangements of the repeating units of the crystalline solids specify a set of operations, which is known today as the symmetry group or space group of the *Bravais lattice*. In short, a Bravais lattice can be introduced as a pure geometrical concept, in which an infinite array of the periodic crystal appears to be exactly the same, no matter from which position the array is viewed.

The most obvious operation in the symmetry group of a Bravais lattice is the *translational symmetry*. It is defined in terms of three non-coplanar basis vectors, denoted by $\hat{a}_1$, $\hat{a}_2$, and $\hat{a}_3$. These vectors are called the *primitive vectors* of the lattice, and are responsible for its generation. The parallelepiped defined by the primitive axes is called a *primitive cell*, and it fills all space under the action of a suitable translation operation. It is also the minimum-volume cell in the Bravais lattice and must contain precisely one lattice point. Note that there is no unique way of choosing a primitive cell for a given Bravais lattice.

If a translation is made between any two locations in the crystal, having identical atomic environments, they can be linked through the fundamental translation vector $\vec{R}$ given by

$$\vec{R} = n_1\hat{a}_1 + n_2\hat{a}_2 + n_3\hat{a}_3, \tag{1.1}$$

where n_1, n_2, and n_3 range through all integer values. The main property of the fundamental translation vector $\vec{R}$ is that the atomic arrangement of the lattice looks the same in every respect, whether viewed from any point $\vec{r}$ or from

$$\vec{r}' = \vec{r} + \vec{R} = \vec{r} + n_1\hat{a}_1 + n_2\hat{a}_2 + n_3\hat{a}_3. \tag{1.2}$$

Apart from the translational symmetry (which has the most important influence on the properties of the crystal), the space group associated with a lattice may also present symmetry operations due to the various rotations and reflections (and combinations of them), which leaves the crystal, as well as its primitive cells, unchanged [6,7]. For instance, for a hypothetical two-dimensional (2D) solid there are five different Bravais lattices, as shown in Fig. 1.1. For 3D solids 14 Bravais lattices are possible (see e.g. Ref. [1]). The points in a Bravais lattice that are closest to a given point in the lattice are called its nearest neighbors, and their number is an invariant property of the lattice.

Since a Bravais lattice is not an arrangement of atoms but a geometrical arrangement of points in the space, it is necessary when defining a more complex *crystal structure* to associate to each point of the Bravais lattice a *basis of atoms*.

1.1. SYMMETRY AND CRYSTAL LATTICES

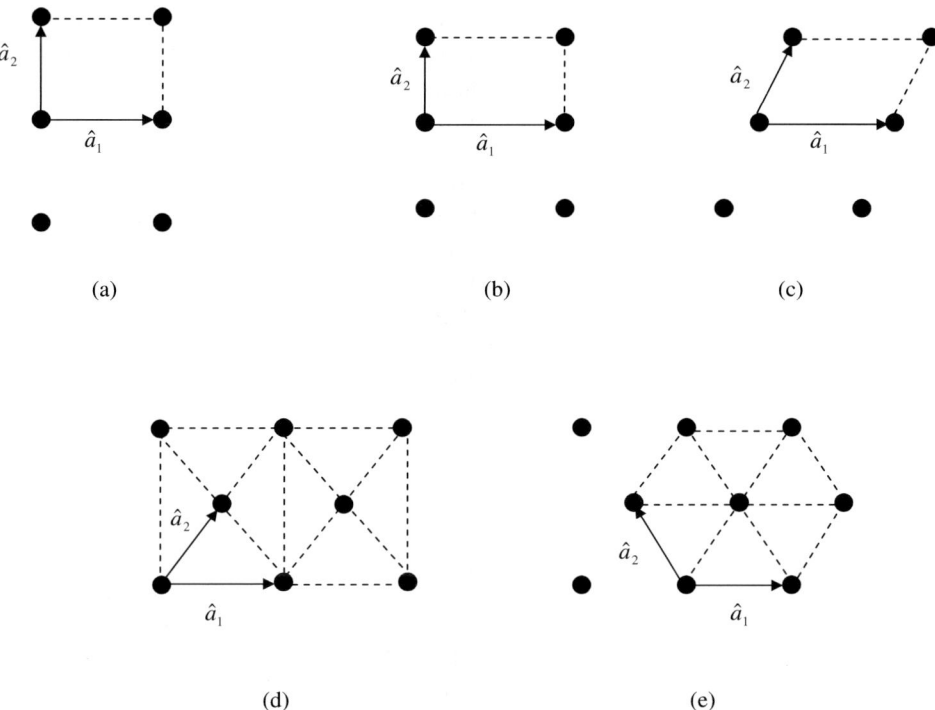

Fig. 1.1. Schematic representation of the five Bravais lattices in 2D: (a) square lattice; (b) rectangular lattice; (c) oblique lattice; (d) centered rectangular lattice; (e) hexagonal lattice.

By a basis of atoms we mean not only the atoms themselves but also their spacing and bond angles, which may form molecules, ions, etc. Of course, the number of atoms in the basis may sometimes simply be unity, as for many metals and the inert gases (He, Ne, Ar, Kr, Xe, and Rn), but it may be larger in general, exceeding 1000 for some inorganic and biochemical structures. Therefore, a crystal structure can be defined as identical copies of the same physical unit, the basis, located at all the points of a Bravais lattice. The geometrical space lattice (defining the Bravais lattice), plus the basis of atoms attached to each lattice point, specifies the full crystal structure [1].

Although the primitive cell is sufficient for characterizing the Bravais lattice, it is sometimes more convenient to work with the so-called *unit cell* of the lattice. The unit cell is the simplest geometrical figure that we can select from a Bravais lattice. Depending on its geometrical arrangement, it may or may not coincide with the primitive cell and therefore it is not always the minimum-volume cell of the Bravais lattice. However, the more straightforward geometrical appearance of the unit cell compensates by far this feature. For example, the body-centered cubic lattice, one of the most studied 3D Bravais lattices, is more easily visualized as a cubic structure with two atoms than its primitive counterpart, which is a much

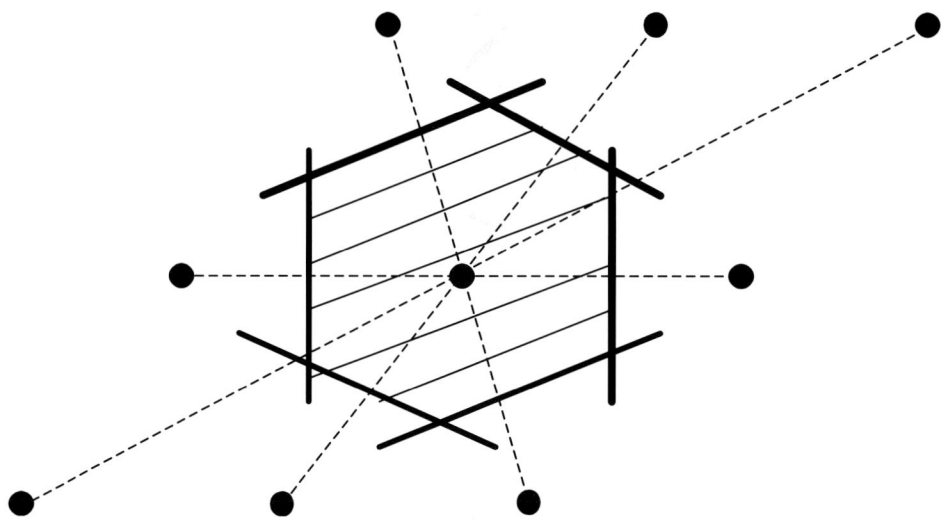

Fig. 1.2. Construction of a primitive Wigner–Seitz cell.

complicated rhombohedron of edge $a\sqrt{3}/2$, with a being the side of the cube, and angle $109° \, 28'$ between adjacent edges.

On the other hand, there is a clever way to construct a primitive cell of any Bravais lattice, the so-called Wigner–Seitz primitive cell [8]. In Fig. 1.2 we show how to draw a Wigner–Seitz cell for a 2D Bravais lattice. The procedure is quite simple: (a) draw lines to connect a given lattice point to all its nearby lattice points; (b) at the midpoint and normal to these lines, draw new lines. The simplest volume enclosed in this way is the Wigner–Seitz primitive cell. It can be shown that in 2D the Wigner–Seitz cell is always a hexagon, with the obvious exceptions of the square and rectangular lattices. As we will see in the next paragraph, the Wigner–Seitz primitive cell plays an important role in the determination of the Brillouin zones.

We often need to describe a particular crystallographic plane or a particular direction within a real 3D crystal. Although a plane can be specified by any three points lying in it, provided the points are not collinear, it is more useful for structural analysis to describe it in terms of its so-called *Miller indices*. For planes the recipe is very simple. First, find the intercepts on the axes $\hat{a}_1$, $\hat{a}_2$, and $\hat{a}_3$, expressed in multiples of the lattice constant. Then, take the reciprocal of these numbers, scaling them (if necessary) to the smallest three integers having the same ratio. The result is displayed in parentheses as (hkl). When a plane cuts an axis (say the $\hat{a}_1$-axis) on the negative side, it is conventional to employ the designation as $(\bar{h}kl)$. Fig. 1.3 shows some examples of planes for cubic lattices, with their Miller notations.

A similar convention is used to specify a particular direction normal to a plane in the real lattice. To avoid confusion with the crystallographic planes, however,

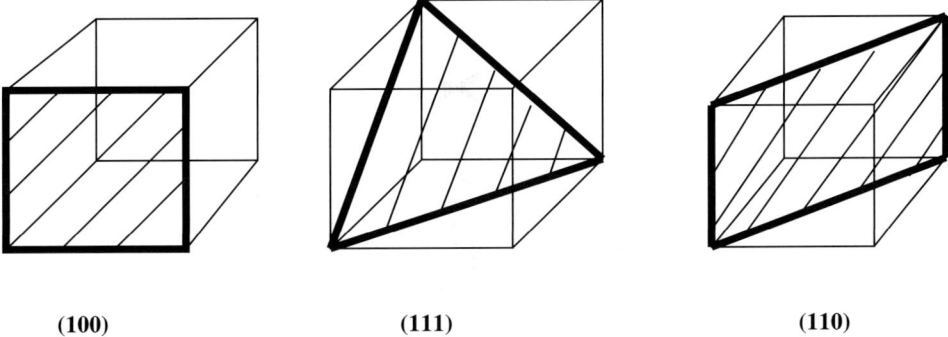

Fig. 1.3. Miller indices for three lattice planes in a simple cubic Bravais lattice.

square brackets, i.e. $[hkl]$, are used instead of parentheses. For instance, the body diagonal of a simple cubic lattice lies in the [111] direction.

1.2 Reciprocal Lattices and Brillouin Zones

A consequence of the translational symmetry of a crystal is that some of the physical properties can be described by a multiply periodic function (denoted here by F), which satisfies the condition

$$F(\vec{r}+\vec{R}) = F(\vec{r}) \qquad (1.3)$$

for all points $\vec{r}$ in space and for all translation lattice vectors $\vec{R}$. On expanding $F(\vec{r})$ in a Fourier series in 3D, we have

$$F(\vec{r}) = \sum_{\vec{Q}} g(\vec{Q}) \exp(i\vec{Q}\cdot\vec{r}). \qquad (1.4)$$

It then follows straightforwardly from Eqs. (1.3) and (1.4) that

$$\exp(i\vec{Q}\cdot\vec{R}) = 1, \qquad (1.5)$$

which implies that $\vec{Q}\cdot\vec{R} = 2\pi \times$ integer. The infinite set of all $\vec{Q}$ vectors that satisfy these conditions defines the *reciprocal lattice*. $\vec{Q}$ is a *reciprocal lattice vector*, and it has the dimension of wavevector (inverse length). Thus multiplication of the set of all reciprocal vectors $\vec{Q}$ by $\hbar$ converts reciprocal space into momentum space. Since the crystal (or Bravais) lattice is in real or ordinary space, the reciprocal (or Fourier) lattice is, apart from a multiplicative constant, in momentum space.

Defining the reciprocal lattice as

$$\vec{Q} = h\hat{b}_1 + k\hat{b}_2 + l\hat{b}_3, \qquad (1.6)$$

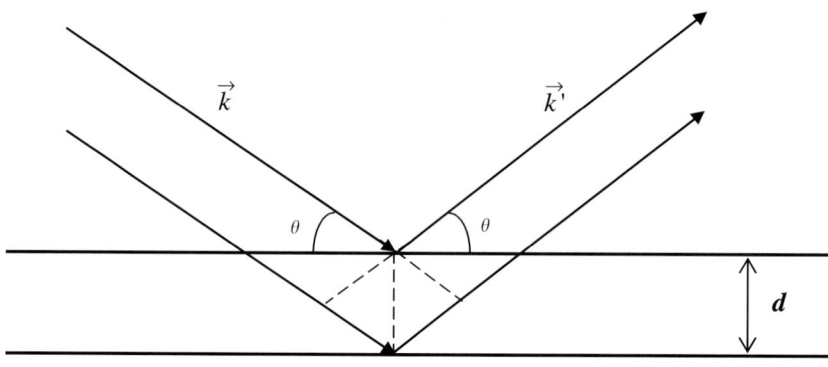

Fig. 1.4. Specular reflection of a parallel beam of X-rays from two adjacent parallel atomic planes separated by a distance d.

where $\hat{b}_1, \hat{b}_2$, and $\hat{b}_3$ are the primitive vectors of the reciprocal lattice, one can easily prove (see e.g. Ref. [9]) that

$$\hat{b}_1 = 2\pi(\hat{a}_2 \times \hat{a}_3)/V, \quad \hat{b}_2 = 2\pi(\hat{a}_3 \times \hat{a}_1)/V, \quad \hat{b}_3 = 2\pi(\hat{a}_1 \times \hat{a}_2)/V, \qquad (1.7)$$

where V is the volume of the primitive real crystal lattice, i.e. $V = |\hat{a}_1 \cdot (\hat{a}_2 \times \hat{a}_3)|$.

Some important examples of the reciprocal lattices are related to the real cubic lattices. From Eq. (1.7), it is easy to show that the simple-cubic (sc) real lattice of side a has, as its reciprocal lattice, another sc structure but with side $2\pi/a$. The face-centered cubic (fcc) Bravais lattice has a reciprocal lattice represented by a body-centered cubic (bcc) lattice with side $4\pi/a$. As one can prove that the reciprocal lattice of a reciprocal lattice is always the original direct lattice [2], it follows that the bcc Bravais lattice has a fcc lattice, with side $4\pi/a$, as its reciprocal lattice.

It is equally important to be able to visualize the reciprocal lattice as it is to visualize the real Bravais lattice. For example, a diffraction pattern of a crystal is a map of the reciprocal lattice of the crystal, while a microscope image is a map of the real crystal structure. To understand the former, consider the well-known Bragg diffraction condition [10] corresponding to the specular reflection of a parallel beam of X-rays from the two successive planes of atoms in a crystal, as depicted in Fig. 1.4. The path difference for rays reflected from adjacent planes is $2d\sin\theta$, where θ is the angle that the X-ray beam makes with the parallel planes. This should be an integer multiple of the wavelength of the incident radiation, say $n\lambda$, to allow constructive reflection of the incident radiation. The concept of reciprocal lattice allows a novel and interesting perspective of the Bragg condition: instead of considering the wavelength of the radiation, we are concerned with the wavevectors $\vec{k}$ and $\vec{k}'$ of the X-ray beam undergoing reflection. Provided that the scattering process is elastic, there is no change in the magnitude of the wavevectors, i.e. $|\vec{k}| = |\vec{k}'| = 2\pi/\lambda$. In this way, it can easily be proved [3] that the diffraction condition is

$$\Delta \vec{k} = \vec{k} - \vec{k}' = \vec{Q}, \qquad (1.8)$$

where $\vec{Q}$ is again a reciprocal lattice vector. The above result can be re-expressed as

$$\vec{k} \cdot \vec{Q}/2 = (|\vec{Q}|/2)^2. \qquad (1.9)$$

This latter expression is often used in the construction of *Brillouin zones* in reciprocal space (or momentum space). It gives a vivid geometrical interpretation of the diffraction condition, as can be visualized by the geometrical construction due to Ewald [11] (for details see e.g. Ref. [2]).

A *first Brillouin zone* is defined as the primitive Wigner–Seitz cell in the reciprocal lattice. As the name suggests, there are also higher-order Brillouin zones, which are primitive cells of a different type. We do not intend to discuss them because they are beyond the context of this book (for details see Ref. [12]). Although the terms Brillouin zone and Wigner–Seitz cell refer to a similar geometrical construction, the former term is applied only in the reciprocal momentum $\vec{k}$-space, while the latter is always related to the real $\vec{r}$-space. For a sc 3D structure with sides a, the first Brillouin zone lies in the region $-\pi/a \leq k_j \leq \pi/a$, with the subscript j equal to the Cartesian axes x, y, or z. On the other hand, because the reciprocal of the bcc lattice is the fcc structure, the first Brillouin zone of a bcc lattice is just the fcc Wigner–Seitz cell (a regular rhombic dodecahedron). Conversely, it follows that the first Brillouin zone of an fcc lattice is a rhombohedron.

1.3 Bulk, Surface, and Superlattice Excitations

The symmetry of the crystal lattice leads directly to some consequences for the elementary excitations (or waves) that may propagate in the crystal. The most important result, in the context of this book, is called *Bloch's theorem*. It is discussed extensively in standard references on solid-state physics (see e.g. Ref. [13]), applied group theory (see e.g. Ref. [14]), and energy band theory (see e.g. Ref. [15]), and so we shall not prove it here. A simple statement of the theorem for an infinite 3D crystal lattice can be made as follows.

Suppose any excitation is described by a spatial amplitude $\psi(\vec{r})$ at position $\vec{r}$. This might be the wave function, obtained from the Schrödinger equation, in the case of an electron. Alternatively, for example, in the case of a phonon or lattice vibration $\psi(\vec{r})$ would be an atomic displacement satisfying its linearized equation of motion. In all such cases, Bloch's theorem states that

$$\psi(\vec{r}) = \exp(i\vec{k} \cdot \vec{r}) U_{\vec{k}}(\vec{r}), \qquad (1.10)$$

where $\vec{k}$ is an arbitrary (real) constant vector with dimensions of reciprocal length. Also, $U_{\vec{k}}(\vec{r})$ is a periodic function satisfying $U_{\vec{k}}(\vec{r}+\vec{R})=U_{\vec{k}}(\vec{r})$, where $\vec{R}$ is the fundamental translation vector of the lattice.

It is important to note that the wavevector $\vec{k}$ in Eq. (1.10) is not uniquely defined. In fact, because of the spatial periodicity of $U_{\vec{k}}(\vec{r})$ and the property stated in Eq. (1.5), the wavevector $\vec{k}+\vec{Q}$ will equally satisfy Bloch's theorem. Hence it is sufficient to consider only those excitation wavevectors that are within the first Brillouin zone of the lattice. Likewise it may be shown that the frequency $\omega(\vec{k})$ of the excitation, which depends in general on the wavevector $\vec{k}$ through a dispersion relation, satisfies

$$\omega(\vec{k}+\vec{Q}) = \omega(\vec{k}). \tag{1.11}$$

All the symmetry properties discussed above refer to infinite crystal lattices, and Bloch's theorem simply shows that the corresponding *bulk excitations* with wavevector $\vec{k}$ have a wave-like behavior in all 3D, arising from the $\exp(i\vec{k}\cdot\vec{r})$ factor in Eq. (1.10). However, if the surfaces of a crystal are taken into account, then the symmetry is reduced and Bloch's theorem must be modified (see e.g. Ref. [16]).

As a simple illustration, let us consider the case of a semi-infinite crystal so that there is just one surface. We assume an sc lattice with a (001) surface, taking the surface plane to be $z=0$ and the crystal to fill the half-space $z\geq 0$. In this case, there are 2D translation symmetry operations as before in the xy-plane (assuming the crystal to be infinite in the x- and y-direction), but no translational symmetry in the z-direction. Denoting $\vec{r}_\parallel = (x,y)$, the 2D analog of Eq. (1.10) becomes (see Ref. [16])

$$\psi(\vec{r}_\parallel, z) = \exp(i\vec{k}_\parallel \cdot \vec{r}_\parallel) U_{\vec{k}_\parallel}(\vec{r}_\parallel, z), \tag{1.12}$$

where $\vec{k}_\parallel = (k_x, k_y)$ is a 2D wavevector for the excitation, and the amplitude function $U_{\vec{k}_\parallel}(\vec{r}_\parallel, z)$ must now satisfy $U_{\vec{k}_\parallel}(\vec{r}_\parallel + \vec{R}_\parallel, z) = U_{\vec{k}_\parallel}(\vec{r}_\parallel, z)$, with $\vec{R}_\parallel$ being a 2D symmetry operation for translation in the xy-plane.

For an excitation of a semi-infinite crystal in this example, the essential difference compared with the infinite case is that it is required to have a wave-like behavior only with respect to the direction parallel to the surface, hence corresponding to the 2D wavevector $\vec{k}_\parallel$. The z-dependence of the excitation amplitude has to be determined by solving for $U_{\vec{k}_\parallel}(\vec{r}_\parallel, z)$ in any specific physical example, but a few generalizations may be made. If solutions exist of the form $\exp(ik_z z)$ with k_z real, these just correspond again to a bulk excitation, with k_z being the third component of the wavevector. On the other hand, solutions of the same form as above, but with k_z *complex* and having a positive imaginary part, would correspond to a *surface excitation* since the amplitude would decay with distance into the crystal. In some simple cases, as we shall see in later chapters, surface excitations correspond to $k_z = i\alpha$ (with α real and positive), so the z-dependence of the excitation amplitude is proportional to $\exp(-\alpha z)$. In other cases, the spatial localization of a surface mode may be more complicated.

Bulk and surface (or interface) excitations may exist in more complex physical systems where the boundaries play a more significant role, such as thin films and multilayers with several parallel planar surfaces (or interfaces) [16]. In general, as the number of different crystal layers in a system is increased, the excitation spectrum will become more complicated. A notable and interesting exception is when the layers are built up in a periodic fashion to form a *superlattice*. As an example, let us consider an effectively infinite superlattice made up by stacking vertically, in an alternating fashion, films of material A (thickness a) and material B (thickness b). This artificial structure now has a new symmetry operation corresponding to translations through multiples of the length $L = a + b$ along the z-direction. There will be a new Brillouin zone associated with this periodicity and the excitations will reflect this property, giving rise to the so-called *superlattice excitations*. The dynamical behavior of superlattices, both periodic and quasiperiodic, will be a major topic throughout this book. It is convenient, first,to introduce some basic properties of crystal excitations in bulk materials.

1.4 Phonon: Quantum of the Lattice Vibrations

A *phonon* is the quantum of energy associated with a lattice vibration or elastic wave. To exemplify this concept, we consider a set of N identical ions of mass m distributed along a monatomic 1D Bravais lattice whose translation vector is $\vec{R} = na\hat{z}$, with n being an integer and a denoting the distance between two adjacent ions. The vibrational motion is assumed here to be confined to the z-direction (along the chain).

Let u_n be the displacement of the ion that oscillates from its equilibrium position $z = na$ along the linear chain (see Fig. 1.5). The number N is taken to be sufficiently large that end effects can be ignored (i.e. the chain is effectively infinite). Assuming that only neighboring ions interact, Newton's equation of motion for the nth ion yields

$$m\partial^2 u_n/\partial t^2 = C[(u_{n+1} - u_n) - (u_n - u_{n-1})], \tag{1.13}$$

where C is the (elastic) force constant between the ions. Considering now only normal modes propagating (with angular frequency ω) in the chain, it follows that the solutions for u_n can be represented by plane waves:

$$u_n = u \exp[i(kna - \omega t)] \tag{1.14}$$

in accordance with Bloch's theorem, the 1D case of Eq. (1.10). Substituting into Eq. (1.13) gives

$$\omega^2 = (2C/m)(1 - \cos ka) = (4C/m)\sin^2(ka/2). \tag{1.15}$$

The dispersion relation for $\Omega = \omega/(4C/m)^{1/2}$ versus the dimensionless wavevector ka is plotted in Fig. 1.6 for its first Brillouin zone, i.e. for $-\pi \leq ka \leq \pi$. As we can see, it is symmetrical between k and $-k$, i.e. waves to the left and to the

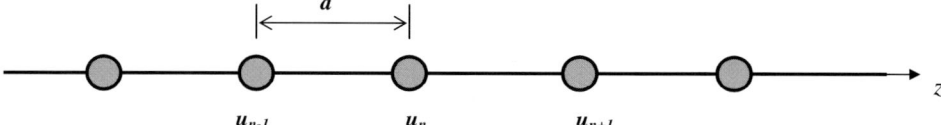

Fig. 1.5. Monatomic linear chain formed by N ions of mass m separated by a lattice parameter a.

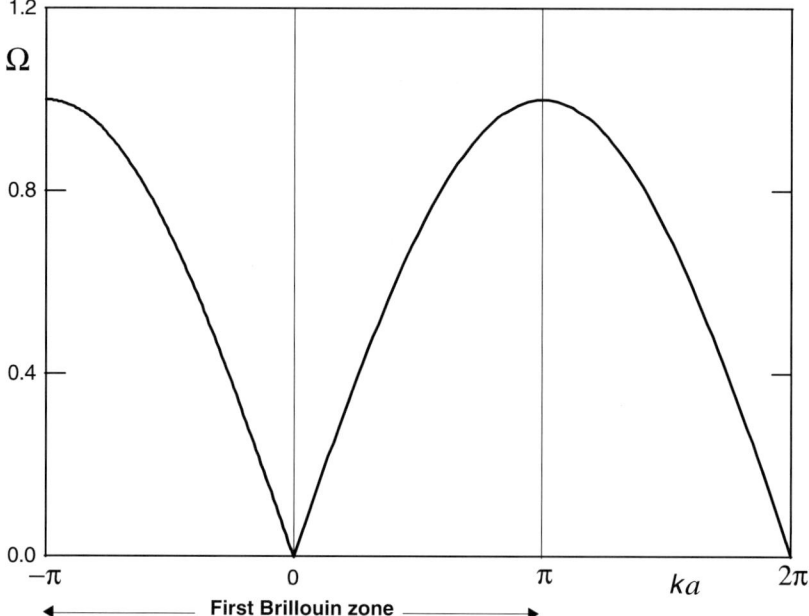

Fig. 1.6. Phonon dispersion relation in the first Brillouin zone for a monatomic linear chain.

right are identical. Although it is not shown in the figure, this dispersion relation profile repeats for each interval of $2\pi/a$, as expected from the previous section. Note that when ka tends to zero, Ω is proportional to $|k|$, and its group velocity, found from $d\omega/dk$, vanishes at the boundaries of the Brillouin zone ($ka = \pm\pi$).

We next consider a 1D Bravais lattice with *two* types of alternating ions with masses m_1 and m_2 per primitive cell (the diatomic chain depicted in Fig. 1.7). It has $2N$ ions (N of each type), and all ions are assumed to be coupled by the same force constant C to their neighbors. Newton's equation of motion is now slightly different for each ion type, namely

$$m_1\,\partial^2 u_n/\partial t^2 = C[(v_n - u_n) - (u_n - v_{n-1})],$$
$$m_2\,\partial^2 v_n/\partial t^2 = C[(u_{n+1} - v_n) - (v_n - u_n)]. \qquad (1.16)$$

1.4. PHONON: QUANTUM OF THE LATTICE VIBRATIONS

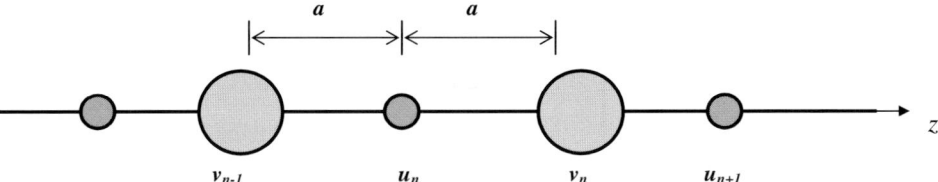

Fig. 1.7. Diatomic linear chain formed by $2N$ ions of masses m_1 and m_2 separated by lattice parameter a.

Again we expect from symmetry that each ion has its normal mode represented by a plane wave similar to Eq. (1.14), but with different amplitudes u and v for the two masses. Thus Eq. (1.16) leads to

$$-\omega^2 m_1 u = Cv[1 + \exp(-ika)] - 2Cu,$$
$$-\omega^2 m_2 v = Cu[1 + \exp(ika)] - 2Cv. \qquad (1.17)$$

This pair of equations for the amplitudes has non-trivial solutions, provided the determinant of the coefficient vanishes. This gives two positive values of ω satisfying

$$\omega^2 = C(m_1^{-1} + m_2^{-1}) \pm C\left[(m_1^{-1} + m_2^{-1})^2 - 4\sin^2(ka/2)/m_1 m_2\right]^{1/2}, \qquad (1.18)$$

and the ratio of amplitudes is

$$\frac{u}{v} = \frac{2C\cos(ka)}{2C - m_1 \omega^2} = \frac{2C - m_2 \omega^2}{2C\cos(ka)}. \qquad (1.19)$$

It is easy to see that for $ka = \pm\pi$ (the Brillouin zone boundaries), the solutions for ω^2 from Eq. (1.18) are $2C/m_1$ and $2C/m_2$. Also, as $ka \to 0$ (close to the Brillouin zone center), the two solutions are approximately

$$\omega^2 = 2C(m_1^{-1} + m_2^{-1}),$$
$$\omega^2 = [2C/(m_1 + m_2)]k^2 a^2. \qquad (1.20)$$

For each value of k there are now *two* separate solutions, which repeat at every wavevector interval of $2\pi/a$. These two branches of the dispersion relation are illustrated in Fig. 1.8. The lower branch has the same qualitative form as the single branch found in the preceding case (the monatomic Bravais lattice). This branch is known as the *acoustic branch* because its dispersion relation for small ka is of the form $\omega = vk$, and is characteristic of a sound wave (which is a longitudinal mode).

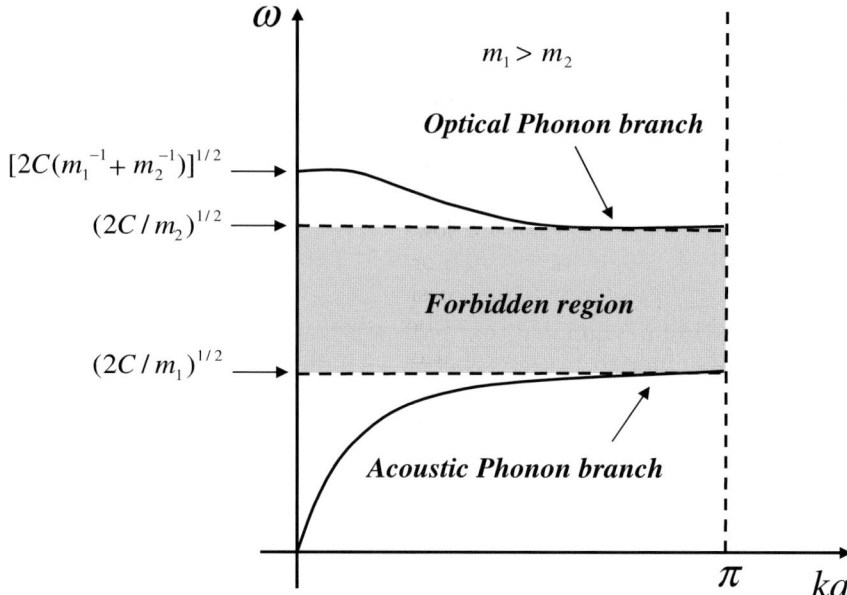

Fig. 1.8. Acoustic and optical phonon branches in the first Brillouin zone for a diatomic linear chain.

The upper branch is known as the *optical branch* because the transverse long-wavelength optical mode in ionic crystals can interact with the electromagnetic radiation, thereby giving rise to the so-called *phonon-polariton* mode, which we will discuss later.

The classification of the vibrational modes into acoustic and optical branches is still applicable for a 3D solid with a polyatomic basis. Suppose we have a crystal with p atoms in each unit cell and a large number N of such cells form the crystal. In this case, there will be $3pN$ degrees of freedom and hence $3pN$ normal modes. Of these branches, $3N$ will be acoustic branches ($2N$ transverse, or TA modes, and N longitudinal, or LA modes) having the property that $\omega(k) \propto k$ as $k \to 0$. The remaining $3(p-1)N$ branches are transverse (TO) and longitudinal (LO) optical branches, having the property that $\omega(k)$ tends to a non-zero constant as $k \to 0$. However, not all of them will be optically active in the sense of interacting strongly with the radiation field.

1.5 Plasmon: Quantum of the Plasma Oscillations

A *plasma oscillation* in a metal is a collective longitudinal excitation of the conduction electron gas. The term "plasma" was suggested in 1929 by Langmuir [17] to describe the collective electrical properties that he noted in an ionized gas. Since that time, many of the phenomena observed for a gaseous plasma (a medium with

1.5. PLASMON: QUANTUM OF THE PLASMA OSCILLATIONS

equal concentrations of positive and negative charges, where at least one charge type is mobile) can be reproduced in the electron "gas" of a metal or a semiconductor. Thus the field of plasma physics in condensed-matter systems has become well established, with several technological applications.

A *plasmon* is a quantum associated with a plasma oscillation. While it has relatively few direct observations from the experimental point of view, a notable exception is the observation of energy losses, in multiples of $\hbar\omega_p$ (where ω_p is the plasma frequency, which we discuss below), when electrons are fired through thin metallic film or by reflecting an electron or a *photon* (the quantum of the radiation field) from a film [18]. However, the possibility of their excitation in any process involving an electron gas should always be borne in mind.

The frequency of the plasma oscillation can be found using a very simple argument. Suppose that an electron in the electron gas (classical or quantized) of a solid oscillates about its equilibrium position under the effect of an electrical restoring force, defined by the equation

$$m d^2 \vec{r}/dt^2 = -e\vec{E}. \tag{1.21}$$

Here e (m) is the electronic charge (mass), $\vec{r}$ is the electron displacement in 3D, and $\vec{E}$ is the electric field vector. For oscillations with a time dependence of the type $\exp(-i\omega t)$, the polarization field, defined as the dipole moment of the electrons per unit volume, is given by

$$\vec{P}(\omega) = -ne\vec{r} = -ne^2 \vec{E}/m\omega^2, \tag{1.22}$$

where n is the electron concentration. The dielectric function associated with the electron gas oscillation is then given (in SI units) by

$$\epsilon(\omega) = 1 + P(\omega)/\epsilon_0 E(\omega) = 1 - \omega_p^2/\omega^2, \tag{1.23}$$

where the plasma frequency ω_p is defined by $\omega_p^2 = ne^2/\epsilon_0 m$. If the positive ion background has a dielectric constant ϵ_∞, assumed to be essentially constant up to frequencies well above ω_p, then the plasmon dielectric function becomes

$$\epsilon(\omega) = \epsilon_\infty (1 - \omega_p^2/\omega^2). \tag{1.24}$$

We observe, from Eq. (1.24), that the frequency-dependent dielectric function $\epsilon(\omega)$ is negative for $\omega < \omega_p$ and positive for $\omega > \omega_p$. The behavior of this dielectric function in terms of the frequency ratio ω/ω_p is plotted in Fig. 1.9. The frequency dependence has important consequences for the propagation of an electromagnetic wave through the plasma. In particular, at low frequencies when $\epsilon(\omega)$ is negative, it follows from Maxwell's equations that no radiation can propagate (the radiation field falls exponentially inside the solid). On the other hand, when $\epsilon(\omega)$ is positive

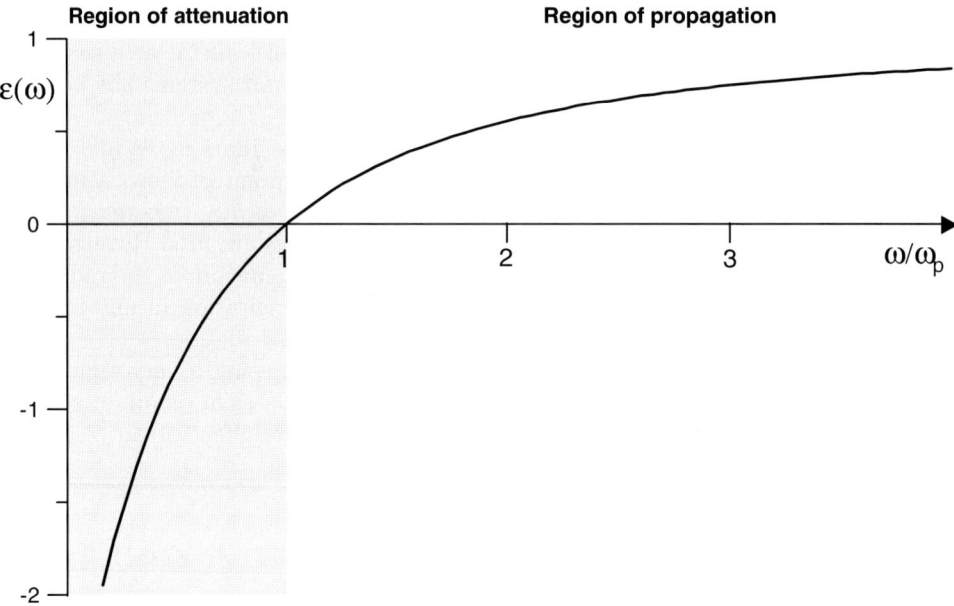

Fig. 1.9. Behavior of the dielectric function for an electron gas, as defined by Eq. (1.24) with $\epsilon_\infty = 1$, as a function of ω/ω_p.

for $\omega > \omega_p$ the electromagnetic radiation can propagate, and the medium should become transparent.

The above situation will be discussed more thoroughly in later chapters on polaritons. However, we may briefly comment here that the transverse electromagnetic mode in the plasma is defined by $k^2 = \epsilon(\omega)\omega^2/c^2$. Using Eq. (1.24) in the case of $\epsilon_\infty = 1$, the expression can easily be rearranged as

$$\omega^2 = \omega_p^2 + c^2 k^2, \tag{1.25}$$

where c is the velocity of light in vacuum. This is the dispersion relation for a mixed mode, which is just the so-called *plasmon-polariton* formed by coupling the transverse electromagnetic radiation to the plasma. Also, from basic electromagnetism, the zeros of the dielectric function determine the frequencies of the longitudinal electromagnetic modes. In the case of Eq. (1.24), this yields simply $\omega = \omega_p$. It is just the longitudinal oscillation mode of an electron gas at the plasma frequency (see Fig. 1.10), which also represents the cut-off frequency of the transverse mode discussed before.

1.6 Exciton: Bound Electron–Hole Pair

The optical characterization of semiconductors provides us with one of the richest sources of information about their electronic properties. In many semiconductors,

1.6. EXCITON: BOUND ELECTRON–HOLE PAIR

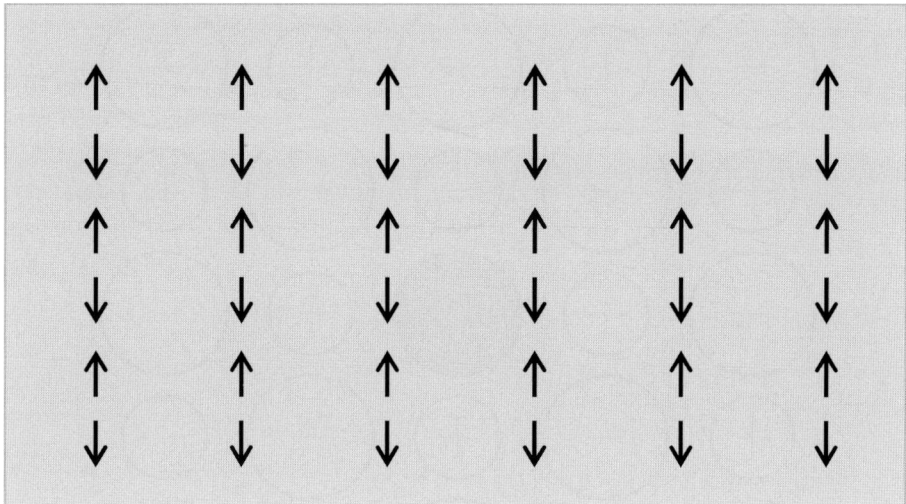

Fig. 1.10. Longitudinal plasma oscillation. The arrows indicate the direction of displacement of the electrons.

photons (light quanta) can interact with the lattice vibrations (phonons) and with the electrons localized on defects. This makes optical techniques useful for studying these excitations and forms the basis of many technological applications such as lasers, light emitting diodes, and photodetectors [19–21].

A widely used approach to view these optical absorption processes is one in which an incident radiation field (a photon) excites an electron–hole pair inside the semiconductor. The attraction between the electron and the hole causes their motion to be correlated, and the resulting bound electron–hole pair is known as an *exciton*. It may move through the crystal transporting excitation energy but not charge. As a consequence, it does not contribute directly to the electrical conductivity. Typically, excitons have been studied in two limiting cases. For strong electron–hole attraction, as found in ionic crystals, the electron and the hole are tightly bound to each other. This excitation is known as a *Frenkel exciton* [22,23]. On the other hand, it is the case for most semiconductors that the Coulomb interaction is strongly screened by the valence electrons via the large dielectric constant. As a result, the electron–hole pair is then only weakly bound, giving rise to a *Wannier–Mott exciton* [24,25].

The Frenkel exciton differs from a single electronic excitation in that it is neutral and carries no current: it does, however, carry energy. It has an associated wavevector $\vec{k}$ that lies within the first Brillouin zone. Evidence of its existence in molecular crystals was obtained by observing the separation between the different $\vec{k} = 0$ Frenkel excitons arising from a single parent, the so-called Davydov splitting [26,27]. The fact that it is a mobile entity, capable of moving from point to point in the crystal, was demonstrated rather directly by Simpson in 1956 [28]. Indeed, it can move through a crystal with much higher mobility than is the case for

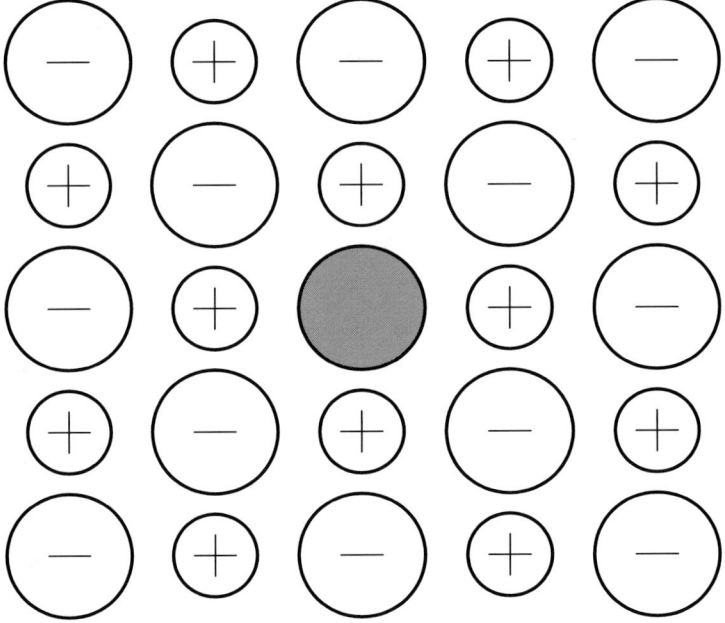

Fig. 1.11. Schematic representation of a tightly bound electron–hole pair forming a Frenkel exciton.

vacancies, interstitials, or substitutional impurities. For most purposes it can be visualized as being at any instant located on or near a single atom, in the sense that the hole is usually on the same atom as the electron. It is considered to be a more complex manifestation of the electronic band structure than a crystal defect. Thus, a Frenkel exciton is essentially an excited state of a single atom, with the excitation hopping from one atom to another by virtue of the coupling between neighbors: it is depicted in Fig. 1.11. In a simple theoretical model (see e.g. Ref. [1]), the Frenkel exciton can be considered as a system of N identical atoms (or molecules) in a crystal, in which one atom is raised to an excited state of energy E and has the ability to hop to neighboring atoms. The problem is somewhat analogous to that of N coupled harmonic oscillators and has, as a solution, a Bloch-type wave function of the same form as in Eq. (1.10).

Because an exciton is a two-particle system, the general problem of calculating its energy levels and wave functions is intrinsically more complicated than the corresponding one-electron problem. However, when the electron–hole pair is not tightly bound, and the particles are separated by many interatomic spacings (as in the case of the Wannier–Mott exciton), its properties can be calculated using an effective mass approximation [29–31].

Within this approximation, the electron and the hole are considered as two moving particles, having the effective masses of the conduction band (m_c) and the valence band (m_v), but separated by many lattice distances. Since the difference

1.6. EXCITON: BOUND ELECTRON–HOLE PAIR

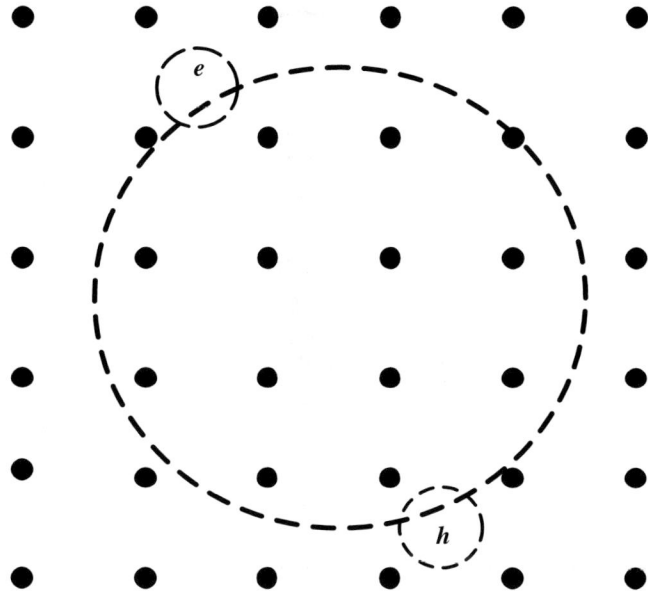

Fig. 1.12. Schematic representation of a weakly bound electron–hole pair separated by many lattice distances forming a Wannier–Mott exciton.

between these effective masses is not as large as that between the electron and the proton, the Wannier–Mott exciton is more analogous to *positronium*, an electron–positron pair. A pictorial representation of the Wannier–Mott exciton is given in Fig. 1.12. The attractive Coulomb force is reduced by the dielectric constant ϵ of the intervening medium. Clearly, this is just the hydrogen-like atom problem, with the reduced mass of the hydrogen atom being replaced by the electron–hole reduced effective mass m^*, where $(m^*)^{-1} = m_c^{-1} + m_v^{-1}$, and the dielectric constant of the medium taken into account. Thus, the energy of the exciton will be described by an infinite series of discrete hydrogenic bound states plus the kinetic energy due to the motion of its center of mass, i.e.

$$E_n(\vec{k}) = -\frac{e^4 m^*}{2\hbar^2 \epsilon^2 n^2} + \frac{\hbar^2 k^2}{2M}, \tag{1.26}$$

where $n = 1, 2, \ldots$ is a label for the different states, $M = m_c + m_v$ is the exciton mass, and $\hbar \vec{k}$ is the momentum of the center of mass. The lowest exciton energy $E_{n=1}$ is lower than the energy $E_c - E_v = E_g$ of the non-interacting electron and hole. Pursuing this analogy with the hydrogen atom, the Bohr radius of the exciton is given by

$$a_{ex} = \epsilon \hbar^2 / m^* e^2 = \epsilon (m/m^*) a_0, \tag{1.27}$$

where $a_0 = \hbar/me^2 = 5.29 \times 10^{-11}$ m denotes the Bohr radius of hydrogen.

The binding energy of the ground state ($n=1$) of the exciton, relative to the free electron–hole pair energy at the band edge, is often called the *exciton Rydberg*:

$$E_1(0) = -\frac{e^4 m^*}{2\hbar^2 \epsilon^2} = -\frac{e^2}{2\epsilon a_{ex}} = -\epsilon^2 (m^*/m) E_0, \qquad (1.28)$$

where E_0 is the ionization energy of hydrogen (13.54 eV). Eq. (1.28) provides a convenient way of estimating the exciton Bohr radius from its ground state energy.

The above model of excitons is useful for understanding qualitatively their effect on the optical spectra. However, as expected, it is not accurate enough for the quantitative interpretation of experimental results, where a more complex view of the electrons in the conduction band as well as the holes in the valence band should be taken into account.

1.7 Magnon: Quantum of the Spin Wave

Nature presents us with various materials having different types of magnetic ordering, as a consequence of their lattice structures and the interactions between magnetic moments (spins) of the atoms. Among these, ferromagnetic materials are a specially simple case since all spins are preferentially ordered parallel to one another (in the low-temperature limit) and there is spontaneous magnetization present even in the absence of an external magnetic field [32]. When one or more spins are reversed (at $T \neq 0$), the system will be in an excited state and a low-lying elementary excitation, the *spin wave*, will move in a wave-like manner through the system of spins (see Fig. 1.13). A *magnon* is the quantized unit of the spin wave energy, defined by analogy with the phonon as a quantized lattice vibration.

Spin waves were postulated in 1930 by Felix Bloch [33]. Since that time, they have been extensively studied theoretically and experimentally (for reviews see Refs. [34–36]). The interest in studying these excitations is due to several factors. First, as with other excitations, spin waves play an important role in determining the thermodynamic properties of a material [37]. Second, they are affected by external factors, such as the radiation field, being probed by several experimental techniques, like microwave excitation [38,39], spectroscopic measurements, light scattering [40–42], among others. More recently, the non-linear properties of spin waves have gained much attention [43,44].

As a simple example, we now give a derivation of the magnon dispersion relation for a bulk ferromagnet by using a spin Hamiltonian of the Heisenberg type [45,46], namely

$$H = -\frac{1}{2} \sum_{i,j} J_{ij} \vec{S}_i \cdot \vec{S}_j - g\mu_B H_0 \sum_i S_i^z, \qquad (1.29)$$

1.7. MAGNON: QUANTUM OF THE SPIN WAVE

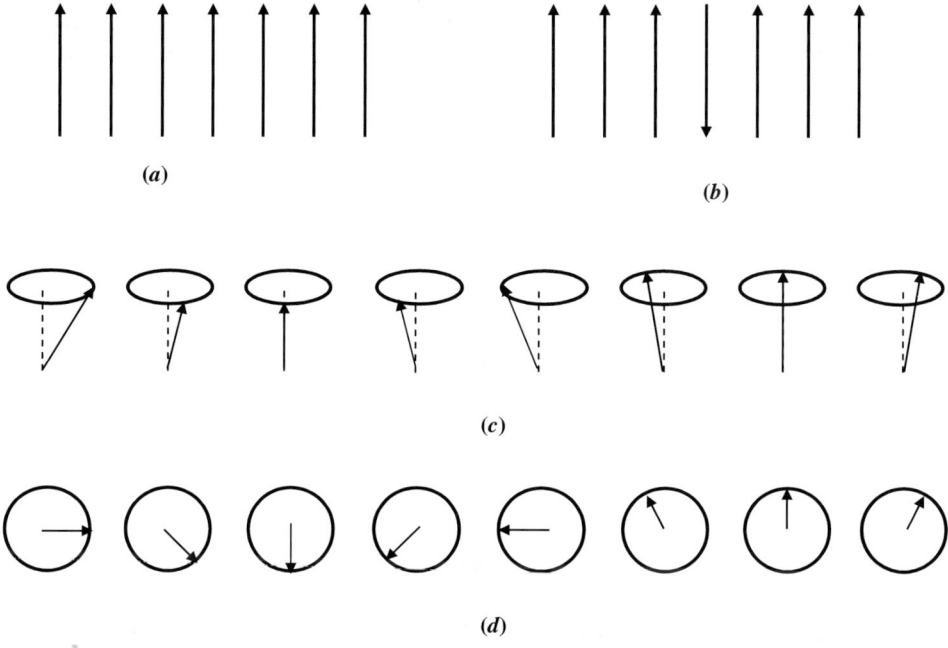

Fig. 1.13. (a) Classical picture of the ground state of a simple ferromagnet, showing all spin vectors aligned; (b) a possible excitation state, with just one spin reversed; (c) the low-lying elementary excitations, called *spin waves*, with the ends of vectors precessing conically; (d) the wave-like propagation of the spins as viewed from above.

where the sums are over all magnetic sites i and nearest neighbors j. Here, J_{ij} is the exchange coupling term between neighboring spins at sites i and j, g is the usual Landé factor, μ_B is the Bohr magneton, and H_0 is a static external magnetic field pointing in the z-direction. From elementary quantum mechanics (e.g. Refs. [47,48]) the equation of motion for the ladder spin operators $S_i^{\pm} = S_i^x \pm iS_i^y$ has the form (we adopt units in which $\hbar = 1$)

$$i\, dS_i^{\pm}/dt = [S_i^{\pm}, H]. \tag{1.30}$$

Using Eq. (1.29) this gives

$$i\, dS_i^{\pm}/dt = g\mu_B H_0 S_i^{\pm} \pm \sum_j J_{ij}(S_j^z S_i^{\pm} - S_i^z S_j^{\pm}) \tag{1.31}$$

at each site i. As there is no coupling term between S_i^+ and S_i^-, due to the absence of any anisotropic coupling terms in the magnetic Heisenberg Hamiltonian, we can work arbitrarily here with either one of these operators.

The product of spin operators in Eq. (1.31) may be simplified by means of the *random-phase approximation* (RPA) [49,50]. This consists of replacing the spin operator S^z in Eq. (1.31) by its thermal average $\langle S_z \rangle$ (which by symmetry is the same at all sites). Also, restricting our calculation to the low-temperature regime $T \ll T_c$ (where T_c is the Curie temperature), we can consider all the spins fully ordered, i.e. $\langle S_z \rangle = S$. Then the dispersion equation for a bulk spin wave can be found, within this linearization approximation, by solving

$$i\, dS_i^+/dt = g\mu_B H_0 S_i^+ + S \sum_j J_{ij}(S_i^+ - S_j^+). \tag{1.32}$$

We therefore define wavevector Fourier transforms by

$$S_j^+ = N^{-1/2} \sum_{\vec{k}} S_{\vec{k}}^+ \exp(i\vec{k} \cdot \vec{r}),$$

$$J_{ij} = N^{-1} \sum_{\vec{k}} J(\vec{k}) \exp[i\vec{k} \cdot (\vec{r_i} - \vec{r_j})], \tag{1.33}$$

where N is the number of magnetic sites, and we seek normal mode solutions with a time dependence like $\exp(i\omega t)$. This leads to $\omega = \omega_B(\vec{k})$, where the bulk (ferromagnetic) magnon frequency is

$$\omega_B(\vec{k}) = g\mu_B H_0 + S[J(0) - J(\vec{k})]. \tag{1.34}$$

For example, in the case of a simple cubic lattice with lattice parameter a and exchange coupling J, one has explicitly

$$J(\vec{k}) = 2J(\cos k_x a + \cos k_y a + \cos k_z a), \tag{1.35}$$

yielding

$$\omega_B(\vec{k}) = g\mu_B H_0 + 2SJ(3 - \cos k_x a - \cos k_y a - \cos k_z a). \tag{1.36}$$

This dispersion relation depends on the direction of the wavevector $\vec{k}$. We give below some specific values of the frequency for some particular values of $\vec{k}$:

$$\begin{aligned}
\omega_B(\vec{k}) &= g\mu_B H_0 & &\text{for } \vec{k}a = (0,0,0) \text{ (Brillouin zone center)}, \\
\omega_B(\vec{k}) &= g\mu_B H_0 + 4JS & &\text{for } \vec{k}a = (\pi,0,0), \\
\omega_B(\vec{k}) &= g\mu_B H_0 + 8JS & &\text{for } \vec{k}a = (\pi,\pi,0), \\
\omega_B(\vec{k}) &= g\mu_B H_0 + 12JS & &\text{for } \vec{k}a = (\pi,\pi,\pi).
\end{aligned} \tag{1.37}$$

The spectrum for three different directions of $\vec{k}$ is depicted in Fig. 1.14. Notice that at long wavelengths $ka \ll 1$, the above dispersion relation gives

$$\omega_B(\vec{k}) = g\mu_B H_0 + 2JSa^2 k^2 + O(k^4). \tag{1.38}$$

Therefore, for the magnon case in zero applied magnetic field, the frequency is proportional to k^2 when $k \to 0$. By contrast, the frequency of an acoustic phonon is proportional to k.

1.7. MAGNON: QUANTUM OF THE SPIN WAVE

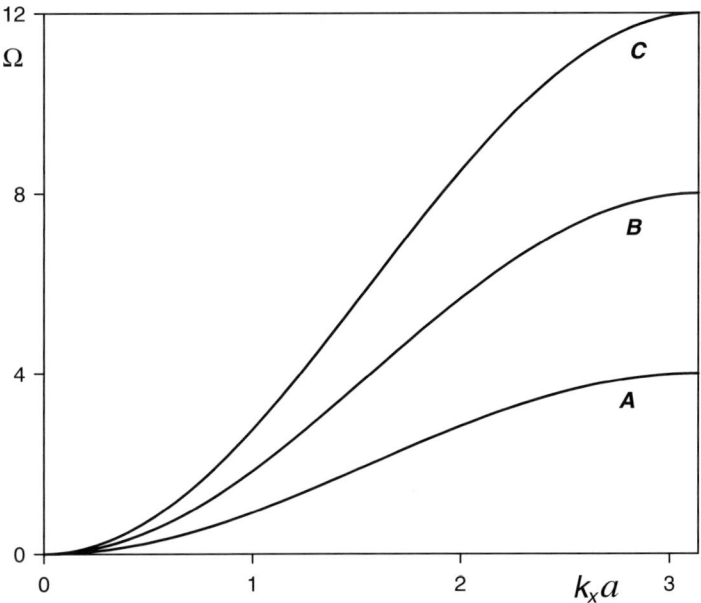

Fig. 1.14. Dispersion relation for bulk spin waves of a Heisenberg ferromagnet with nearest-neighbor interactions plotted as a characteristic frequency $\Omega = (\omega_B - g\mu_B H_0)/JS$ against the dimensionless component $k_x a$ of the wavevector $\vec{k}$ for three different directions: A, [100]; B, [110]; C, [111].

On the experimental side, the technique of inelastic neutron scattering has provided a very effective and direct method to determine the spin-wave energies as a function of wavevectors across the entire Brillouin zone (see e.g. Ref. [51]), just as in the phonon case. In an inelastic scattering process, a neutron may create or destroy a magnon. Diffraction of neutrons by a magnetic crystal also allows the determination of the distribution, direction, and order of the magnetic moments. If the incident neutron has wavevector $\vec{k}_n$ and is scattered to $\vec{k}'_n$ with the creation of a magnon of wavevector $\vec{k}$, then by conservation of crystal momentum (see Fig. 1.15)

$$\vec{k}_n = \vec{k}'_n + \vec{k} + \vec{Q}, \tag{1.39}$$

where $\vec{Q}$ is a reciprocal lattice vector. Also, by conservation of energy

$$\hbar^2 k_n^2 / 2M_n = \hbar^2 k_n'^2 / 2M_n + \hbar \omega_k, \tag{1.40}$$

where M_n denotes the mass of a neutron and $\hbar\omega_k$ is the energy of the magnon created in the process.

In solids with a more complex magnetic order and more complicated coupling terms (such as magnetic dipole–dipole interactions), the spin-wave spectrum is

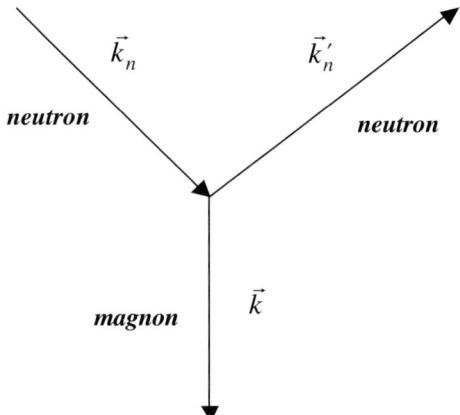

Fig. 1.15. Kinematics of the inelastic scattering of a neutron with creation of a magnon.

more complicated and may consist of many branches [52]. Apart from ferromagnets, two other important types of ordered magnetic materials are antiferromagnets and ferrimagnets. In both cases, the exchange interaction J_{ij} between neighboring spins is of the opposite sign from that for a ferromagnet, favoring an antiparallel alignment of spins. Thus, there are two (or more) interpenetrating sublattices of different spin types. In the case of an antiferromagnet, the two sublattices have equal (but oppositely directed) magnetic moments, and so there is no overall net magnetization in the material. In ferrimagnets the spins are such that the magnetic moments of the sublattices are unequal and so do not cancel. We shall mention these types of magnetic structures (and others) in later chapters.

References

[1] C. Kittel, Introduction to Solid State Physics, 7th ed., Wiley, New York, 1995.
[2] N.W. Ashcroft and N.D. Mermin, Solid State Physics, Saunders College Publishing, New York, 1976.
[3] J.S. Blakemore, Solid State Physics, 2nd ed., Cambridge University Press, Cambridge, 1985.
[4] L. Pauling, The Nature of the Chemical Bond, 3rd ed., Cornell University Press, Ithaca, 1960.
[5] W. Harrison, Electronic Structure and the Properties of Solids: The Physics of the Chemical Bond, Dover, New York, 1989.
[6] V. Heine, Group Theory in Quantum Mechanics, Pergamon, Oxford, 1960.
[7] M. Tinkham, Group Theory and Quantum Mechanics, McGraw-Hill, New York, 1964.
[8] F. Seitz, Modern Theory of Solids, McGraw-Hill, New York, 1940.
[9] J.M. Ziman, Principles of the Theory of Solids, Cambridge University Press, Cambridge, 1972.
[10] W.L. Bragg, Proc. Cambridge Phil. Soc. 17 (1913) 43.
[11] P.P. Ewald, Ann. Phys. 64 (1921) 253.

[12] H. Jones, The Theory of Brillouin Zones and Electronic States in Crystals, North-Holland, Amsterdam, 1962.
[13] R.E. Peierls, Quantum Theory of Solids, Oxford University Press, Oxford, 1964.
[14] B.S. Wherrett, Group Theory for Atoms, Molecules and Solids, Prentice-Hall, London, 1986.
[15] J. Callaway, Energy Band Theory, Academic Press, New York, 1964.
[16] M.G. Cottam and D.R. Tilley, Introduction to Surface and Superlattice Excitations, 2nd. ed., Institute of Physics Publishing, Bristol, 2004.
[17] I. Langmuir, J. Am. Chem. Soc. 38 (1916) 2221.
[18] C.J. Powell and J.B. Swan, Phys. Rev. 115 (1959) 869.
[19] C. de Witt, A. Blandin and C. Cohen-Tannoudji, Eds., Quantum Optics and Electronics, Les Houches Lectures, Gordon and Breach, New York, 1965.
[20] S.M. Sze, Semiconductor Devices, Wiley, New York, 1985.
[21] P.Y. Yu and M. Cardona, Fundamentals of Semiconductors, Springer-Verlag, Heidelberg, 1996.
[22] J. Frenkel, Phys. Rev. 37 (1931) 17.
[23] J. Frenkel, Phys. Rev. 37 (1931) 1276.
[24] G.H. Wannier, Phys. Rev. 52 (1937) 191.
[25] N.F. Mott, Trans. Faraday Soc. 34 (1938) 500.
[26] A.S. Davydov, Zh. Eksp. Teor. Fiz. 18 (1948) 210.
[27] A.S. Davydov, Tr. Inst. Fiz. Akad. Nauk. Ukr. S.S.R. 3 (1951) 36.
[28] O.J. Simpson, Proc. Roy. Soc. (London) A 238 (1956) 402.
[29] W. Kohn in: F. Seitz and D. Turnbull, Eds., Advances in Solid State Physics, Vol. 5, Academic Press, New York, 1957.
[30] G.H. Wannier, Elements of Solid State Theory, Cambridge University Press, Cambridge, 1959.
[31] L.J. Sham and T.M. Rice, Phys. Rev. 144 (1966) 708.
[32] E.P. Wohlfarth, Ferromagnetic Materials, North-Holland, Amsterdam, 1982.
[33] F. Bloch, Z. Physik 61 (1930) 206.
[34] J. van Kranendonk and J.H. Van Vleck, Rev. Mod. Phys. 30 (1958) 1.
[35] F. Keffer, Handbuch der Physik 18 (1966) 1.
[36] D.C. Mattis, The Theory of Magnetism, Springer-Verlag, Heidelberg, 1981.
[37] R.M. White, Quantum Theory of Magnetism, McGraw-Hill, New York, 1970.
[38] C.P. Slichter, Principles of Magnetic Resonance, Harper and Row, New York, 1963.
[39] M. Sparks, Ferromagnetic Relaxation Theory, McGraw-Hill, New York, 1964.
[40] W. Hayes and R. Loudon, Scattering of Light by Crystals, Wiley, New York, 1978.
[41] M.G. Cottam and D.J. Lockwood, Light Scattering in Magnetic Solids, Wiley, New York, 1986.
[42] M. Cardona and G. Güntherodt, Eds., Light Scattering in Solids, Vols. I to VI, Springer-Verlag, Heidelberg, 1975-91.
[43] V.S. L'vov, Wave Turbulence under Parametric Excitation, Springer-Verlag, Berlin, 1994.
[44] M.G. Cottam, Ed., Linear and Nonlinear Spin Waves in Magnetic Films and Superlattices, World Scientific, Singapore, 1994.
[45] W. Heisenberg, Z. Phys. 38 (1926) 411; 49 (1928) 619 .
[46] P.A.M. Dirac, Proc. Roy. Soc. (London) A 112 (1928) 661.
[47] A. Messiah, Quantum Mechanics, Vols. 1 and 2, Wiley, New York, 1968.
[48] C. Cohen-Tannoudji, B. Liu and F. Laloë, Quantum Mechanics, Vols. 1 and 2, Wiley, New York, 1977.
[49] R.A. Tahir-Kheli and D. Ter Haar, Phys. Rev. 127 (1962) 88, 95.
[50] W.E. Parry, The Many-Body Problem, Oxford University Press, Oxford, 1973.
[51] G.E. Bacon, Neutron Diffraction, 3rd. ed., Clarendon, Oxford, 1975.
[52] C. Kittel, Quantum Theory of Solids, 2nd. ed., Wiley, New York, 1987.

Chapter 2

Periodic and Quasiperiodic Structures

The remarkable electronic and structural properties occurring in materials science and device physics illustrate how new and exciting physics can arise unexpectedly due to the role of surface and interface states. Whenever new materials become available, whether they are purer than previously available or belong to a new family, or even have an artificial structure, their surface and interfaces lead to the development of new theoretical models and often to new device applications. For example, many years before the pioneering work of Born and von Karman [1,2] on the dynamical properties of infinitely extended crystals in the early 1900s, Lord Rayleigh [3] had shown, on the basis of the theory of elasticity, that a semi-infinite, elastically isotropic medium bounded by a single, stress-free plane surface could sustain surface vibrational modes of an acoustic character, the so-called *Rayleigh waves*. Much later, it was discovered that in addition to these waves, surface modes of an optical character can exist in a crystal with more than one atom in its primitive cell.

One reason why studies of surface and interfaces became more feasible in recent times was the rapid development of highly sophisticated growth techniques. Their industrial applications have, in turn, led to an increased development of these techniques. For example, Ge and Si single semiconductor crystals are amongst the purest elemental materials available today, as a result of years of perfecting their growth techniques [4].

Nowadays, developments in crystal-growth techniques have made materials science even more versatile. Advances in the fabrication of multilayer materials make it possible to reveal novel features of such structures. The techniques include modern layer-growth methods, such as molecular-beam epitaxy (MBE) and metal-organic chemical vapor deposition (MOCVD), among others [5–7], as well as characterization tools such as X-ray scattering, low-energy electron diffraction (LEED), neutron diffraction, etc. They have made possible the fabrication of layered materials with sharp, high-quality interfaces, and with dimensions comparable to the electron mean free path and the de Broglie wavelength. They allow also for an intriguing new class of artificial *periodic* (or superlattice) and *quasiperiodic* layered materials, in that their macroscopic properties are readily subject to design or control by varying the thickness or composition of the constituent films. In fact, some of these properties may be unique to the multilayer structure and provide the potential for device applications [8].

2.1 Periodic Structures

Tunneling of a particle through a potential barrier is one of the most studied phenomena in the quantum theory of matter, playing an important role in many semiconductor devices. In particular, the tunnel diode or *Esaki diode* discovered by Esaki [9] in 1958 involves tunneling through a forward-biased heavily doped (degenerate) junction in germanium. One important characteristic of the Esaki diode was that it exhibited negative differential resistance (NDR), making possible its application as a high-frequency oscillator [10]. In solid-state physics NDR can be employed to design an AC amplifier which, when coupled with a properly designed positive feedback circuit, can be made into an oscillator. Thus, an important application of materials exhibiting NDR is in the construction of high-frequency oscillators (typically for the microwave range of frequencies). This effect is observed in some semiconductors when the conduction band structure has special properties. An example is GaAs, where the conduction band minimum occurs at the Brillouin zone center and there are also conduction band minima at other points of the Brillouin zone, separated from the band edge by about 0.3 eV.

As a way of achieving this condition in any semiconductor, Esaki and Tsu [11] proposed in 1970 the fabrication of an *artificial periodic structure* consisting of alternate layers of two dissimilar semiconductors, with layer thickness of the order of nanometers. They called this synthetic structure a *superlattice* (SL), and they suggested that the artificial periodicity would "fold" the original Brillouin zones of the constituent bulk materials into smaller Brillouin zones or "mini-zones", for the symmetry reasons already discussed in Section 1.3 of Chapter 1. The artificial periodicity length could be made less than the electron mean free path and the de Broglie wavelength. They envisioned two types of synthesized SLs: doping and compositional, according to whether the SL potential was introduced by means of a periodic variation of impurities or composition, respectively, during epitaxial growth. In 1973, Tsu and Esaki [12] suggested that NDR can also be achieved in an SL, but it was only several years later that NDR was observed in a $GaAs/Ga_{1-x}Al_xAs$ SL [13]. By then, SLs had already opened up a new area of interdisciplinary investigations in the field of materials science and device physics, becoming a lively research area with a broad range of applications.

With the development of more sophisticated growth techniques, it is now possible to fabricate not only the SL structures envisioned by Esaki and Tsu, but also many other kinds of semiconductor structures often with layer thicknesses in the nanometer scale (and therefore referred to generally as *nanostructures*). One reason why interest in nanostructures has intensified recently is that their electronic, optical, vibrational, and magnetic properties are all profoundly modified as a result of their reduced dimensionality, their length scales, and their artificial symmetries.

An SL is a kind of planar or 2D nanostructure in the sense that it still has translational symmetry in directions parallel to the plane interfaces. Other examples of 2D nanostructures are the so-called quantum well (QW) and multiple quantum well (MQW) structures formed from two semiconductor materials. A

QW is a double heterojunction system [14] in which a thin layer of one material (e.g. GaAs) is sandwiched between two thick layers of another material (e.g. $Ga_{1-x}Al_xAs$ for a suitable Al concentration $x \sim 0.3$). In this case, the interfaces are abrupt and epitaxial (i.e. the materials are lattice matched), so the electron states are those of a square-well potential corresponding to the thickness of the GaAs layer (which represents the region of lower potential). The MQWs are similar in construction to the SLs proposed by Esaki and Tsu, in that they consist of a large number of heterojunctions between alternating layers of two materials (e.g. GaAs and $Ga_{1-x}Al_xAs$). However, the quantum well separations (the thickness of the $Ga_{1-x}Al_xAs$ barriers in this example) are large enough to inhibit electrons from tunneling from one well to another [14]. By contrast, in the SLs of Esaki and Tsu, the electrons tunnel through the barriers, so that they see the alternating layers as a periodic potential in addition to the crystal potential.

Nowadays, the term SL tends to be used rather widely to apply to any multilayer structure that is formed from two or more materials such that there is a periodicity (and hence a translation symmetry for the structure) in the growth direction perpendicular to the planar interfaces. Examples are repeats of a basic "unit cell" AB to form the two-component SL corresponding to $\cdots ABABABAB \cdots$, or repeats of ABC to form a three-component SL corresponding to $\cdots ABCABCABC \cdots$ (where A, B, and C represent layers of different materials). There is still a 2D symmetry parallel to the planar interfaces that is consistent with the lattice structure of the constituent materials.

Other structures with even lower dimensionality have also recently been fabricated and successfully studied (see e.g. Ref. [15]). These include the 1D nanostructures referred to as *quantum wires* and the nanometer-sized crystallites known as *quantum dots* (effectively zero-dimensional nanostructures), as well as thin films with a variety of nanoscale patterning applied to a surface (e.g. by etching or ion-beam techniques).

Fig. 2.1 shows a schematic illustration of an infinite binary SL consisting of an alternating $\cdots ABABABAB \cdots$ structure. Here medium A, representing the well (GaAs, for instance), has thickness a, while medium B, representing the barrier ($Ga_{1-x}Al_xAs$ for example), has thickness b. Each unit cell is labelled by an index n ($n =$ any integer) and has length $L = a+b$. There will be a new "mini" Brillouin zone associated with this length (see Section 1.3) with a wavevector component corresponding to $-\pi/L \leq k \leq \pi/L$.

The energy-band diagram of the above semiconductor structure is depicted in Fig. 2.2. Since the band gap of the well A (E_{gA}) is smaller in this case than that of the barrier B (E_{gB}), the conduction and valence band edges of A and B do not align with each other. The difference between their band edges is known as the band offset, producing the potential responsible for confining the carriers (electron and/or holes) in one layer only. Thus the control and understanding of this band offset is crucial in the fabrication of quantum confinement devices. While our current understanding of what determines the band offset of two dissimilar semiconductors is still incomplete, great progress has nevertheless been made in the fabrication techniques to control the shape of the band discontinuity [16].

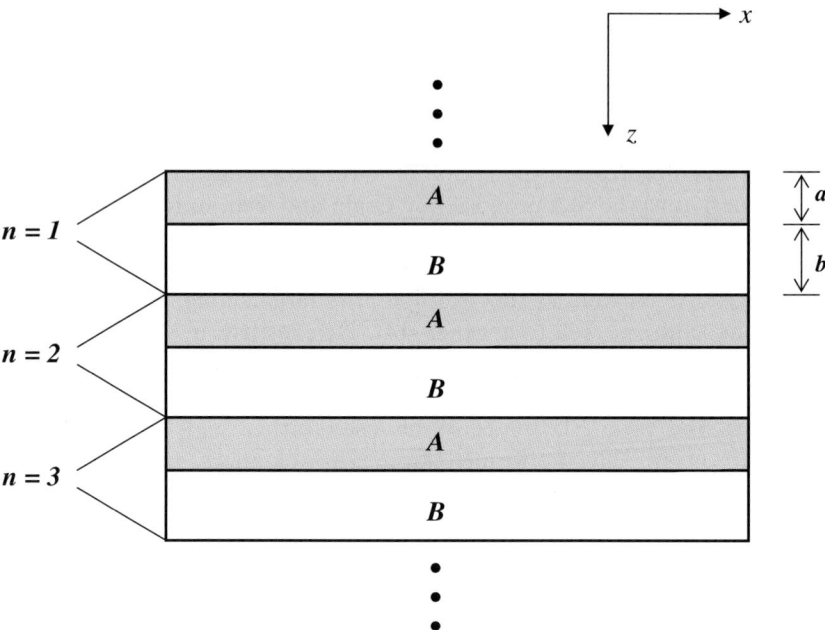

Fig. 2.1. Geometry of a two-component SL of infinite extent, composed of layers of media A and B. The unit cells of the structure are labelled by an integer n.

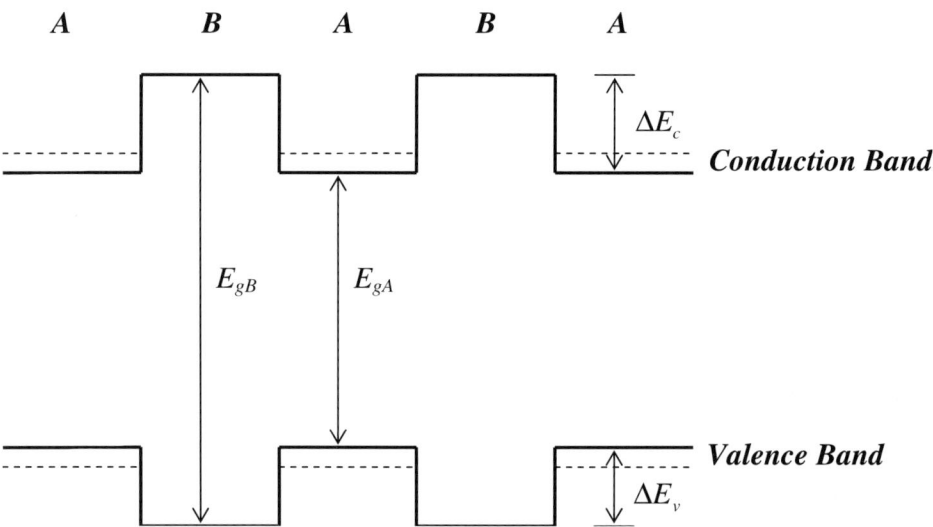

Fig. 2.2. Schematic energy-band diagram showing the regions of confinement of electrons or holes (dashed lines) in a periodic binary SL formed by two semiconductors A and B with band gaps E_{gA} and E_{gB}, respectively. Also shown are the conduction (ΔE_c) and valence (ΔE_v) band offset.

2.1. PERIODIC STRUCTURES

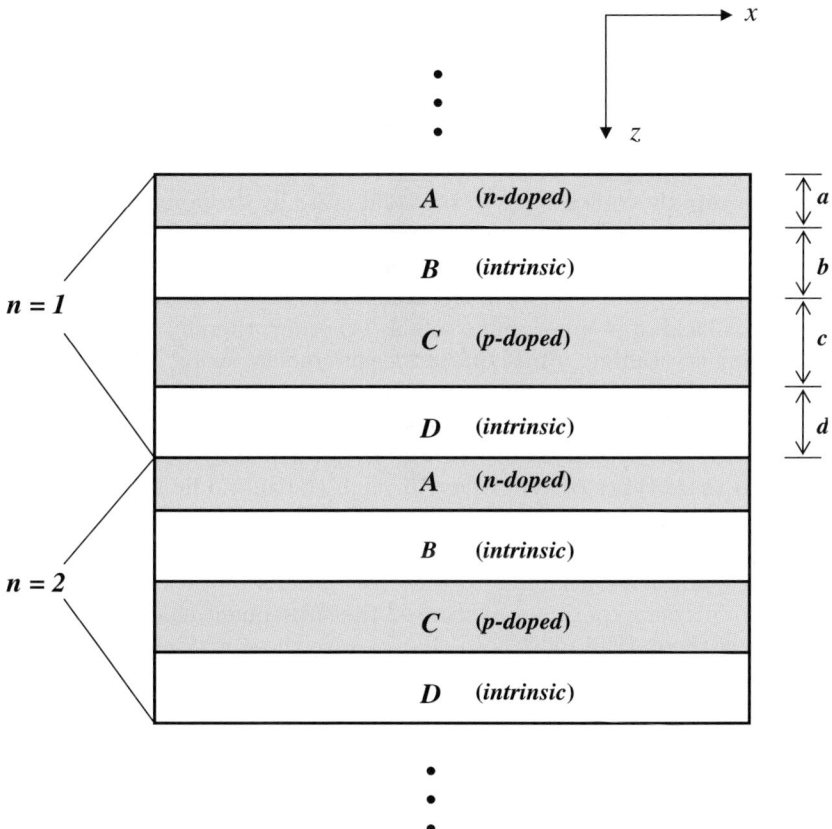

Fig. 2.3. Schematic illustration of a *nipi* semiconductor SL of period L. The unit cells of the SL are again indexed by an integer n.

One may also find more complex SL structures involving differently doped semiconductors. An example is when one layer is an n-type semiconductor while another layer is a p-type semiconductor, and there is an intrinsic layer (denoted by the letter i) at the junctions between them, thereby forming the so-called doping SL or *nipi* SL [17]. This layer structure is shown in Fig. 2.3, where materials A and C are n-doped and p-doped semiconductors with thicknesses a and c, respectively, while materials B and D are intrinsic semiconductors (or insulators) whose thicknesses are b and d, respectively. The unit cells now have length $L = a + b + c + d$ and are designated by an index n as before.

For historical reasons, we have introduced the concept of SLs in terms of the electronic properties of semiconductor materials. However, periodic SLs and their excitations have nowadays been studied for a wide range of different non-magnetic and magnetic materials (see e.g. Ref. [18]), as we shall discuss in subsequent chapters.

2.2 Quasiperiodic Structures

The subject of *quasicrystals* first achieved prominence in 1984, following the report by Schechtman and collaborators [19] of metallic Al–Mn alloys showing very surprising and interesting electron diffraction data. They mixed Al and Mn in a roughly six-to-one proportion and heated the mixture until it melted. The mixture was then rapidly cooled back to the solid state by dropping the liquid on to a cold spinning wheel, a process known as melt spinning. When the solidified alloy was examined, using an electron microscope, a novel structure was revealed. It exhibited five-fold symmetry, which is forbidden in ideal crystals, and a long-range order, which is lacking in amorphous solids. Its order, therefore, was neither truly amorphous nor crystalline. Subsequent measurements, using X-ray scattering at much higher resolution, led to electron diffraction patterns showing not only five-fold but also icosahedral symmetries, forbidden by the rules of crystallography (for reviews see Refs. [20–24]). Theoretical studies developed by Levine and Steinhardt [25] explained these types of symmetry through the aperiodic 2D and 3D Penrose tilings [26] in their diffraction patterns. Tiling is the geometrical operation that results in filling a space with an arrangement of regular polyhedra. Their predictions were, indeed, qualitatively similar to the observations by Schechtman et al. [19]. In addition to further experimental studies, the subsequent challenge has been the development of theoretical models to characterize these artificial structures.

Although the term *quasicrystal* is more appropriate when applied to natural compounds or artificial alloys, in 1D there is no difference between this and the *quasiperiodic* structures formed by the incommensurate arrangement of periodic unit cells. The particular mathematical sequences (Fibonacci, Thue–Morse, etc.) that define the quasiperiodic structures will be discussed in the following section of this chapter. An appealing motivation for studying such structures is that they exhibit a highly fragmented energy spectrum displaying a self-similar pattern. Indeed, from a strictly mathematical perspective, it has been proven that their spectra are *Cantor sets* in the thermodynamic limit [27].

A fascinating feature of these quasiperiodic structures is that they exhibit collective properties not shared by their constituent parts and distinct also compared with the periodic SLs. Furthermore, the long-range correlations induced by the construction of these systems are expected to be reflected to some degree in their various spectra (as in light propagation, electronic transmission, density of states, polaritons, etc.), defining a novel description of disorder [28,29]. Indeed, theoretical transfer matrix treatments (see later) can be used to show that these spectra are fractals [30].

The presence and nature of long-range correlations in such systems preclude using canonical approaches like perturbation theory, where one first separates a small localized piece of the system, treating the rest as a perturbation a posteriori. This approach typically does not work for the cases under consideration here, because the behavior of the overall macroscopic system is quite distinct from the behavior of its separate small pieces, due to the long-range correlations. Fortunately, the presence of long-range correlations itself provides the key to circumventing this

2.2. QUASIPERIODIC STRUCTURES

difficulty, namely that these systems are normally robust to wide modifications on a microscopic scale. An important consequence of this robustness, where many systems that are distinct within a microscopic scale may exhibit the same critical behavior, is that one can classify the various systems in a few *universality classes* (for details see Ref. [31]). For an analogy, we may consider the topic of continuous phase transitions: the critical behavior is known to depend only upon global properties, namely the geometric dimension of the system and the symmetries of its order parameter. Yet it is insensitive to the details of the microscopic interactions between the atoms or molecules [32]. A striking example is the use of the Ising model of interacting spins to describe water: the Ising classical spins oriented *up* (or *down*) are taken to indicate the presence (or absence) of a molecule at sites on a lattice, while the complicated interactions between these molecules are replaced by a nearest-neighbor effective exchange coupling. In spite of its simplicity, this model reproduces quite well many aspects of the behavior of water near its critical temperature [33,34].

The pioneering experimental works of Merlin and collaborators on non-periodic Fibonacci [35–37] and Thue–Morse [38] GaAs–AlAs SLs have generated a large amount of research activity in the field of quasicrystals. Basically, these systems involve defining two distinct building blocks, each of them containing the required physical information, and then having them ordered in a desired manner. For instance, they can be described in terms of a series of generations that obey a particular recursion relation. Thus, they can be regarded as intermediate systems between periodic crystals and random amorphous solids [39], and this is one of the features that makes them of particular interest to study.

On the theoretical side, the spectra of many types of elementary excitations in quasiperiodic structures have been extensively studied by numerous groups. In all cases the spectra were found to be Cantor-like with critical eigenfunctions. For electronic systems, exact eigenfunctions were found only at the special null energy value. However, there are infinitely many eigenvalues in the energy spectrum, although they are rare for the electron chaotic orbits [40,41]. An important issue is to understand the wave functions corresponding to these chaotic orbits. We note that it does not necessarily follow that the wave functions themselves are chaotic, because the orbits represent only selected points on the lattice [42]. In addition, there may be a discrete set of extended states. Similar types of behavior can be found for the phonon problem, and indeed a one-to-one correspondence may be established between certain phonon and electronic spectra (for a review see Ref. [43]). A quite complex *fractal energy spectrum*, which can be considered as a basic signature, is a common feature of these systems. Several different mathematical techniques, including renormalization group theory [44], the transfer matrix method [45], and chaotic Hamiltonian systems [46], to mention just a few, have been successfully applied, leading to remarkable results. For example, for the spectrum in the Thue–Morse case, it is known that the structure factor is composed of a sequence of delta-function peaks [38]. However, there are some conflicting issues. Some authors [47] argue that the results in the case of the electronic properties of the Thue–Morse sequence should depend on the kind of model, whereas others [48] have shown that the trace map of the transfer matrix does not depend on the particular model employed.

Another important motivation for studying these structures comes from recognizing that the localization of electronic states, one of the most active fields in condensed-matter physics, could occur not only in disordered systems but also in deterministic quasiperiodic systems [49,50]. Localization due to electronic properties was studied for a tight-binding Schrödinger equation in 1D by several groups [51–53]. On the other hand, calculations of plasmon-polariton spectra were also reported by Albuquerque and Cottam [54–56]. These latter excitations could provide an excellent way to probe experimentally these localized states, because the localization is essentially due to the wave nature of the electronic states and thus could be found in any wave phenomenon. Recently, there have been several experiments on the localization of photons [57–60], as well as phonons [61], in random media. Furthermore, there are distinct advantages to studying localization using a classical wave equation instead of via the quantum mechanical electronic problem. Indeed, the latter usually deals with other types of interactions, such as the spin–orbit coupling, electron–phonon coupling, and electron–electron interactions, among others, which make the problem more complex and even intractable.

In disordered dielectric materials, experimental proof of the complete localization of light waves (as indicated by a vanishing diffusion coefficient) is difficult. Recently, an unusually small optical diffusion coefficient, consistent with an onset of localization, has been realized in transmission and scattering experiments with microwave signals in random mixtures of Al and Teflon spheres [62], where theoretically predicted scaling properties of the transmission with sample thickness were verified.

2.3 Examples of Quasiperiodic Structures

The quasiperiodic structures considered in this book are of the type generally known as *substitutional sequences*. The sequences generated by substitutions have been studied in several areas of mathematics [63–65], computer science [66,67], and cryptography [68]. The more recent applications in physics have been outlined in the last section. The sequences are characterized by the nature of their Fourier spectrum, which can be dense pure point (as for Fibonacci sequences) or singular continuous (as for Thue–Morse and Double-period sequences) [69].

We start with some general mathematical considerations and terminology. First, we give the definition of a substitutional sequence of the type used here. Consider a finite set ξ (here $\xi = \{A, B\}$, for example, with A and B being two different building blocks) called an *alphabet*, and denote by ξ^* the set of all "words" of finite length (such as $AABAB$) that can be written in this alphabet. Now let us define ζ as a map from ξ to ξ^* by specifying that ζ acts on a word by substituting each letter (e.g. A) of this word by its corresponding image, denoted by $\zeta(A)$. A sequence is then called a substitutional sequence if it is a fixed point of ζ, i.e. if it remains invariant when each letter in the sequence is replaced by its image under ζ.

2.3. EXAMPLES OF QUASIPERIODIC STRUCTURES

These substitutional sequences are described in terms of a series of generations that obey particular inflation rules. Let $a_1, a_2, ..., a_g$ be g basic units, and define this pattern as stage n of the sequence. Then the next stage $n+1$ of the sequence is obtained inductively from stage n by the inflation rule $\vec{a} \to \bar{M}\vec{a}$, where $\vec{a}$ represents the column vector $(a_1, a_2, ..., a_g)^t$ with t denoting the transpose. Also $\bar{M} = (m_{ij})$ is a $g \times g$ matrix with non-negative integer matrix elements. The matrix $\bar{M}$ and its successive applications fully determine the sequence. At each stage, a_i is replaced by $m_{i1}a_1$, followed by $m_{i2}a_2$,..., etc., for $i = 1, 2..., g$. For example, for the case of the Fibonacci lattice to be discussed shortly, we would have $g = 2$ and it turns out that we would operate with the 2×2 substitution matrix $\bar{M}$,

$$\bar{M} = \begin{pmatrix} 1 & 1 \\ 1 & 0 \end{pmatrix}, \tag{2.1}$$

on the vector $(a_1, a_2)^t$ at each stage. This gives, in terms of the building blocks A and B, the substitution rules $A \to AB$, $B \to A$ which will then generate the whole sequence, provided we start with AB as the first compound "word" of the sequence. Similar procedures can be identified to generate other quasiperiodic sequences.

We now proceed to give explicit definitions of the main substitutional sequences to be used here. Other examples will be mentioned later in the book, as required.

2.3.1 Cantor

Probably the most well-known and simple deterministic fractal geometry is the *triadic* Cantor sequence [70]. This set is obtained through the repetition of a simple rule: divide any given segment into three equal parts, then eliminate the central one (we may call this the *inbound* Cantor sequence), and continue this process. For example, if we start algebraically with the closed set $S_0 = [0, 3]$ of all numbers from 0 to 3 and remove its open middle third, we are left with the pair of closed intervals $[0, 1]$ and $[2, 3]$ representing S_1. The open middle thirds in each of these intervals would be removed again to produce four smaller intervals representing S_2, and so on. After many stages, we would have a large number of small intervals, separated by gaps of various sizes.

For applications to the building blocks of multilayered structures, it is more appropriate to consider instead the so-called *outbound* Cantor sequence. This has its nth stage defined in terms of the previous stage by the rule $S_n = S_{n-1} B_n S_{n-1}$, with initial conditions taken as $S_0 = A$ and $S_1 = AB_1 A$. In this case B_n for the nth sequence stage differs from the basic $B_1 (\equiv B)$ for the first stage only by its thickness $d_{B_n} = 3^{n-1} d_{B_1}$. We can also construct the same sequence rather more straightforwardly by the substitutional transformations $A \to ABA$, $B \to BBB$.

The resulting Cantor generations are therefore

$$S_0 = A; \quad S_1 = ABA; \quad S_2 = ABABBBABA; \quad \text{etc.} \tag{2.2}$$

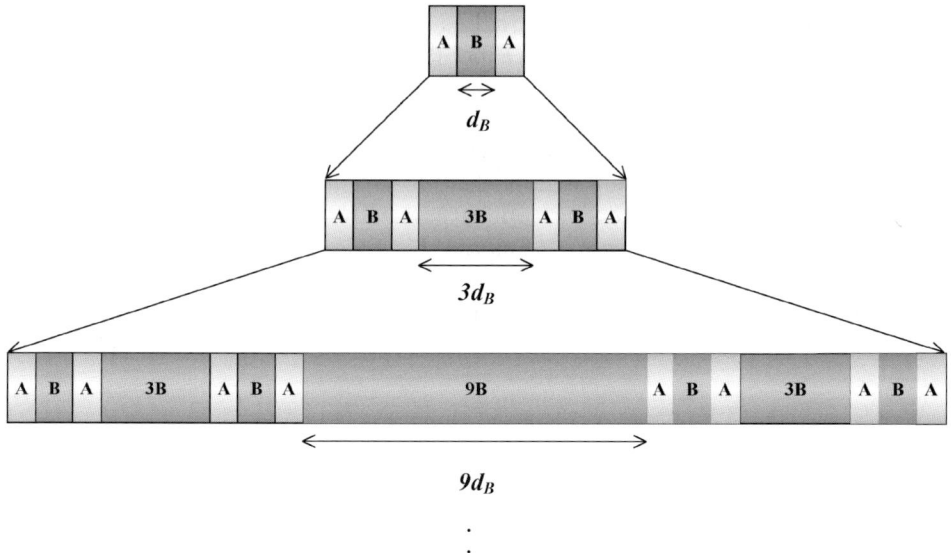

Fig. 2.4. Schematic illustration of the Cantor quasiperiodic structure.

and are represented more clearly by the diagrammatic expansion scheme shown in Fig. 2.4.

The *fractal* properties of the quasiperiodic structures will turn out to be of relevance in later chapters for a description of the excitation spectrum. Roughly speaking, fractals can be thought of as complex geometric shapes with fine structure at arbitrary small scales. It is obvious how this fine structure comes about in the above example of the Cantor sequence because of the repeated division into smaller intervals. Some general accounts of fractals and their properties are to be found in several books, e.g. Refs. [70–72]. The concept of a *fractal dimension* is often useful in relation to the *self-similarity* property, i.e. again roughly, if we magnify a tiny part of the fractal we will see features reminiscent of the whole. If, from a d-dimensional object (a "box") of size l, some number N conformal copies of reduced size lr (with $0 < r < 1$) are produced and the process is repeated a large number of times, then the fractal dimension D_0 can be defined (see Ref. [73]) by the relation $N = \exp(-rD_0)$, or equivalently

$$D_0 = \ln(N)/\ln(1/r). \tag{2.3}$$

For the present example of the Cantor sequence, we have $N = 2$ and $r = \frac{1}{3}$, so it has a fractal dimension equal to a non-integer, namely $\ln 2/\ln 3 \simeq 0.63$. This is less than its geometric dimension $d = 1$.

2.3.2 Fibonacci

The Fibonacci sequence is the oldest example of an aperiodic chain of numbers. It was developed by Leonardo de Pisa (whose nickname was Fibonacci, which

2.3. EXAMPLES OF QUASIPERIODIC STRUCTURES

means son of Bonacci) in 1202 as a result of his investigation on the growth of a population of rabbits. The successive Fibonacci numbers are generated by adding together the two previous numbers in the sequence, after specifying suitable initial conditions.

For our purposes, a Fibonacci structure can be realized experimentally by juxtaposing the two basic building blocks A and B in such a way that the nth stage of the process S_n is given by the recursive rule $S_n = S_{n-1}S_{n-2}$, for $n \geq 2$, starting with $S_0 = B$ and $S_1 = A$. It has the property of being invariant under the transformations $A \to AB$ and $B \to A$.

The Fibonacci generations are (see Fig. 2.5a)

$$S_0 = B; \quad S_1 = A; \quad S_2 = AB; \quad S_3 = ABA; \quad S_4 = ABAAB; \quad \text{etc.} \qquad (2.4)$$

In this case the number of building blocks increases in accordance with the Fibonacci number F_n defined by the rule $F_n = F_{n-1} + F_{n-2}$ (with $F_0 = F_1 = 1$). Also the ratio between the number of building blocks A and the number of building blocks B in the sequence tends to the golden mean number $\tau = (1 + \sqrt{5})/2 \simeq 1.62$ for large generation number n. This particular irrational number is related to five-fold symmetries (e.g. it is twice the ratio of the distance between the center-vertex and the center mid-edge in a pentagon). It is interesting to note that *all* the Fibonacci numbers can be generated from the golden mean number through the relation $F_n = [\tau^n - (-\tau)^{-n}]/\sqrt{5}$. This means that a sequence of *rational* numbers (namely the integer-valued Fibonacci numbers) can be obtained from powers of *irrational* numbers!

There are variations of the above sequence leading to *generalized Fibonacci structures* that involve different relationships between the number of building blocks A and the number of building blocks B (thus generalizing also the golden mean number). In these cases, the nth stage of the structure S_n is taken to be generated by the sequence given recursively as

$$S_{n+1} = S_n^p S_{n-1}^q \qquad (2.5)$$

with, as before, $S_0 = B$ and $S_1 = A$. Here the indexes p and q are arbitrary positive integer numbers and $n \geq 1$. The above notation means that S_n^p represents p adjacent repetitions of the stack S_n. This type of "inheritance" is normal in iterative processes and frequently produces self-similar structures that are the basis of fractal configurations. When $p = q = 1$ (the simplest possible case), we have just the well-known Fibonacci sequence discussed previously. Equivalently, the generalized Fibonacci sequences can also be generated by the substitutional relation

$$B \to A, \quad A \to A^p B^q, \qquad (2.6)$$

where A^p (or B^q) represents a string of p A-blocks (or q B-blocks). The total number of blocks in S_n is equal to the generalized Fibonacci number denoted by

F_n, and given now by the recurrence relation

$$F_n = pF_{n-1} + qF_{n-2}, \qquad (2.7)$$

with initial values $F_0 = F_1 = 1$. It follows that the characteristic value $\sigma(p,q)$, defined as being the ratio of F_n to F_{n-1} in the limit of $n \to \infty$, must satisfy the quadratic equation

$$\sigma^2 - p\,\sigma - q = 0. \qquad (2.8)$$

Solving for the positive root gives explicitly

$$\sigma(p,q) = \lim_{n\to\infty} F_n/F_{n-1} = \frac{p + \sqrt{p^2 + 4q}}{2}. \qquad (2.9)$$

This expression generalizes the previous golden-mean result and introduces other types of means, depending on the values of p and q. For instance, for $p = q = 1$, we have $\sigma(1,1) \equiv \sigma_g$ ($\equiv \tau \simeq 1.62$), the well-known golden mean. Similarly, $\sigma(2,1) \equiv \sigma_s$ ($\simeq 2.41$) is the silver mean, $\sigma(3,1) \equiv \sigma_b$ ($\simeq 3.30$) is the bronze mean, and $\sigma(1,3) \equiv \sigma_n$ ($\simeq 2.30$) is the nickel mean.

It is worth briefly mentioning here a mathematical aspect that has interesting implications for the physical properties of a quasiperiodic system. We may note that the expression for σ in Eq. (2.8) is formally equivalent to a result arising when determining the eigenvalues of the substitution matrix $\bar{M}$ introduced earlier in this section. This was exemplified by Grimm and Baake [74] in treating a quantum spin chain with quasiperiodic pair interactions. Essentially they were able to classify the different substitutional sequences based on the irrationality of $\sigma^-(p,q)$, which denotes the *negative* root of Eq. (2.8). They found that if $|\sigma^-(p,q)| < 1$, it is a so-called Pisot–Vijayraghavan (PV) irrational number, and the fluctuations of the physical properties associated with the sequence are relatively well behaved and stable. On the other hand, if $|\sigma^-(p,q)| > 1$, the fluctuations of the physical properties are almost chaotic. For the examples of generalized Fibonacci cases mentioned above, only the nickel-mean sequence is not a PV type and therefore we expect a more chaotic behavior of its physical properties (as found in the specific heat calculations discussed in Ref. [75]).

2.3.3 Thue–Morse

The Thue–Morse sequence first came about as the result of systematic studies of aperiodic chains initiated by Thue [76] in 1906. His results seem to have been rediscovered many times since then, but the most important contribution to applying and understanding this sequence was made in 1921 by Morse [77,78] in the context of topological dynamics. Although there are several ways to define the Thue–Morse sequence, it is easy to prove they are equivalent to each other. Hence, taking its simplest form, the Thue-Morse sequence can be defined by the recursive relations $S_n = S_{n-1} S^+_{n-1}$ and $S^+_n = S^+_{n-1} S_{n-1}$ (for $n \geq 1$), with $S_0 = A$ and $S^+_0 = B$. An alternative (and perhaps more straightforward) way to build up this sequence is through the inflation rules $A \to AB$ and $B \to BA$.

2.3. EXAMPLES OF QUASIPERIODIC STRUCTURES

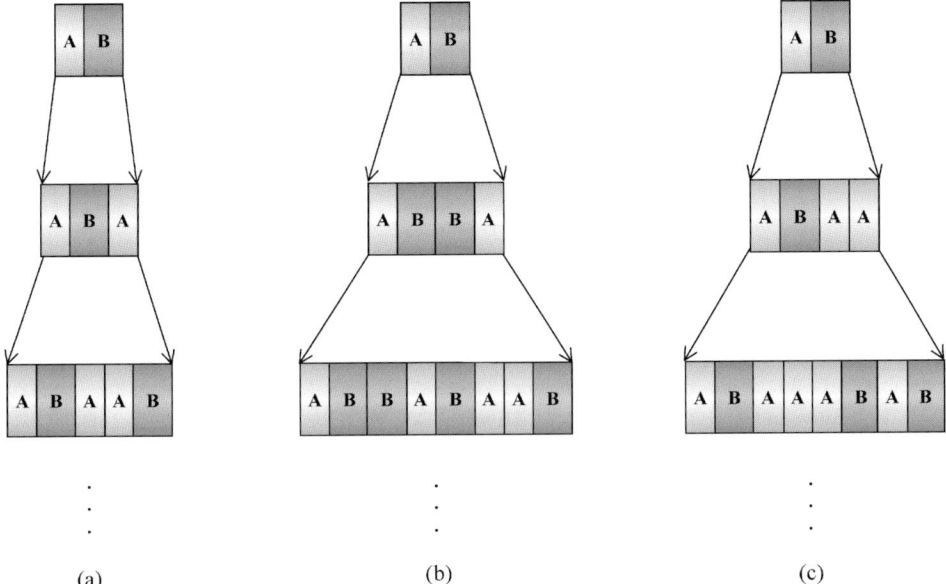

Fig. 2.5. Schematic illustration of the other quasiperiodic structures considered: (a) Fibonacci; (b) Thue–Morse; (c) Double period.

The Thue–Morse generations are (see Fig. 2.5b)

$$S_0 = A; \quad S_1 = AB; \quad S_2 = ABBA; \quad S_3 = ABBABAAB; \quad \text{etc.} \tag{2.10}$$

The number of building blocks in this quasiperiodic system increases with n as 2^n, while the ratio of the number of building blocks A to the number of building blocks B is a constant and equal to unity. In this respect, it is simpler than the Fibonacci case. We notice that B blocks can occur adjacent to one another, whereas this is not possible for the original ($p=q=1$) Fibonacci structure.

2.3.4 Double Period

The final example to be mentioned here, the Double-period sequence, is one of the newest of the aperiodic chains to be investigated. It has its origin in the study of system dynamics [79] and in the applications of lasers to nonlinear optical fibers [80]. Its recursion relation is superficially somewhat similar to the Thue–Morse case: the nth stage is given by $S_n = S_{n-1} S_{n-1}^+$ and $S_n^+ = S_{n-1} S_{n-1}$ for $n \geq 1$, with $S_0 = A$ and $S_0^+ = B$. It is also invariant under the transformations $A \to AB$, $B \to AA$, which makes the distinction clearer.

The Double-period generations are (see Fig. 2.5c)

$$S_0 = A; \quad S_1 = AB; \quad S_2 = ABAA; \quad S_3 = ABAAABAB; \quad \text{etc.} \tag{2.11}$$

The number of building blocks for this sequence increases with n in the same way as in the Thue–Morse sequence, i.e. like 2^n. However, it differs from the previous case in that the ratio between the number of building blocks A to the number of building blocks B is not constant: it tends to 2 as the number of generations goes to infinity. Also, the B blocks always occur singly, as in the Fibonacci case but unlike the Thue–Morse case.

References

[1] M. Born and Th. von Karman, Phys. Z. 13 (1912) 297.
[2] M. Born and Th. von Karman, Phys. Z. 14 (1913) 15.
[3] Lord Rayleigh, Proc. London Math. Soc. 17 (1887) 4.
[4] E.E. Haller and F.S. Goulding in: C. Hilsum, Ed., Handbook on Semiconductors, Vol. 4, Elsevier, Amsterdam, 1993.
[5] L.L. Chang and K. Ploog, Eds., Molecular Beam Epitaxy and Heterostructure, Plenum, New York, 1985.
[6] E.H. Parker, Ed., The Technology and Physics of Molecular Beam Epitaxy, Plenum, New York, 1985.
[7] J. George, Preparation of Thin Films, Dekker, New York, 1992.
[8] M. Jaros, Physics and Applications of Semiconductor Microstructures, Oxford University Press, Oxford, 1989.
[9] L. Esaki, Phys. Rev. 109 (1958) 603.
[10] S. Wang, Fundamentals of Semiconductor Theory and Device Physics, Prentice-Hall, Englewood Cliffs, NJ, 1989.
[11] L. Esaki and R. Tsu, IBM J. Res. Develop. 14 (1970) 61.
[12] R. Tsu and L. Esaki, Appl. Phys. Lett. 22 (1973) 562.
[13] A. Silbille, J.F. Palmier, H. Wang and F. Mollot, Phys. Rev. Lett. 64 (1990) 52.
[14] K. Hess, Advanced Theory of Semiconductor Devices IEEE Press, New Jersey, 1999.
[15] C. Weisbuch and B. Vintner, Quantum Semiconductor Structures, Academic, Boston, 1991.
[16] R. Dingle, W. Weigmann and C.H. Henry, Phys. Rev. Lett. 33 (1974) 827.
[17] K. Ploog and G.H. Döhler, Adv. Phys. 32 (1983) 285.
[18] M.G. Cottam and D.R. Tilley, Introduction to Surface and Superlattice Excitations, 2nd. ed., Institute of Physics Publishing, Bristol, 2004.
[19] D. Schechtman, I. Blech, D. Gratias and J.W. Cahn, Phys. Rev. Lett. 53 (1984) 1951.
[20] P.J. Steinhard and S. Ostlund, The Physics of Quasicrystals, World Scientific, Singapore, 1987.
[21] C. Janot, Quasicrystals, a Primer, Oxford University Press, Oxford, 1993.
[22] F. Hippert and D. Gratias, Eds., Lectures on Quasicrystals, Les Editions de Physique, Les Ulis, 1994.
[23] M. Senechal, Quasicrystals and Geometry, Cambridge University Press, Cambridge, 1995.
[24] D.P. DiVincenzo and P.J. Steinhardt, Eds., Quasicrystals: The State of the Art, 2nd ed., World Scientific, Singapore, 1999.
[25] D. Levine and P.J. Steinhardt, Phys. Rev. Lett. 53 (1984) 2477.
[26] R. Penrose, Bull. Inst. Math. Appl. 10 (1974) 266.
[27] A. Bovier and J.-M. Ghez, Commun. Math. Phys. 158 (1993) 45.
[28] F. Axel and J. Peyriére, C. R. Acad. Sci. de Paris 306 (1988) 179.

REFERENCES

[29] F. Axel and H. Terauchi, Phys. Rev. Lett. 66 (1991) 2223.

[30] E. Maciá and F. Domínguez-Adame, Electrons, Phonons and Excitons in Low Dimensional Aperiodic Systems, Editorial Complutense, Madrid, 2000.

[31] L.E. Reichl, A Modern Course in Statistical Physics, Texas University Press, Austin, 1980.

[32] H.E. Stanley, Introduction to Phase Transitions and Critical Phenomena, Oxford University Press, Oxford, 1971.

[33] J.C. Sartorelli, W.M. Gonçalves and R.D. Pinto, Phys. Rev. B 49 (1994) 3963.

[34] T.J.P. Penna, P.M.C. de Oliveira, J.C. Sartorelli, W.M. Gonçalves and R.D. Pinto, Phys. Rev. E 52 (1995) R2168.

[35] R. Merlin, K. Bajema, R. Clarke, F.-Y. Juang and P.K. Bhattacharya, Phys. Rev. Lett. 55 (1985) 1768.

[36] J. Todd, R. Merlin, R. Clark, K.M. Mohanty and J.D. Axe, Phys. Rev. Lett. 57 (1986) 1157.

[37] K. Bajema and R. Merlin, Phys. Rev. B 36 (1987) 4555.

[38] Z. Cheng, R. Savit and R. Merlin, Phys. Rev. B 37 (1988) 4375.

[39] M. Kohmoto and J.R. Banavar, Phys. Rev. B 34 (1984) 563.

[40] R.E. Prange, D.R. Grempel and S. Fishman, Phys. Rev. B 29 (1984) 6500.

[41] J.M. Luck and D. Petritis, J. Stat. Phys. 42 (1986) 289.

[42] J. Bellisard, D. Bessis and P. Moussa, Phys. Rev. Lett. 49 (1982) 701.

[43] M. Quilichini and T. Janssen, Rev. Mod. Phys. 69 (1997) 277.

[44] S. Ostlund, R. Pandit, D. Rand, H.J. Schellnhuber and E.D. Siggia, Phys. Rev. Lett. 50 (1983) 1873.

[45] M. Kohmoto, L.P. Kadanoff and C. Tang, Phys. Rev. Lett. 50 (1983) 1870.

[46] K. Nakamura, Quantum Chaos: A New Paradigm of Nonlinear Dynamics, Cambridge University Press, Cambridge, 1993.

[47] A. Chakrabarti, S.N. Karmakar and R.K. Moitra, Phys. Rev. Lett. 74 (1995) 1403.

[48] M. Kolár and M.K. Ali and F. Nori, Phys. Rev. B 43 (1991) 1034.

[49] P.A. Lee and T.V. Ramakrishnan, Rev. Mod. Phys. 57 (1985) 287.

[50] J.B. Sokoloff, Phys. Rep. 126 (1985) 189.

[51] S. Ostlund and R. Pandit, Phys. Rev. B 29 (1984) 1394.

[52] J.P. Lu, T. Odagaki and J.L. Birman, Phys. Rev. B 33 (1986) 4809.

[53] F. Nori and J.P. Rodriguez, Phys. Rev. B 34 (1986) 2207.

[54] E.L. Albuquerque and M.G. Cottam, Solid State Commun. 81 (1992) 383.

[55] E.L. Albuquerque and M.G. Cottam, Solid State Commun. 83 (1992) 545.

[56] E.L. Albuquerque, Phys. Lett. A 181 (1993) 409.

[57] Y. Kuga and A. Ishimaru, J. Opt. Soc. Am. A 1 (1984) 831.

[58] M.P. van Albada and A. Lagendijk, Phys. Rev. Lett. 55 (1985) 2692.

[59] P.E. Wolf and G. Maret, Phys. Rev. Lett. 55 (1985) 2696.

[60] M.P. van Albada, M.P. van der Mark and A. Lagendijk, Phys. Rev. Lett. 58 (1987) 361.

[61] S. He and J.D. Maynard, Phys. Rev. Lett. 57 (1986) 3171.

[62] A.Z. Genak and N. Garcia, Phys. Rev. Lett. 66 (1991) 2064.

[63] F.M. Dekking, J. Comb. Theory 27A (1976) 292.

[64] F.M. Dekking, C. R. Acad. Sci. de Paris 285 (1977) 157.

[65] G. Christol, T. Kamae, M. Mendes-France and G. Rauzy, Bull. Soc. Math. (France) 108 (1980) 401.

[66] A. Cobham, Math. Syst. Theory 3 (1969) 186.

[67] A. Cobham, Math. Syst. Theory 6 (1972) 164.

[68] G.T. Herman and G. Hozenberg, Developmental Systems and Languages, North-Holland, Amsterdam, 1975.

[69] M. Queffélec, Substitution Dynamical Systems: Spectral Analysis, Lecture Notes in Mathematics Vol. 1294, Springer-Verlag, Heidelberg, 1987.

[70] B.B. Mandelbrot, The Fractal Geometry of Nature, Freeman, New York, 1982.

[71] P.G. Drazin, Nonlinear Systems, Cambridge University Press, Cambridge, 1992.

[72] S.H. Strogatz, Nonlinear Dynamics and Chaos, Addison-Wesley, Reading, 1994.

[73] G. Nicolis, Introduction to Nonlinear Science, Cambridge University Press, Cambridge, 1995.

[74] U. Grimm and M. Baake, Aperiodic Ising Models, in: The Mathematics of Long-Range Aperiodic Order, Ed., R.V. Moody, Kluwer, Dordrecht, 1997.

[75] P.W. Mauriz, M.S. Vasconcelos and E.L. Albuquerque, Physica A 329 (2003) 101.

[76] A. Thue, Norske Vididensk. Selsk. Skr. I. 7 (1906) 1.

[77] M. Morse, Trans. Am. Math. Soc. 22 (1921) 84.

[78] M. Morse, Am. J. Math. 43 (1921) 35.

[79] J.M. Luck, Phys. Rev. B 39 (1989) 5834.

[80] G. Steinmeyer, D. Jaspert and F. Mitschke, Opt. Commun. 104 (1994) 379.

Chapter 3

Bulk Polaritons

When the electromagnetic radiation propagating through a polarizable dielectric or magnetic crystal excites some internal degrees of freedom of the crystal, it gives rise to a hybrid (or "mixed") mode called a *polariton*. Polaritons, thus, are "quasi-particle" excitations in solids consisting of a photon (the quantum of the electromagnetic field) coupled to an elementary excitation (plasmon, phonon, exciton, magnon, etc.) which polarizes the crystal [1]. Plasmon-polaritons were briefly mentioned in Chapter 1 in the discussion of plasmons (see Section 1.5). The possibility of a coupled vibrational–electromagnetic (or phonon–photon) excitation in solids, giving rise to a *phonon-polariton*, was first predicted by Huang [2,3] in 1951 for cubic ionic crystals of the NaCl type.

The theory of polaritons in bulk materials has been clearly and elegantly discussed by a number of authors (see e.g. Refs. [4–6]). In the case of an electron plasma, for instance, such an excitation will have both a photon and a plasmon content, because the total energy is distributed over the whole system due to the coupling. The resulting mode, the *plasmon-polariton*, may have either a very strong photon or a very strong plasmon component, or may be strongly mixed, depending on the wavevector. Similar considerations apply for the other types of polaritons.

Experimental evidence for the existence of polaritons was provided first by Henry and Hopfield [7] in 1965 for the phonon-like polariton in cubic GaP. Further important experimental investigations of phonon-polaritons in those early days were made by Porto et al. [8] on hexagonal ZnO in 1966 and by Scott et al. [9] on α-quartz in 1967.

The characteristics of these bulk (and also surface) electromagnetic modes in dielectric and semiconductor crystals turn out to be closely related to the dielectric properties of the media and, in particular, to the frequency dependence of the relevant dielectric functions. We have already seen a particular example of this in Section 1.5. It is therefore fundamental to have, first, a proper understanding of the dielectric functions of solids, as presented below, before we introduce the concept of polaritons by outlining their theory for the case of bulk modes in a boundless (infinite) medium.

On the other hand, in magnetic crystals it is the magnetic susceptibility that plays a fundamental role in the understanding of the so-called *magnetic-polaritons*. The determination of the magnetic susceptibility tensor (or matrix) for ferromagnetic and antiferromagnetic crystals is presented in Section 3.4.

3.1 The Frequency-Dependent Dielectric Function

The dielectric function is the response of a system to an external electric field, and it plays an important role in the study of electromagnetically coupled modes, such as the plasmon-, phonon-, and exciton-polaritons [10,11]. For a medium with translational invariance, the position- and time-dependent dielectric function is defined in terms of the electric field $\vec{E}(\vec{r},t)$ and the electric displacement $\vec{D}(\vec{r},t)$ by

$$\vec{D}(\vec{r},t) = \epsilon_0 \int \epsilon(\vec{r}-\vec{r'},t-t')\vec{E}(\vec{r'},t')\, d^3\vec{r'}\, dt'. \qquad (3.1)$$

By translational invariance we mean here that ϵ is a function of the difference $\vec{r}-\vec{r'}$ and not of $\vec{r}$ and $\vec{r'}$ separately. Eq. (3.1) can be written more conveniently in terms of its Fourier transform to wavevector $\vec{k}$ and frequency ω as

$$\vec{D}(\vec{k},\omega) = \epsilon_0 \epsilon(\vec{k},\omega)\vec{E}(\vec{k},\omega). \qquad (3.2)$$

Thus, ϵ is in general a function of both wavevector $\vec{k}$ and frequency ω. However, the polariton regime corresponds to very small wavevector $|\vec{k}|$ (or very long wavelength). Essentially, this is because the photon and the crystal excitation will have comparable energies (as required for the formation of a coupled mode) only at very small wavevector $|\vec{k}|$, due to the large phase velocity of light. This is the regime in which one describes the electromagnetic properties using Maxwell's equations with *retardation* (typically $|\vec{k}| \leq 10^3 \text{ m}^{-1}$).

In this case, the dependence of the dielectric function ϵ on the wavevector $\vec{k}$ (referred to as spatial dependence) can usually be neglected, and so we replace $\epsilon(\vec{k},\omega)$ by $\epsilon(0,\omega)$, from now on simply written as $\epsilon(\omega)$. A notable exception is in the case of exciton-polaritons, where the spatial dependence appears due to the center-of-mass term in the exciton energy (see Section 1.6), and this will be discussed in Section 3.3.

Furthermore, in the case of anisotropic media where vectors $\vec{D}$ and $\vec{E}$ are not necessarily in the same direction, we note that $\epsilon(\omega)$ will be a tensor (or matrix) quantity rather than a scalar. In particular, for a uniaxial material, it will have the form

$$\bar{\epsilon}(\omega) = \begin{pmatrix} \epsilon_\perp(\omega) & 0 & 0 \\ 0 & \epsilon_\perp(\omega) & 0 \\ 0 & 0 & \epsilon_\parallel(\omega) \end{pmatrix}, \qquad (3.3)$$

in terms of the principal axes. The functions $\epsilon_\perp(\omega)$ and $\epsilon_\parallel(\omega)$ describe, respectively, the dielectric response to an electric field transverse and longitudinal to the uniaxis (in this case z).

We now turn our attention to the determination of the dielectric function for an ionic crystal (noting that the electron gas case was already discussed in Section 1.5), using the harmonic approximation [12,13]. The quantum theory

3.1. THE FREQUENCY-DEPENDENT DIELECTRIC FUNCTION

was developed independently by Fano and Hopfield and yields an identical result [14,1].

We consider an infinite diatomic 1D lattice with alternating masses m_1 and m_2, as depicted in Fig. 1.7. As the crystal is ionic, the two sublattices are now associated with opposite electric charges, and thus the polarization vector $\vec{P}$ involves a term proportional to the relative displacement $\vec{u}$, as well as the usual one proportional to the electric field $\vec{E}$, i.e.

$$\vec{P} = \epsilon_0(\alpha \vec{u} + \chi \vec{E}), \tag{3.4}$$

where χ is the electronic susceptibility. Here $\vec{E}$ means the macroscopic electric field, obtained after averaging the local field $\vec{E}_{loc}$ over many unit cells, and α is a constant of proportionality that depends on details of the lattice dynamics.

On the other hand, the equation of motion for the displacement $\vec{u}$ is [12]

$$(-\omega^2 - i\omega\Gamma)\vec{u} = -\omega_T^2 \vec{u} + \beta \vec{E}_{loc}, \tag{3.5}$$

where we have included a damping factor Γ and ω_T denotes the frequency of the transverse optical (TO) phonons. It should be noted that the longitudinal optical (LO) phonons do not couple to the light in the bulk of a crystal. As the relationship between $\vec{E}$ and $\vec{E}_{loc}$ is linear [15], Eq. (3.5) takes the form

$$(\omega^2 + i\omega\Gamma)\vec{u} = \omega_T^2 \vec{u} - \gamma \vec{E}. \tag{3.6}$$

Eqs. (3.4) and (3.6) are easily solved for $\vec{P}$ to give

$$\vec{P} = \epsilon_0 \left[\alpha\gamma(\omega_T^2 - \omega^2 - i\omega\Gamma)^{-1} \hat{u} + \chi \vec{E} \right], \tag{3.7}$$

where $\hat{u}$ is the unit vector in the direction of the displacement vector $\vec{u}$.

Next, using the electromagnetic constitutive equation

$$\vec{D} = \epsilon_0 \vec{E} + \vec{P} = \epsilon_0 \epsilon(\omega) \vec{E}, \tag{3.8}$$

we find that the dielectric function is

$$\epsilon(\omega) = \epsilon_\infty \left[1 + \frac{\omega_L^2 - \omega_T^2}{\omega_T^2 - \omega^2 - i\omega\Gamma} \right], \tag{3.9}$$

where $\epsilon_\infty = 1 + \chi$, and

$$\omega_L^2 - \omega_T^2 = \alpha\gamma/\epsilon_\infty. \tag{3.10}$$

Here ϵ_∞ is the high-frequency dielectric constant (equal to η^2, with η being the refractive index of the medium), and ω_L denotes the frequency of the LO phonon. If damping is neglected, the dielectric function expressed by Eq. (3.9) simplifies to

$$\epsilon(\omega) = \epsilon_\infty \left(\frac{\omega_L^2 - \omega^2}{\omega_T^2 - \omega^2} \right). \tag{3.11}$$

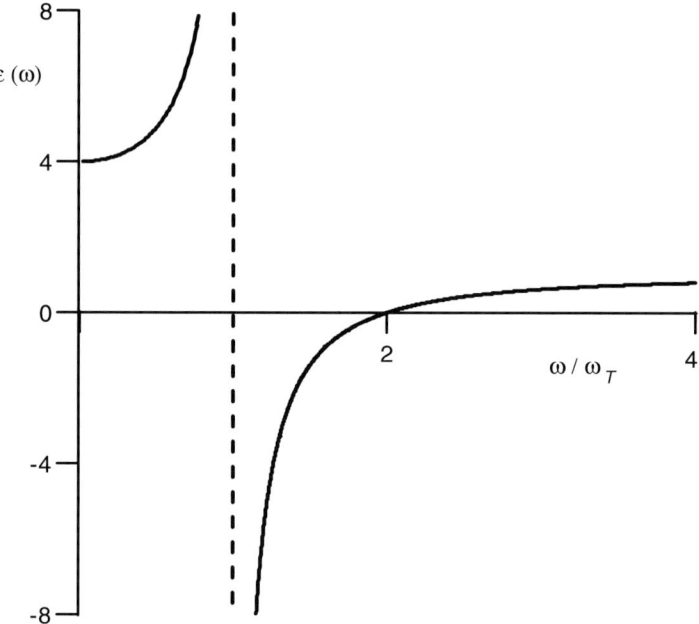

Fig. 3.1. Behavior of the dielectric function for the ionic crystal, defined by Eq. (3.11) for the zero damping case, as a function of the reduced frequency ω/ω_T. Here we have considered $\epsilon_\infty = 1$ and $\omega_L/\omega_T = 2$.

The behavior of this dielectric function as a function of the reduced frequency ω/ω_T is shown in Fig. 3.1.

The zero-frequency value of $\epsilon(\omega)$ is

$$\epsilon(0) = \epsilon_\infty (\omega_L^2/\omega_T^2), \qquad (3.12)$$

which is the Lyddane–Sachs–Teller (LST) relation [16]. Therefore, in the limit of zero damping factor, the zeros of the dielectric function $\epsilon(\omega)$ define the LO phonon frequency ω_L, while the poles of $\epsilon(\omega)$ define the TO phonon frequency ω_T.

In a doped polar semiconductor, the quasi-free electron (or hole) gas and the long-wavelength TO phonons are coupled forming a mixed mode, whose dielectric function is given by (neglecting the damping term)

$$\epsilon(\omega) = \epsilon_\infty \left[1 - \left(\frac{\omega_p^2}{\omega^2}\right) + \left(\frac{\omega_L^2 - \omega^2}{\omega_T^2 - \omega^2}\right) \right]. \qquad (3.13)$$

This has both a plasmon and phonon character, and it can be considered as a generalization of both Eq. (1.24) and Eq. (3.11).

3.2 Bulk Plasmon- and Phonon-Polaritons

The propagation of light in a crystal is governed by Maxwell's equations of electromagnetism. In SI units these can be stated as [17]

$$\nabla \cdot \vec{D} = 0,$$
$$\nabla \times \vec{E} = -\partial \vec{B}/\partial t,$$
$$\nabla \cdot \vec{B} = 0,$$
$$\nabla \times \vec{H} = \partial \vec{D}/\partial t. \qquad (3.14)$$

For plane-wave solutions in a bulk (i.e. effectively infinite) medium, all electromagnetic fields are proportional to $\exp(i\vec{k}\cdot\vec{r} - i\omega t)$, and the Maxwell equation $\nabla\cdot\vec{D}=0$ yields the condition $\epsilon(\omega)\vec{k}\cdot\vec{E}=0$. This implies that

$$\epsilon(\omega) = 0 \quad \text{or} \quad \vec{k}\cdot\vec{E} = 0. \qquad (3.15)$$

The first of the above equations ($\epsilon = 0$), corresponding to the zeros of the dielectric function, gives simply the frequencies $\omega = \omega_p$ and $\omega = \omega_L$ for the case of an electron gas and an ionic medium, respectively (in the absence of damping). These are the longitudinal bulk modes; they turn out to have no surface counterpart, and therefore we do not consider them further.

On the other hand, the second equation ($\vec{k}\cdot\vec{E}=0$) is the transversality condition between the electric field $\vec{E}$ and the wavevector $\vec{k}$. Considering the second and fourth of Maxwell's equations and the constitutive relation $\vec{B}=\mu_0\vec{H}$ in a non-magnetic isotropic medium, we have

$$\nabla \times (\nabla \times \vec{E}) + \mu_0\, \partial^2 \vec{D}/\partial t^2 = 0, \qquad (3.16)$$

which becomes

$$c^2[\nabla(\nabla\cdot\vec{E}) - \nabla^2\vec{E}] + \epsilon(\omega)\, \partial^2\vec{E}/\partial t^2 = 0. \qquad (3.17)$$

Here we have considered the other constitutive equation $\vec{D}=\epsilon_0\epsilon(\omega)\vec{E}$ and defined the velocity of the light in the vacuum as $c^2=(\mu_0\epsilon_0)^{-1}$. Then for plane-wave $\exp[i(\vec{k}\cdot\vec{r}-\omega t)]$ propagation of the electric field, the dispersion relation equation for the transverse mode is easily found to be

$$k^2 = \epsilon(\omega)\omega^2/c^2. \qquad (3.18)$$

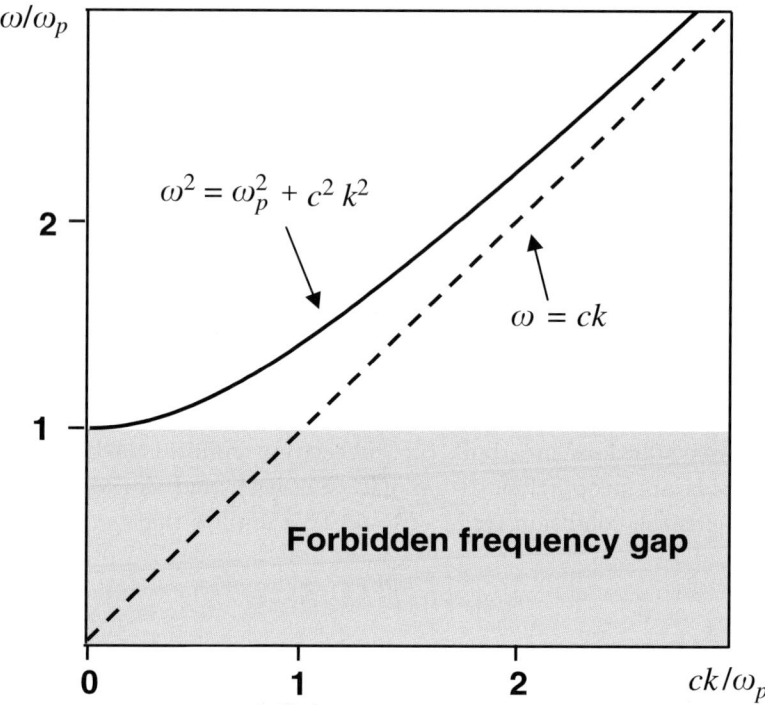

Fig. 3.2. Bulk plasmon-polariton dispersion curve (full line) plotted as ω/ω_p versus ck/ω_p, considering the positive ion background to have a dielectric constant $\epsilon_\infty = 1$. The dashed line represents the light line $\omega = ck$ in the vacuum.

For an electron gas, with the dielectric function given by Eq. (1.24), the above result can be rearranged to give Eq. (1.25); its asymptotic and limiting forms are

(a) k is real for $\omega \geq \omega_p$, with $k \to \omega/c$ as $\omega \to \omega_p$;

(b) k is purely imaginary for $\omega < \omega_p$, and in this region the wave decays exponentially with a characteristic length equal to $1/k$;

(c) $|k| \to \omega_p/c$ when $\omega \to 0$.

Fig. 3.2 depicts a plot of this dispersion curve, which represents the propagation of a transverse electromagnetic wave in a plasma (the so-called *bulk plasmon-polariton*). The spectrum has just one plasmon-polariton branch, for propagation in the region $\omega > \omega_p$. The shaded area $\omega < \omega_p$ corresponds to a forbidden frequency gap. The light line (for an uncoupled photon) is here represented by the dashed line $\omega = ck$.

For the phonon case, with the dielectric function now given by Eq. (3.11), the asymptotic and limiting forms of Eq. (3.18), neglecting the damping term, are

(a) $k \simeq \epsilon(0)^{1/2}\omega/c$ when $\omega \ll \omega_T$;

3.2. BULK PLASMON- AND PHONON-POLARITONS

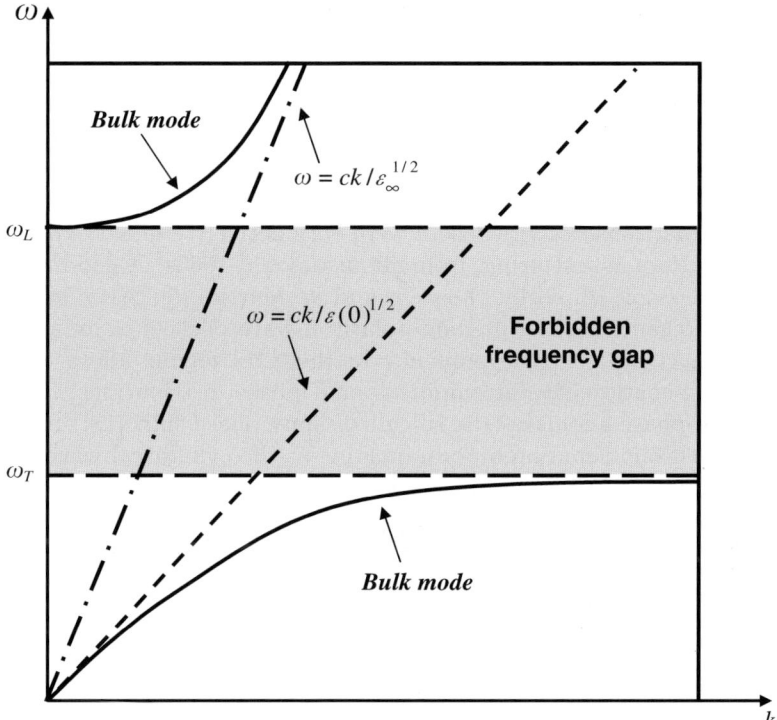

Fig. 3.3. Bulk phonon-polariton dispersion curve (full lines) plotted as ω versus k. The dashed line represents the asymptotic curve $\omega = ck/\epsilon(0)^{1/2}$, while the chain-dotted line represents the asymptotic curve $\omega = ck/\epsilon_\infty^{1/2}$, respectively. The shaded area between ω_T and ω_L is a forbidden frequency gap.

(b) $k \to \infty$ when $\omega \to \omega_T$;

(c) k is purely imaginary for $\omega_T < \omega < \omega_L$. This is the region where the surface phonon-polariton mode propagates (see the next chapter);

(d) $k \to 0$ when $\omega \to \omega_L$;

(e) $k \simeq \epsilon_\infty^{1/2} \omega/c$ when $\omega \to \infty$.

In Fig. 3.3 we show a plot of this dispersion relation curve, which represents the propagation of a transverse electromagnetic wave coupled to a transverse optical phonon mode in a polar crystal. The resulting mixed mode is called a *bulk phonon-polariton*. As we can see, by contrast to the plasmon-polariton case, for every value of the wavevector k there are two solutions for ω, giving rise to the two branches depicted in Fig. 3.3. When the wavevector k approaches zero, one solution (known as the *lower branch*) approaches $\omega^2 = c^2 k^2/\epsilon(0)$ (the dashed line in Fig. 3.3), while the other one (the *upper branch*) approaches the value of the LO frequency ω_L.

Thus, the frequency of the transverse oscillations in the limit of zero wavevector (for the upper branch) becomes degenerate with the longitudinal oscillation frequency. On the other hand, when $k \to \infty$ the dispersion of the upper branch is given by $\omega^2 = c^2 k^2/\epsilon_\infty$ (the chain-dotted line in Fig. 3.3), while the lower branch approaches the value of the TO frequency ω_T. Since the longitudinal oscillations cannot couple to the transverse electromagnetic radiation field, they have no dispersion, a fact that is represented by the horizontal straight line passing through ω_L.

The bulk polaritons described here were very actively studied experimentally, particularly by Raman scattering, from about the mid-1960s. A good reference for this work is the book edited by Burstein and de Martini [5]. Also on the experimental side, the generation of phonon-polariton wave packets in uniaxial LiTaO$_3$ crystals with a typical carrier frequency in the THz regime allows direct measurement of the spatiotemporal amplitude and phase distributions. Under these conditions, the phase anomaly (the so-called Gouy phase shift [18]) may be visualized directly through spatiotemporal imaging as the cylindrical wave propagates through its focus [19]. Furthermore, the dipolar Cherenkov radiation (radiation emitted by charged particles travelling through matter at speeds larger than the phase velocity of light in the medium), in the range of infrared-active phonons, is identical to that of phonon-polaritons produced by impulsive laser excitation [20,21].

3.3 Bulk Exciton-Polaritons

An *exciton-polariton* is a propagating mode in a dielectric or semiconductor crystal in which the electromagnetic wave is coupled with the polarization wave of excitons (the electron–hole pair discussed in Section 1.6). The polarization wave in the present case is associated with the non-zero electric dipole moments of the excitons. As excitons travel in the crystal, they radiate electromagnetic waves which, in turn, can excite excitons. In principle, there is no way to separate the exciton wave from the electromagnetic wave. Thus, introducing an exciton–photon interaction does not necessarily mean that energy will be lost by photons inside the crystal. In this polariton picture, energy is converted from photons to excitons and vice versa in an "exchange" process [22].

Exciton-polaritons were first predicted by Pekar [23] in 1957 by proposing that the kinetic energy term $\hbar^2 k^2/2M$ in the dispersion equation (1.26) for the total energy of an exciton (arising due to the motion of the exciton's center of mass) could play a decisive role in the optics of a crystalline condensed-matter system. Such an effect implies an explicit wavevector dependence in the dielectric response function, leading to the so-called *spatial-dispersion* effects in the optical region of exciton resonance [24,25].

The "signature" of the spatial-dispersion effect, as pointed out by Hopfield and Thomas [26], is the non-local relationship between the polarization $\vec{P}(\vec{r},t)$ and an applied electric field $\vec{E}(\vec{r},t)$, acting as a driving force. The differential equation of motion, deduced by considering the simple model of a single electric-dipole-active exciton resonance (often denoted as the "dielectric

3.3. BULK EXCITON-POLARITONS

approximation"), is [26]

$$\left[(\partial^2/\partial t^2) + \omega_0^2 - D\nabla^2 + \Gamma(\partial/\partial t)\right]\vec{P}(\vec{r},t) = S\vec{E}(\vec{r},t), \qquad (3.19)$$

where ω_0 is the resonant frequency of the uncoupled exciton (the band-gap frequency minus the frequency corresponding to the binding energy), and $D = \hbar\omega_0/M$ with $M = m_c + m_v$ being the total exciton mass (the sum of electron and hole effective band masses). Also, Γ is the damping coefficient and $S = 4\pi\alpha_0\omega_0^2$ is the exciton oscillator strength at $\omega = 0$ and $k = 0$, with α_0 denoting the dipole matrix element for optical excitation of the exciton [27]. It is helpful to emphasize that Eq. (3.19) is a semi-macroscopic equation of motion. It is obtained as a long wave-length approximation to an exciton Schrödinger equation with dipole moment (or polarization) defined via the exciton eigenfunction in a bounded medium [28].

Considering the macroscopic constitutive relation for the exciton polarization, i.e.

$$\vec{P}(\vec{k},\omega) = \chi(\vec{k},\omega)\vec{E}(\vec{k},\omega), \qquad (3.20)$$

and taking into account harmonic plane-wave propagation in the excitonic medium, Eqs. (3.19) and (3.20) lead to

$$\chi(\vec{k},\omega) = S/(\omega_0^2 + Dk^2 - \omega^2 - i\omega\Gamma). \qquad (3.21)$$

The non-local exciton-polariton dielectric function is then given by

$$\epsilon(\vec{k},\omega) = \epsilon_\infty + \chi(\vec{k},\omega), \qquad (3.22)$$

where ϵ_∞ is the background dielectric constant of the crystal. The exciton-polariton dielectric function thus differs from the corresponding phonon form by the spatial-dispersion term Dk^2.

The above "dielectric approximation" to the susceptibility has the merit of being a simple, analytical expression that contains some of the essential physics of spatially dispersive media when the exciting laser frequency is close to an exciton resonance. It presumably applies best well inside the crystal, away from the surface, where evanescent waves have decayed and surface-related phenomena are unimportant.

Eq. (3.22) plus Maxwell's equations constitute the complete set of equations one needs to find the normal modes of the crystal in the resonance frequency region. For the isotropic case there exist the two usual types of modes, whose dispersion relations are given implicitly by

$$\epsilon(\vec{k},\omega) = (ck/\omega)^2 \quad \text{(transverse modes)}, \qquad (3.23)$$

$$\epsilon(\vec{k},\omega) = 0 \quad \text{(longitudinal modes)}. \qquad (3.24)$$

The first of these equations yields the following exciton-polariton dispersion relation for the transverse modes:

$$Dk^4 - (\Omega^2 + D\epsilon_\infty\omega^2/c^2)k^2 + (\omega^2/c^2)(\epsilon_\infty\Omega^2 - S) = 0. \qquad (3.25)$$

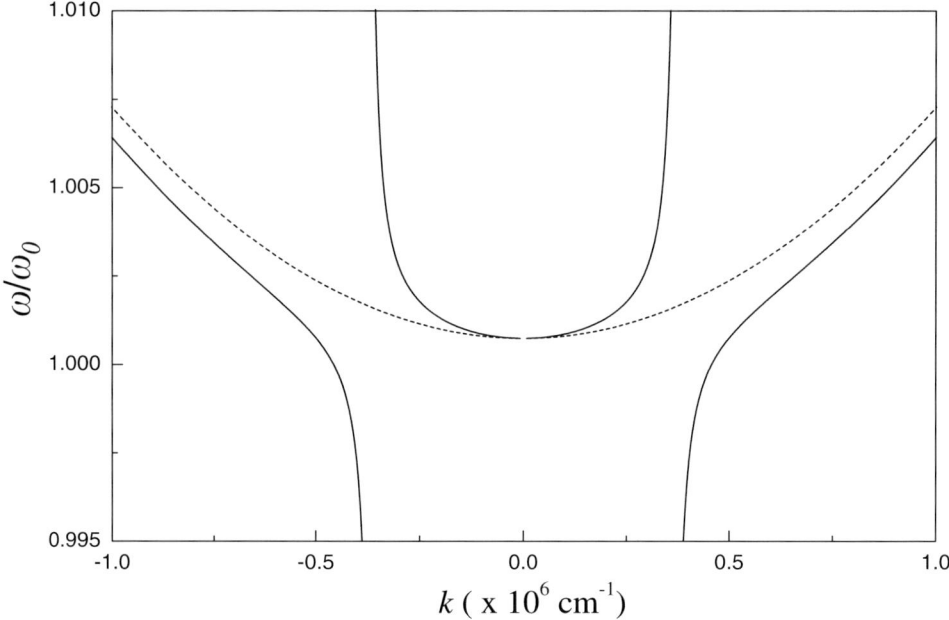

Fig. 3.4. The exciton-polariton curve for CdS calculated with the damping term $\Gamma = 0$. The full curves represent the two transverse branches; the dotted curve represents the longitudinal branch.

This gives rise to the two transverse branches in the exciton-polariton frequency spectrum. The second equation (3.24) yields, for the sole longitudinal mode,

$$Dk^2 = \Omega^2 - S/\epsilon_\infty. \tag{3.26}$$

In the above expressions we have denoted $\Omega^2 = \omega^2 - \omega_0^2 + i\omega\Gamma$.

Fig. 3.4 shows the exciton-polariton dispersion curves for frequencies in the vicinity of the uncoupled exciton resonance ω_0, neglecting the damping term. We have plotted the reduced frequency ω/ω_0 against the wavevector k (in units of 10^6 cm^{-1}). It differs from the corresponding phonon-polariton spectrum in one vitally important way: because of the kinetic energy term $\hbar^2 k^2/2M$, the exciton branches curve upwards, so that the familiar "forbidden band" between the LO and TO phonons does not exist in the exciton case. This curvature leads to the k-dependence in the dielectric function already discussed. Far below resonance ($\omega \ll \omega_0$) there is just one exciton-polariton branch, which is fundamentally photon-like (the first transverse mode). As the frequency increases, this branch takes on more exciton character, as can be seen from Fig. 3.4, until at and above ω_0 it is strongly exciton-like. Two more polariton branches then arise in the region. One of them (the other transverse mode) quickly takes on predominantly photon character, while the other (the longitudinal mode, shown dashed) is always primarily exciton-like.

3.3. BULK EXCITON-POLARITONS

The exciton-polariton spectrum shown in Fig. 3.4 is independent of any boundary conditions, being a strict consequence of Maxwell's electromagnetic equations and the constitutive relations in the medium. If we are only concerned with the spectrum in a bulk dispersive medium, the boundary is irrelevant. However, in a bounded dispersive medium the extra k-dependence in the exciton-polariton dielectric function leads to a major problem related with the totality of boundary conditions which are necessary to describe the exciton-polariton theory. Indeed, in the presence of spatial dispersion, an incident light wave with frequency close to the uncoupled exciton frequency ω_0 can excite more than two waves in the crystal when it strikes the crystal surface. Therefore, in order to calculate the optical spectra in this case, unlike in classical optics, additional boundary conditions (or ABCs) are required beyond the usual Maxwell boundary conditions.

As an example to highlight this, we consider the simple problem of a normally incident plane wave from vacuum, propagating in the z-direction towards a non-local crystal. The incident electric field is $E_0 \exp(ik_0 z)$ and the corresponding reflected wave is $E_R \exp(-ik_0 z)$, with $k_0 = \omega/c$. Two transverse modes, with wavevectors k_{T1} and k_{T2} defined by Eq. (3.25), will be transmitted in the medium; their electric fields are $E_{T1} \exp(ik_{T1} z)$ and $E_{T2} \exp(ik_{T2} z)$. From Maxwell's equations, the corresponding magnetic fields are $(k_0 E_0/\omega) \exp(ik_0 z)$, $-(k_0 E_R/\omega) \exp(-ik_0 z)$, $(k_{T1} E_{T1}/\omega) \exp(ik_{T1} z)$, and $(k_{T2} E_{T2}/\omega) \exp(ik_{T2} z)$, respectively. Hence there are the following electric field amplitudes to deal with: E_0 (the incident field), E_R (the reflected field), plus E_{T1} and E_{T2} (the transmitted fields). As the amplitude of the incident field is externally controllable, it is necessary to find three ratios (E_R/E_0, E_{T1}/E_0, and E_{T2}/E_0). However, the Maxwell boundary conditions give only two equations, namely:

(a) continuity of the tangential component of the electric field, giving

$$1 + (E_R/E_0) = (E_{T1}/E_0) + (E_{T2}/E_0); \tag{3.27}$$

(b) continuity of the normal component of the magnetic field, giving

$$k_0[1 - (E_R/E_0)] = k_{T1}(E_{T1}/E_0) + k_{T2}(E_{T2}/E_0). \tag{3.28}$$

Thus, the problem of determining the additional unknown ratio can only be solved if there is one more boundary condition.

The simplest form of ABC was proposed by Pekar [23,29] by considering that the macroscopic polarization vector vanishes at the boundary of the dispersive medium, i.e.

$$\vec{P} = 0 \quad \text{at} \quad z = 0, \tag{3.29}$$

where the z-axis is considered to be directed along the normal of the crystal surface (at $z = 0$). Although this is a natural and appealing procedure, this boundary condition is not a mathematical consequence of the assumed susceptibility. On the other hand, Ginzburg [30] proposed a phenomenological approach, in which a spatially dispersive susceptibility was postulated for the bounded medium by writing $\epsilon(\vec{k},\omega)$ in a power series up to bilinear terms in $\vec{k}$. Together with Agranovich [31], he analyzed further his proposed ABC starting from essentially the same constitutive relation for the exciton polarization, i.e.

$$(-\omega^2 + \beta + \gamma\nabla + \alpha\nabla^2)\vec{P} = \lambda\vec{E}, \qquad (3.30)$$

and they deduced that the general form of the ABC at the boundary should be

$$\vec{P} + \Gamma\vec{E} = 0 \quad \text{at } z = 0, \qquad (3.31)$$

with the term Γ frequency independent.

At about this time, Hopfield and Thomas [26] realized that the form of ABC depends essentially on the behavior of the potential $U(z)$ for the exciton near the crystal boundary. To simplify the problem, they idealized the surface as being exciton-free (in the so-called *dead-layer* model). In this model the potential $U(z)$ is replaced by an infinite potential barrier at a finite distance d above the crystal, i.e. they assumed

$$U(z) = \begin{cases} 0 & \text{for } d < z < \infty, \\ \infty & \text{for } 0 < z < d. \end{cases} \qquad (3.32)$$

The dead-layer region $0 < z < d$ is characterized by a frequency-independent dielectric constant, which is assumed to be the same as the background dielectric constant of the crystal (ϵ_∞). The usual Maxwell's boundary conditions are then applied at the boundary vacuum-dead-layer (at $z = d$), and Pekar's ABC, Eq. (3.29), is applied at the boundary crystal-dead-layer at $z = 0$. Physically, the dead-layer thickness should be of the order of the exciton Bohr radius a_{ex}. Attempts have been made to reveal the origin of the dead layer using a model of an image force [32,33].

Further improvements were made afterwards. Among others, Agarwal et al. [34,35] reformulated the problem of the electrodynamics of non-local media in a differential equation approach extended to higher order through the non-local (integral) constitutive relation. Maradudin and Mills [36] developed a similar differential equation formulation, while Birman and Sein [37] presented an integral equation method in the framework of a polarization approach and the extinction theorem applied to non-local media [38,39]. Other successful approaches to the ABC problem were developed by Zeyher et al. [40] in the course of investigating the non-translationally invariant susceptibility, and by Skettrup [41] using an interesting variational approach based on properties of the Lagrangian for matter in a non-local medium.

To date, the ABCs proposed can be cast generally into the form

$$\gamma\vec{P} + \beta\, d\vec{P}/dz = 0 \qquad (3.33)$$

at the interface. Some limiting cases are $\gamma = 1, \beta = 0$ (which leads to $\vec{P} = 0$) and $\gamma = 0, \beta = 1$ (meaning $d\vec{P}/dz = 0$).

3.3. BULK EXCITON-POLARITONS

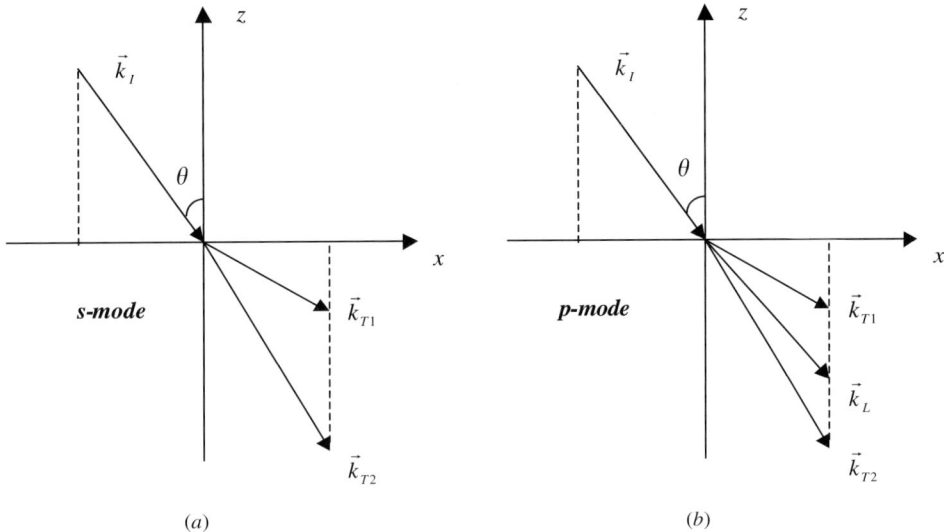

Fig. 3.5. Geometry showing the incident light wave (from vacuum, for example), along with the reflected and transmitted (in a spatial-dispersive medium) light waves: (a) s-polarization, (b) p-polarization.

Calculations of the reflection and transmission spectra in the vicinity of an excitonic resonance using this ABC and its limiting cases were made by Tilley [42] and Albuquerque and Gonçalves da Silva [43]. Their results show a set of exciton anomalies and provide information about the excitonic states. In Fig. 3.5 we show schematically the geometry of the propagation of the light waves, s- and p-polarized, in a non-local crystal. We note that, for an incident s-polarized light wave, two transverse modes with wavevectors given by Eq. (3.25) will propagate, while in the p-polarization case an additional longitudinal mode, whose wavevector is defined by Eq. (3.26), will be excited.

The most important experimental technique to investigate the optical properties of exciton-polaritons is that of resonance Brillouin scattering (RBS) proposed by Brenig et al. [44] in 1972. Using this spectroscopic method, it is possible to determine the basic exciton and photon parameters, such as the exciton's mass M, the damping term Γ for exciton decay, the resonance energy $\hbar\omega_0$, and the oscillator strength S for the exciton–photon coupling. Besides, the technique gives information about the dynamics of the exciton–phonon coupling which produces the scattering interaction (for details see Refs. [45–47]).

An additional advantage of RBS is that only low excitation intensities are required. Thus, there is little danger of altering the optical properties of the crystal in some unforeseen manner as with an intense laser beam. However, only dipole-active exciton-polariton branches may be populated. Thus, transitions involving states with weak or non-existent oscillator strengths are not observed. Also, with RBS it is possible to probe the exciton-polariton dispersion curve at wavevectors that are a significant fraction of Brillouin zone values. RBS experiments are performed at low temperatures to sharpen the exciton resonance and to reduce phonon-related broadband background emission. To observe intense Brillouin peaks over a wide range, competing relaxation processes must be reduced to a minimum. Even small concentrations of impurities can produce strong impurity emission peaks that can overlap exciton resonances. Generally, the best available, pure, single crystals are investigated. RBS can sometimes be observed in poorer quality samples near the exciton resonance, but usually only weakly.

To probe the exciton-polariton dispersion curve by this inelastic process, we consider that the initial and final polariton states have energy $\hbar\omega_i(\vec{k}_i)$ and $\hbar\omega_s(\vec{k}_s)$, with $\hbar\vec{k}_i$ and $\hbar\vec{k}_s$ being their momentum, respectively. Also we let $\Omega(\vec{k}) = v_S|\vec{k}|$ be the dispersion relation for the acoustic phonon. The kinematics of the Brillouin scattering process are based on the conservation of energy and momentum, i.e.

$$\omega_s(\vec{k}_s) = \omega_i(\vec{k}_i) \pm \Omega(\vec{k}), \tag{3.34}$$

$$\vec{k}_s = \vec{k}_i \pm \vec{k}, \tag{3.35}$$

where the minus and plus signs indicate Stokes (phonon emission) and anti-Stokes (phonon absorption) processes, respectively. The scattering experiment in the resonant regime is, for experimental reasons, performed in a backscattering geometry, so that if $\vec{k}_i = k_i\hat{z}$, then $\vec{k}_s = -k_s\hat{z}$.

Fig. 3.6 provides a visualization of the kinematics for this process. In the region $\omega \ll \omega_0$, which is non-resonant, there is relatively small frequency dispersion. In this regime, the photon-like excitation has essentially a vertical slope (the exciton-polariton travels at essentially infinite velocity compared to the phonon) and the transitions are made joining points that lie in the parallel, almost vertical lines, with downward (upward) arrows corresponding to the Stokes (anti-Stokes) processes. This non-resonant Brillouin scattering spectroscopy was studied by Cummins and Schoen [48]. As the exciton-polariton frequency increases towards the resonant frequency of the uncoupled exciton ω_0, qualitative differences occur due to the dispersive properties of the polariton dielectric function. Several Stokes and anti-Stokes peaks are allowed. Details can be found in Tilley [42] for normal incidence back-scattering and in Albuquerque and Gonçalves da Silva [43] for general angles of incidence (see also Chapter 11).

3.4 Magnetic Susceptibility

The magnetic susceptibility tensor plays a major role in the determination of the magnetic-polariton dispersion relation (just as its counterpart, the dielectric tensor, does for the non-magnetic plasmon-, phonon-, and exciton-polaritons). It describes the response of a magnetic crystal to the presence of an external magnetic field $\vec{H}$.

The susceptibility tensor can be found using the torque equation of magnetic resonance theory [49–51]. We shall consider here the case of a two-sublattice antiferromagnetic material. As well as this being required for later applications,

3.4. MAGNETIC SUSCEPTIBILITY

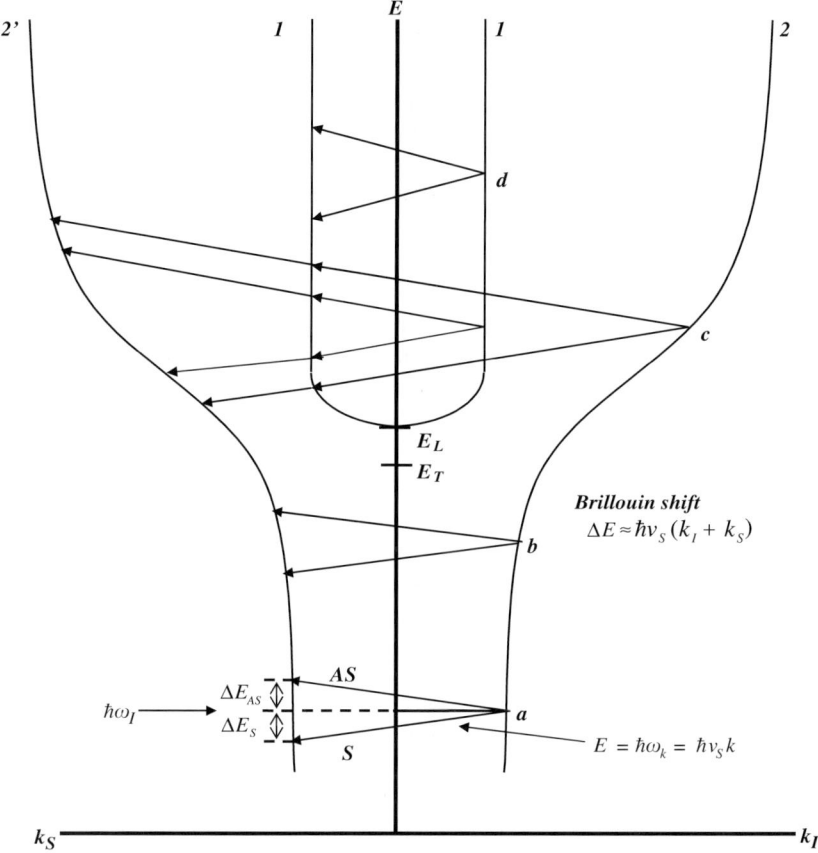

Fig. 3.6. The kinematics of the resonant Brillouin scattering to probe exciton-polaritons. The arrows illustrate the dynamics of the Stokes (downward) and anti-Stokes (upward) processes.

we can straightforwardly take a special case (i.e. the reduction to one sublattice) to deduce the susceptibility tensor appropriate to a ferromagnetic material. Furthermore, although we will not show it here, the antiferromagnetic calculation is easily generalized to two-sublattice ferrimagnets.

We consider therefore a two-sublattice antiferromagnetic material with uniaxial anisotropy, such that the total (instantaneous) sublattice magnetizations $\vec{M}_1$ and $\vec{M}_2$ have their average values directed along the $+z$ and $-z$ directions, respectively. Also, we write M and $-M$ for the *static* components M_{1z} and M_{2z}, and H_A and $-H_A$ for the effective anisotropy fields, respectively. The required susceptibility is the dynamic linear response of $\vec{M}_1 + \vec{M}_2$ to an applied external magnetic field $\vec{H}$ of frequency ω.

The torque equation (without damping) yields

$$d\vec{M}_i/dt = \gamma(\vec{M}_i \times \vec{H}_i^{eff}), \quad i = 1, 2, \tag{3.36}$$

where the gyromagnetic factor γ is negative (with $|\gamma| = g\mu_B/\hbar$, where g is the Landé factor and μ_B is the Bohr magneton). The effective magnetic field in the above equation is

$$\vec{H}_1^{eff} = \vec{H}_0 + \vec{H}_{A1} - \lambda \vec{M}_2 + (\tfrac{1}{3})(\vec{M}_1 + \vec{M}_2) + \vec{h}, \quad (3.37)$$

with a similar expression for $\vec{H}_2^{eff}$ with the subscripts 1 and 2 interchanged. Here $\vec{H}_0$ is the applied static magnetic field in the z-direction and $\vec{H}_{Ai}$ are static anisotropy fields in the $\pm z$-direction for each sublattice. The term $\lambda \vec{M}_2$ represents the inter-sublattice exchange coupling; the intra-sublattice exchange is not included here since it gives a zero contribution to Eq. (3.36). The term $(\tfrac{1}{3})(\vec{M}_1 + \vec{M}_2)$ is a Lorentz term describing local field correction; the need to include it in this susceptibility calculation was pointed out by Harris [52]. The final term $\vec{h}$ in Eq. (3.37) is the fluctuating part of the field, and it leads to fluctuating components in $\vec{M}_1$ and $\vec{M}_2$.

The next step is to linearize Eq. (3.36) in terms of $\vec{h}$. The components M_{ix}, M_{iy} are of first order, but the deviation of M_{iz} from their static values is easily shown to be of second order. Thus, only the transverse components, M_{ix} and M_{iy}, enter in the linearized theory. It is convenient to use a rotating wave representation with

$$M_i^\pm = M_{ix} \pm i M_{iy}, \quad i = 1, 2, \quad (3.38)$$

$$h^\pm = h_x \pm i h_y. \quad (3.39)$$

The equations of motion for $M_i^\pm$ and $h^\pm$ then take the form

$$(\omega + \omega_{12}) M_1^+ + \omega_{11} M_2^+ = -\gamma M h^+, \quad (3.40)$$

$$\omega_{22} M_1^+ - (\omega - \omega_{21}) M_2^+ = -\gamma M h^+, \quad (3.41)$$

with

$$\omega_{11} = -\gamma M(\lambda - \tfrac{1}{3}), \quad (3.42)$$

$$\omega_{12} = \omega_{11} - \gamma(H_0 + H_A), \quad (3.43)$$

$$\omega_{21} = \omega_{11} + \gamma(H_0 - H_A), \quad (3.44)$$

$$\omega_{22} = -\omega_{11}. \quad (3.45)$$

The antiferromagnetic resonance frequencies are obtained from the condition that the coefficient matrix for Eqs. (3.40) and (3.41) is singular. This condition yields

$$\omega = \gamma H_0 \pm [(\omega_{11} - \gamma H_A)^2 - \omega_{11}^2]^{1/2}. \quad (3.46)$$

Thus, there are two resonance frequencies, as might be expected in a two-sublattice material. We shall denote these by ω_1 and ω_2, corresponding to the $+$ and $-$

3.4. MAGNETIC SUSCEPTIBILITY

signs, respectively, in the above equation. Note that these modes are degenerate in magnitude when $H_0 = 0$ and they are equal to the antiferromagnetic resonance (AFMR) frequency.

The sublattice susceptibilities are defined by $M_i^\pm = \chi_i^\pm h^\pm$ and, from Eqs. (3.40) and (3.41), they are found to be

$$\chi_1^+(\omega) = -\gamma M(\omega_{11} + \omega - \omega_{21})/D_1, \tag{3.47}$$

$$\chi_2^+(\omega) = -\gamma M(-\omega_{11} + \omega + \omega_{12})/D_1, \tag{3.48}$$

with

$$D_1 = (\omega - \omega_1)(\omega - \omega_2). \tag{3.49}$$

The corresponding χ_i^- ($i = 1, 2$) are obtained from the above equations by replacing ω by $-\omega$.

The total magnetic susceptibilities (in the rotating wave representation) are

$$\chi^\pm = \chi_1^\pm + \chi_2^\pm. \tag{3.50}$$

Transforming back to the original coordinate frame, the magnetic susceptibility tensor has the gyromagnetic form

$$\bar{\chi}(\omega) = \begin{pmatrix} \chi_a(\omega) & i\chi_b(\omega) \\ -i\chi_b(\omega) & \chi_a(\omega) \end{pmatrix}, \tag{3.51}$$

with

$$\chi_a(\omega) = 2\gamma M(\omega_{11} - \gamma H_A)(\omega^2 - \omega_1\omega_2)/D_2, \tag{3.52}$$

$$\chi_b(\omega) = 2\gamma M\omega(\omega_{11} - \gamma H_A)(\omega_1 + \omega_2)/D_2, \tag{3.53}$$

where

$$D_2 = (\omega^2 - \omega_1^2)(\omega^2 - \omega_2^2). \tag{3.54}$$

Therefore, we see that the antiferromagnetic susceptibility tensor has poles at $\pm\omega_1$ and $\pm\omega_2$. There is again a simplification in the case of zero applied field $H_0 = 0$ because $\omega_2 = -\omega_1$. In fact, we have $\chi_b(\omega) = 0$ in this limit.

For the case of a ferromagnetic crystal, there is a simplification to the above calculation for the susceptibility tensor, because we now have only one sublattice of spins and the inter-sublattice coupling terms are absent. The required equation of motion is therefore

$$(\pm\omega + \omega_{12})M_1^\pm = -\gamma M h^\pm, \tag{3.55}$$

where the ω_{12} parameter is now redefined as

$$\omega_{12} = -\gamma(H_{A1} + H_0). \tag{3.56}$$

Instead of Eq. (3.50) this leads to

$$\chi^\pm(\omega) = -\gamma M/(\omega_{12} \pm \omega), \tag{3.57}$$

and the gyromagnetic tensor has the form of Eq. (3.51) with

$$\chi_a(\omega) = -\gamma M \omega_{12}/(\omega_{12}^2 - \omega^2), \qquad (3.58)$$

$$\chi_b(\omega) = -\gamma M \omega/(\omega_{12}^2 - \omega^2). \qquad (3.59)$$

The ferromagnetic susceptibility tensor has only simple poles at $\omega = \pm \omega_{12}$, so that ω_{12} is the ferromagnetic resonance (FMR) frequency. Very often, $H_A \ll H_0$, so that the FMR frequency becomes simply $-\gamma H_0$.

3.5 Bulk Magnetic-Polaritons

In this section we are concerned with the coupling between the electromagnetic field and a magnetic crystal characterized by a non-vanishing magnetic susceptibility tensor $\chi_{ij}(\omega)$. In particular, we are interested in this interaction for frequencies in the vicinity of a pole of $\chi_{ij}(\omega)$. This will give rise to the so-called *magnetic-polariton* modes, where the susceptibility components play an analogous role to the frequency-dependent dielectric function in the previous examples of non-magnetic polaritons. The magnetic-polaritons have very long wavelengths in comparison to the crystal lattice parameter, so it is usually sufficient to consider macroscopic equations of motion for the magnetization in the crystal to describe the polariton mode (for a review see e.g. Ref. [53]).

The electromagnetic field is described by Maxwell's equations including retardation, as given by Eq. (3.14). Using the second and fourth of these equations, we get

$$\nabla \times (\nabla \times \vec{H}) + \epsilon_0 \epsilon \, \partial^2 \vec{B}/\partial t^2 = 0, \qquad (3.60)$$

where ϵ is the dielectric constant of the medium. Now, employing the definition $\vec{B} = \mu_0(\vec{H} + \vec{M})$, where $\vec{M}$ is the magnetization vector of the crystal, we obtain from the above equation

$$c^2[\nabla(\nabla \cdot \vec{H}) - \nabla^2 \vec{H}] = -\epsilon(\partial^2/\partial t^2)(\vec{H} + \vec{M}). \qquad (3.61)$$

Assuming a plane-wave solution for $\vec{H}$ and $\vec{M}$, and taking into account that the Maxwell equation $\nabla \cdot \vec{B} = 0$ implies

$$\nabla \cdot \vec{H} = -\nabla \cdot \vec{M}, \qquad (3.62)$$

one gets

$$\vec{k}(\vec{k} \cdot \vec{M}) + k^2 \vec{H} = \epsilon(\omega/c)^2(\vec{H} + \vec{M}). \qquad (3.63)$$

Choosing the wavevector $\vec{k}$ to lie in the xz-plane and to make an angle θ with the z-axis (the direction of static magnetization), we have

$$k_x = |\vec{k}| \sin\theta \;\; \text{and} \;\; k_x^2 + k_z^2 = k^2. \qquad (3.64)$$

3.5. BULK MAGNETIC-POLARITONS

Therefore we can rewrite Eq. (3.63) in component form as

$$H_x = (-k_x^2 + \epsilon\omega^2/c^2)M_x(k^2 - \epsilon\omega^2/c^2)^{-1}, \tag{3.65}$$

$$H_y = (\epsilon\omega^2/c^2)M_y(k^2 - \epsilon\omega^2/c^2)^{-1}. \tag{3.66}$$

Defining $M^\pm = M_x \pm iM_y$, as before, and $H^\pm = H_x \pm iH_y$, Eqs. (3.65) and (3.66) yield

$$H^+ = \xi_1 M^+ - \xi_2 M^-, \tag{3.67}$$

$$H^- = -\xi_2 M^+ + \xi_1 M^-, \tag{3.68}$$

where

$$\xi_1 = (\tfrac{1}{2})[2(\epsilon\omega^2/c^2) - k_x^2](k^2 - \epsilon\omega^2/c^2)^{-1}, \tag{3.69}$$

$$\xi_2 = (\tfrac{1}{2})k_x^2(k^2 - \epsilon\omega^2/c^2)^{-1}. \tag{3.70}$$

Eqs. (3.67) and (3.68) can then be combined with the susceptibility relations of the previous section, i.e.

$$M^\pm - \chi^\pm H^\pm = 0, \tag{3.71}$$

to yield the polariton dispersion relation

$$(1 - \xi_1\chi^+)(1 - \xi_1\chi^-) = \xi_2^2 \chi^+ \chi^-. \tag{3.72}$$

This is a general dispersion relation for bulk magnetic polaritons suitable for any ordered magnetic materials (ferro-, antiferro-, or ferrimagnets) provided the appropriate susceptibilities are inserted.

The case of a propagation wavevector along the static field direction (the z-axis), meaning $k_x = \xi_2 = 0$, is of particular importance. As can be seen from Eqs. (3.67) and (3.68), the modes then become separated into two pure circularly polarized components; this was pointed out for antiferromagnets by Bose et al. [54]. The simplified dispersion relation is

$$(1 - \xi_1\chi^+)(1 - \xi_1\chi^-) = 0, \tag{3.73}$$

where each factor inside the brackets above corresponds to a particular circular polarization.

However, for propagation in a general direction, Eqs. (3.67) and (3.68) give a mixing of the left and right circularly polarized modes. We note that when the in-plane wavevector k_x becomes sufficiently large (i.e. $k_x \gg \epsilon^{1/2}\omega/c$), the polariton dispersion relation becomes

$$(\chi^+ + \chi^-)\sin^2\theta = 2, \tag{3.74}$$

since in this limit

$$\xi_1 = -\xi_2 = (\tfrac{1}{2})\sin^2\theta. \tag{3.75}$$

Considering now the particular case of a ferromagnet, Eq. (3.74) can be rewritten as

$$(\omega_0^2 - \omega^2)/\omega_0\omega_M = \sin^2\theta, \tag{3.76}$$

employing the notation $\omega_0 = -\gamma H_0$ and $\omega_M = -\gamma M$. Therefore the dispersion relation is given by

$$\omega^2 - \omega_0^2 + 2\xi_1\omega_0\omega_M = (\xi_1^2 - \xi_2^2)\omega_M^2, \tag{3.77}$$

where we have neglected the anisotropy field. Eq. (3.77) is equivalent to a result first derived by Auld [55].

Fig. 3.7 shows a schematic representation of the low-frequency bulk ferromagnetic polariton dispersion curve, as described by Eq. (3.77), together with its extension to larger wavevectors (see Sections 7.1 and 7.2 of Chapter 7), for propagation along the static field ($\theta = 0$). It is convenient to divide this figure into three regions (according to the wavevector), each having distinctive characteristics:

(a) The electric and magnetic fields are of comparable magnitude, and Maxwell's equations in their full form, including retardation, must be used. We call this the polariton region, or region of electromagnetic propagation, and it forms the topic of the present section.

(b) The electric field is negligible compared with the magnetic field so that the displacement current can be ignored and the magnetic field is derived from a magnetostatic potential. This means that ξ_1 and ξ_2 are given by Eq. (3.75). The frequencies found in this magnetostatic approximation are the same as those of the long-wavelength magnons derived from a microscopic Hamiltonian that includes magnetostatic dipole–dipole interactions. This case will be discussed in Section 7.2.

(c) The exchange energy (initially giving rise to a term proportional to k^2) has a significant effect on the shape of the dispersion curve, and the lower branch bends upward for large values of k. The dipole–dipole effects eventually become small compared with the exchange. This case will be discussed in Section 7.1.

Finally, we turn to the polariton modes of the uniaxial two-sublattice antiferromagnet [56]. At low temperatures one may assume zero static susceptibility in the z-direction [57], implying $M_1 = -M_2$ as assumed before. We define $\omega_0 = -\gamma H_0$ (as in the ferromagnetic case) and $\omega_M = -\gamma M_1 = \gamma M_2$. The anisotropy field and effective exchange field, where $H_E = (\lambda - \tfrac{1}{3})M_1$, correspond to the frequencies $\omega_A = -\gamma H_A$ and $\omega_E = -\gamma H_E$, respectively.

3.5. BULK MAGNETIC-POLARITONS

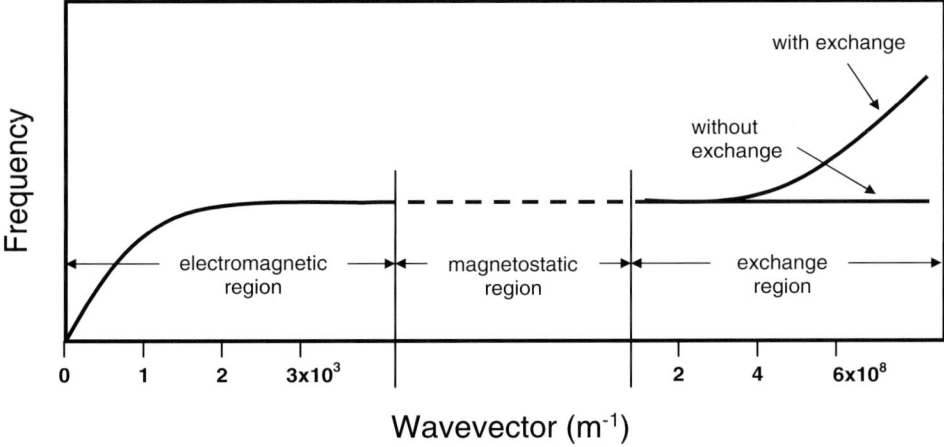

Fig. 3.7. Schematic representation of the low-frequency bulk ferromagnetic polariton propagating along the direction of the static field $\vec{H}_0$ ($\theta = 0$). We have plotted the frequency ω versus the wavevector. The evolution into the magnetostatic and exchange regions for larger wavevectors is shown.

In the magnetostatic limit, where $k_x \gg \epsilon^{1/2}\omega/c$, Eqs. (3.74) and (3.75) yield

$$\omega^2 = \omega_0^2 + \omega_A \omega_x \pm [4\omega_0^2 \omega_A \omega_x + \omega_A^2 \omega_M^2 \sin^4 \theta]^{1/2}, \quad (3.78)$$

where

$$\omega_x = \omega_A + 2\omega_E + \omega_M(\sin^2\theta - \tfrac{2}{3}). \quad (3.79)$$

In the absence of an external magnetic field (when $\omega_0 = 0$), Eq. (3.78) gives the dipole–dipole correction to the long-wavelength magnon frequencies originally derived by Loudon and Pincus [58]. In the general case, when retardation effects are included, the general magnetic-polariton dispersion relation, Eq. (3.72), yields

$$\left[(\omega+\omega_0)^2 - \omega_{AF}^2 + 2\omega_M\omega_A\xi_1\right]\left[(\omega-\omega_0)^2 - \omega_{AF}^2 + 2\omega_M\omega_A\xi_1\right] = 4\omega_M^2\omega_A^2\xi_2^2, \quad (3.80)$$

where

$$\omega_{AF} = [\omega_A^2 + 2\omega_A\omega_E]^{1/2} \quad (3.81)$$

gives the AFMR frequency for $H_0 = 0$.

The calculations that we have presented so far are all concerned with the solution of homogeneous equations of motions coupled to an electromagnetic field. It is more informative, although algebraically harder, to find the linear-response functions (or Green functions) of the magnetic crystal to a fictitious external magnetic field $\vec{H}_{ext}$, as was carried out for ferromagnetic [59] and antiferromagnetic [60] materials. The required mathematical formalism is outlined in Section A.3 of the Appendix.

References

[1] J.J. Hopfield, Phys. Rev. 112 (1958) 1555.
[2] K. Huang, Nature 167 (1951) 779.
[3] K. Huang, Proc. Roy. Soc. London, A208 (1951) 352.
[4] R. Loudon, Adv. Phys. 13 (1964) 423.
[5] E. Burstein and F. de Martini, Eds., Polaritons, Pergamon, Oxford, 1974.
[6] D.L. Mills and E. Burstein, Rep. Prog. Phys. 37 (1974) 817.
[7] C.H. Henry and J.J. Hopfield, Phys. Rev. Lett. 15 (1965) 964.
[8] S.P.S. Porto, B. Tell and T.C. Damen, Phys. Rev. Lett. 16 (1966) 450.
[9] J.F. Scott, L.E. Cheesman and S.P.S. Porto, Phys. Rev. 162 (1967) 834.
[10] F. Stern, Phys. Rev. Lett. 18 (1967) 546.
[11] A.L. Fetter, Ann. Phys. (NY) 88 (1974) 1.
[12] M. Born and K. Huang, Dynamical Theory of Crystal Lattices, Oxford Univ. Press, Oxford, 1954.
[13] A.A. Maradudin, E.W. Montroll, G.H. Weiss and I.P. Ipatova, Theory of Lattice Dynamics in the Harmonic Approximation, Academic Press, New York, 1971.
[14] U. Fano, Phys. Rev. 103 (1956) 1202.
[15] R.J. Elliott and A.F. Gibson, An Introduction to Solid State Physics and its Applications, Macmillan, London, 1974.
[16] R.H. Lyddane, R.G. Sachs and E. Teller, Phys. Rev. 59 (1941) 673.
[17] J.D. Jackson, Classical Electrodynamics, 3rd. ed., Wiley, New York, 1999.
[18] S.M. Feng and H.G. Winful, Opt. Lett. 26 (2001) 485.
[19] T. Feurer, S. Stoyanov, D.W. Ward and K.A. Nelson, Phys. Rev. Lett. 88 (2002) 257402.
[20] T.E. Stevens, J.K. Wahlstrand, J. Kuhl and R. Merlin, Science 291 (2001) 627.
[21] J.K. Wahlstrand, T.E. Stevens, J. Kuhl and R. Merlin, Physica B 316 (2002) 55.
[22] V.M. Agranovich and V.L. Ginzburg, Spatial Dispersion in Crystal Optics and the Theory of Excitons, 2nd. ed., Springer-Verlag, Berlin, 1981.
[23] S.I. Pekar, Zh. Eksp. Teor. Fiz. 33 (1957) 1022 [Sov. Phys.-JETP 6 (1958) 785].
[24] V.M. Agranovich and A.A. Rukhadze, Sov. Phys.-JETP 7 (1959) 685.
[25] V.L. Ginzburg, Propagation of Electromagnetic Waves in a Plasma, Pergamon, London, 1960.
[26] J.J. Hopfield and D.J. Thomas, Phys. Rev. 132 (1963) 563.
[27] E.L. Ivchenko in: Excitons, Eds., E.I. Rashba and M.D. Sturge, North-Holland, Amsterdam, 1982.
[28] J.L. Birman in: Excitons, Eds., E.I. Rashba and M.D. Sturge, North-Holland, Amsterdam, 1982.
[29] S.I. Pekar, Zh. Eksp. Teor. Fiz. 34 (1958) 1176 [Sov. Phys.-JETP 7 (1958) 813].
[30] V.L. Ginzburg, Zh. Eksp. Teor. Fiz. 34 (1958) 1593.
[31] V.M. Agranovich and V.L. Ginzburg, Sov. Phys.-Uspekhi 5 (1962) 323, 675.
[32] J.J. Hopfield, J. Phys. Soc. Japan 21 (1966) 77.
[33] S. Sakoda, J. Phys. Soc. Japan 40 (1976) 152.
[34] G.S. Agarwal, D.N. Pattanayak and E. Wolf, Phys. Rev. Lett. 27 (1971) 1022.
[35] G.S. Agarwal, D.N. Pattanayak and E. Wolf, Phys. Rev. B 8 (1973) 4768.
[36] A.A. Maradudin and D.L. Mills, Phys. Rev. B 7 (1973) 2787.
[37] J.L. Birman and J.J. Sein, Phys. Rev. B 6 (1972) 2482.
[38] J.J. Sein, Phys. Lett. A 32 (1970) 141.

REFERENCES

[39] J.J. Sein, Optics Commun. 2 (1970) 170.
[40] R. Zeyher, J.L. Birman and W. Brening, Phys. Rev. B 6 (1972) 4613.
[41] T. Skettrup, Phys. Status Solidi (b) 60 (1973) 695.
[42] D.R. Tilley, J. Phys. C 13 (1980) 781.
[43] E.L. Albuquerque and C.E.T. Gonçalves da Silva, J. Phys. C 18 (1985) 665.
[44] W. Brening, R. Zeyher and J.L. Birman, Phys. Rev. B 6 (1972) 4617.
[45] E. Koteles in: Excitons, Eds., E.I. Rashba and M.D. Sturge, North-Holland, Amsterdam, 1982.
[46] P. Yu in: Excitons, Ed., K. Cho, Springer, Berlin, 1979.
[47] R.G. Ulbrich and C. Weisbuch, in: Advances in Solid State Physics, Festkörperprobleme XVIII (1978) 217.
[48] H.Z. Cummins and P.E. Schoen, in: Laser Handbook Vol. 2, Eds., F.T. Arrechi and E.O. Schultz, North Holland, Amsterdam, 1987.
[49] S. Foner in: Magnetism Vol. I, Eds., G.T. Rado and H. Suhl, Academic, New York, 1963.
[50] E.F. Sarmento and D.R. Tilley, in: Electromagnetic Surface Modes, Ed., A.D. Boardman, Wiley, New York, 1982.
[51] F.A. Oliveira, A.F. Khater, E.F. Sarmento and D.R. Tilley, J. Phys. C 12 (1979) 4021.
[52] A.B. Harris, Phys. Rev. 143 (1966) 353.
[53] M.I. Kaganov, N. Pustylnik and T.I. Shalaeva, Usp. Fiz. Nauk 167 (1997) 191.
[54] M.S. Bose, E.N. Foo and M.A. Zuniga, Phys. Rev. B 12 (1975) 3855.
[55] B.A. Auld, J. Appl. Phys. 31 (1960) 1642.
[56] C. Manohar and G. Venkataraman, Phys. Rev. B 5 (1972) 1993.
[57] J.A. Eisele and F. Keffer, Phys. Rev. 96 (1954) 929.
[58] R. Loudon and P. Pincus, Phys. Rev. 132 (1963) 673.
[59] E.F. Sarmento and D.R. Tilley, J. Phys. C 9 (1976) 2943.
[60] E.F. Sarmento and D.R. Tilley, J. Phys. C 10 (1977) 795.

Chapter 4

Surface Plasmon- and Phonon-Polaritons

It has been known for some time that electromagnetic waves can propagate along an interface between two media, provided that at least one of the media is dispersive. *Surface polaritons* are those "mixed" electromagnetic modes that exist at crystalline surfaces and may be excited by various methods [1–3]. They are typically localized to within a few wavelengths of a surface, in the sense that their amplitude is a maximum at the surface and decays (usually exponentially) away from it. The specific properties of the surface polaritons depend on the characteristics of the materials, normally as described by their dielectric function. Many aspects of surface polaritons have been already extensively studied, e.g. their behavior in slab (or thin-film) geometries, non-planar geometries, effects of anisotropy, damping and external magnetic fields, etc. Reviews are to be found in e.g. Refs. [4–7].

The aim in this chapter is to present a discussion of the propagation of these electromagnetic modes (with particular reference to the non-magnetic cases of plasmon- and phonon-polaritons) at single and double interfaces. We restrict our attention to planar interfaces between the two media, establishing the formalism in a way that will facilitate our generalizations to superlattices and other multilayers in the subsequent chapters.

4.1 Single-Interface Modes: Isotropic Media

Consider the geometry depicted in Fig. 4.1, where we assume that there is a single interface at $z=0$ separating two isotropic media. Medium A occupies the half-space $z>0$ and is characterized by a dielectric function $\epsilon_A(\omega)$, while medium B is in $z<0$ and has a dielectric function $\epsilon_B(\omega)$. Also the direction of propagation of the mode is taken along the x-axis (with no loss of generality).

The solution of the electromagnetic wave equation inside any medium has the form

$$\vec{E}_J(\vec{r}, t) = (E_{xJ}, 0, E_{zJ}) \exp(ik_x x - i\omega t) \exp(ik_{zJ} z), \quad (4.1)$$

where $J = A, B$ and k_x is the common wavevector in the x-direction. Here we have assumed a transverse magnetic (TM) electromagnetic mode solution, i.e. a mode where the magnetic field is perpendicular to the plane of propagation of the radiation field (also known as p-polarization), as the case of interest. This choice is made because it can be easily proved that there is no surface mode for the

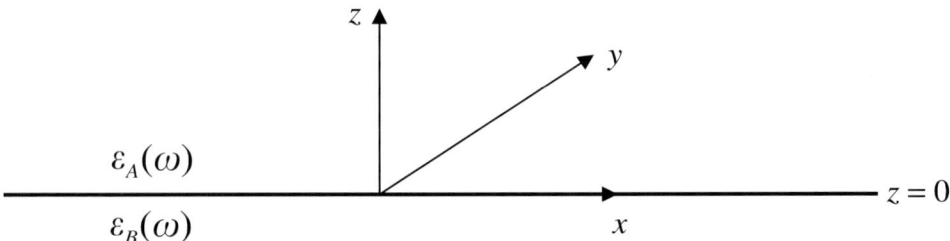

Fig. 4.1. Geometry used to specify the propagation of surface polaritons at a single interface. Media A and B occupy the half-spaces $z>0$ and $z<0$, and have dielectric functions $\epsilon_A(\omega)$ and $\epsilon_B(\omega)$, respectively.

transverse electric (TE) electromagnetic mode propagation, i.e. a mode where the electric field is perpendicular to the plane of propagation of the radiation field (also known as s-polarization) [8]. Substitution of Eq. (4.1) into the electromagnetic wave equation given in Eq. (3.17) yields

$$k_{zJ} = [\epsilon_J(\omega)\omega^2/c^2 - k_x^2]^{1/2}. \tag{4.2}$$

For localized modes, k_{zJ} must necessarily be complex in both media, which implies the inequality $k_x^2 > \epsilon_J(\omega)\omega^2/c^2$ if $\epsilon_J(\omega)$ is real.

Now applying the usual electromagnetic boundary conditions, namely the continuity of the tangential component of $\vec{E}$ and the normal component of $\vec{D}$ at the interface $z=0$, we find

$$k_{zA}/k_{zB} = \epsilon_A(\omega)/\epsilon_B(\omega). \tag{4.3}$$

This, together with Eq. (4.2), yields the dispersion relation for the *surface polaritons*, i.e.

$$k_x^2 = (\omega^2/c^2)\left[\epsilon_A(\omega)\epsilon_B(\omega)\right]\left[\epsilon_A(\omega) + \epsilon_B(\omega)\right]^{-1}. \tag{4.4}$$

Both $\epsilon_A(\omega)$ and $\epsilon_B(\omega)$ are real in the absence of damping, and from the localization condition we can write $k_{zA} = i\alpha_A$ and $k_{zB} = -i\alpha_B$, where α_A and α_B are real and positive, and the signs account for the appropriate limits as z tends to $\pm\infty$. It therefore follows from Eq. (4.3) that $\epsilon_A(\omega)$ and $\epsilon_B(\omega)$ must have opposite signs at any frequency ω corresponding to a surface polariton. Furthermore, taking into account Eq. (4.4), we have

$$\epsilon_A(\omega) + \epsilon_B(\omega) < 0. \tag{4.5}$$

We introduce the terminology that the medium with negative $\epsilon(\omega)$ is called the *surface-active* medium, while the medium with positive $\epsilon(\omega)$ is the *surface-inactive* medium.

As a first example of these general results, let us consider medium A as vacuum ($\epsilon_A = 1$) and medium B with $\epsilon_B(\omega)$ in the plasma form of Eq. (1.24), i.e.

4.1. SINGLE-INTERFACE MODES: ISOTROPIC MEDIA

$\epsilon_B(\omega) = \epsilon_\infty(1 - \omega_p^2/\omega^2)$. Medium B could, for example, be an n-doped GaAs. The dispersion relation for the *surface plasmon-polariton* frequency ω_{SPL} is obtained from Eq. (4.4) as

$$\omega_{SPL}^2(k_x) = (\tfrac{1}{2}\epsilon_\infty)\left[(1+\epsilon_\infty)c^2 k_x^2 + \epsilon_\infty \omega_p^2 - \Delta_{SPL}^{1/2}\right], \tag{4.6}$$

where

$$\Delta_{SPL} = \left[(1+\epsilon_\infty)c^2 k_x^2 + \epsilon_\infty \omega_p^2\right]^2 - (2\epsilon_\infty c k_x \omega_p)^2. \tag{4.7}$$

This dispersion relation is illustrated in Fig. 4.2, where a reduced frequency ω/ω_p is plotted against the reduced in-plane wavevector ck_x/ω_p. We note several features in this dispersion curve, as summarized below:

(a) the surface plasmon-polariton exists in the frequency interval

$$0 < \omega_{SPL}(k_x) < \omega_p/(1+\epsilon_\infty^{-1})^{1/2}; \tag{4.8}$$

(b) it lies entirely to the right of the light line in the vacuum, i.e. the line $\omega = ck_x$;

(c) it has a photon-like behavior ($\omega \simeq ck_x$) at small wave number, changing to a plasmon-like behavior for large wave number;

(d) at large values of k_x it approaches the asymptotic value $\omega_{SPL}(\infty) = \omega_p/(1+\epsilon_\infty^{-1})^{1/2}$;

(e) in this type of plot (against k_x) the bulk plasmon-polaritons appear as a *band* (shown shaded) at higher frequencies.

As a second example, we consider that medium B is a polar medium, whose real dielectric function is described by Eq. (3.11), with the damping factor $\Gamma = 0$, and medium A is still vacuum. The dispersion relation, obtained from Eq. (4.4), is now

$$\omega_{SPH}^2(k_x) = (\tfrac{1}{2}\epsilon_\infty)\left[(1+\epsilon_\infty)c^2 k_x^2 + \epsilon_\infty \omega_L^2 - \Delta_{SPH}^{1/2}\right], \tag{4.9}$$

where

$$\Delta_{SPH} = \left[(1+\epsilon_\infty)c^2 k_x^2 + \epsilon_\infty \omega_L^2\right]^2 - 4\epsilon_\infty c^2 k_x^2 (\omega_T^2 + \epsilon_\infty \omega_L^2) \tag{4.10}$$

with ω_L (ω_T) being the longitudinal (transverse) optical phonon frequency. In this case the so-called *surface phonon-polariton* mode occupies the frequency interval

$$\omega_T < \omega_{SPH}(k_x) < \left[(\epsilon_\infty \omega_L^2 + \omega_T^2)/(1+\epsilon_\infty)\right]^{1/2}. \tag{4.11}$$

We note that it formally reduces to the surface plasmon-polariton case if the limits of $\omega_T \to 0$ and $\omega_L \to \omega_p$ are taken. Fig. 4.3 shows the surface phonon-polariton dispersion relation for an intrinsic GaAs crystal with vacuum outside, where we

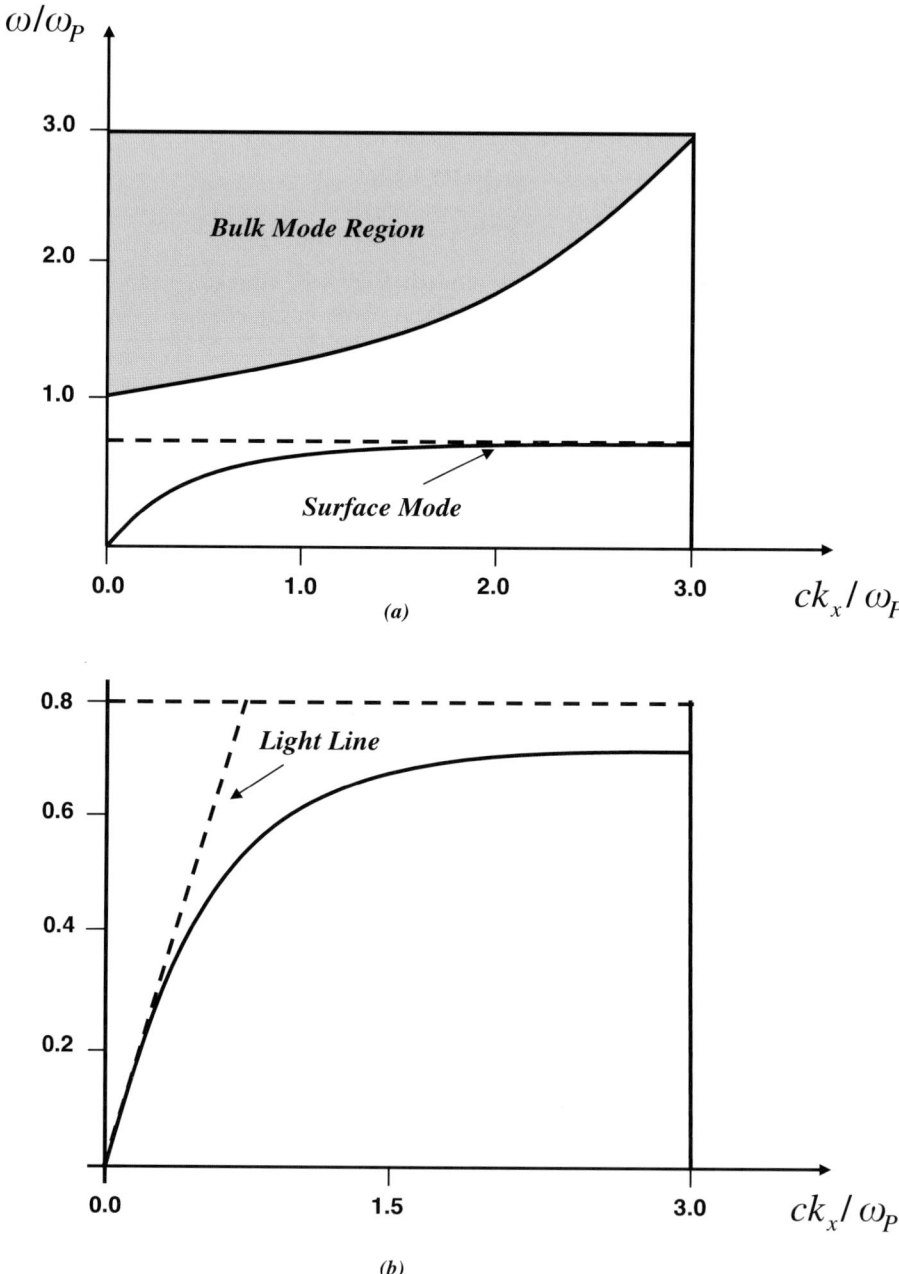

Fig. 4.2. (a) Plasmon-polariton dispersion curve for a semi-infinite sample of n-doped GaAs with vacuum outside. (b) Expanded scale for the low-frequency surface mode. The curve is asymptotic (at large k_x) to the value $\omega/\omega_p = 0.71$. The almost vertical dotted line is the light line in vacuum.

4.1. SINGLE-INTERFACE MODES: ISOTROPIC MEDIA

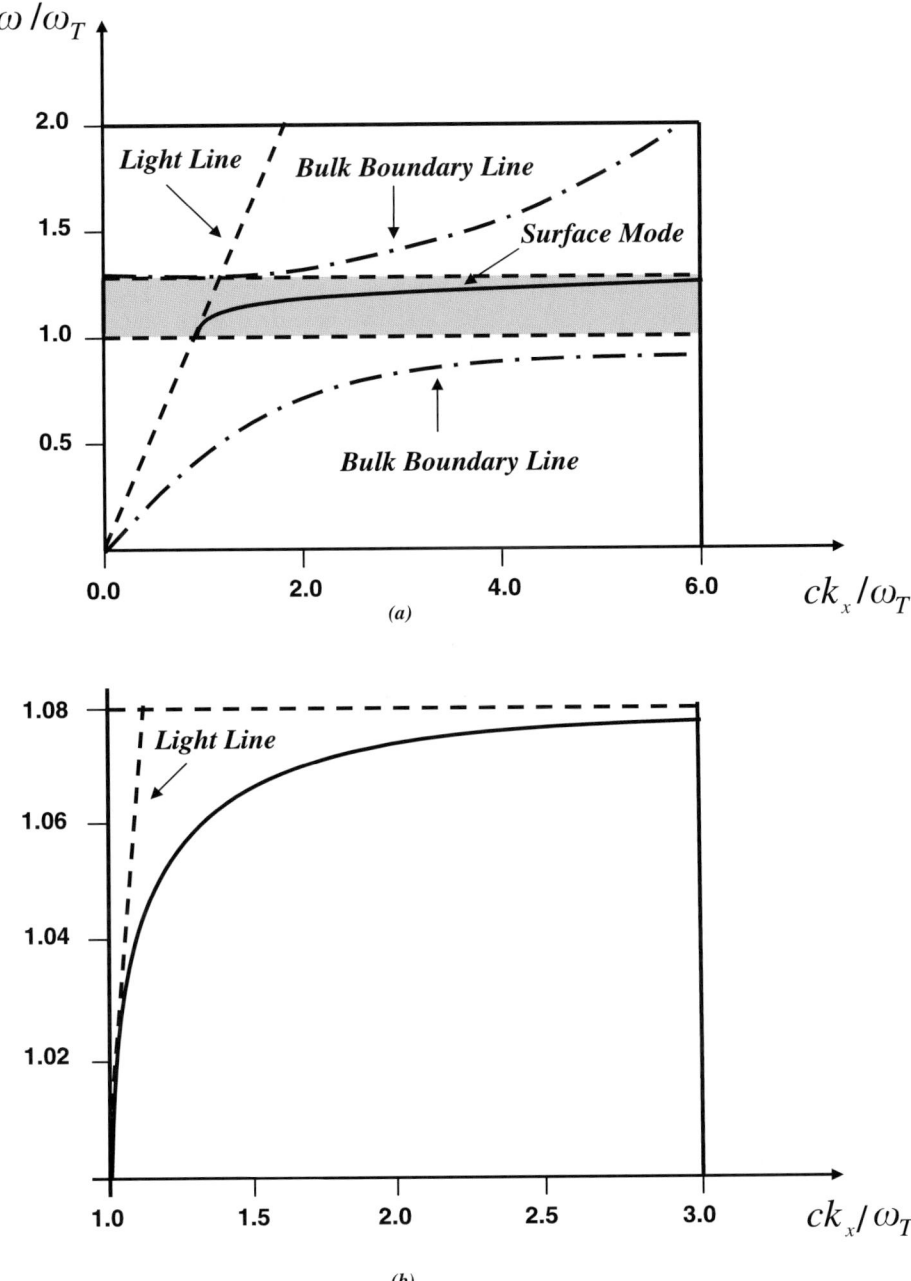

Fig. 4.3. (a) Phonon-polariton dispersion curve for an intrinsic GaAs/vacuum interface. The surface mode region is shown shaded here. (b) Expanded scale for the surface mode. The curve is asymptotic to the value $\omega/\omega_T = 1.0$. The almost vertical dotted line is the light line in vacuum.

have plotted a reduced frequency ω/ω_T against the reduced in-plane wavevector ck_x/ω_T. The dispersion curve starts on the light line at ω_T and, as the in-plane wavevector $k_x \to \infty$, approaches the asymptotic value

$$\omega_{SPH}(\infty) = \left[(\epsilon_\infty \omega_L^2 + \omega_T^2)/(1+\epsilon_\infty)\right]^{1/2}. \qquad (4.12)$$

The regime where the retardation effects (due to the non-zero propagation time of an electromagnetic signal) are unimportant is called the *electrostatic limit*. This corresponds to the condition $k_x \gg \omega/c$, which means that the wavelength $2\pi/k_x$ of the surface wave is much smaller than the free-space wavelength $2\pi c/\omega$ of the light. In this limit (effectively the limit of $c \to \infty$), we can deal with the solutions of Laplace's equation for the scalar potential instead of the full electromagnetic wave equation (3.17). Then, a plane-wave solution of this equation, which is localized at the interface $z=0$ and travels in the x-direction, has the form

$$\begin{aligned}\phi_A(\vec{r},t) &= A_1 \exp(-\alpha_A z)\exp(ik_x x - i\omega t), & z>0, \\ \phi_B(\vec{r},t) &= A_2 \exp(\alpha_B z)\exp(ik_x x - i\omega t), & z<0.\end{aligned} \qquad (4.13)$$

The substitution of the above expressions for the scalar potential into Laplace's equation, which is $\nabla^2 \phi(\vec{r},t)=0$, yields simply $\alpha_J^2 = k_x^2$ ($J=A,B$). The electric fields, which are obtained using $\vec{E}_J(\vec{r},t) = -\nabla \phi_J(\vec{r},t)$, are then given by

$$\begin{aligned}\vec{E}_A(\vec{r},t) &= (-ik_x, 0, k_x)\phi_A(\vec{r},t), & z>0, \\ \vec{E}_B(\vec{r},t) &= (-ik_x, 0, -k_x)\phi_B(\vec{r},t), & z<0.\end{aligned} \qquad (4.14)$$

Therefore E_x and E_z (in either medium) are equal in magnitude but are out of phase with each other by $\pi/2$.

Imposing the standard boundary conditions, which now take the form that ϕ and $\epsilon \partial \phi/\partial t$ are continuous across the interface $z=0$, we find that $A_1 = A_2$ and

$$\epsilon_A(\omega) + \epsilon_B(\omega) = 0, \qquad (4.15)$$

which is the unique condition for such a wave to exist. The surface excitation, whose potential is given by Eq. (4.13) and whose frequency is obtained from Eq. (4.15), is called the *unretarded surface polariton*.

As a final example, we mention the case of a surface polariton associated with the plasmon and optical phonon modes, which can exist in a doped polar semiconductor. The appropriate dielectric function now contains both plasmon and optical phonon contributions, and is given by

$$\epsilon(\omega) = \epsilon_\infty \left[1 + \left(\frac{\omega^2 - \omega_L^2}{\omega^2 - \omega_T^2}\right) - \left(\frac{\omega_p^2}{\omega^2}\right)\right]. \qquad (4.16)$$

Considering medium B to be an n-doped semiconductor material, and medium A to be vacuum as before, and taking into account Eqs. (4.4) and (4.16), the surface polariton spectrum for this case consists of two branches, as expected. One is similar to the surface plasmon-polariton case (depicted in Fig. 4.2) while the other is related to the surface phonon-polariton case (shown in Fig. 4.3). Bryksin et al. [9] have, in fact, been able to observe this spectrum, finding good agreement with the theoretical predictions, although the measured frequencies of the plasmon-like surface polaritons are somewhat lower than the calculated ones.

4.2 Single-Interface Modes: Anisotropic Media

There are many crystals whose optical properties are found to be anisotropic, and so an extension of the results of the previous section is required when one (or both) of the media is taken to be anisotropic [10–13].

Suppose now that medium A is isotropic, with dielectric constant $\epsilon_A > 0$ and independent of frequency, while the second medium has its optical properties characterized by a dielectric tensor $\epsilon_B(\omega)$, whose principal axes (x', y', z') are at an arbitrary orientation with respect to the surface-related axes (x, y, z). Wallis et al. [14] have developed the theory for the most general form of this dielectric tensor $\epsilon_B(\omega)$, while considering $\epsilon_A = 1$. A simpler case, which can be treated analytically and is still of experimental interest, is the special case in which the interface is parallel to a principal axis of the anisotropic dielectric tensor, and the in-plane wavevector k_x is parallel to a second principal axis (taken to be the x-axis). Denoting the principal values of the anisotropic dielectric tensor along the x-axis and the axis perpendicular to the surface by $\epsilon_\parallel(\omega)$ and $\epsilon_\perp(\omega)$, respectively, the dispersion relation for the surface polariton is found to be [15,16]

$$\epsilon_\parallel(\omega)/\epsilon_A = -\alpha_B/\alpha_A, \qquad (4.17)$$

where we now define

$$\alpha_A^2 = k_x^2 - \epsilon_A \omega^2/c^2, \qquad (4.18)$$

$$\alpha_B^2 = [\epsilon_\parallel(\omega)/\epsilon_\perp(\omega)][k_x^2 - \epsilon_\perp(\omega)\omega^2/c^2]. \qquad (4.19)$$

Using Eqs. (4.17)–(4.19) we can solve for k_x to obtain

$$k_x^2 = \frac{\omega^2}{c^2} \frac{\epsilon_A \epsilon_\perp(\omega)[\epsilon_A - \epsilon_\perp(\omega)]}{\epsilon_A^2 - \epsilon_\parallel(\omega)\epsilon_\perp(\omega)}, \qquad (4.20)$$

which is a generalization of Eq. (4.4).

The solutions of Eq. (4.20) correspond to surface polaritons only if either one of the conditions described below can be satisfied:

(a) $\epsilon_\parallel(\omega)$ and $\epsilon_\perp(\omega)$ are both negative, with $k_x^2 > \epsilon_A(\omega/c)^2$;

(b) $\epsilon_\parallel(\omega)$ is negative and $\epsilon_\perp(\omega)$ is positive, with $\epsilon_A < (ck_x/\omega)^2 < \epsilon_\perp(\omega)$.

Case (a) describes the generalization of the surface polaritons discussed in the previous section. They were called *type I* surface polaritons by Bryksin et al. [17].

Case (b), called *type II* surface polaritons [17], represents a new type of surface polariton that does not exist in the unretarded limit. It exists only for a limited range of values of k_x and also requires $\epsilon_\perp(\omega) > \epsilon_A$. It is a *photon-induced* surface polariton.

Another example of interest is the special case when the y-axis is a principal axis of the tensor $\epsilon_B(\omega)$, i.e. $y \equiv y'$. As before, we consider a mode propagating in the x-direction and p-polarization electromagnetic modes. It can be shown that the dispersion relation equation for the surface polariton is

$$k_x^2 = \frac{\omega^2}{c^2} \frac{\epsilon_A[\epsilon_x'\epsilon_z' - \epsilon_A\epsilon_{zz}]}{\epsilon_x'\epsilon_z' - \epsilon_A^2}. \tag{4.21}$$

Here ϵ_x' and ϵ_z' are the principal values of ϵ_B in the $x'z'$-plane and ϵ_{zz} is the appropriate component of ϵ_B in the (x, y, z) system of axes.

One application of these results is to ferroelectrics (i.e. materials that have a spontaneous polarization, or electric dipole moment, below a critical temperature). These can be characterized, in the simplest approximation, as having a uniaxial dielectric tensor of the form quoted in Eq. (3.3) in terms of its principal axes. Cottam et al. [18] present some discussion of the possibilities for using surface polaritons to investigate the behavior of the tensor element $\epsilon_\parallel$. Another application was made to α-quartz by Falge and Otto [19]. They studied both the ordinary surface polaritons, which occur when the in-plane wavevector k_x is perpendicular to the easy-optical axis c, which is in turn parallel to the surface, and the extraordinary surface polaritons, which occur when the easy-optical axis c is parallel to the surface and to the in-plane wavevector k_x. Both type I and II surface polaritons were observed, with excellent agreement with the theoretical curves.

4.3 Charge-Sheet Modes

In some circumstances it is possible to have a thin sheet of mobile electrons trapped at an interface between two media. Examples are a charge sheet at the surface of liquid helium (where the charges are trapped in the weak image potential) and a charge sheet at a semiconductor heterojunction such as that between GaAs and $Al_xGa_{1-x}As$. Further discussion can be found in Ref. [20]. By employing a model in which these sheets are treated as a 2D electron (or hole) gas at the interface between two isotropic dielectric media, we now extend the results of the isotropic media to predict the existence of a surface-polariton-like mode.

We use the same geometry and coordinate axes as in Fig. 4.1, except that we now include a 2D charge-sheet plasma localized at $z = 0$. Eqs. (4.1) and (4.2) still apply, but the form of the boundary conditions must be modified. We may assume continuity of the tangential electric fields ($E_{Ax} = E_{Bx}$) as before, while

4.4. THIN FILMS

the magnetic field component H_y now satisfies

$$H_{By} - H_{Ay} = j_x. \tag{4.22}$$

Here j_x is the current density in the charge sheet and is driven by the alternating electrical field at the interface. Ignoring collisions in the electron gas, the classical response equations give

$$j_x = \sigma E_{Ax}, \quad \sigma = ine^2/m\omega, \tag{4.23}$$

where n denotes the number of charges per unit area at the interface, and the remaining notations are the same as before. Using the above boundary conditions, we obtain

$$\epsilon_A/\alpha_A + \epsilon_B/\alpha_B = \Omega_c/\omega^2, \tag{4.24}$$

where Ω_c is a characteristic frequency defined by

$$\Omega_c = ne^2/\epsilon_0 mc. \tag{4.25}$$

Eq. (4.24) generalizes the previous result in Eq. (4.3) and provides the surface-polariton dispersion relation first derived by Nakayama [21]. It simplifies in the electrostatic limit (where α_A and α_B are both replaced by k_x) to give

$$\omega = [\Omega_c c k_x/(\epsilon_A + \epsilon_B)]^{1/2}, \tag{4.26}$$

which had been obtained earlier by Stern [22] using linear-response theory.

We note that the surface modes predicted by Eqs. (4.24) and (4.26) can exist even when ϵ_A and ϵ_B are both positive, in contrast to the case described in Section 4.1, where ϵ_A and ϵ_B were required to have opposite signs. An example of the predicted dispersion relation, Eq. (4.24) in the case of $\epsilon_A = \epsilon_B > 0$, is depicted in Fig. 4.4. Here we have plotted the reduced frequency Ω/Ω_c against ck_x/Ω_c, with $\Omega \equiv \epsilon_A^{1/2}\omega$.

4.4 Thin Films

Next we extend some of the calculations of the previous sections to the three-layer system. This introduces effects involving *two* interfaces, and there is a length parameter L corresponding to the distance apart of the interfaces.

Fig. 4.5 shows the structure consisting of a dielectric thin film of dielectric function $\epsilon_B(\omega)$, sandwiched between two semi-infinite bounding media of dielectric functions $\epsilon_A(\omega)$ and $\epsilon_C(\omega)$. The two interfaces are characterized by the planes $z = 0$ and $z = -L$, where we have taken the z-axis normal to the interfaces, as before. For a p-polarized electromagnetic mode, the electric fields in the three media are

$$\vec{E} = [E_{xA}, 0, (-k_x/k_{zA})E_{xA}]\exp(ik_{zA})\exp(ik_x x - i\omega t) \quad \text{for } z > 0, \tag{4.27}$$

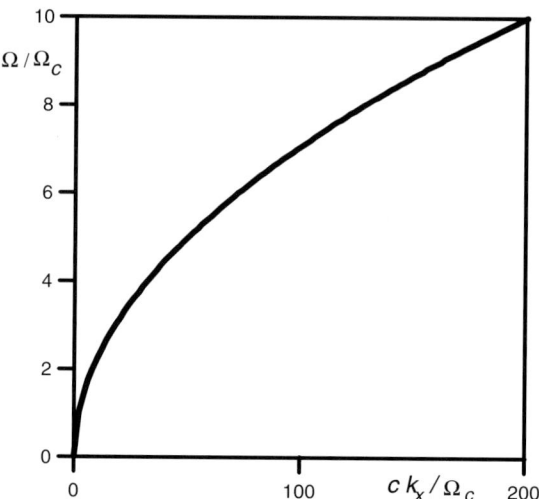

Fig. 4.4. Charge-sheet dispersion curve corresponding to Eq. (4.24) for the case where $\epsilon_A = \epsilon_B$ and positive.

$$\vec{E} = [E_{B+}, 0, (-k_x/k_{zB})E_{B-}]\exp(ik_x x - i\omega t) \quad \text{for } 0 > z > -L, \tag{4.28}$$

$$\vec{E} = [E_{xC}, 0, (-k_x/k_{zC})E_{xC}]\exp[ik_{zC}(z+L)]\exp(ik_x x - i\omega t) \quad \text{for } z < -L \tag{4.29}$$

denoting $E_{B\pm} = A\exp(i\theta_B) \pm B\exp(-i\theta_B)$ and $\theta_B = k_{zB}(z + L/2)$. Here we have used the condition $\nabla \cdot \vec{D} = 0$ to relate the z- and x-components of the electric field $\vec{E}$ in each medium. Also k_{zj} ($j = A, B, C$) is the z-component of the wavevector k_j, while k_x is their common in-plane x-component. The localization condition for surface modes now requires Im $k_{zA} > 0$ and Im $k_{zC} < 0$, yielding

$$k_x^2 > \epsilon_J \omega^2/c^2, \quad J = A, C. \tag{4.30}$$

Next, using the standard electromagnetic boundary conditions at the two interfaces, one finds four homogeneous equations in the amplitudes E_{xA}, A, B, and E_{xC}, whose solvability condition gives the desired dispersion relation for the surface polaritons, i.e.

$$\frac{\epsilon_A(\omega)k_{zB} + \epsilon_B(\omega)k_{zA}}{\epsilon_B(\omega)k_{zA} - \epsilon_A(\omega)k_{zB}}\exp(-2ik_{zB}L) = \frac{\epsilon_C(\omega)k_{zB} + \epsilon_B(\omega)k_{zC}}{\epsilon_B(\omega)k_{zC} - \epsilon_C(\omega)k_{zB}}. \tag{4.31}$$

We observe that, on taking the limit of $L \to \infty$, Eq. (4.31) yields the two equations

$$\epsilon_B(\omega)k_{zA} - \epsilon_A(\omega)k_{zB} = 0, \tag{4.32}$$

$$\epsilon_C(\omega)k_{zB} + \epsilon_B(\omega)k_{zC} = 0, \tag{4.33}$$

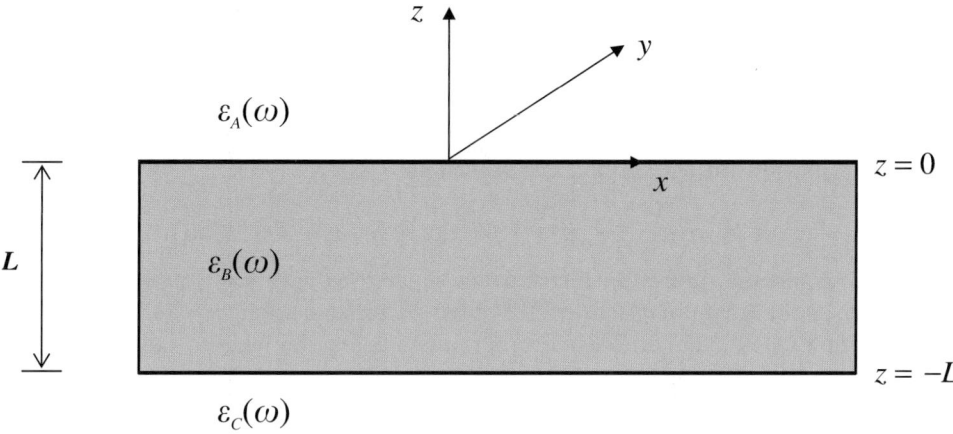

Fig. 4.5. Schematic representation for the propagation of surface polaritons in thin films. The bounding media A and C occupy the regions $z > 0$ and $z < -L$, and have dielectric functions $\epsilon_A(\omega)$ and $\epsilon_C(\omega)$, respectively. The film (medium B) has thickness L and dielectric function $\epsilon_B(\omega)$.

which correspond to the dispersion relations of uncoupled surface polaritons propagating at the $z=0$ and $z=-L$ interfaces, respectively.

We note that, for finite L and in the absence of damping, k_{zA} and k_{zC} must be purely imaginary, but k_{zB} can be either real or imaginary. For the case of imaginary k_{zB}, the surface polariton at one interface is perturbed due to the presence of the other interface and, as L is decreased, the fields of the two surface polaritons overlap and perturb one another. On the other hand, for the case of real k_{zB}, the z-dependence of the fields in medium B is oscillatory, so that it behaves like a waveguide, giving rise to the so-called *guided-wave polaritons* [23].

Guided optical waves consist of light waves trapped by total internal reflection in a region of a dielectric medium that is characterized by a higher refractive index than its surrounding parts. The most common system is a thin film, with thickness of the order of an optical wavelength, deposited on a substrate with a refractive index lower than that of the film. These waves are propagating transverse optical resonances of the film coupled via the boundaries to evanescent fields in the adjacent media. Above a cut-off frequency (which depends on the film thickness and refractive index ratios), there are no restrictions on the frequencies of the guided modes and hence a full spectrum of modes is possible [24–26]. These are examples of modes that form the basis for the integrated optics technology [27]. Note that, although the surface polariton modes are p-polarized electromagnetic waves, guided-wave polaritons can in fact be found in both p- and s-polarizations.

It was in such a three-layer system that the first experimental observations of surface polaritons by Raman scattering were carried out [28]. The theory of Raman scattering applicable to these experimental results was presented shortly afterwards [29–31].

The general result for the dispersion relation in Eq. (4.31) is rather complicated, since many parameters are involved. However, a special case of Eq. (4.31) is the symmetric geometry, in which we have $\epsilon_A(\omega) = \epsilon_C(\omega) \equiv \epsilon_A$ and $\epsilon_B(\omega) \equiv \epsilon(\omega)$, so the film has a plane of reflection symmetry at $z = -L/2$. The localization condition becomes $k_{zA} = -k_{zC} = i\alpha_A$, with $\alpha_A > 0$ given by Eq. (4.18), and the dispersion relation Eq. (4.31) has two solutions:

$$\exp(ik_{zB}L) = \pm [\epsilon_A k_{zB} + i\epsilon(\omega)\alpha_A]/[\epsilon_A k_{zB} - i\epsilon(\omega)\alpha_A]. \tag{4.34}$$

The above equation applies both to surface polaritons with k_{zB} purely imaginary ($k_{zB} = i\alpha_B$) and to p-polarized guided-wave polaritons with k_{zB} real. The two solutions in Eqs. (4.34) can then be reorganized, for the surface polariton case, into the forms

$$\epsilon(\omega)/\epsilon_A = -(\alpha_B/\alpha_A)\tanh(\alpha_B L/2), \tag{4.35}$$

$$\epsilon(\omega)/\epsilon_A = -(\alpha_B/\alpha_A)\coth(\alpha_B L/2). \tag{4.36}$$

It is now easy to see that, as L tends to infinity, the hyperbolic functions in Eqs. (4.35) and (4.36) tend to unity. Then, both equations have the same asymptotic behavior, recovering the dispersion relation for the single-interface mode. Furthermore, as might be expected, the two surface polariton modes defined by Eqs. (4.35) and (4.36) are essentially a bonding and an anti-bonding combination. In Fig. 4.6 we plot the two surface polaritons for the case of an LiF slab surrounded by vacuum and three sets of parameter values [32]. We have plotted the reduced frequency ω/ω_T, with ω_T being the transverse optical phonon frequency in LiF, against the dimensionless wavevector k_x/k_T, with $k_T = \omega_T/c$. As we can see, the upper mode for an unsupported film with $k_T L = 0.1$ (curve a) has the property that as k_x increases the frequency increases at first, then decreases to the asymptotic value, with the group velocity being zero at the maximum. Its lower mode is depicted in curve b, with no equivalent mode with a metal substrate. In curve c we see the profile of the upper mode for $L \to \infty$ for any substrate. The relatively minor effect of a good (rather than perfect) metal substrate is seen in curve d. By contrast, curves e and f show the large effect due to a substrate with a large constant dielectric function.

For p-polarized guided-wave polaritons, the two solutions of Eq. (4.34), now considering k_{zB} to be real, are

$$\epsilon(\omega)/\epsilon_A = (k_{zB}/\alpha_A)\tanh(k_{zB}L/2), \tag{4.37}$$

$$\epsilon(\omega)/\epsilon_A = -(k_{zB}/\alpha_A)\coth(k_{zB}L/2). \tag{4.38}$$

The task of solving these equations numerically is similar to that of finding the odd- and even-parity modes in a quantum-mechanical one-dimensional square well, or finding the dispersion curves for the surface acoustic Love waves [33].

So far, we have been concerned just with p-polarized modes. However, for applications of dielectric waveguides [34], the guided-wave modes with s-polarization

4.4. THIN FILMS

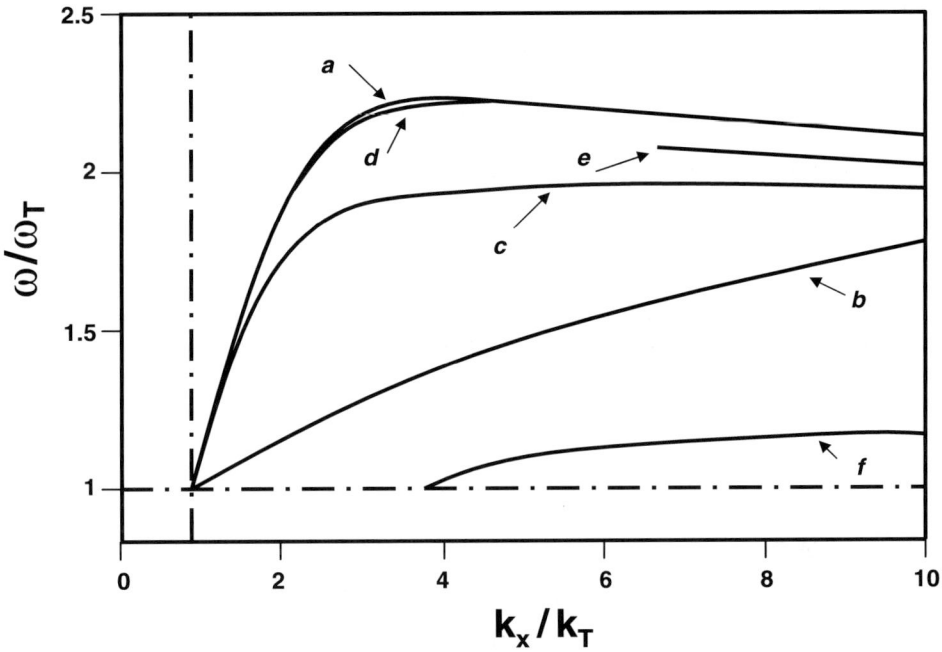

Fig. 4.6. The two surface phonon-polaritons (upper and lower modes) for a LiF thin film on various substrates, including retardation effects. The curves corresponds to: *a* upper branch for an unsupported film with $k_T L = 0.1$; *b* lower branch for an unsupported film with $k_T L = 0.1$; *c* upper mode for $L \to \infty$ for any substrate; *d* potassium substrate for $k_T L = 0.05$ and $\omega_p = 3.8\,\text{eV}$; *e* upper mode for a silicon substrate with $\epsilon_C = 11.7$ and $k_T L = 0.1$; *f* lower mode for a silicon substrate (after Oliveira et al. [32]).

[35] are the more important. The calculation of their dispersion relation is very similar to that for the *p*-polarized modes. In fact, they are again described by Eq. (4.31) provided we make the formal substitutions

$$\epsilon_B(\omega) k_{zA} / \epsilon_A(\omega) k_{zB} \to k_{zB}/k_{zA}, \tag{4.39}$$

$$\epsilon_B(\omega) k_{zC} / \epsilon_C(\omega) k_{zB} \to k_{zB}/k_{zC}. \tag{4.40}$$

Thus, for the symmetric three-layer geometry described before, the *s*-polarized guided-wave polaritons are given by

$$(k_{zB}/\alpha_A) \tanh(k_{zB} L/2) = 1, \tag{4.41}$$

$$(k_{zB}/\alpha_A) \coth(k_{zB} L/2) = -1. \tag{4.42}$$

In Fig. 4.7 we illustrate the dispersion relation of guided-wave polaritons for the case where a dielectric film is surrounded by vacuum, considering both *p*- and *s*-polarization [36].

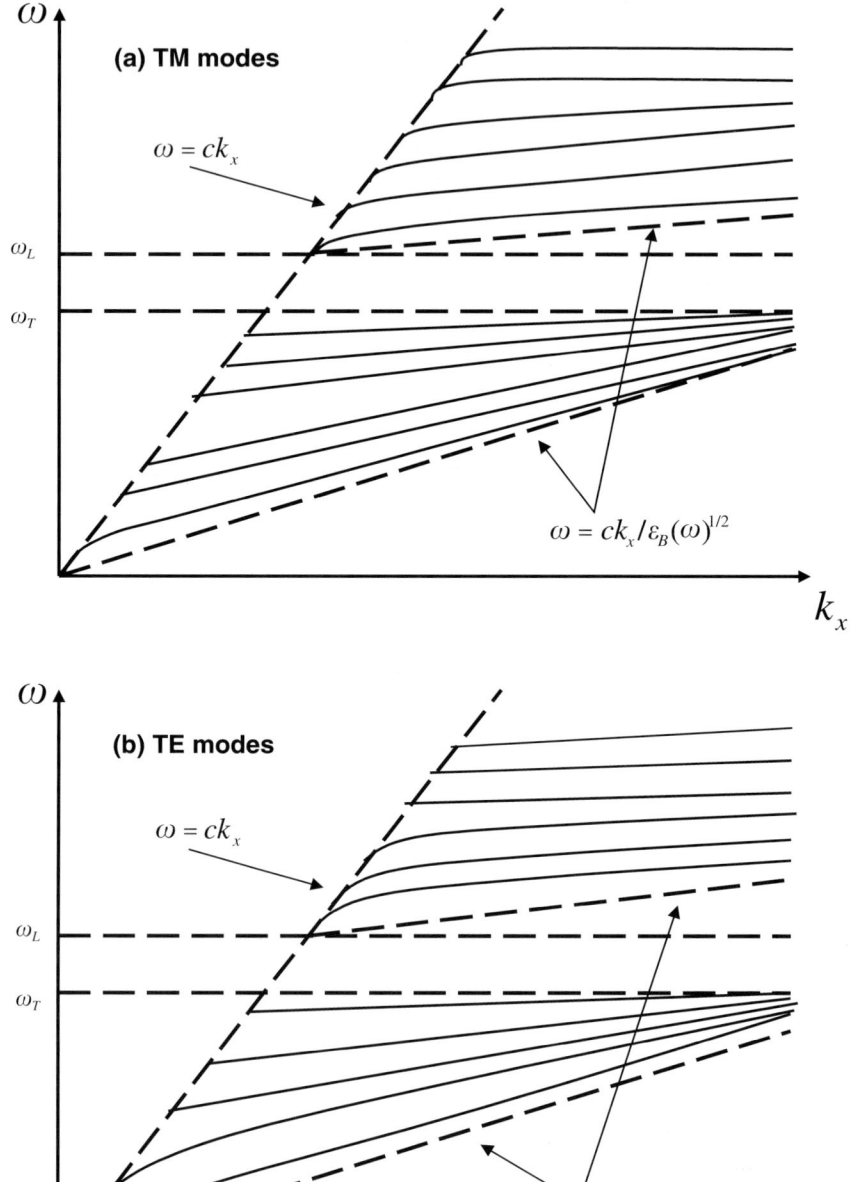

Fig. 4.7. Dispersion relations for guided-wave polaritons with (a) *p*-polarization and (b) *s*-polarization (after Ushioda and Loudon [36]).

4.4. THIN FILMS

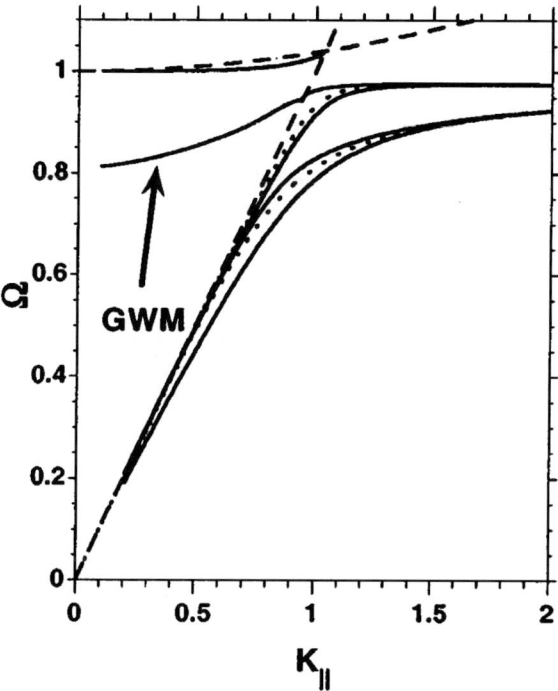

Fig. 4.8. Plasmon-polariton dispersion curve for two doped GaAs films separated and surrounded by vacuum. We have plotted the reduced frequency $\Omega = \omega/\omega_p$ against the reduced wavevector $K_\parallel = ck_x/\omega_p$. Dashed lines indicate the light line and the boundary of the bulk-mode region. The dotted lines depict the single-active-film surface plasmon modes. The guided-wave mode (GWM) is labelled by an arrow (after Gilmore and Johnson [37]).

All the discussions above were made for a single thin-film geometry. Recently, Gilmore and Johnson [37] found that in a structure consisting of three thin films on a substrate, where two of the thin films have free charge carriers and are separated by a static dielectric, an unusual guided-wave plasmon-polariton can be supported in a normally forbidden region of the surface polariton dispersion. The modes emerge from both bulk and surface collective-mode bands and evolve into a ladder of solutions in the forbidden region. The polariton spectrum is shown in Fig. 4.8 for two doped GaAs films separated and surrounded by vacuum.

So far we have assumed, for simplicity, that the dielectric functions are real in all the media, i.e. we have ignored damping. In general the dielectric function may be complex and for an electron gas, instead of the expression given by Eq. (1.24), it has the form

$$\epsilon(\omega) = \epsilon_\infty [1 - \omega_p^2/(\omega^2 + i\omega\Gamma)], \qquad (4.43)$$

where Γ is a damping constant. Fukui et al. [38] and Sarid [39] carried out numerical calculations for the localized plasmon-polariton modes in metallic films with a dielectric function as in Eq. (4.43). In both cases they made use of the general dispersion relation, Eq. (4.31), and its symmetrical forms, Eqs. (4.35) and (4.36). Fukui et al. considered k_x to be real and calculated Im ω, while Sarid took ω to be real and calculated instead Im k_x. These different assumptions corresponded to different experimental applications in the two cases.

The important conclusion in both papers concerned the behavior of the higher-frequency plasmon-polariton mode. It was found that the lifetime (Im $\omega)^{-1}$ in the former case, or similarly the decay length (Im $k_x)^{-1}$ in the latter case, increased as the film thickness L decreased. This mode was then called the *long-range surface plasmon* mode. In the symmetric geometry ($\epsilon_A = \epsilon_C$ for the media bounding the film), the long-range surface plasmon is the mode with the antisymmetric variation of the electric field across the film thickness, while the lower-frequency plasmon mode has a symmetric distribution for the electric field. The long decay length of the upper-frequency mode can be seen as arising from the fact that, as the film thickness decreases, a smaller proportion of the mode energy is transported within the film [40]. The experimental confirmation of these predictions was obtained by a number of groups (see e.g. the work of Craig et al. [41]). Also, the experimental realization of highly efficient optical elements built up from metal nanostructures to manipulate surface plasmon-polaritons propagating along a silver/polymer interface was achieved more recently [42].

Some extensions of the theory to surface phonon-polaritons and to surface plasmon–phonon polaritons were also carried out (see e.g. Refs. [43,44]) by using different expressions for the dielectric function $\epsilon(\omega)$ of the film with damping included. The long-range surface plasmons may be of interest for applications in non-linear optics, where the long-range nature of the modes is advantageous [45]. Furthermore, localization and waveguiding of surface plasmon-polaritons propagating along the gold film surface covered with randomly located scatterers were recently used in strongly scattering non-absorbing random media for guiding electromagnetic waves [46,47].

Finally in this section, we turn our attention to the phonon-polariton propagation in anisotropic thin films. The frequency-dependent dielectric function is assumed to have a diagonal form, as in Eq. (3.3) with

$$\epsilon_\|(\omega) = \epsilon_{\|\infty} \prod_{i=1}^{m_\|} \frac{[(\omega_{\|i}^L)^2 - \omega^2]}{[(\omega_{\|i}^T)^2 - \omega^2]}, \tag{4.44}$$

in terms of the frequencies $\omega_{\|i}^L$ and $\omega_{\|i}^T$ of the LO and TO phonons, respectively. Here the index i labels the reststrahlen bands, and $m_\|$ is the number of these bands. For example, $m_\| = 4$ and $m_\perp = 8$ in the case of α-quartz [19]. There is an analogous expression for $\epsilon_\perp(\omega)$.

We consider the special case of a symmetric geometry, in which the anisotropic film is surrounded by vacuum, i.e. $\epsilon_A(\omega) = \epsilon_C(\omega) = 1$. Taking into account that the electric field inside the film now has the form

$$\vec{E} = [E_{B+}, 0, (-k_x \epsilon_\| / k_{zB} \epsilon_\perp) E_{B-}] \exp(i k_x x - i\omega t) \quad \text{for } 0 > z > -L, \tag{4.45}$$

4.4. THIN FILMS

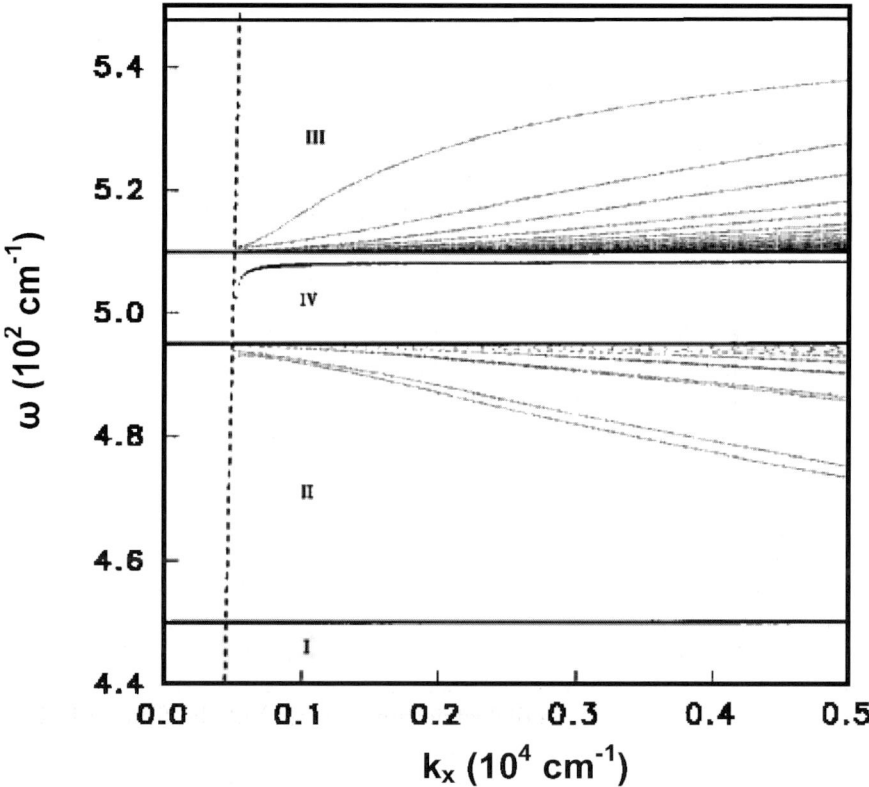

Fig. 4.9. Surface phonon-polariton mode in a film of α-quartz (thickness 4 μm) surrounded by vacuum, showing frequency (in units of 10^2 cm^{-1}) plotted against the wavevector k_x (in units of 10^4 cm^{-1}). The optical axis of the crystal is oriented parallel to the x-axis. Here the short-dashed line is the light line $k_x = \omega/c$. The notation for the different regions is explained in the main text (after Costa Filho et al. [48]).

instead of Eq. (4.28), and that

$$-ik_{zB} = \alpha_B = (\epsilon_\parallel/\epsilon_\perp)^{1/2}[k_x^2 - \epsilon_\perp(\omega/c)^2]^{1/2}, \tag{4.46}$$

instead of Eq. (4.19), the surface phonon-polariton dispersion relation is given by

$$\exp(2\alpha_B L) = [(\alpha_B - \alpha_A \epsilon_\parallel)/(\alpha_B + \alpha_A \epsilon_\parallel)]^2, \tag{4.47}$$

where we define $\alpha_A = [k_x^2 - (\omega/c)^2]^{1/2}$.

Fig. 4.9 shows a typical region of the surface phonon-polariton spectrum calculated for a film of α-quartz with its principal axis c (the optical axis) parallel to the x-axis. We used the dielectric function defined in Eq. (4.44), with parameter values as quoted in Ref. [19]. The thick horizontal lines correspond to optical phonon frequencies $\omega_{\perp 4}^T \simeq 450$ cm^{-1}, $\omega_{\parallel 2}^T \simeq 495$ cm^{-1}, $\omega_{\perp 4}^L \simeq 510$ cm^{-1}, and

$\omega^L_{\|2} \simeq 548$ cm^{-1}. Several different regions of surface phonon-polariton behavior can be identified between the horizontal lines as follows [48]:

(a) Region I corresponds to $\epsilon_\perp > 0$ and $\epsilon_\| > 0$. In this case there is no solution for either $k_x < \omega/c$ or $k_x > \epsilon_\perp^{1/2}\omega/c$, since α_A and α_B are either real or purely imaginary quantities. On the other hand, for $\omega < ck_x < \epsilon_\perp^{1/2}\omega$, α_A is real although α_B is purely imaginary, and there can be guided modes confined in the film.

(b) Region II has $\epsilon_\perp < 0$ and $\epsilon_\| > 0$. Here α_B is purely imaginary whatever the value of k_x, which implies guided modes for all $k_x > \omega/c$.

(c) Region III corresponds to $\epsilon_\perp > 0$ and $\epsilon_\| < 0$. There is no solution for $k_x < \omega/c$. For the range $\omega < ck_x < \epsilon_\perp^{1/2}\omega$, α_B is real and therefore the so-called virtual surface phonon-polariton modes [49], which would not occur in an isotropic medium, are allowed to propagate. These modes usually terminate at a finite value of k_x. For $k_x > \epsilon_\perp^{1/2}\omega/c$, α_B is purely imaginary, and guided modes can propagate in the film.

(d) Region IV corresponds to $\epsilon_\perp < 0$ and $\epsilon_\| < 0$. We have the propagation of real surface phonon-polariton modes, i.e. those that are analogous to the surface phonon-polariton modes in an isotropic medium, for $k_x > \omega/c$.

In addition to the deformed real surface phonon-polariton branches and the appearance of virtual surface phonon-polariton modes, the anisotropy gives rise to a pronounced directional dependence. This difference is clearly illustrated in Fig. 4.10, which is also for α-quartz but with the optical axis now in the z-direction perpendicular to the surfaces [48].

4.5 Experimental Studies

The observation of surface polaritons cannot be made directly by conventional optical absorption measurements, for the following reasons. On the one hand, conservation of energy requires that the frequency of the light that is incident from the vacuum must equal the frequency $\omega_S(k_x)$ of the surface polariton, i.e.

$$k^2 = k_x^2 + k_z^2 = \omega_S^2(k_x)/c^2. \tag{4.48}$$

However, for the polariton field to decay with the distance from the vacuum/sample interface, the surface polariton must be confined to the region of the ωk_x-plane satisfying $k_x^2 > \omega_S^2/c^2$ (see Section 4.1). This is compatible with Eq. (4.48) only if $k_z^2 < 0$, which means that the light incident on the sample must be attenuated in the z-direction. It is for this reason that the surface polariton mode is described as *non-radiative*.

This restriction on k_z can be achieved using the method of attenuated total reflection (ATR), developed by Otto [50–52]. In this method a prism of dielectric constant ϵ_p is placed above the crystal and is separated from it by a gap of thickness

4.5. EXPERIMENTAL STUDIES

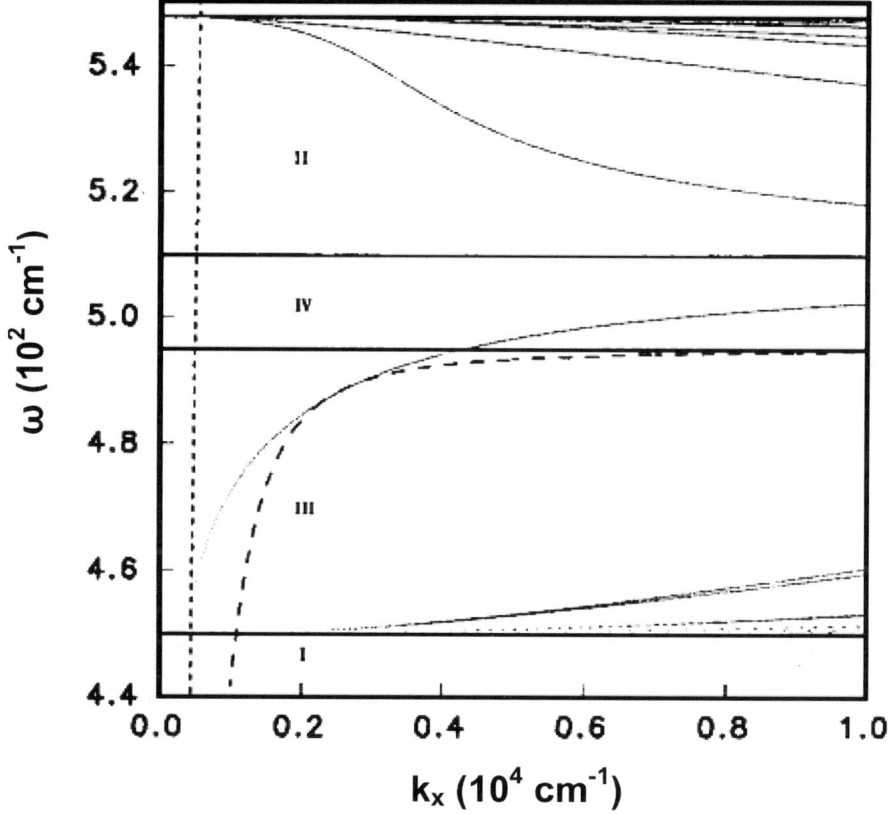

Fig. 4.10. As in Fig. 4.9, but with the optical axis in the z-direction. The long-dashed line represents the curve $k_x = \epsilon_\perp^{1/2} \omega/c$ (after Costa Filho et al. [48]).

d and dielectric constant ϵ_g (see Fig. 4.11). Both the dielectric constants ϵ_p and ϵ_g are assumed to be positive and constant in the frequency range where the surface polariton exists, and the materials are chosen so that $\epsilon_p > \epsilon_g$. The sample is the surface-active medium and has a dielectric function $\epsilon_S(\omega)$. We assume also that the angle of incidence θ on the interface prism/gap is greater than the critical angle θ_C for total internal reflection in the prism, i.e. $\theta > \theta_C$ with

$$\theta_C = \sin^{-1}(\epsilon_g/\epsilon_p). \qquad (4.49)$$

In the absence of the sample, the incident light would be totally reflected. However, with the arrangement as in Fig. 4.11, the evanescent mode (which has decreasing amplitude and a purely imaginary z-component of the wavevector) can excite a surface mode in the gap between the prism and the sample at the gap/sample interface. The energy associated with this surface mode produces an attenuation of the total reflection. In other words, the in-plane wavevector, which

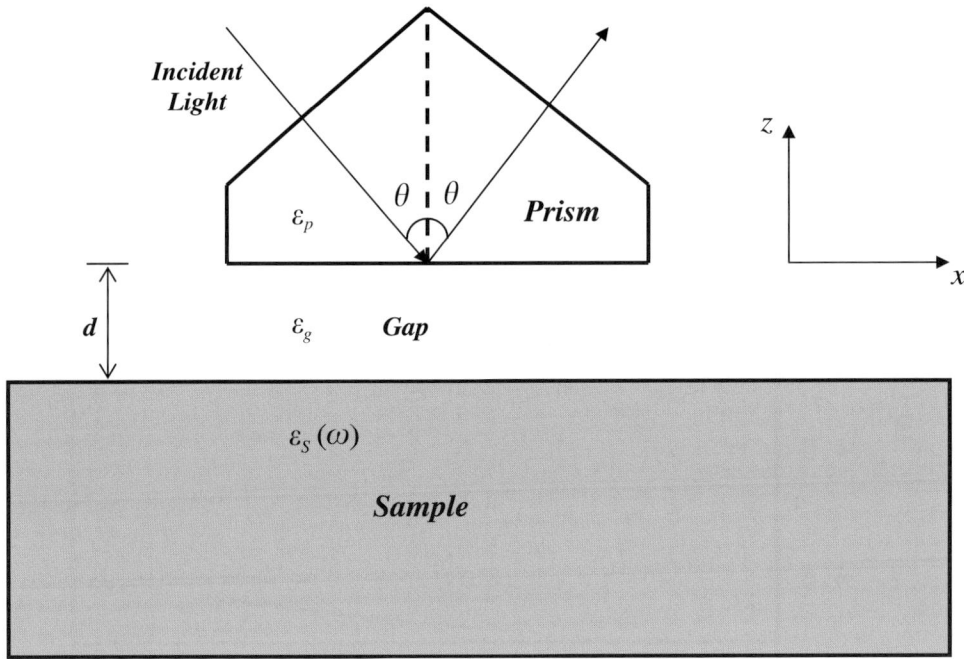

Fig. 4.11. Schematic representation of an ATR experiment in the configuration proposed by Otto [50–52].

is common to all three media and defined by

$$k_x = \epsilon_p^{1/2}(\omega/c)\sin\theta, \tag{4.50}$$

can be made to satisfy the condition

$$k_x > \epsilon_p^{1/2}(\omega/c) \tag{4.51}$$

necessary for the excitation of a surface polariton mode at the gap/sample interface, provided ϵ_p and θ are chosen to be sufficiently large.

In Otto's ATR experiment, surface polaritons were excited at a metallic surface, thus probing the surface plasmon-polaritons. Subsequently, Ruppin [53–55] proposed a modification of Otto's geometry for the excitation of polaritons in an insulating slab, probing the surface phonon-polaritons. In either form of the experiment, the ATR method was related to the excitation of bound modes, providing an extremely sensitive and direct way of measuring the wavevector range, where the surface polariton has a significant photon component.

The main technical problems in applying the ATR method are the control and uniformity of the gap thickness [56]. It is evident that if it is not carefully chosen, the surface polariton's dispersion curves obtained could not be the free surface ones, being perturbed by the presence of the prism [57,58]. Nevertheless, it has

4.5. EXPERIMENTAL STUDIES

provided the most detailed experimental results available concerning the dispersion relations of the surface polaritons.

The first observation of surface phonon-polaritons was reported by Bryksin et al. for NaCl films [59]. These authors subsequently investigated the same type of excitation in films of KBr, NaF, LiF, CdF$_2$, and CaF$_2$ [60]. Surface phonon-polaritons have been successfully observed also in both the long- [61] and short-period [62] superlattices that we shall discuss in later chapters. In addition, experiments have been reported for the surface plasmon-polaritons in n-doped Ge [63] and in a multiple δ-doped GaAs sample [64,65]. However, we should point out that the simple model described in this chapter gives rather poor agreement quantitatively with experiments made on the alkali halides, due to anharmonic effects and phonon–phonon interactions [66,67].

An alternative to the ATR method, as a means of studying surface polaritons, is the use of a laser beam as an energy source and prism couplers for launching and detecting the surface polaritons. This procedure was used by Schoenwald et al. [68] to investigate surface polaritons propagating along a surface of copper.

Another experimental method that is somewhat related to the ATR method is the observation of the reflectivity from a surface upon which a grating has been ruled. Surface polaritons can then be excited without the aid of a prism, even at normal incidence of the light. The presence of the grating relaxes the condition of wavevector conservation for the in-plane component k_x, whose effective value is now given by

$$k_x = (\omega_I/c)\sin\theta + 2\pi n/d, \quad n = 1, 2, \ldots, \qquad (4.52)$$

where ω_I is the incident frequency, θ is the angle of incidence, d is the spacing of the grating, and n is an integer. For d sufficiently small, one can have $ck_x > \omega_I$, even when n is a small integer. The reflectivity exhibits a dip whenever the frequency of a surface polariton is equal to ω_I for some value of n, corresponding to the excitation of the surface polariton by the incident field. This procedure was used by Marschall et al. [69,70], to obtain the spectrum of plasmon-polaritons in n-doped InSb. However, the method has the potential drawback that the surface may be perturbed in a rather complicated way, and the effect of this perturbation on the surface polariton spectrum is difficult to determine precisely.

A completely different technique for the excitation of surface polaritons is Raman scattering. For example, Evans et al. [71] employed near-forward scattering of light through an undoped GaAs thin film on a sapphire substrate. Further theoretical analysis by Mills and Maradudin [72] confirmed that surface polaritons were indeed excited.

More details about the experimental techniques that were briefly covered in this section will be given in later chapters, particularly Chapter 11. However, it is useful to mention here the review articles by Sambles et al. [73] as an introductory account of ATR and the book by Kawata [74] for more recent developments. The role of surface polaritons in surface-enhanced Raman scattering (SERS) is discussed, for example, by Moskovits [75].

References

[1] V.M. Agranovich and D.L. Mills, Eds., Surface Polaritons, North-Holland, Amsterdam, 1982.
[2] A.D. Boardman, in: Electromagnetic Surface Modes, Ed., A.D. Boardman, Wiley, New York, 1982.
[3] P. Halevi, in: Electromagnetic Surface Modes, Ed., A.D. Boardman, Wiley, New York, 1982.
[4] V.M. Agranovich and T.A. Leskova, Fiz. Tverd. Tela 19 (1977) 804 [Sov. Phys.-Solid State 19 (1977) 465].
[5] E.N. Economu and K.L. Ngai, Adv. Chem. Phys. 29 (1974) 265.
[6] R.J. Bell, R.W. Alexander, C.A. Ward and I.L. Tyler, Surf. Sci. 48 (1975) 253.
[7] G. Borstel and H.J. Falge, in: Electromagnetic Surface Modes, Ed., A.D. Boardman, Wiley, New York, 1982.
[8] A.A. Maradudin, R.F. Wallis and L. Dobrzynski, Surface Phonons and Polaritons, Garland STPM Press, New York, 1980.
[9] V.V. Bryksin, D.N. Mirlin and I.I. Reshina, Solid State Commun. 11 (1972) 695.
[10] G. Borstel, H. Falge and A. Otto, in: Springer Tracts in Modern Physics, Vol. 74, 1974, p. 107.
[11] G. Borstel and H. Falge, Phys. Status Solidi (b) 83 (1977) 11.
[12] G. Borstel and H. Falge, Appl. Phys. 16 (1978) 211.
[13] D.N. Mirlin, in: Surface Polaritons, Eds., V.M. Abranovich and D.L. Mills, North-Holland, Amsterdam, 1982.
[14] R.F. Wallis, J.J. Brion, E. Burstein and A. Hartstein, Phys. Rev. B 9 (1974) 3424.
[15] A.A. Maradudin in: Advances in Solid State Physics, Festkörperprobleme XXI (1981) 25.
[16] R.F. Wallis, in: Dynamical Properties of Solids, Vol. 2, Eds., G.K. Horton and A.A. Maradudin, North-Holland, Amsterdam, 1975.
[17] V.V. Bryksin, D.N. Mirlin and I.I. Reshina, Zh. Eksp. Teor. Fiz. Pis. Red. 16 (1972) 445 [JETP Lett. 16 (1972) 315].
[18] M.G. Cottam, D.R. Tilley and B. Zeks, J. Phys. C 17 (1984) 1793.
[19] H.J. Falge and A. Otto, Phys. Status Solidi (b) 56 (1973) 523.
[20] M.W. Cole, Rev. Mod. Phys. 46 (1974) 451.
[21] M. Nakayama, J. Phys. Soc. Jpn. 36 (1974) 393.
[22] F. Stern, Phys. Rev. Lett. 18 (1967) 546.
[23] L. Wendler, Phys. Status Solidi (b) 123 (1984) 469.
[24] D.B. Anderson and J.T. Boyd, Appl. Phys. Lett. 19 (1971) 266.
[25] H. Ito, N. Uesugi and H. Inaba, Appl. Phys. Lett. 25 (1974) 385.
[26] W. Sohler and H. Suche, Appl. Phys. Lett. 33 (1978) 518.
[27] P.K. Tien, Rev. Mod. Phys. 49 (1977) 361.
[28] J. Prieur and S. Ushioda, Phys. Rev. Lett. 34 (1975) 1012.
[29] Y.J. Chen, E. Burstein and D.L. Mills, Phys. Rev. Lett. 34 (1975) 1516.
[30] V.M. Agranovitch and T.A. Leskova, Fiz. Tverd. Tela 17 (1975) 1367 [Sov. Phys.-Solid State 17 (1975) 880].
[31] J.S. Nkoma, R. Loudon and D.R. Tilley, J. Phys. C 7 (1974) 3547.
[32] F.A. Oliveira, M.G. Cottam and D.R. Tilley, Phys. Status Solidi (b) 107 (1981) 737.
[33] E.L. Albuquerque, R. Loudon and D.R. Tilley, J. Phys. C 13 (1980) 1775.
[34] D. Marcuse, Light Transmission Optics, 2nd ed., Van Nostrand, New York, 1982.
[35] L. Wendler, Phys. Status Solidi (b) 128 (1985) 425.
[36] S. Ushioda and R. Loudon in: Surface Polaritons, Eds., V.M. Abranovich and D.L. Mills, North Holland, Amsterdam, 1982.

[37] M.A. Gilmore and B.L. Johnson, J. Appl. Phys. 93 (2003) 4497.
[38] M. Fukui, V.C.Y. So and R. Normandin, Phys. Status Solidi (b) 91 (1979) K61.
[39] D. Sarid, Phys. Rev. Lett. 47 (1981) 1927.
[40] L. Wendler and R. Haupt, J. Appl. Phys. 59 (1986) 3289.
[41] A.E. Craig, G.A. Olson and D. Sarid, Opt. Lett. 8 (1983) 380.
[42] H. Ditlbacher, J.R. Krenn, G. Schider, A. Leitner and F.R. Aussenegg, Appl. Phys. Lett. 81 (2002) 1762.
[43] L. Wendler and R. Haupt, Phys. Status Solidi (b) 137 (1986) 286.
[44] L. Wendler and R. Haupt, J. Phys. C 19 (1986) 1871.
[45] M. Fukui and G.I. Stegeman, in: Electromagnetic Surface Modes, Ed., A.D. Boardman, Wiley, New York, 1982.
[46] S.I. Bozhevolnyi, J. Erland, K. Leosson, P.M.W. Skovgaard and J.M. Hvam, Phys. Rev. Lett. 86 (2001) 3008.
[47] S.I. Bozhevolnyi, V.S. Volkov and K. Leosson, Phys. Rev. Lett. 89 (2002) 186801.
[48] R.N. Costa Filho, M.G. Cottam, E.L. Albuquerque and G.A. Farias, Phys. Rev. B 54 (1996) 2949.
[49] A. Hartstein, E. Burstein, J.J. Brion and R.F. Wallis, Solid State Commun. 12 (1973) 1083.
[50] A. Otto, Z. Physik 216 (1968) 398.
[51] A. Otto, in: Advances in Solid State Physics, Festkörperprobleme XIV (1974) 1.
[52] A. Otto, in: Optical Properties of Solids: New Developments, Ed., B.O. Seraphim, North-Holland, Amsterdam, 1976.
[53] R. Ruppin, Solid State Commun. 8 (1970) 1129.
[54] R. Ruppin, Surf. Sci. 34 (1973) 20.
[55] R. Ruppin and R. Englman, Rep. Prog. Phys. 33 (1970) 149.
[56] F. Abeles, in: Electromagnetic Surface Excitations, Eds., R.F. Wallis and G.I. Stegeman, Springer-Verlag, Heildelberg, 1986.
[57] K.L. Kliewer and R. Fuchs, Adv. Chem. Phys. 17 (1974) 355.
[58] A. Otto, in: Polaritons, Eds., E. Burstein and F. de Martini, Pergamon, Oxford, 1974.
[59] V.V. Bryksin, Y.M. Gerbshtein and D.N. Mirlin, Fiz. Tverd. Tela 13 (1972) 2125 [Sov. Phys.-Solid State 13 (1972) 1779].
[60] V.V. Bryksin, Y.M. Gerbshtein and D.N. Mirlin, Fiz. Tverd. Tela 14 (1972) 543, 3368 [Sov. Phys.-Solid State 14 (1972) 453, 2849].
[61] A.R. El Gohary, T.J. Parker, N. Raj, D.R. Tilley, P.J. Dobson, D. Hilton and C.T.B. Foxon, Semicond. Sci. Technol. 4 (1986) 388.
[62] M. Haraguchi, M. Fukui and S. Muto, Phys. Rev. B 41 (1990) 1254.
[63] A.S. Barker Jr, Phys. Rev. Lett. 28 (1972) 892.
[64] T. Dumelow, T.J. Parker, D.R. Tilley, R.B. Beall and J.J. Harris, Solid State Commun. 77 (1991) 253.
[65] T. Dumelow, A.A. Hamilton, T.J. Parker, D.R. Tilley, B. Samson, S.R.P. Smith, R.B. Beall and J.J. Harris, Superlattices and Microstruct. 9 (1991) 517.
[66] B. Fisher, I.L. Tyler and R.J. Bell, in: Polaritons, Eds., E. Burstein and F. de Martini, Pergamon, Oxford, 1974.
[67] B. Fisher, N. Marschall and H.J. Queisser, Surf. Sci. 34 (1973) 50.
[68] J. Schoenwald, E. Burstein and J.M. Elson, Solid State Commun. 12 (1973) 185.
[69] N. Marschall, B. Fischer and H.J. Queisser, Phys. Rev. Lett. 17 (1966) 379.
[70] N. Marschall and B. Fischer, Phys. Rev. Lett. 28 (1972) 811.
[71] D.J. Evans, S. Ushioda and J.D. McMullen, Phys. Rev. Lett. 31 (1973) 369.
[72] D.L. Mills and A.A. Maradudin, Phys. Rev. Lett. 31 (1973) 372.
[73] J.R. Sambles, G.W. Bradberry and F. Yang, Contemp. Phys. 32 (1991) 173.
[74] S. Kawata, Ed., Near-Field Optics and Surface Plasmon Polaritons, Springer, Berlin, 2001.
[75] M. Moskovits, Rev. Mod. Phys. 57 (1985) 783.

Chapter 5

Plasmon-Polaritons in Periodic Structures

We now begin extending some of the polariton results to structures that involve multiple layers and interfaces, going beyond the single-film case of the previous chapter. In doing this, we shall focus on the case of plasmon-polaritons, although the formalism and many of the results carry over rather straightforwardly to other non-magnetic polaritons. For simplicity, we first discuss *periodic* superlattices, and then in Chapter 6 we generalize to other types of multilayer systems that are generated using the quasiperiodic sequences.

Plasmons were already defined in Section 1.5 as the quanta associated with the collective plasma-like excitations in an interacting electron gas, such as occurs in metals or semiconductors. Plasmon-polaritons, formed by coupling the transverse electromagnetic radiation to the plasma, were discussed in Section 3.2 for bulk media and in Chapter 4 for single- and double-interface geometries. An extensive review that emphasizes the behavior of plasmon-polaritons in periodic superlattices can be found in Ref. [1].

We recall that a periodic superlattice is a multilayer system composed of layers of two (or more) different materials, built up so as to give an overall periodicity to the structure (see Section 1.3). Semiconductor superlattices were first proposed by Esaki and Tsu [2] in 1970, and since that time there has been a considerable and increasing interest in their physical properties (both in semiconductors and other materials). Many of the advances have been due to improvements in fabrication and growth technologies, and the investigations are spurred on by the novel physical properties of superlattices and their potential for device applications [3]. They are of great importance in a variety of fields, more recently in the laser devices area, e.g. quantum cascade lasers based on the intraband transitions of GaAs/AlAs superlattices [4].

The existence of electromagnetic collective excitations in a superlattice, such as bulk and surface polaritons, can be understood in the following way. The excitation of a polariton within a material layer produces electromagnetic fields that extend outside its boundaries, and these fields can couple with elementary excitations of the entire superlattice. Through the use of Bloch's theorem (see Section 1.3), one finds that this coupling creates a set of collective excitations in the superlattice. This collective mode is characterized by a wave component that is normal to the interfaces and can transmit energy normal to the layers of the superlattice structure. Taking Q to denote the wavevector normal to the interfaces, and L as the unit cell length of the superlattice, it is sufficient to consider the bulk polariton when Q lies within the new Brillouin zone associated with the periodic length L, i.e. $0 \leq QL \leq \pi$. Since L may be much larger than the microscopic atomic periodicity length within each of the superlattice constituents, it follows that the zone-boundary value π/L is smaller than for the microscopic Brillouin zones. This artificially introduced *mini-Brillouin zone* leads to striking effects on the excitation spectrum of the superlattice. Indeed, as we shall show shortly, the dispersion curves for the excitations are now "folded back" into these mini-Brillouin zones. This results in a splitting into mini-bands separated by mini-gaps at the zone center and zone edges. The theoretical analysis of these effects in infinite superlattices was pioneered by Rytov [5], who considered acoustic phonons in layered elastic media. However, in a finite or semi-infinite superlattice we will show that surface (i.e. localized) polariton modes may exist with frequencies above, below, and in between the bulk bands [6–8].

It is the aim of this chapter to describe the plasmon-polariton spectrum in a periodic superlattice, including the effects of an external magnetic field that gives rise to the so-called *magnetoplasmon-polaritons*. Then, as we shall see later, it will be straightforward to extend the results to more complex layered structures.

5.1 Two-Component Superlattices

To form a periodic semiconductor superlattice, we consider two different building blocks, A and B (see Fig. 5.1), which are arranged in an alternating way $ABAB\cdots$. For the present application, we assume that the building block A (B) consists of a two-dimensional electron gas (2DEG) layer with a carrier concentration n_A (n_B), supported by a dielectric layer A (B). The layers A and B are characterized in general by frequency-dependent dielectric functions $\epsilon_A(\omega)$ and $\epsilon_B(\omega)$, and have thicknesses a and b, respectively.

For simplicity, we shall ignore the effects of the charge layers throughout the rest of this section, i.e. we set $n_A = n_B = 0$. The additional effects of the charge layers are then taken into account in Section 5.2.

5.1.1 Infinite Superlattices

To find the bulk polariton modes we consider an infinitely extended structure with the Cartesian axes chosen such that the z-axis is normal to the plane of the layers (the xy-plane), just as in Fig. 4.1. Let us assume that the propagation of the electromagnetic wave is TM (or p-polarized), and is characterized by the electric and magnetic fields in the form

$$\vec{E}(\vec{r},t) = (E_x, 0, E_z)\exp(ik_x x - \omega t), \tag{5.1}$$

$$\vec{H}(\vec{r},t) = (0, H_y, 0)\exp(ik_x x - \omega t). \tag{5.2}$$

Here k_x is the in-plane component of the wavevector (taken to be in the x-direction without loss of generality, assuming media A and B both to be isotropic). Within

5.1. TWO-COMPONENT SUPERLATTICES

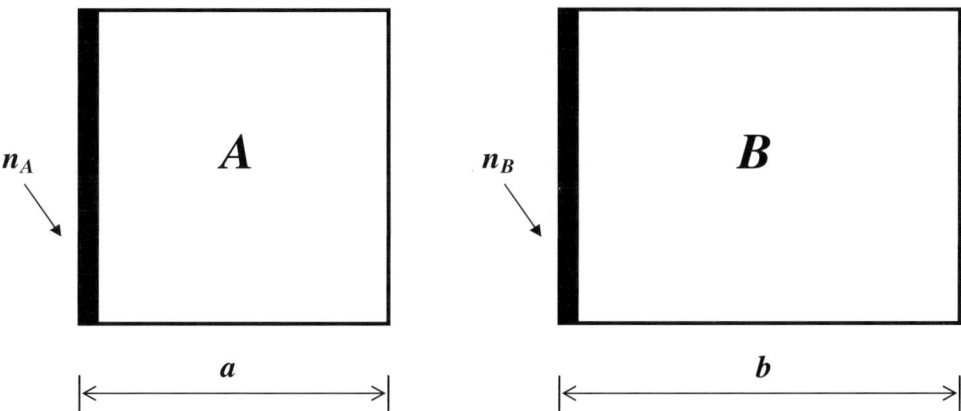

Fig. 5.1. The two building blocks A and B used here to characterize the periodic and quasiperiodic structures.

each layer, the electric field must satisfy the usual wave equation:

$$\nabla \times \left[\nabla \times \vec{E}(\vec{r},t)\right] = -\mu_0 \epsilon_0 \epsilon_j(\omega) \partial^2 \vec{E}(\vec{r},t)/\partial t^2, \tag{5.3}$$

where the label j denotes A or B. Also, from Maxwell's equations,

$$\nabla \times \vec{H}(\vec{r},t) = \epsilon_0 \epsilon_j(\omega) \partial \vec{E}(\vec{r},t)/\partial t. \tag{5.4}$$

We define the nth unit cell AB as that which extends from $z = nL$ to $z = (n+1)L$, where $L = a + b$ is the periodic length and n is any integer. From Eqs. (5.1)–(5.4) it is easy to deduce that the components of the electric and magnetic fields within layer A of the nth unit cell ($nL < z < nL + a$) have the form

$$E_x(z) = A_{1A}^n \exp(-\alpha_A z) + A_{2A}^n \exp(\alpha_A z), \tag{5.5}$$

$$E_z(z) = (ik_x/\alpha_A)[A_{1A}^n \exp(-\alpha_A z) - A_{2A}^n \exp(\alpha_A z)], \tag{5.6}$$

$$H_y(z) = [-i\omega\epsilon_0\epsilon_A(\omega)/\alpha_A][A_{1A}^n \exp(-\alpha_A z) - A_{2A}^n \exp(\alpha_A z)]. \tag{5.7}$$

Here we have either

$$\alpha_A = [k_x^2 - \epsilon_A(\omega)\omega^2/c^2]^{1/2} \quad \text{if } k_x^2 > \epsilon_A(\omega)\omega^2/c^2 \tag{5.8}$$

or

$$\alpha_A = i[\epsilon_A(\omega)\omega^2/c^2 - k_x^2]^{1/2} \quad \text{if } k_x^2 < \epsilon_A(\omega)\omega^2/c^2 \tag{5.9}$$

so that α_A is either real or purely imaginary.

The results for the fields within layer B of the nth cell [$nL + a < z < (n+1)L$] are identical to those for layer A, provided we replace the coefficients A_{1A} and A_{2A} by A_{1B} and A_{2B}, respectively, and we replace α_A by α_B, which is defined in an analogous manner to Eqs. (5.8) and (5.9).

Next, we use the standard boundary conditions for the electromagnetic fields, namely the continuity of $E_x(z)$ and $H_y(z)$, at the two types of interface of the nth unit cell, i.e. the interfaces corresponding to $z = nL + a$ and $z = (n+1)L$. This gives the following equations:

$$A_{1A}^n f_A + A_{2A}^n \bar{f}_A = A_{1B}^n + A_{2B}^n, \tag{5.10}$$

$$\xi_A(A_{1A}^n f_A - A_{2A}^n \bar{f}_A) = \xi_B(A_{1B}^n - A_{2B}^n), \tag{5.11}$$

$$A_{1B}^n f_B + A_{2B}^n \bar{f}_B = A_{1A}^{n+1} + A_{2A}^{n+1}, \tag{5.12}$$

$$\xi_B(A_{1B}^n f_B - A_{2B}^n \bar{f}_B) = \xi_A(A_{1A}^{n+1} - A_{2A}^{n+1}), \tag{5.13}$$

where, as mentioned, we are not considering the presence of the 2DEG at the interfaces. We have introduced the shorthand notation

$$\xi_j = \epsilon_j(\omega)/\alpha_j, \tag{5.14}$$

$$f_j = \exp(-\alpha_j d_j) \text{ and } \bar{f}_j = 1/f_j, \tag{5.15}$$

with $j = A, B$ and $d_j = a, b$.

Although there are different theoretical methods available to study these systems, like the surface Green function matching (SGFM) [9–11] and the interface response theory [12,13], we find it more convenient to make use of a *transfer-matrix formalism* [14–16] by defining for each medium the two-component column vector

$$|A_j^n\rangle = \begin{bmatrix} A_{1j}^n \\ A_{2j}^n \end{bmatrix}. \tag{5.16}$$

Now we can rewrite Eqs. (5.10)–(5.13) in matrix form as

$$\bar{M}_A |A_A^n\rangle = \bar{N}_B |A_B^n\rangle, \tag{5.17}$$

$$\bar{M}_B |A_B^n\rangle = \bar{N}_A |A_A^{n+1}\rangle, \tag{5.18}$$

where we have defined the 2×2 matrices

$$\bar{M}_j = \begin{pmatrix} f_j & \bar{f}_j \\ \xi_j f_j & -\xi_j \bar{f}_j \end{pmatrix}, \tag{5.19}$$

$$\bar{N}_j = \begin{pmatrix} 1 & 1 \\ \xi_j & -\xi_j \end{pmatrix}. \tag{5.20}$$

Using Eqs. (5.17) and (5.18) it is easy to deduce that

$$|A_A^{n+1}\rangle = \bar{T}|A_A^n\rangle, \quad \bar{T} = \bar{N}_A^{-1} \bar{M}_B \bar{N}_B^{-1} \bar{M}_A. \tag{5.21}$$

Here the matrix $\bar{T}$ is called a *transfer matrix* because it relates the electric (and hence the magnetic) field amplitudes at any point with coordinate z in cell n to the equivalent point (i.e. with coordinate $z + L$) in cell $n + 1$.

5.1. TWO-COMPONENT SUPERLATTICES

Taking into account the translational symmetry of the system by using Bloch's ansatz, i.e. the 1D analog of Eq. (1.10), we have

$$|A_A^{n+1}\rangle = \exp(iQL)|A_A^n\rangle, \tag{5.22}$$

where Q is the Bloch wavevector and L is the size of the superlattice unit cell. From this, together with Eq. (5.21), we obtain the following eigenvalue equation (with its corresponding inversion equation):

$$\bar{T}|A_A^n\rangle = \exp(iQL)|A_A^n\rangle, \tag{5.23}$$

$$\bar{T}^{-1}|A_A^n\rangle = \exp(-iQL)|A_A^n\rangle. \tag{5.24}$$

Consequently, by combining these results, we have

$$[\cos(QL)\bar{I} - (\tfrac{1}{2})(\bar{T} + \bar{T}^{-1})]|A_A^n\rangle = 0, \tag{5.25}$$

where $\bar{I}$ is the 2×2 unit matrix.

Since Eq. (5.25) has been deduced for any general vector $|A_A^n\rangle$ of the superlattice, we must have

$$\cos(QL) = (\tfrac{1}{2})(\bar{T} + \bar{T}^{-1}). \tag{5.26}$$

Furthermore, as $\bar{T}$ is a unimodular matrix, its determinant is equal to unity, and so the dispersion relation for the *superlattice bulk plasmon-polariton* is simply given by

$$\cos(QL) = (\tfrac{1}{2})Tr(\bar{T}). \tag{5.27}$$

This is an important general result, which will be useful in many contexts.

If retardation effects can be ignored, i.e. $k_x^2 \gg \epsilon_j \omega^2/c^2$, then $\alpha_A = \alpha_B = k_x$ and the bulk dispersion relation, Eq. (5.27), reduces to the following simpler explicit form [17]:

$$\cos(QL) = \cosh(k_x a)\cosh(k_x b) + f(\omega)\sinh(k_x a)\sinh(k_x b), \tag{5.28}$$

where

$$f(\omega) = (\tfrac{1}{2})[\epsilon_A(\omega)\epsilon_B^{-1}(\omega) + \epsilon_B(\omega)\epsilon_A^{-1}(\omega)]. \tag{5.29}$$

In general, it is necessary to solve Eq. (5.28) numerically, but we can get some insight into the solutions by defining the positive quantity $C(k_x, Q)$ as

$$C(k_x, Q) = \frac{\cosh(k_x a)\cosh(k_x b) - \cos(QL)}{\sinh(k_x a)\sinh(k_x b)}. \tag{5.30}$$

The solutions of Eq. (5.28) then correspond to

$$\epsilon_A(\omega)/\epsilon_B(\omega) = -C(k_x, Q) \pm [C^2(k_x, Q) - 1]^{1/2}, \tag{5.31}$$

where all the frequency dependence is on the left-hand side. There is the additional condition that $C(k_x, Q) \geq 1$.

As an example, let us suppose the medium A is described by the plasmon dielectric function as in Eq. (1.24), while medium B has a dielectric constant $\epsilon_{\infty B}$. Solving Eq. (5.31), we find two solutions $\omega_\pm$ for the frequencies of the coupled plasmon modes, i.e.

$$\omega_\pm^2(k_x, Q) = \omega_{pA}^2 \left[1 - s\left(-C(k_x, Q) \pm [C^2(k_x, Q) - 1]^{1/2} \right) \right]^{-1}, \quad (5.32)$$

where s denotes the ratio $\epsilon_{\infty B}/\epsilon_{\infty A}$. Thus, we expect the infinite superlattice described by this model to have two characteristic branches of excitations given by Eq. (5.32): a lower or *acoustic branch*, with frequency ω_-, and an upper or *optical branch*, with frequency ω_+. We present some numerical illustrations of this later.

5.1.2 Semi-Infinite Superlattices

We now introduce an external surface to the superlattice by considering it truncated at $z=0$ and with the half-space $z<0$ filled by a transparent medium C, whose frequency-independent dielectric constant is denoted by ϵ_C. This semi-infinite superlattice no longer possesses full translational symmetry in the z-direction through multiples of the unit cell thickness L, and therefore we may no longer assume Bloch's ansatz as in the bulk case. On the other hand, this new interface between the material C and the superlattice allows the appearance of another class of solution, namely electromagnetic modes that are localized in the vicinity of the superlattice surface. Going from one unit cell to the next, their amplitude decays with distance from the plane $z=0$. For these superlattice *surface modes*, we therefore have, instead of Eq. (5.22),

$$|A_A^{n+1}\rangle = \exp(-\beta L)|A_A^n\rangle, \quad (5.33)$$

with $\mathrm{Re}(\beta) > 0$, as the condition for a localized mode. Therefore, Eq. (5.27) still holds provided we formally replace Q by the complex quantity $i\beta$ to give

$$\cosh(\beta L) = (1/2)Tr(T). \quad (5.34)$$

Since we also have to consider the extra boundary conditions for the new interface at $z=0$, this imposes a further constraint that enables us eventually to determine the attenuation factor β, as described below.

The relevant electromagnetic fields in the region occupied by medium C ($z<0$) must have the form

$$E_x(z) = C\exp(\alpha_C z), \quad (5.35)$$

$$H_y(z) = (i\omega\epsilon_0\epsilon_C/\alpha_C)C\exp(\alpha_C z), \quad (5.36)$$

where C is constant, and

$$\alpha_C = [k_x^2 - \epsilon_C \omega^2/c^2]^{1/2} \quad \text{with} \quad k_x^2 > \epsilon_C \omega^2/c^2. \quad (5.37)$$

5.1. TWO-COMPONENT SUPERLATTICES

Next, from the boundary conditions at $z = 0$ (the continuity of the x-component of the electric field and the y-component of the magnetic field), and assuming that layer A is the outermost layer in the superlattice, we find from Eqs. (5.5), (5.7), (5.35), and (5.36) that

$$C = A^0_{1A} + A^0_{2A}, \tag{5.38}$$

$$\xi_C C = -\xi_A (A^0_{1A} - A^0_{2A}), \tag{5.39}$$

where we denote $\xi_C = \epsilon_C/\alpha_C$. Then, using Eq. (5.23) with the formal replacement of Q by $i\beta$, we have for the vector $|A^0_A\rangle$ in the case of a surface mode

$$\bar{T}|A^0_A\rangle = \exp(-\beta L)|A^0_A\rangle. \tag{5.40}$$

Here $|A^0_A\rangle$ is given by Eq. (5.16) for $n=0$ and $j=A$. Eliminating the unknown coefficients C, A^0_{1A}, and A^0_{2A} from Eqs. (5.38)–(5.40), we obtain

$$T_{11} + T_{12}\lambda = T_{22} + T_{21}\lambda^{-1}, \tag{5.41}$$

where T_{mn} (with $m, n = 1, 2$) denote elements of the transfer matrix T, and λ is a parameter characteristic of the surface:

$$\lambda = (\xi_A + \xi_C)/(\xi_A - \xi_C). \tag{5.42}$$

Eq. (5.41) represents an implicit dispersion relation for the surface polariton modes. Once it is solved, we can obtain a value for β that must satisfy Eq. (5.34) together with the requirement $\text{Re}(\beta) > 0$ to ensure localization.

Fig. 5.2 shows the dispersion relation for the unretarded case described by Eq. (5.32) (bulk modes) and Eqs. (5.34) and (5.41) (surface modes). In this example we have considered medium A to be Al, with a plasmon-type frequency-dependent dielectric function as in Eq. (1.24) with $\omega_{pA} = 15$ eV. For medium B we assume the physical parameters of Al_2O_3, whose frequency-independent dielectric function is $\epsilon_B = 3$. The damping is neglected and we assume the external medium C to be vacuum ($\epsilon_C = 1$). As we can see, the bulk modes fall into two well-defined branches ω_+ and ω_- as expected, separated by a gap that tends to become narrower when $k_x a$ increases [17]. The bands are bounded by the curves for $QL=0$ and $QL=\pi$. The surface modes lie between and above the bulk bands in this case. The former has βL purely real and positive, and it merges with the ω_+ bulk band when $k_x a \simeq 1.5$. On the other hand, the latter is associated with β having the form $\beta L = i\pi + \chi$, where χ is real and positive; it has the value 6.37 at $k_x a = 5.0$. It is interesting to note that for large values of $k_x a$, this surface branch tends to the frequency appropriate to an Al/vacuum interface (10.6 eV). As $k_x a$ decreases, it merges with the ω_+ branch.

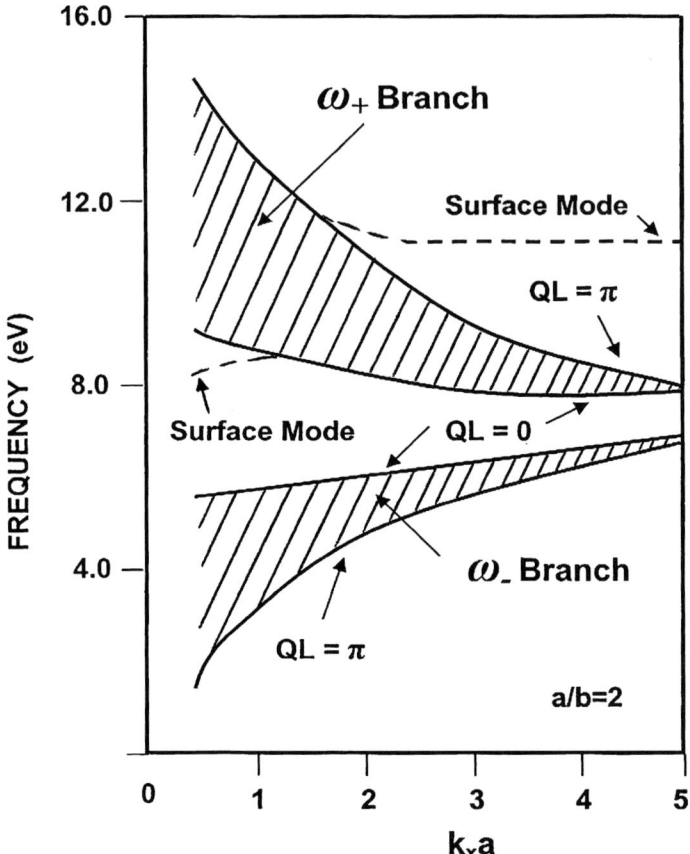

Fig. 5.2. Plasmon-polariton spectrum as given by Eq. (5.32) (unretarded bulk modes) and Eqs. (5.34) and (5.41) (surface modes) in a two-component superlattice, for $a/b = 2$. The energy (in eV units) is plotted against the dimensionless in-plane wavevector $k_x a$. The shaded areas represent the two bulk mode regions (ω_+ and ω_- branches), while the two surface modes are represented by the dotted lines just above and below the high-frequency bulk mode region (after Camley and Mills [17]).

As a second example, in Fig. 5.3 we illustrate the frequencies of the superlattice plasmon-polaritons as a function of the ratio a/b of layer thicknesses, taking the dimensionless in-plane wavevector $k_x a = 1.0$ [17]. The unretarded limit is once again assumed. We have considered here $\epsilon_B = 1$ with the other physical parameter the same as used in Fig. 5.2. As we can now see, the surface mode exists only when a is greater than b, and lies between the bulk bands. The ω_+ and ω_- bulk modes, shown shaded in Fig. 5.3, touch each other for $a = b$; otherwise they are separated by a gap. They are bounded by the curves $QL = 0$ and $QL = \pi$.

5.1. TWO-COMPONENT SUPERLATTICES

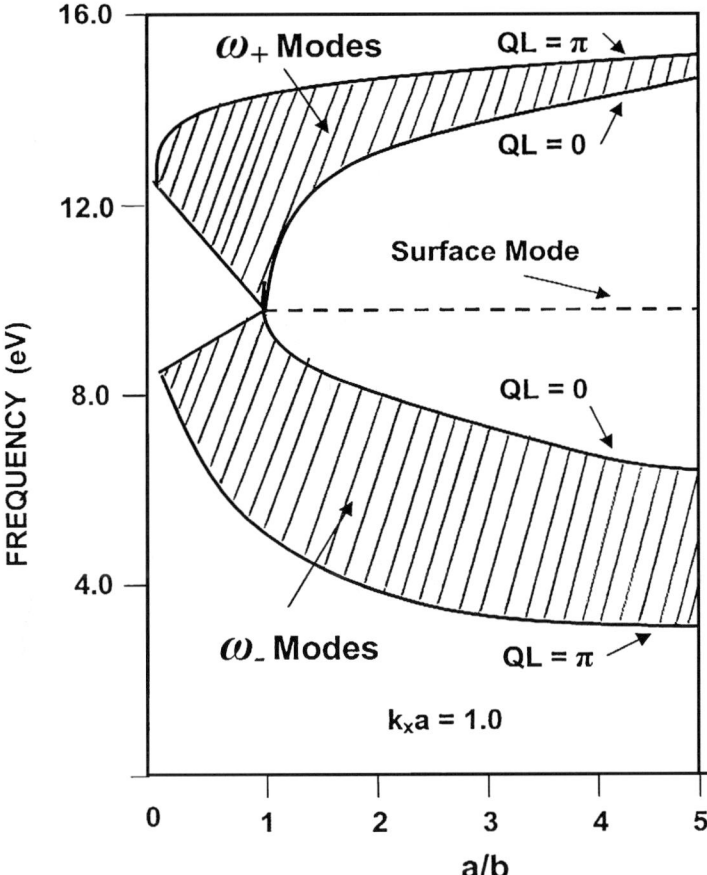

Fig. 5.3. Unretarded plasmon-polariton dispersion relation for a periodic two-component superlattice as a function of the thickness ratio a/b, considering $k_x a = 1.0$. The bulk bands are shaded and limited by the curves $QL=0$ and $QL=\pi$ (after Camley and Mills [17]).

5.1.3 Finite Superlattices

We consider now the finite two-component superlattice structure, shown in Fig. 5.4, obtained from the infinite superlattice by truncating it at $z=0$ and $z=pL$, with p being a positive integer and $L=a+b$ the size of the superlattice unit cell. It is bounded by the isotropic media E and F, as shown, which have dielectric constants ϵ_E and ϵ_F, respectively.

Just as in the semi-infinite case, we can no longer use Bloch's theorem to relate the amplitude in one layer to that in another one through the envelope function $\exp(imQL)$, with m being the difference in labels of the cells involved. Instead, by an extension of results in the previous subsection where there was just one surface, we now have to employ the envelope functions $\exp(-m\beta L)$ and $\exp[-(p-m)\beta L]$,

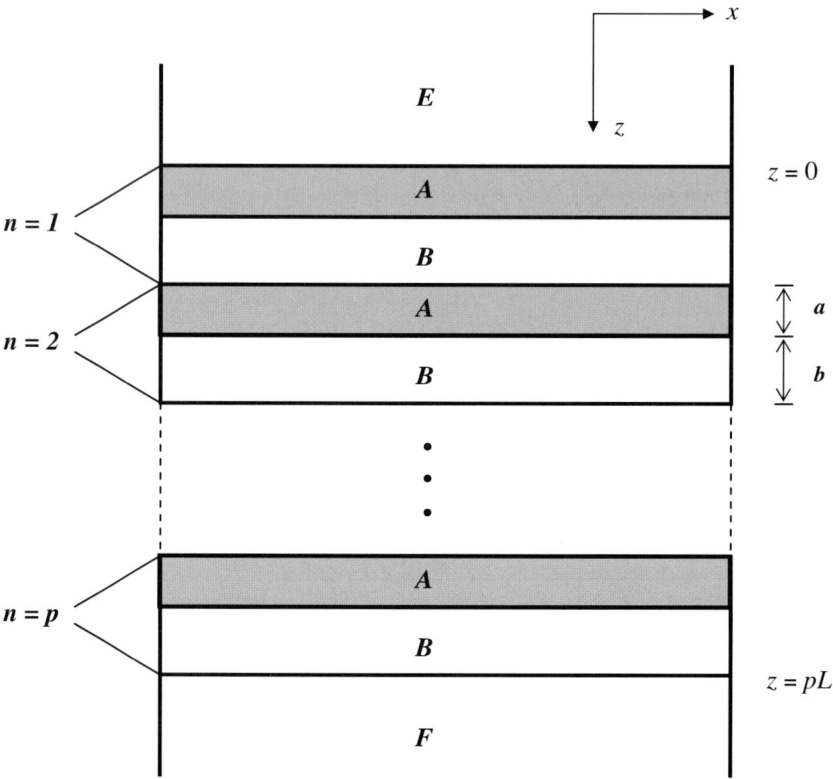

Fig. 5.4. Geometry of a superlattice of finite extent, composed of alternating layers of materials A and B, bounded by materials E and F above and below the structure, respectively.

which are defined to correspond to localization of a surface mode at the top and bottom surfaces, respectively. The case of the bulk mode would simply correspond to substituting β by $-iQ$.

Let us assume, to simplify the algebra of the problem, that the electromagnetic-field coefficients A_{jA} and A_{jB} ($j=1,2$) in layers A and B, respectively, which are related to the envelope function $\exp(-m\beta L)$, are independent of the analogous coefficients A'_{jA} and A'_{jB} associated with the envelope function $\exp[-(p-m)\beta L]$. This assumption enables us to relate these coefficients in the eigenvalue equation of the transfer matrix $\bar{T}$ at the nth cell, i.e.

$$[\bar{T} - \exp(-\beta L)]|A_A^n\rangle = 0, \tag{5.43}$$

$$[\bar{T} - \exp(\beta L)]|A_A'^n\rangle = 0. \tag{5.44}$$

Hence we deduce that

$$A_{2A}^n = K A_{1A}^n, \quad A_{2A}'^n = K' A_{1A}'^n, \tag{5.45}$$

5.1. TWO-COMPONENT SUPERLATTICES

with
$$K = [\exp(-\beta L) - T_{11}]/T_{12}, \qquad (5.46)$$

while K' is given by a similar expression to K but with β replaced by $-\beta$ in Eq. (5.46).

From the electromagnetic boundary conditions for the B–A interface at $z = (n+1)L$ inside any cell n of the superlattice, we obtain (henceforth in this section dropping the superscript n for convenience)

$$A_{1B}f_B + A_{2B}\bar{f}_B = \exp(-\beta L)(A_{1A} + A_{2A}), \qquad (5.47)$$

$$A'_{1B}f_B + A'_{2B}\bar{f}_B = \exp(\beta L)(A'_{1A} + A'_{2A}), \qquad (5.48)$$

$$\xi_B(A_{1B}f_B - A_{2B}\bar{f}_B) = \exp(-\beta L)\xi_A(A_{1A} - A_{2A}), \qquad (5.49)$$

$$\xi_B(A'_{1B}f_B - A'_{2B}\bar{f}_B) = \exp(\beta L)\xi_A(A'_{1A} - A'_{2A}), \qquad (5.50)$$

where ξ_j, f_j and $\bar{f}_j$ ($j = A, B$) are defined by Eqs. (5.14) and (5.15).

The relevant electromagnetic fields for the region $z < 0$ can be written as

$$E_x(z) = C_E \exp(\alpha_E z), \qquad (5.51)$$

$$H_y(z) = i\omega\epsilon_0 \xi_E C_E \exp(\alpha_E z), \qquad (5.52)$$

while for the region $z > pL$ they are

$$E_x(z) = C_F \exp(-\alpha_F z), \qquad (5.53)$$

$$H_y(z) = -i\omega\epsilon_0 \xi_F C_F \exp(-\alpha_F z). \qquad (5.54)$$

Here we have introduced the notation

$$\alpha_E = (k_x^2 - \epsilon_E \omega^2/c^2)^{1/2}, \quad k_x^2 > \epsilon_E \omega^2/c^2, \qquad (5.55)$$

$$\xi_E = \epsilon_E/\alpha_E, \qquad (5.56)$$

with similar definitions for α_F and ξ_F.

Next, imposing the standard electromagnetic boundary conditions at the E–A interface ($z = 0$), we have

$$C_E = A_{1A} + A_{2A} + A'_{1A} + A'_{2A}, \qquad (5.57)$$

$$\xi_E C_E = \xi_A(-A_{1A} + A_{2A} - A'_{1A} + A'_{2A}). \qquad (5.58)$$

Likewise, at the B–F interface ($z = pL$), the boundary conditions yield

$$C_F \exp(-p\alpha_F L) = \theta^-(A_{1B}f_B + A_{2B}\bar{f}_B) + \theta^+(A'_{1B}f_B + A'_{2B}\bar{f}_B), \qquad (5.59)$$

$$\xi_F C_F \exp(-p\alpha_F L) = \xi_B[\theta^-(A_{1B}f_B - A_{2B}\bar{f}_B) + \theta^+(A'_{1B}f_B - A'_{2B}\bar{f}_B)], \qquad (5.60)$$

where we denote $\theta^{\pm} = \exp[\pm(p-1)\beta L]$.

Using Eq. (5.45) and Eqs. (5.47)–(5.50), we can now re-express Eqs. (5.57)–(5.60) in the matrix form below:

$$\begin{bmatrix} 1+K & 1+K' & -1 & 0 \\ \xi_A(1-K) & \xi_A(1-K') & \xi_E & 0 \\ (1+K)\exp(-p\beta L) & (1+K')\exp(p\beta L) & 0 & -\exp(-p\alpha_F L) \\ \xi_A(1-K)\exp(-p\beta L) & \xi_A(1-K')\exp(p\beta L) & 0 & -\xi_F\exp(-p\alpha_F L) \end{bmatrix}$$

$$\times \begin{bmatrix} A_{1A} \\ A'_{1A} \\ C_E \\ C_F \end{bmatrix} = 0. \tag{5.61}$$

The above system of equations has a non-trivial solution if the determinant of the 4×4 coefficient matrix is equal to zero. Therefore, it follows that the implicit dispersion relation for the propagation of plasmon-polaritons in a finite, two-component superlattice is given by [18]

$$\tanh(p\beta L) = \frac{\xi_A(\xi_E + \xi_F)(K - K')}{\xi_A^2(1-K)(1-K') - \xi_E\xi_F(1+K)(1+K') - \xi_A(1-KK')(\xi_E - \xi_F)}. \tag{5.62}$$

This equation is particularly convenient to employ, as far as numerical calculations are concerned, because the number of cells in the finite superlattice, p, appears only in the argument of the hyperbolic tangent function.

As a numerical example, we suppose that the superlattice is composed of alternating layers of Al (as medium A), with a plasmon dielectric function, and Al_2O_3 (as medium B), with a dielectric constant equal to 3.0. The surrounding media E and F are considered to be vacuum ($\epsilon_E = 1$) and SiO_2 ($\epsilon_F = 2.5$), respectively. The plasmon-polariton dispersion curve, obtained from Eq. (5.62), is shown in Fig. 5.5 for the unretarded limit case, taking the number of unit cells $p = 10$ and the ratio of the layer thicknesses $a/b = 2$. One important and new feature of this spectrum is the quantization of the bulk modes due to the finite thickness of the structure. This contrasts with the continuous band of bulk states found for the infinite and semi-infinite superlattices. This quantization effect becomes more pronounced (with a wider separation of the modes) as p becomes smaller. The most striking feature, however, is the existence of a short segment of surface mode, which emerges from the upper bulk-mode branch at $k_x a \simeq 0.5$ and becomes a bulk mode again at $k_x a \simeq 1.3$. In this region, the dispersion curve is quite flat. Above the upper bulk band, there is a high-frequency surface mode which, as in the semi-infinite case, tends to the frequency value of 10.6 eV for large values of $k_x a$. Also there is a low-frequency surface mode that exists for all values of $k_x a$ and has an almost flat behavior at the frequency where $\hbar\omega \simeq 7.8\,\text{eV}$.

Finally, we show in Fig. 5.6 the situation where the thickness ratio is inverted compared with the above case, i.e. $a/b = 0.5$, while other parameters are the same. As we can see, the bulk modes are considerably more compressed than

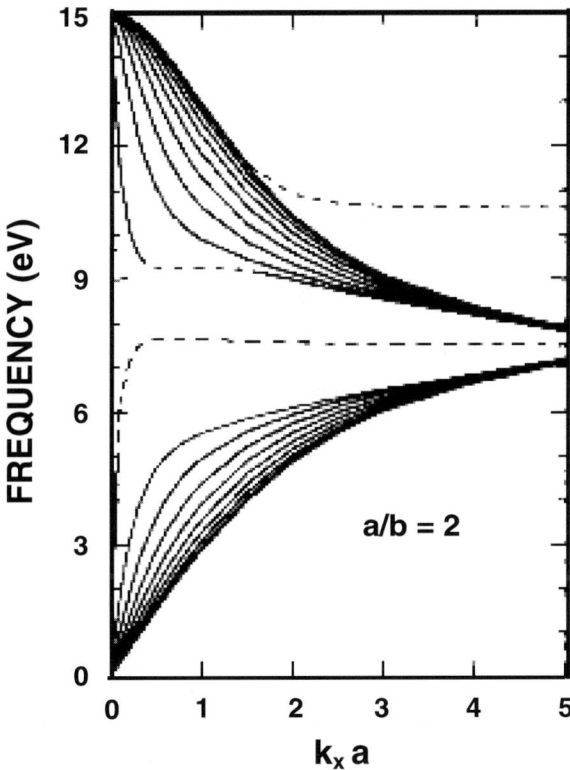

Fig. 5.5. Dispersion curve as a function of the frequency versus the dimensionless in-plane wavevector $k_x a$ for plasmon-polariton in a finite superlattice in the unretarded limit case, taking the ratio of the layer thicknesses $a/b = 2$ and the number of unit cells $p = 10$. Note the quantization of the bulk modes (full lines). The surface modes are shown by broken lines (after Johnson et al. [18]).

in the previous figure. Also, the surface mode that appeared in Fig. 5.5 for $0.5 < k_x a < 1.3$ no longer exists.

5.2 Superlattices with Charge Sheets

We now generalize some of the preceding results to include the effects of charge sheets (2DEG or 2DHG) arising at the interfaces of the superlattice due to the structures in Fig. 5.1. The calculations can be done in a way that is a generalization of the case of modes for a single charge sheet (see Chapter 4, Section 4.3). We consider that at each A–B interface of either an infinite two-component superlattice $\cdots ABABAB \cdots$ structure or a semi-infinite vacuum-$ABABAB \cdots$ structure, there is a 2D electron (or hole) gas layer, whose thickness is negligible compared with the thickness of the layers a and b. Modulation-doped GaAs/Al$_x$Ga$_{1-x}$As

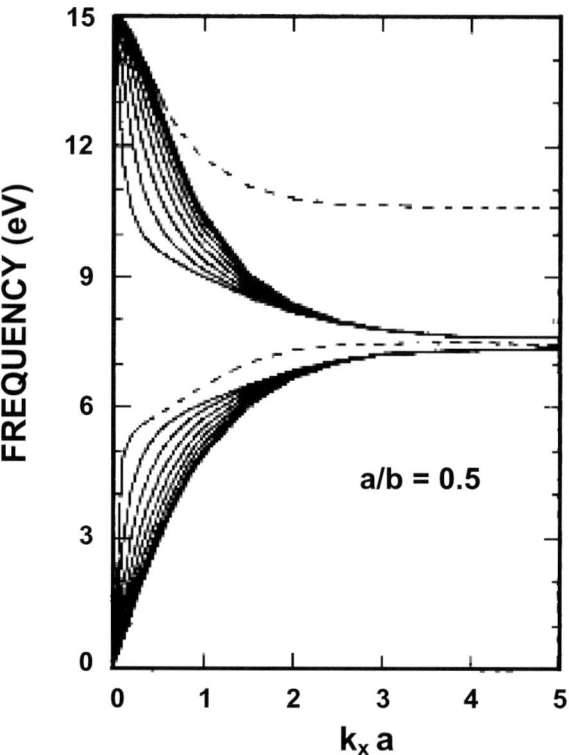

Fig. 5.6. The same as in Fig. 5.5 but for the thickness ratio $a/b = 0.5$ (after Johnson et al. [18]).

superlattices can provide an approximate realization of this model, as we discuss later. For simplicity, we omit the treatment of a finite superlattice.

Compared with the calculations done for the determination of the plasmon-polariton dispersion relation in the absence of charge sheets in the previous section, the main difference here is that we no longer have the y-component of the magnetic field continuous across an interface. Instead, there is a discontinuity due to the presence of a current density j_x at each interface, as given by Eq. (4.23) in a previous example, and the relevant boundary condition is as written in Eq. (4.22). Otherwise, we can use the transfer-matrix formalism as developed earlier in this chapter.

It can easily be proved that with the presence of a 2DEG (or a 2DHG) at each interface of the layers A and B, with a carrier concentration n_j per unit area ($j = A, B$), the plasmon-polariton dispersion relations are given formally by the same expressions as found previously, namely Eq. (5.27) for the bulk modes, and Eq. (5.41) for the surface modes, provided the matrix $\bar{N}_j$ in Eq. (5.20) is replaced by [19]

$$\bar{N}_j = \begin{pmatrix} \frac{1}{\xi_j} - \bar{\sigma} & \frac{1}{-\xi_j} - \bar{\sigma} \end{pmatrix}, \tag{5.63}$$

5.2. SUPERLATTICES WITH CHARGE SHEETS

with
$$\bar{\sigma} = \sigma/i\omega\epsilon_0 = ne^2/m\omega^2\epsilon_0, \qquad (5.64)$$
and we have used Eq. (4.23) which defines σ.

In the special case where ϵ_A and ϵ_B are constants (independent of ω), it is convenient to write the bulk dispersion relation explicitly as [19]

$$(4r/GH)\cos(QL) = \left[\left(\frac{r}{G} + \frac{1}{H}\right)^2 - \left(\frac{\Omega}{\omega}\right)^4\right]\cos(G + sH)$$

$$+ 2\left(\frac{\Omega}{\omega}\right)^2 \left[\left(\frac{r}{G} + \frac{1}{H}\right)\sin(G + sH)\right.$$

$$\left. - \left(\frac{r}{G} - \frac{1}{H}\right)\sin(G - sH)\right]$$

$$- \left[\left(\frac{r}{G} - \frac{1}{H}\right)^2 - \left(\frac{\Omega}{\omega}\right)^4\right]\cos(G - sH), \qquad (5.65)$$

where
$$G = \left[r(\omega a/c)^2 - (k_x a)^2\right]^{1/2}, \qquad (5.66)$$

$$H = \left[(\omega a/c)^2 - (k_x a)^2\right]^{1/2}. \qquad (5.67)$$

We have introduced here the ratios $r = \epsilon_A/\epsilon_B$, $s = b/a$ and Ω is a characteristic frequency defined by $\Omega = (ne^2/m\epsilon_0 a \epsilon_b)^{1/2}$.

Eq. (5.65) simplifies considerably when the dielectric media A and B are identical (i.e. the case of $r = s = 1$). Furthermore, if retardation effects are negligible (i.e. when $k_x^2 \gg \omega^2/c^2$), the bulk plasmon frequencies correspond to [19]

$$\omega = \Omega \left[\frac{k_x a \sinh(k_x a)}{2\cosh(k_x a) - 2\cos(Qa)}\right]^{1/2}, \qquad (5.68)$$

in agreement with calculations using different methods [20–24].

For the surface plasmon-polaritons in a semi-infinite superlattice, the explicit form of the dispersion relation, considering $r = s = 1$ for simplicity, is [19]

$$\left[\mu_S(\mu_S - 1)(k_x a)^2 \sinh(k_x a)\right](\Omega/\omega)^4$$

$$+ (k_x a)\left[r_S(1 - 2\mu_S)\sinh(k_x a) + \cosh(k_x a)\right](\Omega/\omega)^2$$

$$+ (r_S^2 - 1)\sinh(k_x a) = 0, \qquad (5.69)$$

where $r_S = \epsilon_C/\epsilon_B$, with ϵ_C being the dielectric constant of the medium outside the superlattice. Also, we have allowed here for the possibility of the charge sheet at

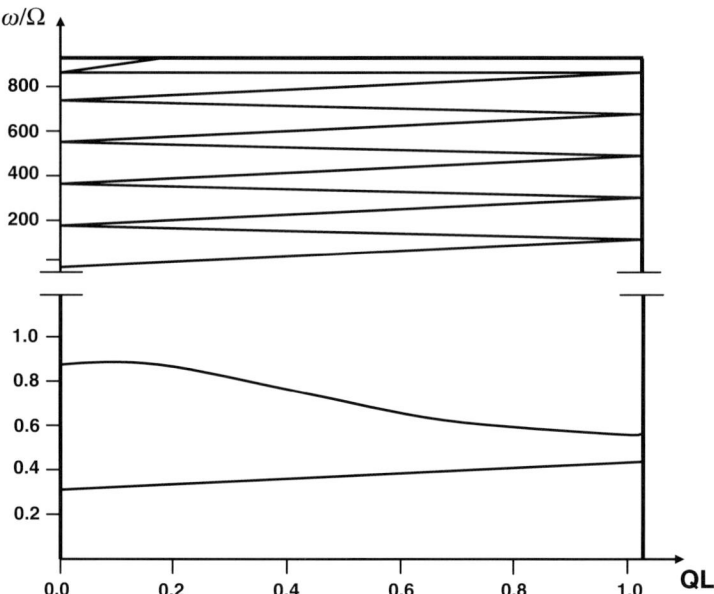

Fig. 5.7. Bulk plasmon-polariton frequencies ω/Ω plotted against QL, showing the effect of retardation for $\omega \gg \Omega$. See the text for parameter values (after Constantinou and Cottam [19]).

the surface having a different carrier density from that at an interior interface; we defined $\mu_S = n_S/n$, where n_S is the carrier concentration per unit area at $z=0$.

In Fig. 5.7 we plot the bulk plasmon-polariton spectrum, according to Eq. (5.65), as a function of QL for a fixed value of the in-plane common wavevector $k_x a = 0.5$. The other physical parameters used are $n = 6 \times 10^{15}$ m^{-2}, $m = 6.4 \times 10^{15}$ kg, $a = 40$ nm, $\epsilon_B = 12.9$, $s = 2$, and $r = 1$. These are appropriate to a layered GaAs/Al$_x$Ga$_{1-x}$As system. For $\omega \leq \Omega$, this dispersion relation gives essentially the spectrum one would expect for unretarded modes, with both the acoustic and the optical branches being well defined. On the other hand, for $\omega \gg \Omega$ where the retardation effects are important, there are higher-frequency solutions of Eq. (5.65) that approximate to

$$\omega = \pm(c/a)QL + 2\pi mc/(1+s)a, \quad m = \text{integer}. \qquad (5.70)$$

This behavior has some resemblance to the folded dispersion curve for phonons in superlattices [25].

Fig. 5.8 shows the dispersion curve for surface plasmon-polaritons, using Eq. (5.69), for several values of the surface parameters r_S and μ_S. We have plotted the reduced frequency ω/Ω against the dimensionless wavevector $k_x a$. The bulk continuum is shown shaded. This dispersion curve is consistent with the results obtained by Giuliani and Quinn [26]. When $\mu_S = 1$, it corresponds to a surface branch existing above or below the bulk-plasmon continuum if $r_S < 1$ or $r_S > 1$,

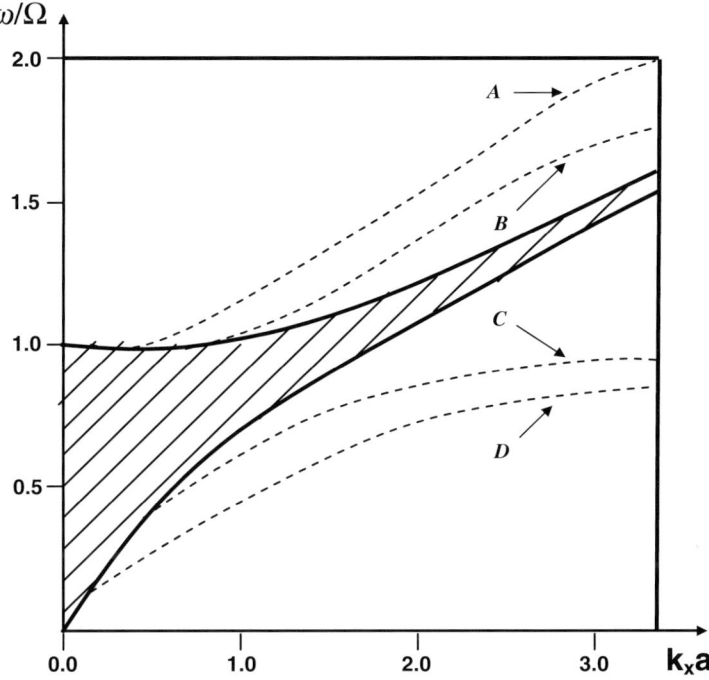

Fig. 5.8. Bulk (shaded region) and surface (dotted lines) plasmon-polariton modes for $a = 40$ nm, $\epsilon_A = 12.9$, and several values of the surface parameters: A, $r_S = 0.08$, $\mu_S = 1.0$; B, $r_S = 0.08$, $\mu_S = 0.75$; C, $r_S = 4.0$, $\mu_S = 1.0$; D, $r_S = 4.0$, $\mu_S = 0.75$ (after Constantinou and Cottam [19]).

respectively. There is then a cut-off wavevector corresponding to the condition

$$k_x a > \ln|(1 + r_S)/(1 - r_S)|. \tag{5.71}$$

However, as μ_S is reduced from unity, it is found that each surface branch moves to a lower frequency and the cut-off value for $k_x a$ is modified. In the limit $\mu_S \to 0$, the surface mode frequency has collapsed to zero.

5.3 Doped Semiconductor Superlattices

Synthetic semiconductor superlattice structures are of both fundamental and technological interest. In the past, most suggestions concerning the growth of such structures focused on either a multilayer heterojunction arrangement or a periodic alternation of the doping of only one semiconductor to form a series of homojunctions (for a review see Ref. [27] and the references therein). However, with the development of more sophisticated techniques to grow artificial structures, the possibility of well-controlled n- and p-doping of otherwise homogeneous semiconductors, including, in particular, the realization of extremely abrupt doping profiles, has been investigated and applied to design devices with high performance and new properties. These doping superlattices are crystallographically only slightly perturbed by a relatively small amount of dopants (typically 10^{17} to 10^{19} cm^{-3} for the most interesting range). Among these *doping superlattices*, one might expect that a superlattice consisting of thin layers of alternate n- and p-doped semiconductors, with an insulating material in between, leading to the so-called *nipi* superlattice, should display novel properties not found in the ordinary cases [28]. For instance, propagation of plasmon-polariton excitations in the periodic *nipi* superlattice was the subject of intense investigation in the 1980s [29,30].

The *nipi* superlattice is an artificial structure composed of doped semiconductors that exhibits many interesting properties. It is formed by a periodic array of n- and p-doped semiconductor layers separated by an insulator layer (or an intrinsic semiconductor material), and it was proposed by Döhler [31]. The layer structure is of the type shown in Fig. 2.3, where materials A and C are now n- and p-doped semiconductors with dielectric functions $\epsilon_A(\omega)$ and $\epsilon_C(\omega)$ and thicknesses a and c, respectively. The other materials B and D are insulators with frequency-independent dielectric constants ϵ_B and ϵ_D and thicknesses b and d, respectively. Each unit cell is of length $L = a + b + c + d$, and is designated by the index n ($n = 1, 2$, etc.). In the nth unit cell, at the interfaces $z = nL$ and $z = nL + a$, it is assumed that there are 2DEG (due to the ionization of impurity levels), while at the interfaces $z = nL + a + b$ and $z = nL + a + b + c$, we have 2DHG. The 2DEG and the 2DHG at each interface are modelled by the presence of a surface current density, σ_e for electrons and σ_h for holes. This is in accordance with Ohm's law, as explained in Section 5.2 (replacing $\bar{\sigma}$ by σ_e and σ_h, respectively).

We assume, as in the previous cases, a p-polarization for the electromagnetic mode and propagation in the x-direction with wavevector k_x. Then using Maxwell's equations together with the electromagnetic boundary conditions at the interfaces of the nth cell of the *nipi* superlattice, we obtain the matrix equations [30]

$$\bar{M}_A |A_A^n\rangle = \bar{N}_B |A_B^n\rangle, \tag{5.72}$$

$$\bar{M}_B |A_B^n\rangle = \bar{N}_C |A_C^n\rangle, \tag{5.73}$$

$$\bar{M}_C |A_C^n\rangle = \bar{N}_D |A_D^n\rangle, \tag{5.74}$$

$$\bar{M}_D |A_D^n\rangle = \bar{N}_A |A_A^{n+1}\rangle, \tag{5.75}$$

where the two-component column vector $|A_A^n\rangle$, as well as the matrices $\bar{M}_j$ and $\bar{N}_j$ (with now $j = A, B, C, D$), are defined as in Eqs. (5.16), (5.19), and (5.63), respectively.

The bulk and surface modes of the plasmon-polariton spectra are still given formally by the expressions in Eqs. (5.27) and (5.41), respectively, but the transfer matrix for the present case is defined by

$$\bar{T} = \bar{N}_A^{-1} \bar{M}_D \bar{N}_D^{-1} \bar{M}_C \bar{N}_C^{-1} \bar{M}_B \bar{N}_B^{-1} \bar{M}_A. \tag{5.76}$$

We note that the simple binary superlattice results of Subsections 5.1.1 and 5.1.2 are a special case of Eq. (5.76) with $\sigma_e = \sigma_h = 0$, and $a = c$, $b = d$. Also the charge-sheet superlattice results of Section 5.2 can be obtained from Eq. (5.76) with $\sigma_e = \sigma_h = \bar{\sigma}$, and $a = c$, $b = d$.

In order to present some numerical results, we consider materials A and C to be Si doped with n and p impurities, and described by a plasmon-type dielectric function with background dielectric constant $\epsilon_\infty = 11.7$ and plasma frequency $\omega_p = 76.5$ THz. The dielectric constants of materials B and D are both assumed to be equal to 12.3, appropriate for SiO$_2$.

The spectra for plasmon-polaritons propagating in a *nipi* superlattice are illustrated in Fig. 5.9, where we have considered all layers to have the same thickness equal to 40 nm. We also assume that the carrier concentrations per unit area, n_e and n_h, have the same absolute value equal to 2×10^{16} m^{-2}. In Fig. 5.9a we have one surface mode (dotted line) localized below the lower bulk band, corresponding to βL purely real and positive, and another one within the gap between the bulk bands which is associated with $\beta L = i\pi + \chi_1$, where χ_1 is a positive quantity. On the other hand, in Fig. 5.9b we have surface modes situated above and below the bulk bands corresponding to βL purely real. The other mode, which exists between the bulk bands, corresponds to $\beta L = i\pi + \chi_2$ ($\chi_2 > 0$) and merges with the upper band for small values of $k_x a$. In both cases the continuum bulk bands are shown shaded, and are bounded by the lines $QL = 0$ and $QL = \pi$.

In Fig. 5.10 we show the effects of varying the layer thicknesses, considering now that all layers have thicknesses equal to 20 nm, with the same values as before for all the other physical parameters. As one can see, the bulk bands (shown shaded) are slightly shifted by comparison with Figs. 5.9a and b, allowing the existence of only three surface branches (compared with a total of five surface modes in the previous case). This behavior arises because as the bulk bands are shifted, they approach the value of the plasma frequency $\omega_{pA} = \omega_{pB}$, which is in turn the limiting frequency for the existence of these modes.

In all the cases reported here, it is found that the retardation effects are important only in the region corresponding to small values of the dimensionless in-plane wavevector $k_x a$, as expected.

5.4 Piezoelectric Superlattices

So far in this chapter, we have considered situations where the propagation of the light through a dielectric or semiconductor medium is governed solely by the use of Maxwell's equations with an appropriate dielectric function $\epsilon(\omega)$. While this is correct in the majority of known crystals, it is not *always* the case, and we consider in this section the example of *piezoelectric materials*. These are crystals characterized by the fact that deforming them produces an electric field, and conversely,

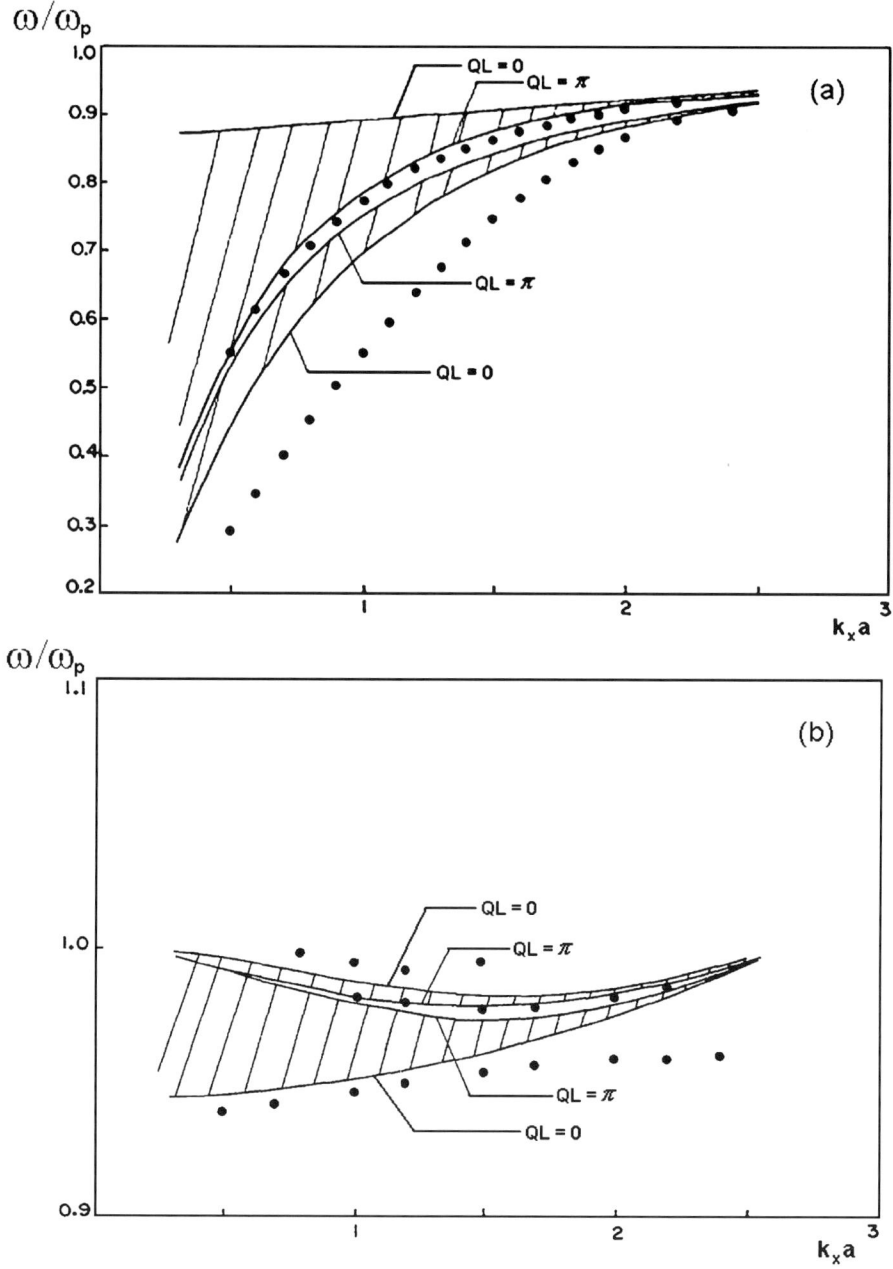

Fig. 5.9. Plasmon-polariton spectra, plotted as ω/ω_p versus the in-plane wavevector $k_x a$ for a *nipi* periodic superlattice showing the bulk (shaded area) and the surface (dots) modes for (a) the two lower bulk bands, and (b) the two upper bulk bands. See the text for the parameter values (after Farias et al. [30]).

5.4. PIEZOELECTRIC SUPERLATTICES

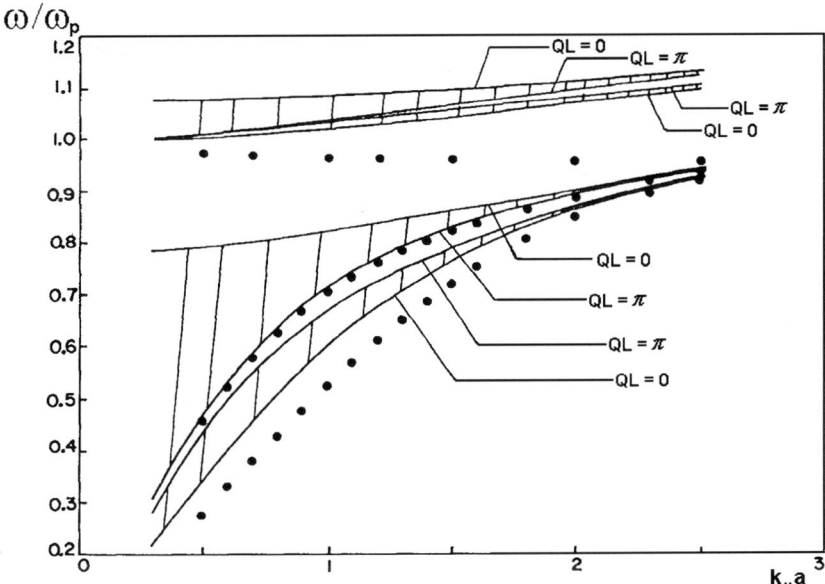

Fig. 5.10. The same as Fig. 5.9, but with layer thicknesses reduced by a factor of two, so that $a = b = c = d = 20$ nm (after Farias et al. [30]).

if one applies an electric field to them, they deform. Therefore, in these materials, the applicable equations to be concerned with are combinations of the elastic equations of motion together with Maxwell's equations. The coupling parameters are the elements of the piezoelectric tensor (or matrix).

Since the first quartz resonator around 1920, piezoelectric materials have been envisaged for many practical applications. One example is in the field of acousto-electronics [32,33], where there are a number of important devices using piezoelectric effects, such as signal processing, spread-spectrum communications, radar and acoustic charge transport, among others [34–38]. Additionally, they give rise to some elastic surface waves, the so-called Bleustein–Gulyaev waves, with no surface counterpart in the purely elastic materials [39,40]. As a result, the study of the propagation of acoustic waves in periodic piezoelectric superlattices has also been the object of much attention in the last two decades [41–45], following the technological trend for seeking more device applications from superlattice structures [46].

In this section we extend our study of the propagation of plasmon-polariton modes in two-component superlattices by considering that one of the superlattice constituents is a piezoelectric material. Following the same formal approach as in Section 5.1, we consider two composite building blocks α and β that are now defined as shown in Fig. 5.11. Each block is composed of two layers of different materials, one of them being a piezoelectric semiconductor crystal with either cubic symmetry (in block α) or hexagonal 6 mm symmetry (in block β). These

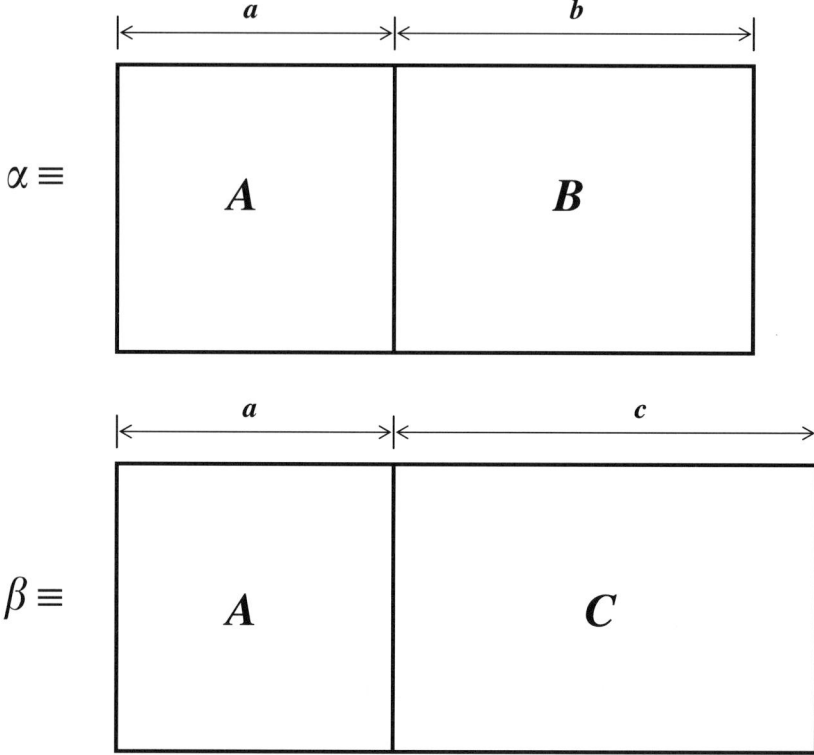

Fig. 5.11. The two building blocks α and β assumed for the piezoelectric two-component periodic superlattice. Here A is a non-piezoelectric insulating material, B is a cubic piezoelectric semiconductor material, and C is a hexagonal 6 mm piezoelectric semiconductor material.

constitute the two important symmetry cases for piezoelectric materials. In both cases, layer A is a non-piezoelectric insulator material. The layers may have different thicknesses and dielectric functions.

5.4.1 Piezoelectric Layer

Inside a single layer of a piezoelectric material (B or C), the piezoelectric coupling is usually weak enough to allow the hybrid wave solutions to behave as a combination of a quasi-acoustic mode (with a phase velocity slightly lower than the uncoupled acoustic mode) and a quasi-electromagnetic mode (with a phase velocity shifted to a slightly higher value than the electromagnetic wave). As the electromagnetic wave has a velocity approximately five orders of magnitude higher than the acoustic wave, we can describe the former in the so-called *static field approximation* where the particle displacement u_j ($j = x$, y, or z) along the coordinate axes r_j is coupled to the electrical potential ϕ, through the piezoelectric

5.4. PIEZOELECTRIC SUPERLATTICES

tensor, by the following set of equations [47–49]:

$$\rho \frac{\partial^2 u_j}{\partial t^2} - C_{ijkl} \frac{\partial^2 u_k}{\partial r_i \partial r_l} - e_{kij} \frac{\partial^2 \phi}{\partial r_i \partial r_k} = 0, \tag{5.77}$$

$$e_{ikl} \frac{\partial^2 u_k}{\partial r_i \partial r_l} - \epsilon_{ik} \frac{\partial^2 \phi}{\partial r_i \partial r_k} = 0. \tag{5.78}$$

Here the indices i, j, k, l can be x, y, or z, and repeated subscripts are implicitly assumed to be summed over. Also C_{ijkl} is the fourth-rank elastic tensor at constant electric field, e_{ikl} is the third-rank piezoelectric tensor, ϵ_{ik} is the second-rank dielectric permittivity tensor, and ρ is the density of the material.

Considering only one layer at present, let us assume that the hybrid wave is propagating in the x-direction with a phase velocity equal to ω/k_x. The solutions of the coupled equations (5.77) and (5.78) can be cast as

$$u_j = \alpha_j \exp(ik_z z)\exp(ik_x x - i\omega t), \quad j = x, y, z, \tag{5.79}$$

$$\phi = \alpha_4 \exp(ikz)\exp(ik_x x - i\omega t), \tag{5.80}$$

where the α coefficients are amplitude factors. Substitution of the above solutions, Eqs. (5.79) and (5.80), into the coupled equations (5.77) and (5.78) yields coupled differential equations for the two pairs (u_x, u_z) and (u_y, ϕ) in both symmetry cases. For (u_y, ϕ) the coupling is due to the presence of the piezoelectric tensor. On solving these coupled equations, and considering only the piezoelectric case of interest, we eventually find

$$u_y = (2k_x k_{Tz}^2)^{-1}\Big\{L(k_1)\Big[B_1 \exp(ik_1 z) - B_2 \exp(-ik_1 z)\Big]$$
$$+(\epsilon_{xx}/e_{x4})L(k_2)\Big[B_3 \exp(ik_2 z) - B_4 \exp(-ik_2 z)\Big]\Big\}, \tag{5.81}$$

$$\phi = (e_{x4}/\epsilon_{xx})\Big[B_1 \exp(ik_1 z) - B_2 \exp(-ik_1 z)\Big]$$
$$+B_3 \exp(ik_2 z) + B_4 \exp(-ik_2 z), \tag{5.82}$$

for the case of *cubic symmetry*. Here B_r ($r=1,2,3,4$) are unknown coefficients to be determined later by applying the boundary conditions, and $k_{1,2} = (k_{+,-})^{1/2}$ with

$$k_{\pm}^2 = (\tfrac{1}{2})\Big[k_{Tz}^2 - k_x^2(1+4p) \pm \Delta\Big], \tag{5.83}$$

$$\Delta = \Big[(k_{Tz}^2 + k_x^2)^2 + 8k_x^2 p(2k_x^2 p - k_{Tz}^2 + k_x^2)\Big]^{1/2}. \tag{5.84}$$

Also we have defined

$$L(k) = k\Big[k^2 + k_x^2(1+4p)\Big], \tag{5.85}$$

$$p = e_{x4}^2/C_{44}\epsilon_{xx}, \tag{5.86}$$

where p is a dimensionless coupling coefficient, and we have used the usual shorthand notation C_{IJ} and e_{iJ} for the elastic and piezoelectric tensors [50]. The z-component of the transverse acoustic wavevector is given by

$$k_{Tz}^2 = (\omega/v_T)^2 - k_x^2, \quad v_T = (C_{44}/\rho)^{1/2}, \tag{5.87}$$

where v_T is the velocity of the transverse acoustic mode. We observe that Eq. (5.83) gives the correct limits for the uncoupled case ($p=0$), namely $k_+^2 = k_{Tz}^2$ and $k_-^2 = -k_x^2$.

On the other hand, considering now the case of *hexagonal symmetry* 6 mm, the solutions of u_y and ϕ are found to have a slightly simpler form:

$$u_y = B_1' \exp(ikz) + B_2' \exp(-ikz), \tag{5.88}$$

$$\phi = (e_{x5}/\epsilon_{xx})\left[B_1' \exp(ikz) + B_2' \exp(-ikz)\right] + B_3' \exp(-k_x z) + B_4' \exp(k_x z). \tag{5.89}$$

Here k is given by

$$k^2 = (k_{Tz}^2 - k_x^2 p')/(1+p'), \tag{5.90}$$

and p' is defined as for p provided we replace e_{x4} by e_{x5}. The unknown B' coefficients can later be found, if needed, using the appropriate boundary conditions.

5.4.2 Superlattice Structure

We now generalize to the case of a two-component periodic piezoelectric superlattice, where one of the layers is a piezoelectric material, the other one being a non-piezoelectric insulating medium. The Cartesian axes are chosen so that the z-axis is normal to the plane of the layers, with the nth superlattice unit cell bounded by the planes $z=(n-1)L$ and $z=nL$, with the size L of the superlattice unit cell denoted by either $a+b$ for the cubic symmetry case, or $a+c$ for the hexagonal symmetry case.

Inside layer A (the non-piezoelectric material), at the nth cell, the uncoupled elastic and electromagnetic differential equations have the well-known solutions

$$u_y = A_1^n \exp(ik_{Tz}z) + A_2^n \exp(-ik_{Tz}z), \tag{5.91}$$

$$\phi = A_3^n \exp(-k_x z) + A_4^n \exp(k_x z). \tag{5.92}$$

On the other hand, inside layer B (the piezoelectric medium), the coupled elastic and electromagnetic differential equations at the nth cell are given by Eqs. (5.81) and (5.82) for cubic symmetry, and by Eqs. (5.88) and (5.89) for hexagonal symmetry.

Now using the standard elastic and electromagnetic boundary conditions [47–49] and employing the transfer-matrix treatment to simplify the algebra as in the previous section, we find the eigenvalue equation

$$\bar{T}|A^n\rangle = \exp(iQL)|A^n\rangle, \tag{5.93}$$

5.4. PIEZOELECTRIC SUPERLATTICES

where $|A^n\rangle$ is the 1×4 column vector formed by the unknown coefficients in Eqs. (5.91) and (5.92), Q is the Bloch wavevector, and the 4×4 transfer matrix $\bar{T}$ is here calculated from

$$\bar{T}_\alpha = \bar{M}_2^{-1}\bar{N}_2\bar{N}_1^{-1}\bar{M}_1 \tag{5.94}$$

or

$$\bar{T}_\beta = \bar{M}'^{-1}_2\bar{N}'_2\bar{N}'^{-1}_1\bar{M}'_1, \tag{5.95}$$

for cubic ($\bar{T}_\alpha$) and hexagonal ($\bar{T}_\beta$) symmetries. The $\bar{M}$ and $\bar{N}$ matrices are defined by [51]

$$\bar{M}_1 = \begin{bmatrix} f_A & \bar{f}_A & 0 & 0 \\ \mu k_{Tz} f_A & -\mu k_{Tz}\bar{f}_A & -p_1 k_x f_x & -p_1 k_x \bar{f}_x \\ 0 & 0 & f_x & \bar{f}_x \\ 0 & 0 & f_x & -\bar{f}_x \end{bmatrix}, \tag{5.96}$$

$$\bar{N}_1 = \begin{bmatrix} L(k_1) & -L(k_1) & L(k_2)/p_2 & -L(k_2)/p_2 \\ k_1 L(k_1) & k_1 L(k_1) & k_2 L(k_2)/p_2 & k_2 L(k_2)/p_2 \\ p_2 & p_2 & 1 & 1 \\ -i\epsilon_B k_1 p_2/\epsilon_A k_x & i\epsilon_B k_1 p_2/\epsilon_A k_x & -i\epsilon_B k_2/\epsilon_A k_x & i\epsilon_B k_2/\epsilon_A k_x \end{bmatrix}. \tag{5.97}$$

The matrix $\bar{M}_2$ can be obtained from $\bar{M}_1$ by dividing the first row by f_A, the second by $\bar{f}_A$, the third by f_x, and the fourth by $\bar{f}_x$. Also we can obtain the matrix $\bar{N}_2$ from $\bar{N}_1$ by multiplying the first row by f_{B1}, the second by $\bar{f}_{B1}$, the third by f_{B2}, and the fourth by $\bar{f}_{B2}$. Here we have used the following definitions:

$$f_A = \exp(ik_{Tz}a) = 1/\bar{f}_A, \qquad f_x = \exp(ik_x a) = 1/\bar{f}_x, \tag{5.98}$$

$$f_{Bj} = \exp(ik_j b) = 1/\bar{f}_{Bj}, \qquad j = 1, 2, \tag{5.99}$$

$$\mu = C_{44A}/C_{44B}, \qquad p_1 = e_{x4}/C_{44B}, \qquad p_2 = e_{x4}/\epsilon_B. \tag{5.100}$$

In the case of hexagonal symmetry the results for the matrices in Eq. (5.95) are [51]

$$\bar{M}'_1 = \begin{bmatrix} f_A & \bar{f}_A & 0 & 0 \\ \mu' k_{Tz} f_A & -\mu' k_{Tz}\bar{f}_A & 0 & 0 \\ 0 & 0 & f_x & \bar{f}_x \\ 0 & 0 & f_x & -\bar{f}_x \end{bmatrix}, \tag{5.101}$$

$$\bar{N}'_1 = \begin{bmatrix} 1 & 1 & 0 & 0 \\ k(1+p') & -k(1+p') & ip'_1 k_x & -ip'_1 k_x \\ p'_2 & p'_2 & 1 & 1 \\ 0 & 0 & \epsilon_B/\epsilon_A & -\epsilon_B/\epsilon_A \end{bmatrix}. \tag{5.102}$$

The matrix $\bar{M}'_2$ can be found in the same way as discussed for the cubic case. Also the matrix $\bar{N}'_2$ can be obtained from $\bar{N}'_1$ by multiplying the first row by

$f_C = \exp(ikc)$, the second by $\bar{f}_C = 1/f_C$, the third by f_x, and the fourth by $\bar{f}_x$. Here p'_1 and p'_2 are defined as in Eq. (5.100) provided we replace e_{x4} by e_{x5} and set $\mu' = C_{44A}/C_{44C}$.

We note, as expected, that when the piezoelectric coupling terms e_{x4} or e_{x5} are set equal to zero, the matrix $\bar{T}$ reduces to a partitioned form

$$\bar{T} = \begin{pmatrix} \bar{T}_1 & | & 0 \\ --- & | & --- \\ 0 & | & \bar{T}_2 \end{pmatrix}, \qquad (5.103)$$

where $\bar{T}_1$ and $\bar{T}_2$ are the two 2×2 matrices that describe the separate spectra of phonon-polaritons (in the elastic approximation) and plasmon-polaritons (in the unretarded limit), respectively, in binary superlattices.

To study the surface modes in a semi-infinite structure, let us consider that the superlattice is truncated at the surface $z = 0$, with vacuum filling the space $z < 0$. The electrical potential in the vacuum region is simply given by

$$\phi = C \exp(k_x z), \qquad (5.104)$$

where C is a constant.

We suppose, for simplicity, that the outermost layer of the superlattice is occupied by the non-piezoelectric medium A, and that its surface can either be metallized (with a thin layer at zero electrical potential) or free. The elastic and electromagnetic boundary conditions give us the following dispersion relation for the surface polariton modes:

$$(T_{11} + T_{12} - T_{21} - T_{22})(T_{33} - T_{44} + T_{34}\lambda - T_{43}\lambda^{-1})$$
$$+ (T_{41} + T_{42})[(T_{13} - T_{23})\lambda^{-1} + T_{14} - T_{24}]$$
$$- (T_{31} + T_{32})[T_{13} - T_{23} + (T_{14} - T_{24})\lambda] = 0, \qquad (5.105)$$

with $\lambda = -1$ for the metallic surface and $\lambda = (\epsilon_A + 1)/(\epsilon_A - 1)$ for the free surface. We note that when the piezoelectric coupling term is equal to zero, the second and third lines of Eq. (5.105) vanish. Then the two uncoupled first terms reduce to the appropriate surface dispersion relations for phonon- and plasmon-polariton modes, respectively.

Next we present some numerical examples to illustrate properties of the polariton dispersion relations in piezoelectric periodic superlattices. In what follows, we have taken physical parameters appropriate to fused silica as layer A, while the piezoelectric semiconductor materials B (cubic symmetry) and C (hexagonal symmetry) were considered to be GaAs and ZnO, respectively. The assumed thicknesses are $a = 20$ nm and $b = c = 40$ nm. Also, the dielectric constants of the materials are $\epsilon_A = 3.8$, $\epsilon_B = 12.5$, and $\epsilon_C = 9.2$. The elastic tensor elements C_{44} are equal to 31.2, 59.4, and 42.47 (all in units of 10^9 N/m^2) for materials A, B, and C, respectively, while the piezoelectric tensor element e_{x4} is equal to 0.154 C/m^2

5.5. MAGNETOPLASMON-POLARITONS IN FINITE AND INFINITE SUPERLATTICES

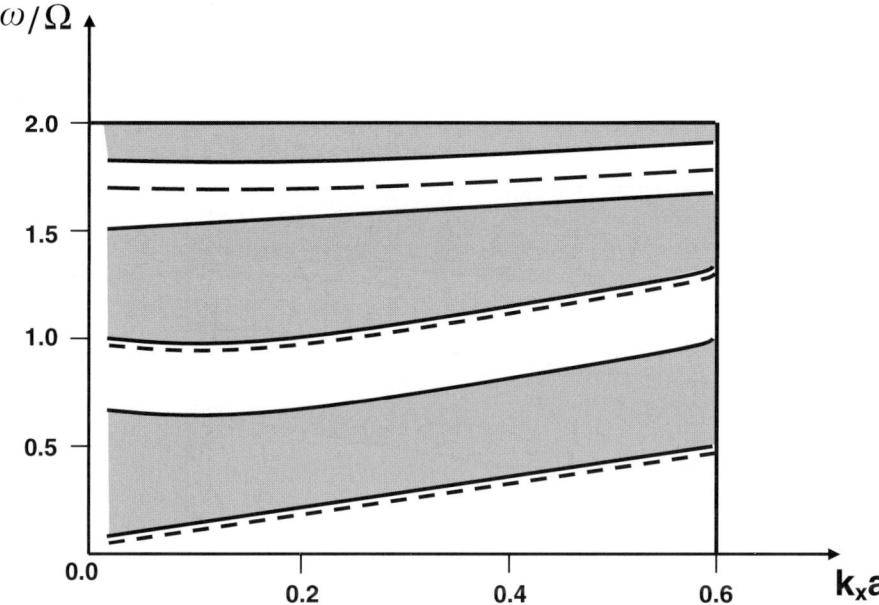

Fig. 5.12. Dispersion relations for bulk bands (shaded) and surface polariton modes (dotted lines) propagating in a cubic piezoelectric superlattice. We have plotted the reduced frequency ω/Ω against the dimensionless wavevector $k_x a$. See text for parameter values (after Albuquerque and Cottam [51]).

(GaAs), and for ZnO we have $e_{x5} = 0.48\,\text{C/m}^2$. Finally, the densities of the layers are 2.2 (material A), 5.3 (material B), and 5.7 (material C), all in units of $10^3\,\text{kg/m}^3$ [51].

Fig. 5.12 shows the polariton spectra for the case of cubic piezoelectric symmetry (i.e. the fused silica/GaAs structure). The outermost interface is considered to be metallized (but for the non-metallized case the spectra are almost identical in this case). The characteristic frequency Ω in the vertical axis is equal to v_{TA}/a, where v_{TA} is the velocity of the acoustic mode that propagates in layer A (the insulator). The shaded bulk bands are bounded by the curves $QL = 0$ and $QL = \pi$. The dotted curves represent the surface modes of a semi-infinite superlattice. For comparison, we show in Fig. 5.13 the corresponding case for the hexagonal 6 mm symmetry. Although a quantitative difference is apparent, there is no striking qualitative difference in the spectra, as a consequence of their common (u_y, ϕ) piezoelectric coupling.

5.5 Magnetoplasmon-Polaritons in Finite and Infinite Superlattices

In this final section we examine the effects of an external applied magnetic field on the spectrum of the superlattice plasmon-polaritons. The magnetic field leads

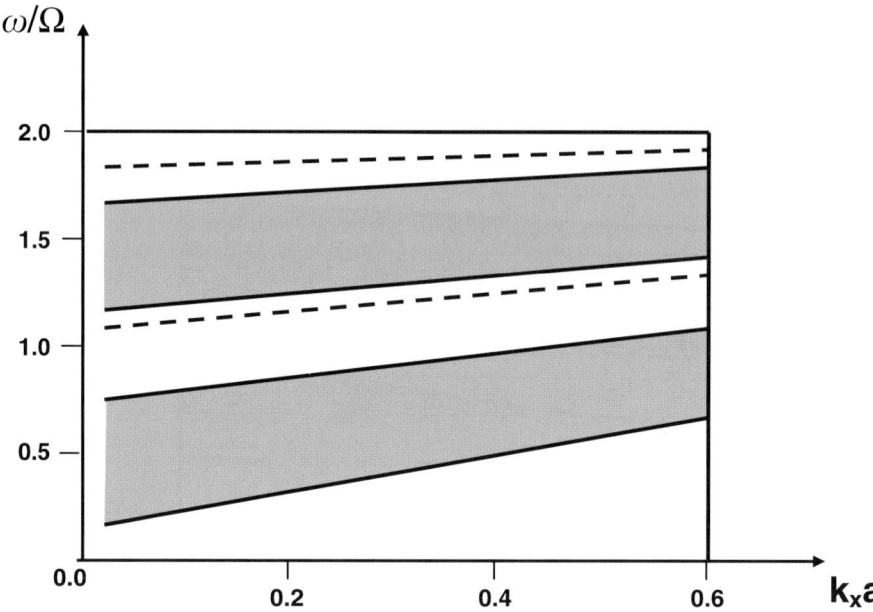

Fig. 5.13. As in Fig. 5.12 but for the hexagonal piezoelectric superlattice (after Albuquerque and Cottam [51]).

to the so-called *magnetoplasmon-polariton* excitations. They may be studied using the same transfer-matrix technique as presented in the earlier sections of this chapter, but with a modified form of the dielectric function $\epsilon(\omega)$. We shall proceed directly to present the results for superlattices; for a description of the earlier work in other geometries (including bulk materials, single films, etc.), the reader is referred to Refs. [52–57] for the plasmon case. We mention Refs. [58–60] for the analogous magnetic field effects in the phonon case.

We begin with a discussion of the bulk and surface magnetoplasmon-polariton modes in binary superlattices as the simplest case. Then we extend these results to the more complex *nipi* superlattices.

Specifically we assume that a uniform external magnetic field H_0 is imposed in the y-direction (parallel to the interfaces), supposing that the orientation of the Cartesian axes for the superlattice is the same as used in Fig. 5.4. The excitations are taken to have in-plane propagation in the x-direction and are described by Bloch wavevector Q and frequency ω.

We consider initially a binary superlattice with alternating layers A and B extending indefinitely in both the positive and negative z-directions. The layers A and B have thicknesses a and b, as well as dielectric functions $\epsilon_A(\omega)$ and $\epsilon_B(\omega)$, respectively. When the external magnetic field is present, the dielectric functions in the media become anisotropic and have off-diagonal terms. Their general form

5.5. MAGNETOPLASMON-POLARITONS IN FINITE AND INFINITE SUPERLATTICES

(for an applied field in the y-direction) is

$$\bar{\epsilon} = \begin{pmatrix} \epsilon_1 & 0 & -i\epsilon_2 \\ 0 & \epsilon_3 & 0 \\ i\epsilon_2 & 0 & \epsilon_1 \end{pmatrix}, \tag{5.106}$$

with

$$\epsilon_1 = \epsilon_\infty \left(1 + \frac{\omega_p^2}{\omega_c^2 - \omega^2}\right), \tag{5.107}$$

$$\epsilon_2 = \frac{\epsilon_\infty \omega_p^2 \omega_c}{\omega(\omega_c^2 - \omega^2)}, \tag{5.108}$$

$$\epsilon_3 = \epsilon_\infty [1 - (\omega_p^2/\omega^2)]. \tag{5.109}$$

As before, ω_p and ϵ_∞ are the plasma frequency and the background dielectric constant, respectively, in the layer under consideration. The external magnetic field effects are given in terms of the cyclotron frequency $\omega_c = eH_0/m$; this represents the frequency of circular motion of the electrons in the plane perpendicular to the applied field. Eqs. (5.106) and (5.107)–(5.109) can be derived by following a hydrodynamic approach, e.g. by adding a term $-e(d\vec{r}/dt \times \vec{H}_0)$ in Eq. (1.21).

Generalizing results in Subsection 5.1.1, the solutions of Maxwell's equations in the layers of the nth unit cell are (for $j = A$ or B)

$$E_{xj}(z) = A_{1j}^n \exp(-\alpha_j z) + A_{2j}^n \exp(\alpha_j z), \tag{5.110}$$

$$E_{zj}(z) = i[\lambda_{1j} A_{1j}^n \exp(-\alpha_j z) + \lambda_{2j} A_{2j}^n \exp(\alpha_j z)], \tag{5.111}$$

$$H_{yj}(z) = -i\omega\epsilon_0 [\xi_{1j} A_{1j}^n \exp(-\alpha_j z) - \xi_{2j} A_{2j}^n \exp(\alpha_A z)]. \tag{5.112}$$

Here α_A and α_B are defined as in Eqs. (5.8) and (5.9), with ϵ_j denoting the following combination of dielectric-function matrix elements:

$$\epsilon_j = \epsilon_{1j}(\omega) - \epsilon_{2j}^2(\omega)/\epsilon_{1j}(\omega). \tag{5.113}$$

Also we have defined (for $r = 1, 2$)

$$\lambda_{rj}(z) = (k_x \epsilon_{1j} \pm \alpha_j \epsilon_{2j})/(k_x \epsilon_{2j} \pm \alpha_j \epsilon_{1j}), \tag{5.114}$$

$$\xi_{rj}(z) = \frac{\epsilon_{2j}[k_x^2(\epsilon_{1j}^2 + \epsilon_{2j}^2) - 2\alpha_j^2 \epsilon_{1j}^2] \pm \alpha_j k_x \epsilon_{1j}(\epsilon_{2j}^2 - \epsilon_{1j}^2)}{k_x(k_x^2 \epsilon_{2j}^2 - \alpha_j^2 \epsilon_{1j}^2)}. \tag{5.115}$$

In Eqs. (5.114) and (5.115), the upper and lower signs are associated with $r = 1$ and $r = 2$, respectively.

Now, using the standard electromagnetic boundary conditions for the nth cell, and following the same steps as in Subsection 5.1.1, we find that the superlattice bulk dispersion relation is as formally quoted in Eq. (5.27). The relevant transfer matrix $\bar{T}$ is formally defined in the same way as in Eq. (5.21), provided we replace ξ_j by ξ_{rj} in the $\bar{M}$ and $\bar{N}$ matrices defined by Eqs. (5.19) and (5.20), respectively, taking $r=1$ (2) for the matrix elements M_{12} and N_{12} (M_{22} and N_{22}). Explicitly, this dispersion relation reads [61]

$$2(Y_{1A} - Y_{2A})(Y_{1B} - Y_{2B})\cos(2QL) = (Y_{1A}f_A - Y_{2A}\bar{f}_A)(Y_{1B}f_B - Y_{2B}\bar{f}_B)$$
$$+ (Y_{1A}\bar{f}_A - Y_{2A}f_A)(Y_{1B}\bar{f}_B - Y_{2B}f_B)$$
$$- (Y_{1A}Y_{2A} + Y_{1B}Y_{2B})(f_A - \bar{f}_A)(f_B - \bar{f}_B),$$

(5.116)

with

$$Y_{rj} = \epsilon_{1j}\lambda_{rj} - \epsilon_{2j}, \quad r = 1, 2, \quad j = A, B, \tag{5.117}$$

and f_j is defined as in Eq. (5.15).

Next, for a semi-infinite superlattice, we can generalize the calculation in Subsection 5.1.2 to include an external magnetic field. We suppose that the superlattice is truncated at the plane $z=0$ with a medium C, which has a dielectric constant ϵ_C independent of the external magnetic field and occupies the region $z \leq 0$. Also we take the first layer of the semi-infinite superlattice to be, as previously, a layer of medium A. Then the dispersion relation for surface magnetoplasmon-polariton modes is again given formally by Eq. (5.41) provided the appropriate changes are made in the elements of the $\bar{T}$-matrix (as described above for the infinite case) and the factor λ is here defined by

$$\lambda = (\xi_{1A} + \xi_C)/(\xi_{2A} - \xi_C). \tag{5.118}$$

When written explicitly, the dispersion relation for the surface mode becomes [61]

$$\left[X_1(1+V_{2A})\bar{f}_A - X_2(1+V_{1A})f_A\right](1-V_{2B})(X_1U_{12} - X_2U_{11})f_B$$
$$= \left\{2(X_1U_{22} - X_2U_{21})\bar{f}_B + \left[X_1(1+V_{2B})U_{12} - X_2(1+V_{2B})U_{11}\right]f_B\right\}$$
$$\times \left[X_1(1-V_{2A})\bar{f}_A - X_2(1+V_{1A})f_A\right]. \tag{5.119}$$

Here we denote

$$V_{rj} = Y_{rj}/Y_{1B}, \tag{5.120}$$

$$U_{rs} = (-1)^r(Y_{rB} - Y_{sA})/(Y_{2B} - Y_{1B}), \tag{5.121}$$

$$X_r = k_x\xi_C + Y_{rA}, \tag{5.122}$$

with r and s equal to 1 or 2.

5.5. MAGNETOPLASMON-POLARITONS IN FINITE AND INFINITE SUPERLATTICES

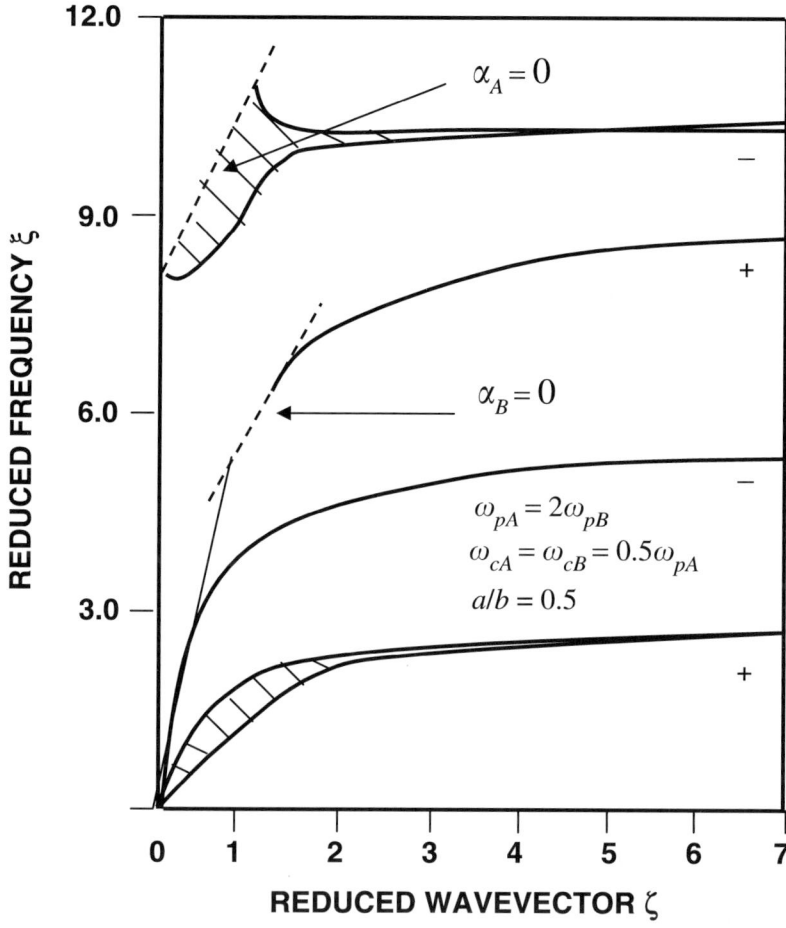

Fig. 5.14. Dispersion curves for magnetoplasmon-polaritons in a two-component semi-infinite superlattice, plotted in terms of a reduced frequency $\xi = \omega/\omega_{pA}$ against a reduced wavevector $\zeta = ck_x/\omega_{pA}$. See the text for parameter values. Note the four surface modes shown here as solid lines with a plus or minus sign attached (after Wallis et al. [61]).

Some dispersion curves are plotted in Fig. 5.14 in terms of a reduced frequency $\xi = \omega/\omega_{pA}$ against a reduced wavevector $\zeta = ck_x/\omega_{pA}$, considering medium C as vacuum [61]. The ratio between the plasma frequencies of medium A and medium B is taken to be 2.0, while their thickness ratio $a/b = 0.5$. The external magnetic field is specified through the cyclotron frequency ω_c, which is taken as half the value of the plasma frequency ω_{pA}. It can be shown that, in the absence of the external magnetic field, there are two surface modes, doubly degenerate, corresponding to propagation in opposite directions (i.e. parallel and antiparallel to the x-axis) for any given magnitude of the wavevector k_x. However, in the presence of the external magnetic field, this degeneracy is removed, leading to the four surface modes (full lines) labelled by the plus and minus signs in Fig. 5.14. Three of them lie in the gap between the bulk bands (which are shown shaded as usual) while the other is in the region below the lower bulk band, merging into it when $\zeta \simeq 3.0$. We note that the upper plus surface mode intersects the curve $\alpha_B = 0$ (shown dashed in Fig. 5.14) at two points, and it ceases to exist as a surface magnetoplasmon-polariton mode between these two points.

As a further extension of the previous case, we now briefly present the magneto-plasmon-polariton spectra for a semi-infinite *nipi* superlattice. In doing this we also allow for the presence of a 2D charge sheet at each interface. The dispersion relations are formally the same as those in the absence of a magnetic field (see Section 5.3), with a transfer matrix defined by Eq. (5.76), provided we make the same replacements described earlier, i.e. ξ_j replaced by ξ_{rj}, in the matrices $\bar{M}$ and $\bar{N}$.

Similar considerations can be applied to the magnetoplasmon-polariton spectra in *nipi* superlattices with a finite number p of unit cells (each with thickness L), bounded by isotropic media E and F whose dielectric constants ϵ_E and ϵ_F are not affected by presence of the external magnetic field H_0. The dispersion relation for the surface magnetoplasmon-polariton modes is now given by [62]

$$\exp(2\beta pL) = \frac{[\vartheta_A - \xi_F(1+K)][\xi_{1A} - K'\xi_{2A} + \xi_E(1+K')]}{[\vartheta'_A - \xi_F(1+K')][\xi_{1A} - K\xi_{2A} + \xi_E(1+K)]}, \qquad (5.123)$$

with

$$\vartheta_A = \xi_{1A} - K\xi_{2A} - \bar{\sigma}_h(1+K), \qquad (5.124)$$

and ϑ'_A is defined similarly to ϑ_A provided we replace K by K' in Eq. (5.124). Also ξ_{1A} and ξ_{2A} are given by Eq. (5.115), ξ_E by Eq. (5.56) (with a similar expression for ξ_F), K and K' by Eq. (5.46), and the surface current density for holes, $\bar{\sigma}_h$, defined by Eq. (5.64).

The dispersion relation, defined by Eq. (5.123), is plotted in Fig. 5.15 with the surface modes represented by dots and the bulk bands shown shaded. We have considered, as in Section 5.3, the layer A and C as n- and p-doped Si. The layers B and D were taken to have diagonal dielectric tensors, with all diagonal elements equal to 12.3, which is suitable for SiO_2. The finite superlattice, whose overall size is taken as $20L$ in this example, is considered to be surrounded by the vacuum. The thicknesses of the layers are $a = 2b = c = 2d = 40$ nm, and the applied external magnetic field is assumed to be such that $\omega_c/2\pi = 38.25$ THz. The quantization of the bulk continuum due to the finite thickness of the structure is apparent. This quantization, as is well known, leads to the modes being more widely spaced as p, the number of superlattice unit cells, becomes smaller. The surface polaritons at the top and bottom surfaces of the finite structure are essentially degenerate in frequency. As the thickness of the superlattice is decreased, the decaying exponential envelopes of these degenerate modes begin to overlap, and the frequencies split into a bonding and antibonding pair. This effect is clearly seen in the intermediate-frequency mode of Fig. 5.15 [62].

5.5. MAGNETOPLASMON-POLARITONS IN FINITE AND INFINITE SUPERLATTICES

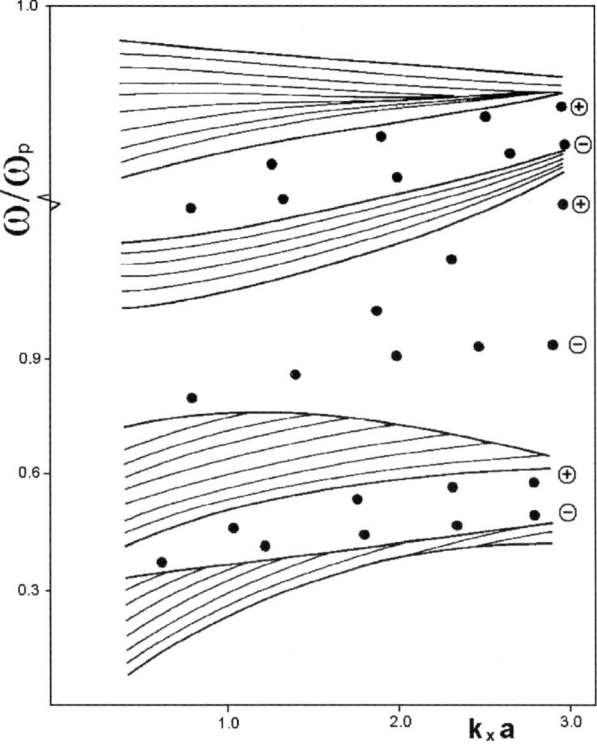

Fig. 5.15. Dispersion curves for magnetoplasmon-polaritons in a *nipi* finite superlattice. We have plotted the reduced frequency ω/ω_p against the dimensionless wavevector $k_x a$. The six magnetoplasmon-polariton surface mode branches are represented here by dots, and are labelled by the $(+)$ and $(-)$ signs. See the text for parameter values (after Albuquerque et al. [62]).

The surface plasmon modes that exist in the gap between the bulk bands are normally doubly degenerate with respect to propagation in opposite directions for the wavevector $\vec{k}$. However, when the magnetic field is switched on, this degeneracy is removed and the surface modes split, giving rise to the six magnetoplasmon-polariton surface modes indicated by the plus and minus signs in Fig. 5.15. Two of these modes are localized between the upper bulk bands (at $\omega/\omega_p \simeq 0.93$) and two more modes are localized between the lower bulk bands for $0.3 < \omega/\omega_p < 0.5$. Note that all four of these modes correspond to $\beta L = i\pi + \chi$, where χ is a positive quantity, and they merge into the bulk bands at $k_x a \simeq 3.2$. The other two modes are localized between the upper and lower bulk branches and correspond to βL being real and positive. They exist in the range $0.7 < \omega/\omega_p < 0.95$. The upper $(+)$ branch enters the upper bulk band at $k_x a \simeq 3.5$, while the lower one $(-)$ terminates when it enters the lower bulk band at $k_x a \simeq 3.5$.

Another important point can be inferred from Eq. (5.123) by rewriting it as

$$\exp(2\beta pL) = N/D, \tag{5.125}$$

where N and D are the numerator and denominator terms, respectively. As $p \to \infty$, i.e. we take the semi-infinite limit, it follows that $\exp(2\beta pL) \to \infty$ for $\beta > 0$ and tends to zero for $\beta < 0$. The different signs correspond to the surface modes localized on the upper and lower surfaces, so $N = 0$ corresponds to the dispersion relation of one of these modes, while $D = 0$ corresponds to the other.

References

[1] E.L. Albuquerque and M.G. Cottam, Phys. Rep. 233 (1993) 68.
[2] L. Esaki and R. Tsu, IBM J. Res. Develop. 14 (1970) 61.
[3] B. Vinter and C. Weisbuch, Quantum Semiconductor Structures, Academic Press, San Diego, 1991.
[4] G. Strasser, S. Gianordoli, L. Hvozdara, W. Schrenk, K. Unterrainer and E. Gomik, Appl. Phys. Lett. 75 (1999) 1345.
[5] S.M. Rytov, Akust. Zh. 2 (1956) 71 [Sov. Phys.-Acoust. 2 (1956) 68].
[6] W. Liu, G. Eliasson and J.J. Quinn, Solid State Commun. 55 (1985) 533.
[7] J.W. Wu, P. Hawrylak, G. Eliasson and J.J. Quinn, Phys. Rev. B 33 (1986) 7091.
[8] Y. Zhu, S. Cai and S. Zhou, Phys. Rev. B 38 (1988) 9941.
[9] M.C. Muñoz, V.R. Velasco and F. Garcia-Moliner, Phys. Rev. B 39 (1989) 1786.
[10] F. Garcia-Moliner and V.R. Velasco, Theory of Single and Multiple Interfaces, World Scientific, Singapore, 1992.
[11] L. Fernández-Alvarez, G. Monsivais, S. Vlaev and V.R. Velasco, Surf. Sci. 369 (1996) 367.
[12] L. Dobrzynski, Surf. Sci. 175 (1986) 1.
[13] L. Dobrzynski, Surf. Sci. Rep. 6 (1986) 119.
[14] R.E. Camley, B. Djafari-Rouhani, L. Dobrzynski and A.A. Maradudin, Phys. Rev. B 27 (1983) 7318.
[15] B. Djafari-Rouhani, L. Dobrzynski, O. Hardouin Duparc, R.E. Camley and A.A. Maradudin, Phys. Rev. B 28 (1983) 1711.
[16] E.L. Albuquerque, P. Fulco and D.R. Tilley, Phys. Status Solidi (b) 146 (1988) 449.
[17] R.E. Camley and D.L. Mills, Phys. Rev. B 29 (1984) 1695.
[18] B.L. Johnson, J.T. Weiler and R.E. Camley, Phys. Rev. B 32 (1985) 6544.
[19] N.C. Constantinou and M.G. Cottam, J. Phys. C 19 (1986) 739.
[20] S. Das Sarma and J.J. Quinn, Phys. Rev. B 25 (1982) 7603.
[21] W.L. Bloss and E.M. Brody, Solid State Commun. 43 (1982) 523.
[22] W.L. Bloss, Solid State Commun. 44 (1982) 363.
[23] G. Qin, G.F. Giuliani and J.J. Quinn, Phys. Rev. B 28 (1983) 6144.
[24] G.F. Giuliani, G. Qin and J.J. Quinn, Surf. Sci. 142 (1984) 433.
[25] M. Babiker, D.R. Tilley, E.L. Albuquerque and C.E.T Gonçalves da Silva, J. Phys. C 18 (1985) 1269.
[26] G.F. Giuliani and J.J. Quinn, Phys. Rev. Lett. 51 (1983) 919.
[27] A. MacDonald in: Interfaces, Quantum Wells and Superlattices, C.R. Leavens and R. Taylor, Eds., Plenum, New York, 1987.
[28] G.H. Döhler, J. Vac. Sci. and Technol. 16 (1979) 851.
[29] B.L. Johnson and R.E. Camley, Phys. Rev. B 38 (1988) 3311.
[30] G.A. Farias, M.M. Auto and E.L. Albuquerque, Phys. Rev. B 38 (1988) 12540.
[31] G.H. Döhler, Phys. Status Solidi (b) 52 (1972) 79.

[32] E. Dieulesaint and D. Royer, Elastic Waves in Solids, Wiley, New York, 1980.
[33] A.A. Chaban, Fiz. Tverd. Tela 32 (1990) 2137 [Sov. Phys.-Solid State 32 (1990) 1241].
[34] D.L. White, J. Appl. Phys. 33 (1962) 2547.
[35] S. Swierkonski, T. Vanduzer and C.W. Turner, IEEE Trans. Sonics Ultrason. SU-20 (1973) 260.
[36] R.A. Becker, R.W. Ralston and P.V. Wright, IEEE Trans. Sonics Ultrason. SU-29 (1982) 289.
[37] M.S. Hoskins, E.G. Bogus and B.J. Hunsinger, IEEE Electron Devices Lett. EDL-4 (1983) 396.
[38] F. Palma, L. Saccani and P. Das, Superlattices and Microstruct. 3 (1987) 181.
[39] J.L. Bleustein, Appl. Phys. Lett. 13 (1968) 412.
[40] Y.V. Gulyaev, JETP Lett. 9 (1969) 37.
[41] M. Grimsditch and F. Nizzoli, Phys. Rev. B 33 (1986) 9891.
[42] L. Fernández and V.R. Velasco, Surf. Sci. 185 (1987) 175.
[43] L. Fernández, V.R. Velasco and F. Garcia-Moliner, Europhys. Lett. 3 (1987) 723; Surf. Sci. 188 (1987) 140.
[44] A. Nougaoui and B. Djafari-Rouhani, Surf. Sci. 185 (1987) 154.
[45] E. Akcakaya and G.W. Farnell, J. Appl. Phys. 64 (1988) 4469.
[46] P. Dhez and C. Weisbuch, Eds., Physics, Fabrication and Applications of Multilayered Structures, Plenum, New York, 1988.
[47] G.W. Farnell and E.L. Adler, in: Physical Acoustics, Vol. IX, Ed., W.P. Mason, Academic Press, New York, 1972.
[48] B.A. Auld, Acoustic Fields and Waves in Solids, Vols. 1 and 2, 2nd ed., Krieger, Malabar-Florida, 1990.
[49] E.L. Albuquerque, Phys. Status Solidi (b) 104 (1981) 667.
[50] J.F. Nye, Physical Properties of Crystals, Oxford Univ. Press, Oxford, 1985.
[51] E.L. Albuquerque and M.G. Cottam, Solid State Commun. 83 (1992) 545.
[52] J.J. Brion, R.F. Wallis, A. Hartstein and E. Burstein, Phys. Rev. Lett. 28 (1972) 1455.
[53] J.J. Quinn and K.W. Chiu, in: Polaritons, Eds., E. Burstein and F. de Martini, Pergamon, New York, 1974.
[54] B.G. Martin, A.A. Maradudin and R.F. Wallis, Surf. Sci. 77 (1978) 416.
[55] M.S. Kushwaha and P. Halevi, Phys. Rev. B 35 (1987) 3879; 36 (1987) 5960; 37 (1988) 2724 (erratum).
[56] P. Halevi and M.S. Kushwaha, in: Electrodynamics of Interfaces and Composite Systems, Eds., R. Barrera and W.L. Mochan, World Scientific, Singapore, 1988.
[57] M.S. Kushwaha, Phys. Rev. B 40 (1989) 1692, 1969.
[58] A.C. Tselis and J.J. Quinn, Phys. Rev. B 29 (1984) 3318.
[59] H.C.A. Oji and A.H. MacDonald, Phys. Rev. B 33 (1986) 3810; 34 (1986) 1371.
[60] L. Wendler and R. Pechstedt, J. Phys.: Condens. Matter, 2 (1990) 8881.
[61] R.F. Wallis, R. Szenics, J.J. Quinn and G.F. Giuliani, Phys. Rev. B 36 (1987) 1218.
[62] E.L. Albuquerque, P. Fulco, G.A. Farias and M.M. Auto, Phys. Rev. B 43 (1991) 2032.

Chapter 6

Plasmon-Polaritons in Quasiperiodic Structures

The spectra of many different types of elementary excitations in quasiperiodic structures have been studied by several groups. In all cases the detailed theory showed the spectra to be Cantor-like with critical eigenfunctions and localized states. Furthermore, a quite complex *fractal energy spectrum*, which can be considered as a basic signature of quasiperiodicity, is a common feature of these systems [1], as mentioned in Chapter 2. On the experimental side much progress has also been made in studying the quasiperiodic structures, mainly to determine their optical properties like the picosecond luminescence measurements [2], photoluminescence excitation spectroscopy [3], and reflectance spectroscopy [4].

In this chapter we present a discussion focusing on the propagation of plasmon-polaritons in quasiperiodic structures. Our geometry is similar to that used in Chapter 5 for periodic structures, i.e. in the optical studies, light comes from a semi-infinite transparent medium (e.g. vacuum), occupying the region $z<0$, and is transmitted into a quasiperiodic structure, bounded by the plane $z=0$ and occupying the region $z>0$ (see Fig. 6.1).

We shall first discuss the polariton spectra, emphasizing their self-similar pattern. Then we present a quantitative analysis of the results, pointing out the distribution of the polariton bandwidths for high generations. This gives a good insight to their localization and their power laws, which are a guide to their *universality classes*. We proceed further to discuss the multifractal properties of the plasmon-polariton spectra using the so-called $f(\alpha)$ function, which describes the distribution of different fractal dimensions of the object upon variation of the singularities of strength α [5]. A brief discussion about more complex quasiperiodic structures of the *nipi* type is then presented. Finally, we provide an analysis for their interesting and distinctive thermodynamic properties.

6.1 Two-Component Quasiperiodic Structures

In this section we extend the method presented in Chapter 5, which was suitable for a periodic superlattice [6], to more complex layered structures, such as those exhibiting deterministic disorders [7–9], i.e. Fibonacci, Thue–Morse, Double-period, and Cantor. All we need is to determine the transfer matrices T associated with

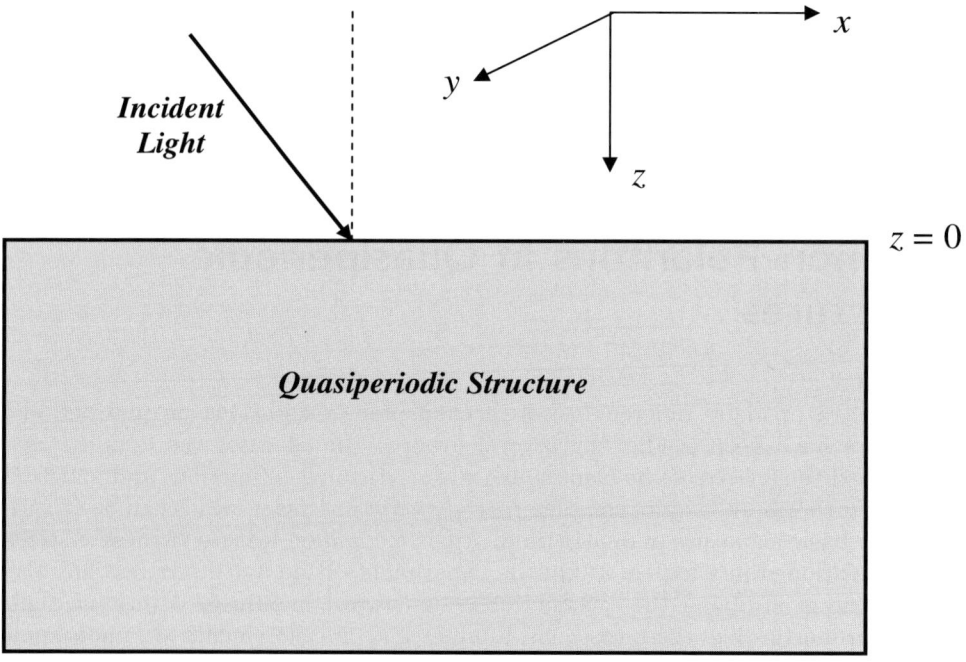

Fig. 6.1. Schematic representation for the plasmon-polariton propagation in quasiperiodic structures.

the building blocks A and B (depicted in Fig. 5.1) for each quasiperiodic structure, as was done for the periodic case and found in other applications in this subject [10–12]. After that one may again use Eqs. (5.27), (5.34), and (5.41) to find the bulk and surface polariton modes.

We begin with Fibonacci structures. As defined in Subsection 2.3.2, the Fibonacci sequence is characterized by the generations

$$S_0 = B;\quad S_1 = A;\quad S_2 = AB;\quad S_3 = ABA;\quad S_4 = ABAAB;\quad \text{etc.} \tag{6.1}$$

It is straightforward, starting with the simplest cases, to prove that their transfer matrices are [13]:

(a) for $S_0 = B$ and $S_1 = A$

$$T_{S_0} = N_B^{-1} M_B;\quad T_{S_1} = N_A^{-1} M_A; \tag{6.2}$$

(b) for $S_2 = AB$

$$T_{S_2} = N_A^{-1} M_B N_B^{-1} M_A; \tag{6.3}$$

(c) for any higher generation (with $n \geq 1$) by application of the rule

$$T_{S_{n+2}} = T_{S_n} T_{S_{n+1}}. \tag{6.4}$$

6.1. TWO-COMPONENT QUASIPERIODIC STRUCTURES

Therefore, from a knowledge of the transfer matrices T_{S_0}, T_{S_1}, and T_{S_2} we can proceed to determine the transfer matrix of any other Fibonacci generation.

Similarly, for the Thue–Morse sequence, whose generations are (see Subsection 2.3.3)

$$S_0 = A; \quad S_1 = AB; \quad S_2 = ABBA; \quad S_3 = ABBABAAB; \quad \text{etc.}, \tag{6.5}$$

the transfer matrices are found to be [13]:

(a) for $S_1 = AB$

$$T_{S_1} = N_A^{-1} M_B N_B^{-1} M_A = N_A^{-1} T_{\beta_1} T_{\alpha_1} N_A, \tag{6.6}$$

where

$$T_{\xi_1} = M_j N_j^{-1} \quad (\xi = \alpha, j = A \text{ or } \xi = \beta, j = B); \tag{6.7}$$

(b) for $S_2 = ABBA$

$$T_{S_2} = N_A^{-1} T_{\alpha_1} T_{\beta_1} T_{\beta_1} T_{\alpha_1} N_A = N_A^{-1} T_{\beta_2} T_{\alpha_2} N_A, \tag{6.8}$$

where

$$T_{\alpha_2} = T_{\beta_1} T_{\alpha_1}; \quad T_{\beta_2} = T_{\alpha_1} T_{\beta_1}; \tag{6.9}$$

(c) for any generation n ($n \geq 1$)

$$T_{S_n} = N_A^{-1} T_{\beta_n} T_{\alpha_k} N_A, \tag{6.10}$$

with

$$T_{\alpha_{n+1}} = T_{\beta_n} T_{\alpha_n}, \tag{6.11}$$

$$T_{\beta_{n+1}} = T_{\alpha_n} T_{\beta_n}. \tag{6.12}$$

As expected, a similar rule holds for the Double-period sequence (see Subsection 2.3.4), where the transfer matrix for the 2nd generation $S_2 = ABAA$ is given by [13]

$$T_{S_2} = N_A^{-1} M_A N_A^{-1} M_A N_A^{-1} M_B N_B^{-1} M_A \tag{6.13}$$

or

$$T_{S_2} = T_{S_0} T_{S_0} T_{S_1}. \tag{6.14}$$

For any higher generation (with $n \geq 1$)

$$T_{S_{n+2}} = T_{S_n} T_{S_n} T_{S_{n+1}}. \tag{6.15}$$

Finally, for completeness, in the case of the so-called Cantor superlattice (see Subsection 2.3.1) the transfer matrix for its nth generation is [14]

$$T_{S_n} = T_A T_n. \tag{6.16}$$

Here we define

$$T_A = N_A^{-1} M_A, \tag{6.17}$$

and

$$T_n = T_{n-1} T_{AB_n} T_{n-1} \quad (n \geq 2), \tag{6.18}$$

with

$$T_{AB_n} = N_A^{-1} M_{B_n} N_{B_n}^{-1} M_A \tag{6.19}$$

and

$$T_1 = T_{AB_1}. \tag{6.20}$$

6.1.1 Numerical Examples

With the transfer-matrix formalism established, we next present some numerical results for the spectra of the bulk and surface polaritons that can propagate in these quasiperiodic structures. We consider medium A to be GaAs, with a plasmon-type frequency-dependent dielectric function. For medium B we assume the physical parameters of SiO_2, whose frequency-independent dielectric function is $\epsilon_B = 12.3$. The other physical parameters used here are $\epsilon_{\infty A} = 12.9$, $m = 6.4 \times 10^{-32}$ kg, $a = b/2 = 40$ nm, $n_A = n_B = 6 \times 10^5$ m^{-2}, and $\omega_{pA} = 4.04$ THz. The damping term in the dielectric function is neglected, and we assume the external medium C to be vacuum ($\epsilon_C = 1$). It is convenient to plot the results in terms of a reduced frequency ω/Ω, where $\Omega = \left(\epsilon_{\infty A} n_A e^2 / m \epsilon_0 a\right)^{1/2}$, versus the dimensionless in-plane wavevector $k_x a$. For GaAs the value of Ω is approximately 23 THz.

The plasmon-polariton spectra for the quasiperiodic Fibonacci (2nd and 4th generations), Thue–Morse (3rd generation), Double-period (3rd generation), and Cantor (2nd generation) superlattices are presented in Figs. 6.2–6.6, respectively. In all cases the surface modes are represented by dashed lines, while the bulk bands are characterized by the shaded areas, which have boundaries corresponding to $QL = 0$ and $QL = \pi$, with Q being the Bloch wavevector and L the size of the unit cells of the quasiperiodic structures.

The plasmon-polariton spectrum for the 2nd Fibonacci generation is depicted in Fig. 6.2. Actually, since it corresponds to an $\cdots ABABAB \cdots$ structure, it is equivalent to the periodic case treated in Chapter 5. In addition, we are including retardation effects in the present numerical example. As we can see, it has two well-defined branches [13]. When $k_x a \ll 1$ the high-frequency bulk branch is in the range $0.18 < \omega/\Omega < 0.6$, while the low-frequency one lies in the region $\omega/\Omega < 0.17$ for the same $k_x a$. The high-frequency surface mode emerges from the bulk band at $\omega/\Omega = 0.6$, and then evolves separately from the bulk band. The low-frequency surface mode, on the other hand, starts at $\omega/\Omega = 0$ and then merges into the bulk

6.1. TWO-COMPONENT QUASIPERIODIC STRUCTURES

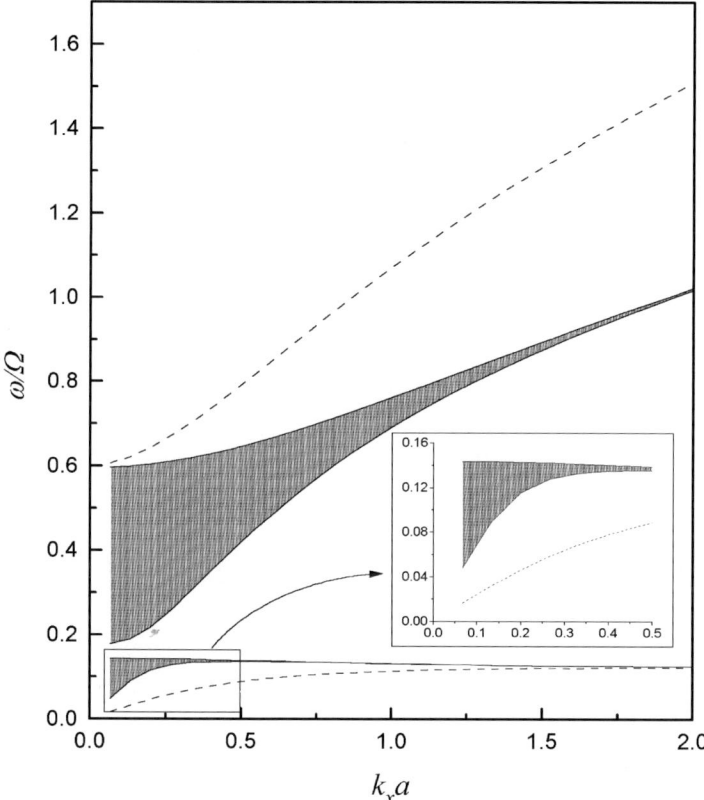

Fig. 6.2. Plasmon-polariton spectrum for the reduced frequency ω/Ω versus $k_x a$ for the 2nd generation of the Fibonacci quasiperiodic structure, which is equivalent to the periodic case. The bulk bands are shaded, while the surface modes are represented by the dashed lines (after Vasconcelos and Albuquerque [13]).

band at $\omega/\Omega \simeq 0.126$ for $k_x a \simeq 2.6$. The inset shows that the spectrum has no self-similarity pattern (in contrast to the cases to be shown later).

In Fig. 6.3 the plasmon-polariton spectrum for the 4th Fibonacci generation is presented. We observe that the number of bulk bands is equal to the corresponding Fibonacci number F_4 (namely, 5), and indeed the number of bulk bands is always equal to the Fibonacci number of the corresponding generation. Two of the surface modes are seen to have a behavior similar to those found in the periodic case: a high-frequency mode, which starts at the top of the bulk band and then propagates apart from it for higher $k_x a$, and a low-frequency mode, which starts (for small $k_x a$) split off below the bulk band, merging with it again at $k_x a \simeq 2.6$. It is interesting that the latter property holds for *all* quasiperiodic structures studied here. The inset provides a qualitative indication of the fractal aspect of the spectrum, in the sense that the inset has the same qualitative features as the main figure.

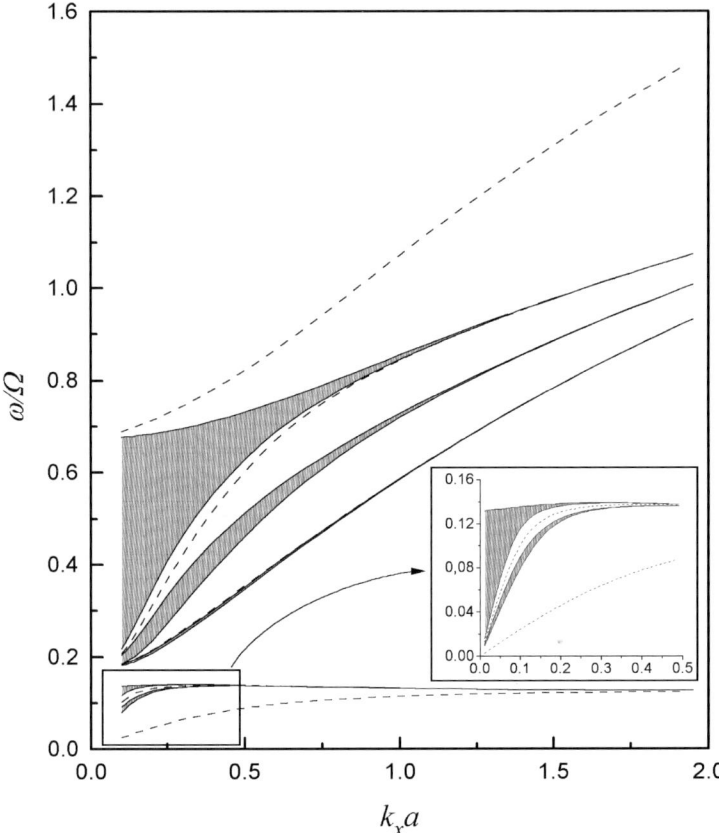

Fig. 6.3. Same as Fig. 6.2, but for the 4th generation of the Fibonacci quasiperiodic structure (after Vasconcelos and Albuquerque [13]).

The Thue–Morse 3rd generation is shown in Fig. 6.4. Here, as in the previous case, we have two well-defined regions for the plasmon-polariton spectrum. The number of bulk bands now increases as $2^{n-1} + 1$, with n being the Thue–Morse generation number. The surface modes lie between the bulk bands. The qualitative self-similarity aspect of the spectrum in this case is also apparent on comparing the inset with the main figure.

The plasmon-polariton spectrum for the 3rd generation of the Double-period structure is shown in Fig. 6.5. In some respects it is similar to the Thue–Morse case, but there is an important difference to notice. The number of bulk bands in the high-frequency region of the spectrum for each generation is equal to the number of A building blocks of the corresponding generation, while the number of bulk bands in the low-frequency region is equal to the number of B building blocks of the same generation. Altogether, however, the number of bulk bands increases as 2^n for this quasiperiodic sequence.

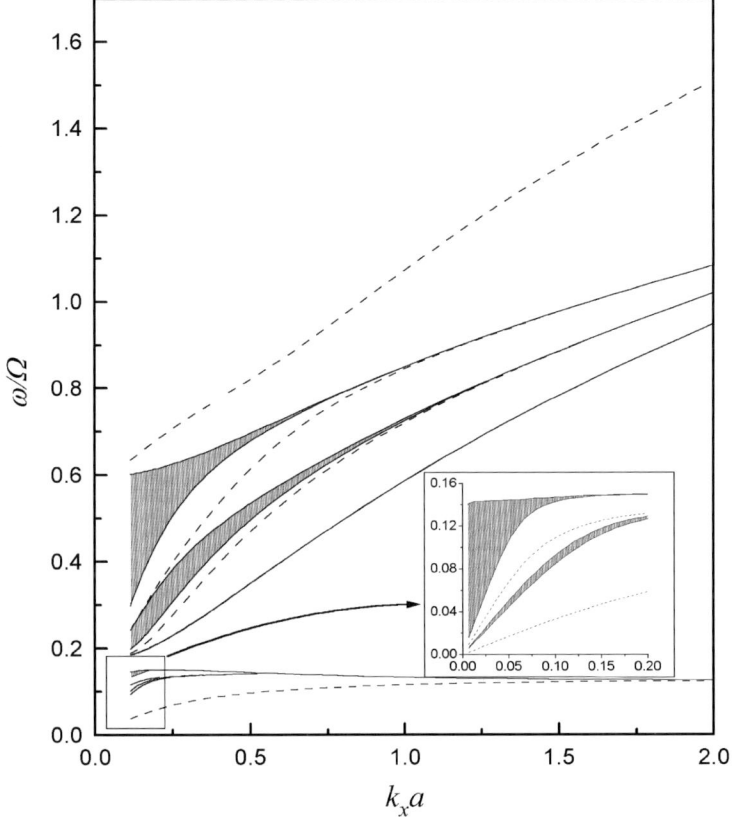

Fig. 6.4. Same as Fig. 6.2, but for the 3rd generation of the Thue–Morse quasiperiodic structure (after Vasconcelos and Albuquerque [13]).

Finally, for completeness, we show in Fig. 6.6 the plasmon-polariton spectrum for the 2nd generation of the Cantor structure. Of course, as commented in Chapter 2, it is not included in the substitutional classes defined by the other structures. The number of bulk bands satisfies the following numerical sequence [14]:

$$S_n = \begin{cases} 7n/2 & \text{for } n \text{ even,} \\ (7n-1)/2 & \text{for } n \text{ odd.} \end{cases} \tag{6.21}$$

It again has a qualitative self-similarity aspect, which is clear from the inset.

6.2 Localization and Scaling Properties

One of the most fascinating aspects of excitations in quasiperiodic structures concerns their localization and connection with fractal behavior. Thus, we now proceed with an analysis of the confinement effects arising from competition between

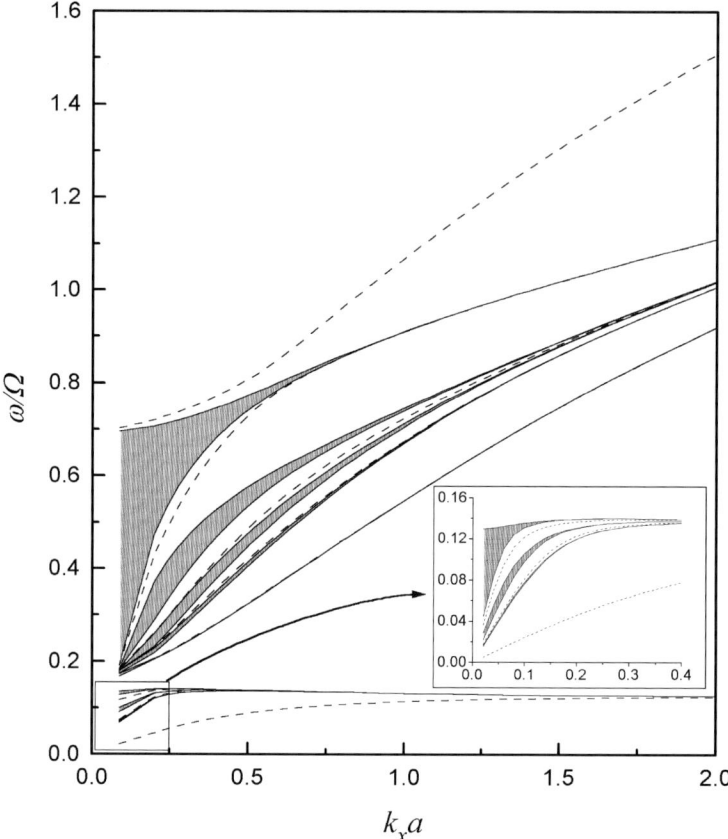

Fig. 6.5. Same as Fig. 6.2, but for the 3rd generation of the Double-period quasiperiodic structure (after Vasconcelos and Albuquerque [13]).

the long-range aperiodic order, which is induced by the quasiperiodic structure, and the short-range disorder [15,16]. To this end, a quantitative analysis will be made of the localization and magnitude of the allowed bandwidth in the plasmon-polariton spectra, which were described in the previous section for quasiperiodic structures. Also, we shall discuss the related scaling behavior as a function of the number of generations of the sequences.

Taking the Fibonacci case, data for the distribution of the bandwidths are shown in Fig. 6.7a for the dimensionless in-plane wavevector $k_x a = 0.25$. One can deduce the forbidden and allowed energy bands as a function of the generation number n up to the 10th generation of the Fibonacci sequence, which means a unit cell with 55 A and 34 B building blocks. We note that, as expected, the allowed band regions exhibit a highly fragmented energy spectrum for large n, as an indication of greater localization of the modes [17,18]. In fact, the total width Δ of the allowed energy regions (which is known as the Lebesgue measure

6.2. LOCALIZATION AND SCALING PROPERTIES

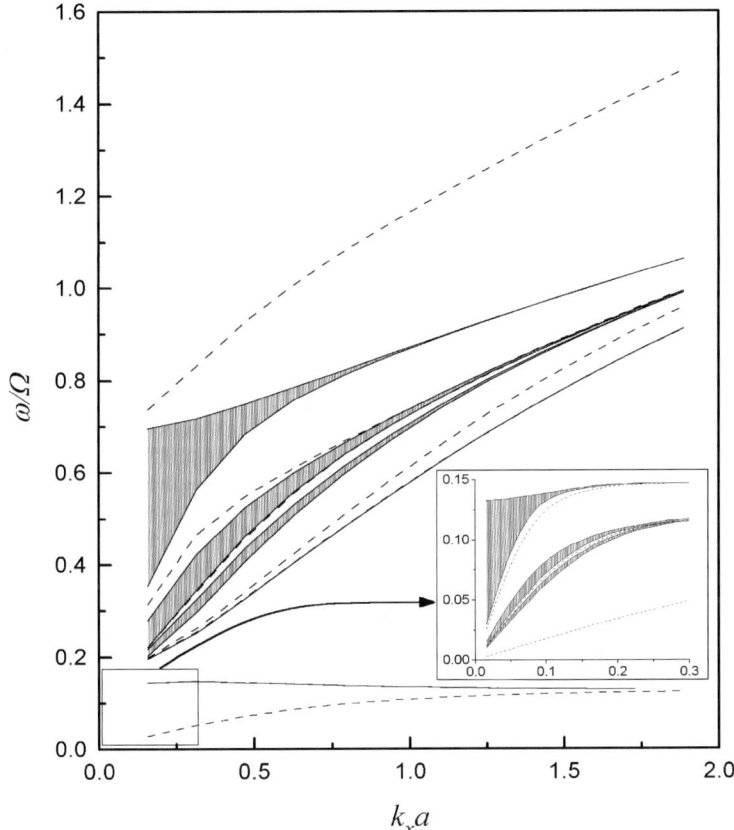

Fig. 6.6. Same as Fig. 6.2, but for the 2nd generation of the Cantor quasiperiodic structure (after Vasconcelos and Albuquerque [13]).

of the energy spectrum) decreases with n as the power law $\Delta \sim F_n^{-\delta}$. Here F_n is the Fibonacci number and the exponent δ (the so-called *diffusion constant* of the spectra) is a function of the common in-plane wavevector $k_x a$. This exponent can be considered as indicating the degree of localization of the excitation [19–21]. In Fig. 6.7b we show a log–log plot of these power laws for three different values of $k_x a$.

In a similar way, Fig. 6.8a shows the forbidden and allowed region of propagation for the plasmon-polaritons in the Thue–Morse quasiperiodic system as a function of its generation number. We again go to the 10th generation of the sequence, which in this case means a unit cell with 2^{10} A and B building blocks. The total allowed bandwidth scales as the power law $\Delta \sim (2^n)^{-\delta}$, where now δ is independent of the wavevector $k_x a$. Indeed, in Fig. 6.8b we see a log–log plot of the width Δ of the allowed regions against 2^n for three different values of $k_x a$, with almost the same value of δ, the small difference probably being due to numerical uncertainties.

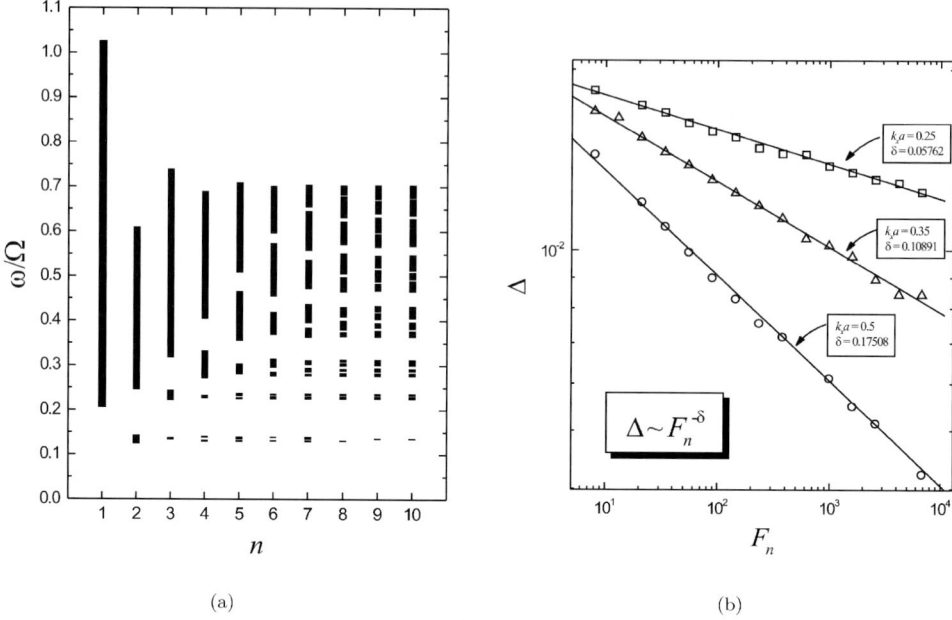

Fig. 6.7. Localization and scaling properties of the plasmon-polaritons in Fibonacci structures: (a) distribution of bandwidths as a function of the generation number n; (b) log–log plot of the total width Δ of the allowed regions against the Fibonacci number (after Vasconcelos and Albuquerque [13]).

Finally, the forbidden and allowed regions of plasmon-polaritons propagating in quasiperiodic Double-period sequences, as a function of the generation number n, are shown in Fig. 6.9a. The scaling behavior of the allowed bandwidth is $\Delta \sim (2^n)^{-\delta}$, where here, as with Fibonacci structures, it is the case that the exponent δ depends on the wavevector $k_x a$, as can be seen in Fig. 6.9b. We surmise that this may be a feature of sequences for which the numbers of A and B blocks are different.

6.3 Multifractal Analysis

We now go beyond the simple scaling results of the previous section to investigate the multifractal properties for the spectra of plasmon-polaritons in quasiperiodic multilayers. Multifractal analysis has proved to be a convenient statistical description for the study of the long-term dynamical behavior of a physical system. A multifractal behavior is, in general, a common property of strange attractors in non-linear systems [22,23]. Thus, before covering applications to quasiperiodic systems, it is useful to mention some general concepts (the details are to be found in textbooks on non-linear dynamics, e.g. see [24]). Loosely speaking, an attractor is a set to which all neighboring trajectories converge, and a strange attractor is

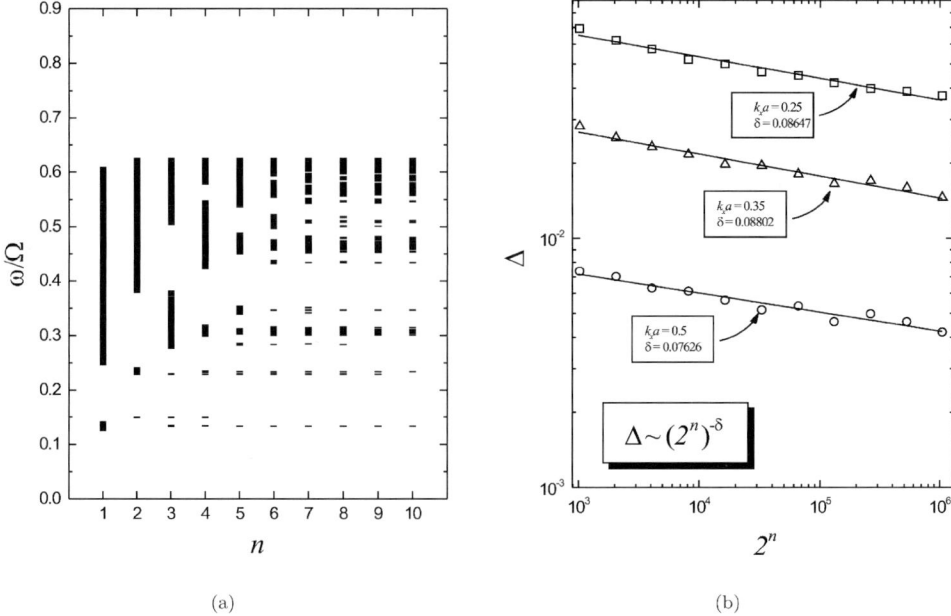

Fig. 6.8. Localization and scaling properties of the plasmon-polaritons in Thue–Morse structures: (a) distribution of bandwidths as a function of the generation number n; (b) log–log plot of the total width Δ of the allowed regions against 2^n (after Vasconcelos and Albuquerque [13]).

one that exhibits a sensitive dependence on the initial conditions. For the simple Cantor sequence there is a uniform scaling and the fractal dimension is a constant, as noted in Subsection 2.3.1. However, in general, this will not be the case and in multifractals the dimension will vary across the attractor. Multifractals differ from monofractals in the sense that they need an infinite set of exponents (rather than a single exponent that suffices for a fractal) to characterize their spectra. In particular, they arise when multiplicative processes are involved in defining the sequence [25,26], and this is the case for most of the sequences that are of interest to us here.

In order to characterize these objects, it is convenient to introduce the function $f(\alpha)$, which is known as the *multifractal spectrum* or the *spectrum of scaling indices*. Loosely, one may think of the multifractal as an interwoven set of fractals of different dimensions α, where $f(\alpha)$ is a measure of their relative strength [27]. The formalism relies on the fact that highly non-uniform probability distributions arise from the non-uniformity of the system. In spite of quasiperiodic systems not being classifiable in the non-linear physics context, they do exhibit multifractality in their spectra. This feature was firstly studied by Kohmoto and collaborators for electrons in a 1D quasiperiodic discontinuous potential [28]. In the case of the Cantor sequence, $f(\alpha)$ would consist of a delta function "spike" at the value

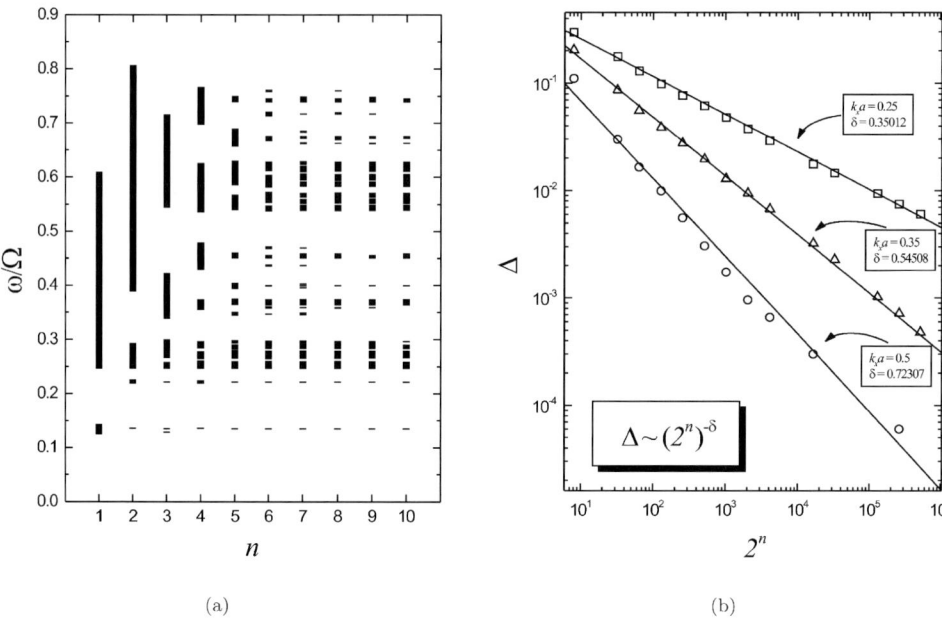

Fig. 6.9. Same as Fig. 6.8, but for Double-period structures (after Vasconcelos and Albuquerque [13]).

$\ln 2 / \ln 3$ corresponding to its fractal dimension. By contrast, we shall show examples later where $f(\alpha)$ is a continuous smooth distribution.

Fractal and multifractal objects have been identified in a multitude of physical situations, ranging from problems of aggregation to the behavior of chaotic dynamical systems [29,30]. The multifractal sets can be characterized on the basis of the generalized dimensions D_q (which is "generalized" in the sense that it is defined in terms of a variable q) and the associated spectrum of singularities given by $f(\alpha)$. In fact, they can be completely described *either* by an infinite number of generalized dimensions D_q *or* by the $f(\alpha)$ function [31,32]. The definition of D_q as a function of q is provided by the expression [33]

$$D_q = (q-1)^{-1} \lim_{N' \to \infty} \left[-\ln \sum_i p_i^q / \ln N' \right] \quad (q \neq 1), \tag{6.22}$$

while for $q = 1$ we have

$$D_1 = \lim_{N' \to \infty} \left[-\sum_i p_i \ln p_i / \ln N' \right]. \tag{6.23}$$

Here $p_i = \int_{\text{box}} d\mu$, with μ being the probability measure of the multifractal set, and $i = 1, 2, \ldots, N'$ (where N' is the number of boxes). Also i is the index of a box

6.3. MULTIFRACTAL ANALYSIS

that belongs to a grid that covers the set and has a linear size given by $\epsilon = 1/N'$. In this formalism the scaling exponent α is defined by [33]

$$\alpha(x) = \lim_{N' \to \infty} \left[-\ln p(x) / \ln N' \right], \tag{6.24}$$

where $p(x)$ is the integral of $d\mu$ over a box centered at x. The $f(\alpha)$ function is then defined by the relation

$$N'(\alpha, \epsilon) \sim \epsilon^{-f(\alpha)}, \tag{6.25}$$

for $\epsilon \to 0$. Here the number of boxes (specified by ϵ) with α between α and $\alpha + \Delta \alpha$ is defined to be $N'(\alpha, \epsilon) \Delta \alpha$.

To evaluate the multifractal spectra, one should first obtain the measures from either real or computer experiments and then use a Legendre transform. The difficulty with this method lies with the Legendre transform itself, depending on the system being considered, and on eventual discontinuities that might arise on the $f(\alpha)$ curves [34]. A new approach for this problem was due to Chhabra and collaborators [35,36], who introduced one of the most efficient algorithms to calculate the $f(\alpha)$ function. It allows the $f(\alpha)$ function to be found with excellent numerical precision, and it is the method that we shall follow here. Let us first define a measure ζ_i by normalizing the local plasmon-polariton bandwidths Δ_i, i.e.

$$\zeta_i = \Delta_i \Big/ \sum_i \Delta_i, \tag{6.26}$$

where the summation is over all bands. Then we construct a parametrized family of normalized measures defined by

$$\mu_i = \zeta_i^q \Big/ \sum_i \zeta_i^q. \tag{6.27}$$

This represents a generalization (to $q \neq 1$) of the original measure ζ_i. The spectrum $f(\alpha)$ is then obtained by varying the parameter q and calculating

$$f(\alpha_q) = \lim_{N' \to \infty} \left[-\sum_i \mu_i \ln \mu_i / \ln N' \right], \tag{6.28}$$

$$\alpha_q = \lim_{N' \to \infty} \left[-\sum_i \mu_i \ln \zeta_i / \ln N' \right]. \tag{6.29}$$

Fig. 6.10a shows the $f(\alpha)$ functions for the 10th generation of the Fibonacci sequence. Here we have considered three different values of the dimensionless in-plane wavevector $k_x a$, and we notice that the spectra are independent of the $k_x a$ values. For comparison, Fig. 6.10b shows the $f(\alpha)$ functions for the Thue–Morse structure. We consider here its 8th generation, whose unit cell is composed of 128 A and 128 B building layers. For this structure we can infer small variations of the spectra for different values of $k_x a$, although the $f(\alpha)$ spectra widths are

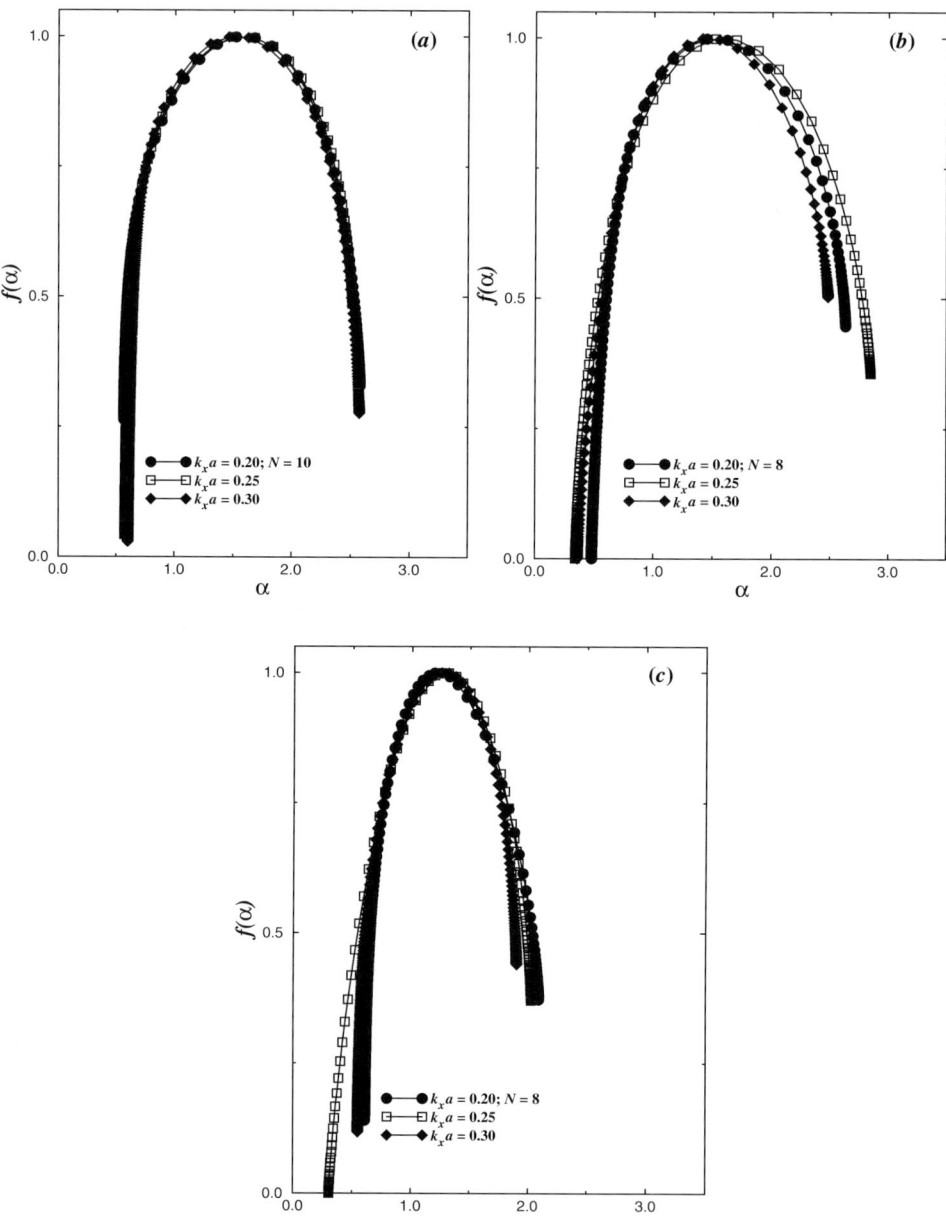

Fig. 6.10. $f(\alpha)$ functions of the plasmon-polariton bandwidths for the quasiperiodic structures: (a) Fibonacci; (b) Thue–Morse; (c) Double-period. The generation number N and the values of the in-plane dimensionless wavevector $k_x a$ are indicated (after Vasconcelos et al. [33]).

almost the same. Next the $f(\alpha)$ functions for the 8th Double-period generation are shown in Fig. 6.10c, where the unit cell now has 161 A and 85 B building layers. As in the Thue–Morse case, we notice only small variations of the spectra for different values of $k_x a$.

For all these cases the extremes α_{min} and α_{max} of the abscissa of the $f(\alpha)$ curves represent the minimum and maximum of the singularity exponent α, which acts as an appropriate weight in the reciprocal space. In fact, $\alpha_{min} = \lim_{N' \to +\infty} D_q$ and $\alpha_{max} = \lim_{N' \to -\infty} D_q$ characterize the scaling properties of the most concentrated and most rarified region of the intensity measure, respectively. The value of $\Delta \alpha \equiv \alpha_{max} - \alpha_{min}$ may be used as a parameter reflecting the randomness of the intensity measure. Also, using general scaling arguments at the edge of chaos within the generalized statistics model [37], it can be proved that $1/(1-q) = 1/\alpha_{min} - 1/\alpha_{max}$, where q is an entropic index [38].

The above multifractal analysis revealed a smooth $f(\alpha)$ function distributed in a finite range $[\alpha_{min}, \alpha_{max}]$ for all the quasiperiodic structures, with a summit at $f(\alpha_0) = 1$ for some value α_0 of α. These investigations clearly demonstrate that all the spectra correspond to highly non-uniform intensity distributions, and therefore they possess the scaling properties of a multifractal.

6.4 Quasiperiodic *nipi* Structures

The plasmon-polariton spectrum of the *nipi* periodic superlattice was studied in Section 5.3. Our purpose here is to extend this calculation to quasiperiodic structures, taking into account two main goals: first we want to determine their fractal-profile spectra, and then their localized modes, as a power law for their allowed bandwidth distribution [39].

To form a *nipi* quasiperiodic superlattice, we consider the four different building blocks as described in Chapter 5, namely A (*n*-type semiconductor), B (insulator), C (*p*-type semiconductor), and D (insulator). The structure can be generated (or "grown") by means of the transformations $A \to AC$, $B \to DC$, $C \to AB$, and $D \to DA$. The building blocks A and C are formed by a two-dimensional electron gas (2DEG) and a two-dimensional holes gas (2DHG), respectively, both supported by layers of doped semiconductor materials, whose thicknesses are a and c, respectively (as described in Section 5.3). The building blocks B and D are insulators, whose thicknesses are b and d. The total number of building blocks in the unit cell increases with 2^{n+1}. The generations of this quasiperiodic structure are

$$S_1 = [ABCD], \qquad S_2 = [ACDCABDA],$$

$$S_3 = [ACABDAABACDCDAAC], \quad \text{etc.} \qquad (6.30)$$

To study the plasmon-polariton modes, we consider an infinite periodic repetition of the unit cell for the nth quasiperiodic *nipi* generation. The Cartesian axes are chosen so that the z-axis is the normal direction to the xy-plane of the layers. We also assume that the propagation of the electromagnetic wave is p-polarized.

The transfer matrices for the *nipi* quasiperiodic superlattice can be formed and defined by using the method of induction. The results are:

(a) for the first generation $S_1 = [ABCD]$, which is the periodic *nipi* structure:

$$T_{S_1} = N_A^{-1} T_{D_1} T_{C_1} T_{B_1} T_{A_1} N_A, \qquad (6.31)$$

where

$$T_{A_1} = M_A N_A^{-1},$$
$$T_{B_1} = M_B N_B^{-1},$$
$$T_{C_1} = M_C N_C^{-1},$$
$$T_{D_1} = M_D N_D^{-1}. \qquad (6.32)$$

(b) for the second generation $S_2 = [ACDCABDA]$:

$$T_{S_2} = N_A^{-1} T_{D_2} T_{C_2} T_{B_2} T_{A_2} N_A, \qquad (6.33)$$

with

$$T_{A_2} = T_{C_1} T_{A_1},$$
$$T_{B_2} = T_{C_1} T_{D_1},$$
$$T_{C_2} = T_{B_1} T_{A_1},$$
$$T_{D_2} = T_{A_1} T_{D_1}. \qquad (6.34)$$

(c) for the nth generation ($n > 2$):

$$T_{S_n} = N_A^{-1} T_{D_n} T_{C_n} T_{B_n} T_{A_n} N_A, \qquad (6.35)$$

where

$$T_{A_n} = T_{C_{n-1}} T_{A_{n-1}},$$
$$T_{B_n} = T_{C_{n-1}} T_{D_{n-1}},$$
$$T_{C_n} = T_{B_{n-1}} T_{A_{n-1}},$$
$$T_{D_n} = T_{A_{n-1}} T_{D_{n-1}}. \qquad (6.36)$$

The basic matrices M_j and N_j, that enter into the above expressions, are defined in Eqs. (5.19) and (5.63).

6.4. QUASIPERIODIC NIPI STRUCTURES

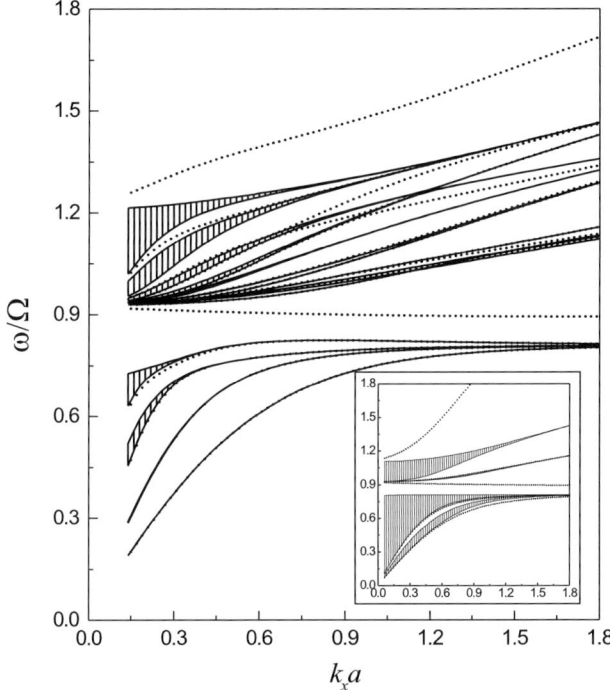

Fig. 6.11. Plasmon-polariton spectra for the reduced frequency ω/Ω vs $k_x a$ for the third generation of the *nipi* quasiperiodic superlattice. The shaded areas represent the bulk bands, while the surface modes are here represented by dashed lines. The inset shows, for comparison, the plasmon-polariton spectra of the periodic *nipi* superlattice (after Vasconcelos and Albuquerque [39]).

Next we present some numerical calculations to illustrate the above theory. Let us consider, as in the previous chapter, that the media A and C are n- and p-doped Si, respectively, with plasmon frequency-dependent dielectric functions whose physical parameters are $\epsilon_\infty = 11.7$ and $\omega_p \equiv \omega_{pA} = \omega_{pC} = 76.5\,\text{THz}$ (the plasma frequency). We assume that the surface carrier concentration of the electrons (equal to that of the holes) at the interfaces is $2 \times 10^{16}\,\text{m}^{-2}$. The dielectric materials B and D are both considered to be SiO_2, whose frequency-independent dielectric constant is $\epsilon_B = \epsilon_D = 12.3$. The thicknesses of the layers are the same, that is, we take $a = b = c = d = 40\,\text{nm}$. For the numerical results, instead of using the frequency ω, we replace it, as before, by the reduced frequency ω/Ω, where Ω is the characteristic frequency already defined in Section 6.1.

Fig. 6.11 shows the plasmon-polariton spectra plotted against the common in-plane dimensionless wavevector $k_x a$ for the 3rd generation of the *nipi* quasiperiodic structure. For comparison, we have shown in the inset the much simpler plasmon-polariton spectra for the periodic *nipi* superlattice with the same choice of parameters. In both cases, the bulk modes occur as bands with edges corresponding to the lines $QL=0$ and $QL=\pi$, in such a way that they alternate from one band to another, following the sequence $0, \pi, \pi, 0, 0, \pi, \ldots$. Also, for relatively small $k_x a$ the bulk bands are wide and are shown shaded, while for large $k_x a$ they are narrow. The surface modes lie between the bulk bands, and are shown as dotted lines. There are, as expected, two well-defined regions in the spectra. However, the 3rd generation *nipi* quasiperiodic structure presents a much richer spectrum: in the high-frequency region there are 11 bulk bands and 11 surface modes distributed in the interval $0.9 \leq \omega/\Omega \leq 1.8$, while in the low-frequency branch four bulk bands and four surface modes lie in the region $\omega/\Omega \leq 0.8$. Altogether, the number of bulk bands grows like $15 \times 3^{n-3}$ for $n > 2$. Of course, as the generation number increases, so does the number of allowed bands, leading to a *fractal* spectra profile that is absent in the periodic case. In this regard, it is similar to the behavior found in other quasiperiodic structures generalized from *two-component* periodic superlattices, as described in Ref. [13] for *non-doped* semiconductor building blocks. The novelty here comes because we have obtained the plasmon-polariton spectra in artificial quasiperiodic structures derived from *four-component* superlattices with *doped* semiconductor building blocks. Furthermore, doped superlattices display quite a number of novel physical properties, such as tunability of the carrier concentration, of the band gap, and of the subband separation. These special features are a consequence of the different origin of the periodic superlattice potentials, due to the presence of the 2DEG and the 2DHG at the interfaces. These are formed due to a mobility enhancement behavior attributed to the spatial separation between electrons/holes and their parent donor impurities: the n- and p-doped semiconductor building layers. Despite all of this, the quasiperiodic aspect of the structure is sufficiently robust to keep the major profile of the spectra unchanged, i.e. they display an intriguing *complex fractal* behavior! We have not consider any coupling effects between the electron and hole gases, since the plasmon-polaritons in the structure are adequately described by propagating Bloch states rather than in terms of hopping between the dynamical 2D states associated with the subband spectrum.

To investigate the effects on the confinement due to competition between the long-range quasiperiodic order and the short-range disorder, we have performed a bandwidth scaling study. To this end we deduce, for a given value of the dimensionless wavevector $k_x a$, the bandwidth distribution of the plasmon-polariton spectra, as shown in Fig. 6.12 for $k_x a = 0.3$. From this figure, one can infer the forbidden and allowed energies as a function of the generation number n up to the 5th generation of the quasiperiodic *nipi* sequence, which means 2^6 building blocks. Notice that, as expected, for large n the allowed band regions get narrower and narrower, as an indication of the modes being more localized. In fact, the total width Δ of the allowed energy regions decreases as the power law $\Delta \sim (2^{n+1})^{-\delta}$, where the exponent δ is the diffusion constant of the spectra. In Fig. 6.13 we show a log–log plot of this power law for four different values of $k_x a$, namely 0.2, 0.3, 0.4, and 0.5. We observe that there is only a slight dependence of δ on the in-plane dimensionless wavevector $k_x a$. According to earlier work [13], such behavior is

6.4. QUASIPERIODIC NIPI STRUCTURES

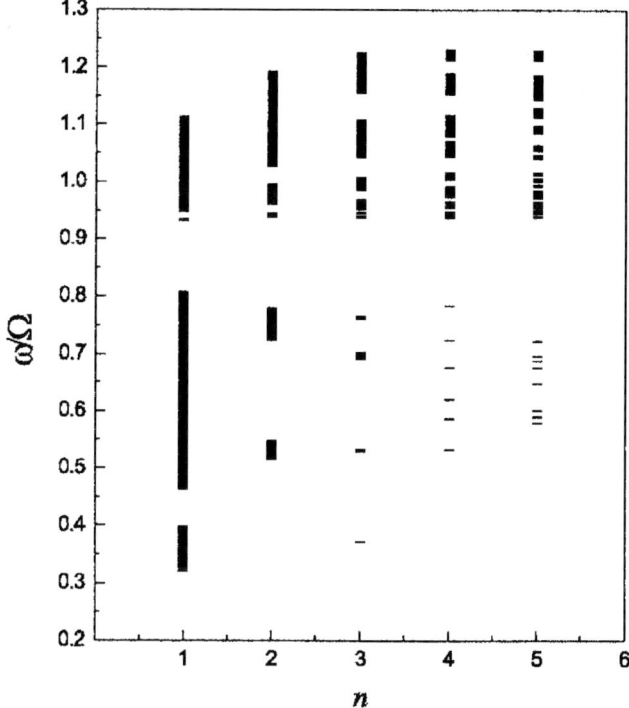

Fig. 6.12. Allowed bandwidth distribution for plasmon-polaritons as a function of the number of generation n, for $k_x a = 0.3$ (after Vasconcelos and Albuquerque [39]).

characteristic of a fractal-like energy spectrum, becoming a Cantor set in the limit $n \to \infty$.

Quasiperiodic systems are essentially a new class of materials in the sense that they have introduced ideas about condensed matter structures with topological long-range order that are distinct from the usual periodicity. Their discovery and understanding has involved strong motivation from both basic science and technical applications. The main purpose of this section has been to show how rich are the spectra and the localization properties of plasmon-polaritons in those quasiperiodic structures generalized from the doped four-component *nipi* periodic superlattices. This is emphasized in Fig. 6.11 where, for comparison, the periodic case is depicted in the inset. The fractal-like behavior of the spectra for high generations is perhaps the clearest signature of the excitations in these systems. We have also shown that the defining rules of the quasiperiodic *nipi* sequence (imposing long-range correlations on the plasmon-polariton propagation) were responsible for the Cantor-like bandwidth structure profile depicted in Fig. 6.12. Furthermore, the spectra have self-similar properties, as exemplified by the power law that governs the scaling of the bandwidth (see Fig. 6.13).

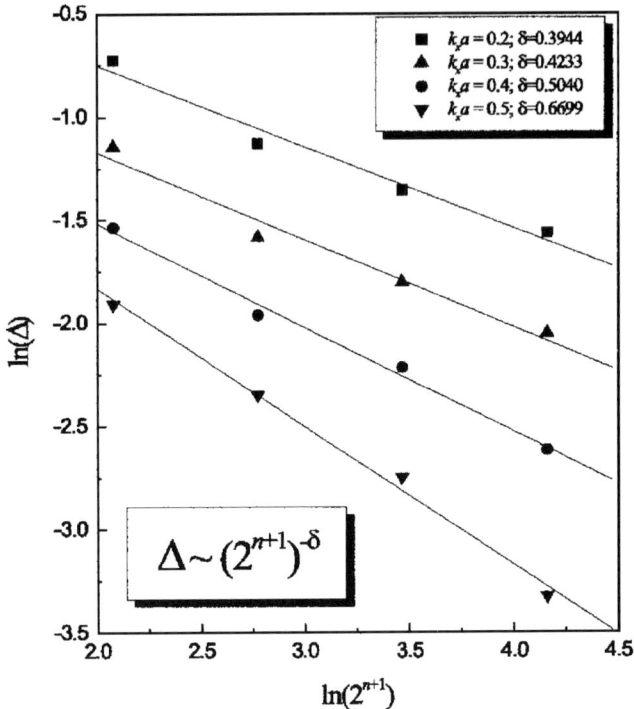

Fig. 6.13. Log–log plot for the total allowed bandwidth (the Lebesgue measure of the energy spectrum) Δ versus 2^{n+1} for four values of $k_x a$, namely 0.2, 0.3, 0.4, and 0.5 (after Vasconcelos and Albuquerque [39]).

The most important experimental techniques used to probe plasmon-polariton spectra are Raman light scattering and attenuated total reflection (ATR). In the case of Raman scattering, one uses a grating spectrometer to detect and analyze the scattered light. The typical shift of the frequency of the scattered light is in the range 0.6–500 meV, which makes this technique very appropriate for probing polariton spectra. On the other hand, ATR spectroscopy is much easier to implement than Raman spectroscopy, but typically it gives less precise results. However, it has been employed with success in a number of experiments (see e.g. Refs. [40,41] for reviews). We shall discuss these experimental methods (along with others) in Chapter 11.

6.5 Thermodynamic Properties

In an interesting development, Tsallis et al. [42] have reported calculations for a mathematical fractal set, namely the geometrical *triadic Cantor set* mentioned in Subsection 2.3.1. They argued that if the energy spectrum of the excitations in a quasicrystal is a Cantor set of zero measure, it can be approximated as the

6.5. THERMODYNAMIC PROPERTIES

fractal object obtained from the Cantor triadic set. Based on this approximation and using the classical Maxwell–Boltzmann statistics, they found an interesting behavior for the specific heat at low temperature: it oscillates around the fractal (or spectral) dimension $\ln 2/\ln 3 \simeq 0.63$ of the triadic Cantor set. In addition, a non-uniform convergence between the so-called banded and discrete models was observed. The results of Tsallis et al. were then extended to the *two-scale* Cantor set [43,44] and for *phonon states* [45,46]. For these more general cases the specific heat was found to exhibit log-periodic oscillations around the fractal dimension, since it was considered that the corresponding spectral staircase satisfies a simple scaling law, whose general solution can be written as a power law times a log-periodic function [47]. These oscillatory features were also studied by other groups with similar conclusions [48,49]. Later Carpena et al. [50], using the properties of *multifractal* spectra, showed under what conditions the oscillatory regime would disappear. Similar results were obtained by Kimball and Frisch [51] for the distribution of normal mode frequencies of fractal-based models. Finally, using a logistic map as an example of a fractal system [52], Curado and Rego–Monteiro [53] examined the thermodynamic properties of the model in the chaotic region of the map.

In this section we extend the model described by Tsallis and collaborators with the aim of studying the thermodynamic properties, in particular the specific heat $C(T)$, of *real* excitation modes (such as polaritons) in quasiperiodic structures. We take into account two different aspects of this problem: (i) instead of a geometrical Cantor set, we consider *multifractal* energy spectra of the *real* excitations; and (ii) we look for connections with the quasiperiodic aspects of these spectra (scaling laws, fractal dimension, etc.) to see if there is some kind of common behavior in the specific heat spectra. We show that when $T \to 0$ the specific heat displays oscillations and when $T \to \infty$ the specific heat goes to zero as T^{-2}, because the energy spectrum under consideration is bounded. Throughout the section we use the classical Maxwell–Boltzmann statistics.

6.5.1 Theoretical Model

Our starting point in developing a theoretical model is to consider the energy spectrum for the general schematic scaled continuous multifractal set depicted in Fig. 6.14. We see that

$$\Delta_1 = \epsilon_2 - \epsilon_1 \implies \epsilon_2 = \epsilon_1 + \Delta_1, \tag{6.37}$$

$$\Delta_2 = \epsilon_4 - \epsilon_3 \implies \epsilon_4 = \epsilon_3 + \Delta_2, \tag{6.38}$$

$$\vdots$$

$$\Delta_i = \epsilon_{2i} - \epsilon_{2i-1} \implies \epsilon_{2i} = \epsilon_{2i-1} + \Delta_i. \tag{6.39}$$

Thus, in this hypothetical representation of the energy spectrum, the case of generation number $n=1$ corresponds to a continuum spectrum going from ϵ_1 to

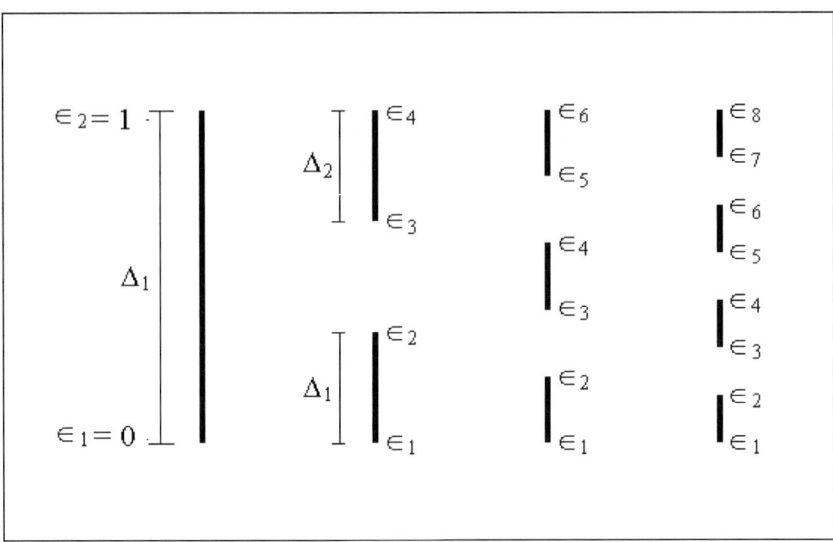

Fig. 6.14. Assumed energy spectrum, showing the first four steps in the construction of a general schematic fractal set.

$(\epsilon_1 + \Delta_1)$. Also $n = 2$ corresponds to a spectrum whose first branch goes from ϵ_1 to $(\epsilon_1 + \Delta_1)$ and the second one goes from ϵ_3 to $(\epsilon_3 + \Delta_2)$, and so on, for increasing n. We take the level density inside each band to be constant, and to be the same for all bands in a given hierarchy. A fractal or multifractal energy spectrum emerges at the $n \to \infty$ limit. This model is quite general and can be applied to *any* real energy excitation spectra in quasiperiodic structures. Obviously the number of bands depends on the sequence chosen. In what follows, we consider a normalization in the frequency spectrum, in such a way that the bands stay within the limits 0 and 1 (in units of $\hbar\omega$).

The partition function is given by

$$Z_N = \int_0^\infty \rho(\epsilon) \exp(-\beta \epsilon) \, d\epsilon, \qquad (6.40)$$

where β is $1/T$ (in units such that Boltzmann's constant k_B is unity) and we take a constant density of states $\rho(\epsilon) = 1$. After some calculations, it is easy to rewrite Eq. (6.40) in terms of a summation as [54]

$$Z_N = \frac{1}{\beta} \sum_{i=1,3,\ldots}^{2N-1} \exp(-\beta \epsilon_i) \left[1 - e^{-\beta \Delta_i} \right], \qquad (6.41)$$

where N is the number of bands of the multifractal spectrum. For example, in the case of the Fibonacci quasiperiodic system, N is just the Fibonacci number F_n. Once we know the partition function, it is possible to calculate the specific heat

6.5. THERMODYNAMIC PROPERTIES

using the standard result that

$$C_N = \beta^2 \partial^2 (\ln Z_N)/\partial \beta^2. \tag{6.42}$$

On taking the natural logarithm of Eq. (6.41), differentiating twice with respect to β, and substituting the result into Eq. (6.42), we obtain

$$C_N = 1 + (\beta f_N/Z_N) - (g_N^2/Z_N^2), \tag{6.43}$$

where

$$f_N = \sum_{i=1,3,\ldots}^{2N-1} \left[\epsilon_i^2 \exp(-\beta\epsilon_i) - \epsilon_{i+1}^2 \exp(-\beta\epsilon_{i+1}) \right] \tag{6.44}$$

and

$$g_N = \sum_{i=1,3,\ldots}^{2N-1} \left[\epsilon_i \exp(-\beta\epsilon_i) - \epsilon_{i+1} \exp(-\beta\epsilon_{i+1}) \right], \tag{6.45}$$

with the summation here being over odd integer values of i. Note that it is necessary only to know the distribution of the energy spectrum of a given multifractal system to calculate Eqs. (6.44) and (6.45), so we can determine the specific heat by using Eq. (6.43).

We will now apply this method to determine the specific heat spectra of the plasmon-polaritons, whose multifractal energy spectra were already determined in Section 6.2.

6.5.2 Specific Heat Profiles

Using the multifractal energy profile depicted in Fig. 6.7a, we show in Fig. 6.15 the corresponding polariton specific heat calculated as a function of the temperature, for several generations of the Fibonacci sequence [54]. For high temperatures ($T \to \infty$), the specific heat for all generation numbers converges and decays as T^{-2}, which is consistent with the triadic case. This asymptotic behavior is mainly due to the fact that we have considered our system bounded. It can be seen that as n increases, the plotted specific heat curves have a negative slope for a larger and larger temperature range. More importantly, the inset of this figure shows the oscillatory behavior of the specific heat for low temperatures. These oscillations are *neither* around the fractal dimensionality of the spectrum *nor* an approximation of the idealized oscillations found in the triadic Cantor set.

For these Fibonacci structures, there are clearly two classes of oscillations, one for the *even* and the other for the *odd* generation numbers of the sequence, the amplitude of the odd oscillations being bigger than the amplitude of the even ones. This surprising behavior is illustrated more clearly in Figs. 6.16 and 6.17 in terms of log–log plots of the specific heat against the temperature for even and odd values of n, respectively. It is intriguing that this peculiar behavior is some kind of *signature* of the Fibonacci structure, with no counterpart in the other quasiperiodic structures considered here. In general, the self-similarity of the spectra is more

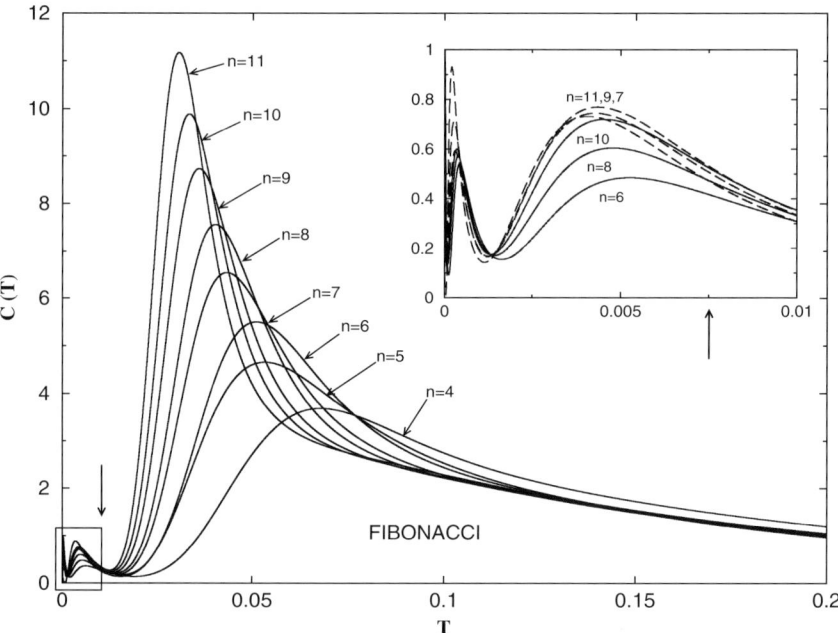

Fig. 6.15. Polariton specific heat versus temperature for the Fibonacci quasiperiodic sequence considered up to its 11th generation (after Mauriz et al. [54]).

pronounced for sequences with a difference of two in the generation number. For example, in the tight-binding Fibonacci spectra one can see that the biggest gaps appearing in the spectrum, and leading to the biggest subbands, occur (for any sequence of length F_n) in F_{n-2} at low energy and in F_{n-1} at high energy (a similar relation holds inside each subband) [55]. We note that the specific heat properties in the log scale are basically controlled by the behavior of the low-energy region at the scale considered (i.e. each oscillation can be considered as a change of scale in the spectrum). In this sense, at a high-scale F_n the low-energy region would be controlled by F_{n-2}; at a smaller scale the low-energy region would be controlled by $F_{(n-2)-2} = F_{n-4}$ and so on.

The log-periodic behavior of the specific heat is shown in Fig. 6.18, where we have plotted $C(T)$ versus $\log(T)$ for the Fibonacci sequence. The curves resemble the triadic case with a mean value d, where $C(T)$ oscillates log-periodically around it, although this value is not related to the fractal dimension for the Fibonacci quasiperiodic structure. The reason is that our polariton spectra are not strictly invariant under changes of scales (as in the triadic case, whose log-periodic behavior can be explained as a natural consequence of discrete scale invariance [56,57]). These oscillations are similar to those found for the tight-binding Fibonacci spectra [58]. Nevertheless, this mean value d can be given approximately by the so-called spectral dimension (the exponent of a power-law fit of the integrated density of states [59,60]), which in this case is equal to 0.5762. It is also associated with

6.5. THERMODYNAMIC PROPERTIES

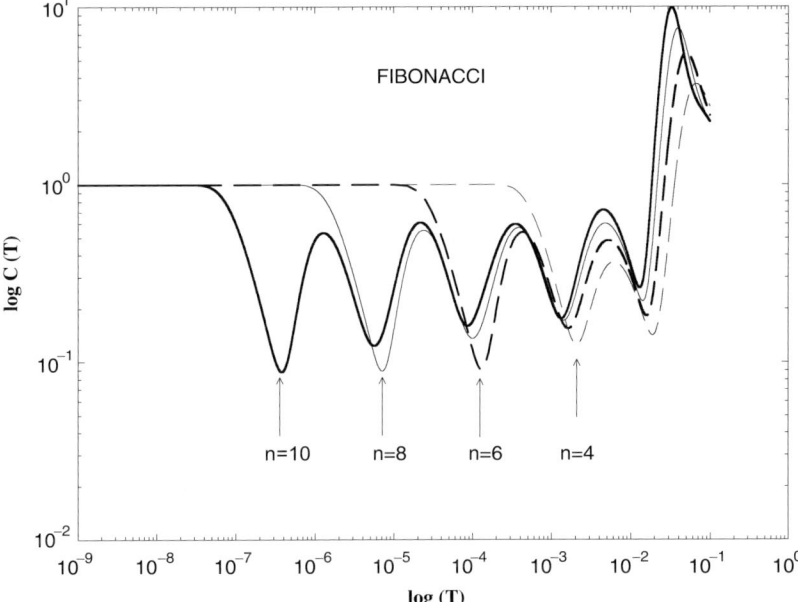

Fig. 6.16. Log–log plot of the polariton specific heat versus temperature for the even generation numbers of the Fibonacci quasiperiodic sequence (after Mauriz et al. [54]).

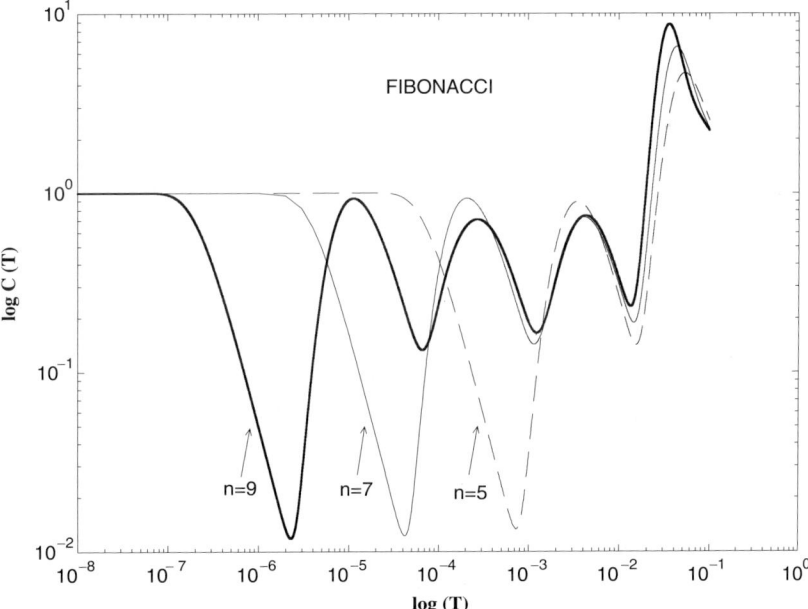

Fig. 6.17. Log–log plot of the polariton specific heat versus temperature for the odd generation numbers of the Fibonacci quasiperiodic sequence (after Mauriz et al. [54]).

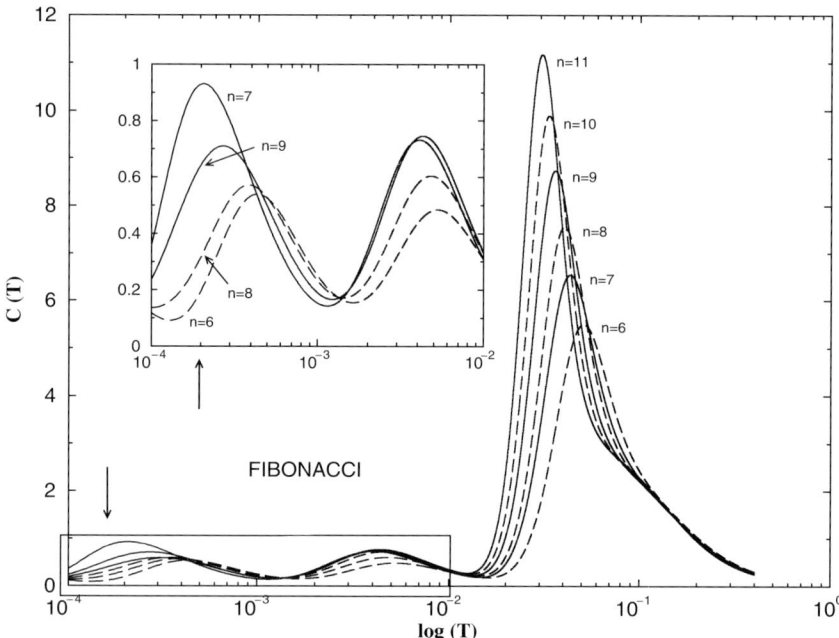

Fig. 6.18. Polariton specific heat plotted versus $\log(T)$ to show the log-periodicity effects in the Fibonacci structure (after Mauriz et al. [54]).

the minimum singularity exponent α in the multifractal $f(\alpha)$ spectrum shown in Fig. 6.10a. Again, although it is not particularly clear in the figure, there is a difference in the behavior of the log-periodicity for even and odd generation numbers of the Fibonacci sequence (see the inset in Fig. 6.18).

In Fig. 6.19 we show the specific heat as a function of temperature for the Thue–Morse sequence. By analogy with what was done in the Fibonacci case, we have considered the polariton multifractal energy profile of the Thue–Morse sequence, as depicted in Fig. 6.8a. Again, in the limit when $T \to \infty$, the specific heat goes to zero as T^{-2} for all generation numbers. Although there are oscillations in the region near $T \to 0$ (shown more clearly in the inset of the figure), they do not have the same type of behavior as in the Fibonacci sequence (where there are two groups of oscillations corresponding to even and odd generation numbers of the sequence).

Fig. 6.20 shows the log–log plot for this same case, where the oscillatory behavior of the specific heat can be better appreciated. The pattern is the same for all generation numbers, with the oscillations becoming more accentuated for large n, i.e. the bigger the number of generations, the greater the number of oscillations.

The log-periodicity of the specific heat profile is shown in Fig. 6.21. From there we can infer that this log-scale shows erratic oscillations, which do not resemble

6.5. THERMODYNAMIC PROPERTIES

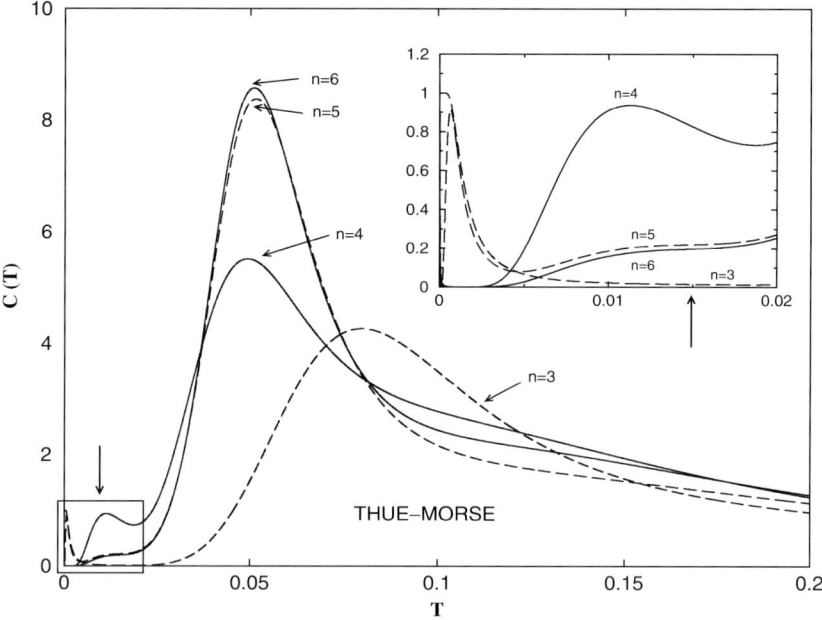

Fig. 6.19. Same as Fig. 6.15 but for the Thue–Morse quasiperiodic sequence considered up to its 6th generation (after Mauriz et al. [54]).

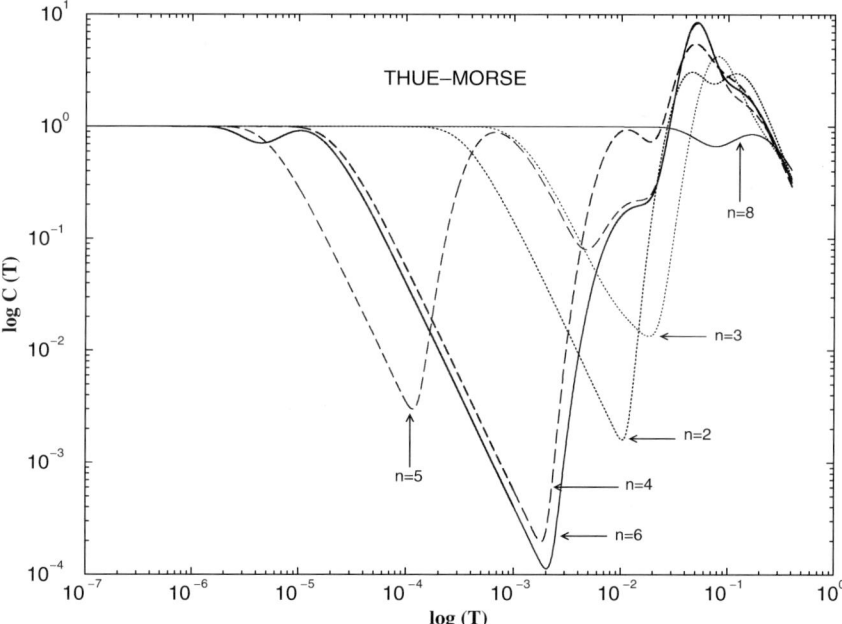

Fig. 6.20. Same as in Fig. 6.16 but for the Thue–Morse sequence (after Mauriz et al. [54]).

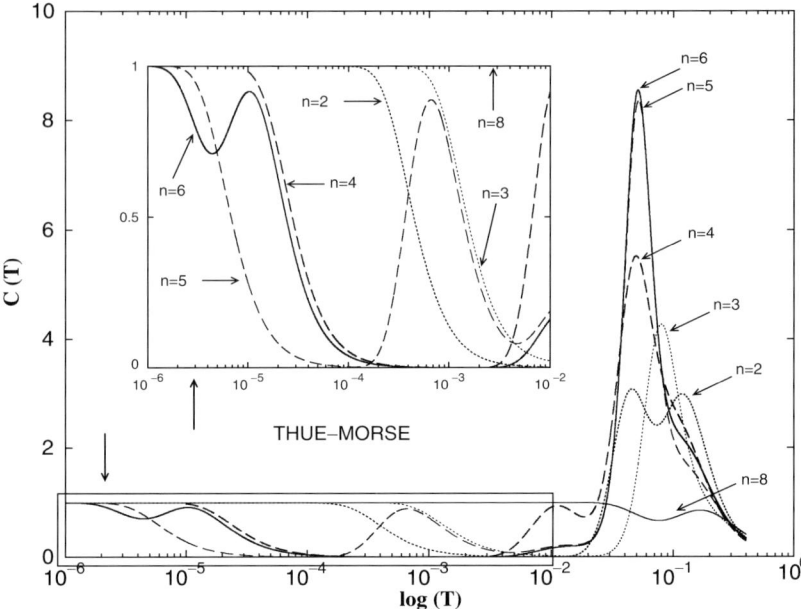

Fig. 6.21. Same as in Fig. 6.18 but for the Thue–Morse sequence (after Mauriz et al. [54]).

the pattern found in either the even and odd generation numbers of the Fibonacci case. Also there is *no* average value around which the specific heat $C(T)$ has a log-periodical oscillation. Therefore, apart from the common asymptotic behavior of the specific heat as $T \to \infty$, there seems to be no other connection (regarding the specific heat spectra) among the quasiperiodic structures considered here.

For completeness, we also describe analytical and numerical calculations of the *surface* plasmon-polariton contribution to the specific heat associated with successive hierarchical sequences of quasiperiodic Fibonacci structures. We take the building blocks that are used to set up these structures as the semiconductor GaAs and the insulator SiO_2. At each interface there is a 2D electron gas, modelled due to the presence of an areal current density of electrons, whose expression is given by Eq. (4.23). The corresponding scaled discrete fractal set is shown in Fig. 6.22, where the allowed frequencies are plotted as a function of the generation number n [61]. We have considered here the dimensionless in-plane wavevector $k_x a = 0.25$. We can infer that a fractal spectrum emerges when n goes to infinity. In the representation of the frequency (or energy) spectrum, it is easy to see that for $n = 2$ there are two discrete levels, the first at $\epsilon_1 = 0.0553$ and the second at $\epsilon_2 = 0.6629$. For $n = 3$ we have three levels at $\epsilon_1 = 0.0554$, $\epsilon_2 = 0.2436$, and $\epsilon_3 = 0.8036$. We have shown the spectra up to $n = 9$, which means a unit cell with 34 building blocks A and 21 building blocks B; this should be enough to model the Fibonacci quasiperiodicity.

6.5. THERMODYNAMIC PROPERTIES

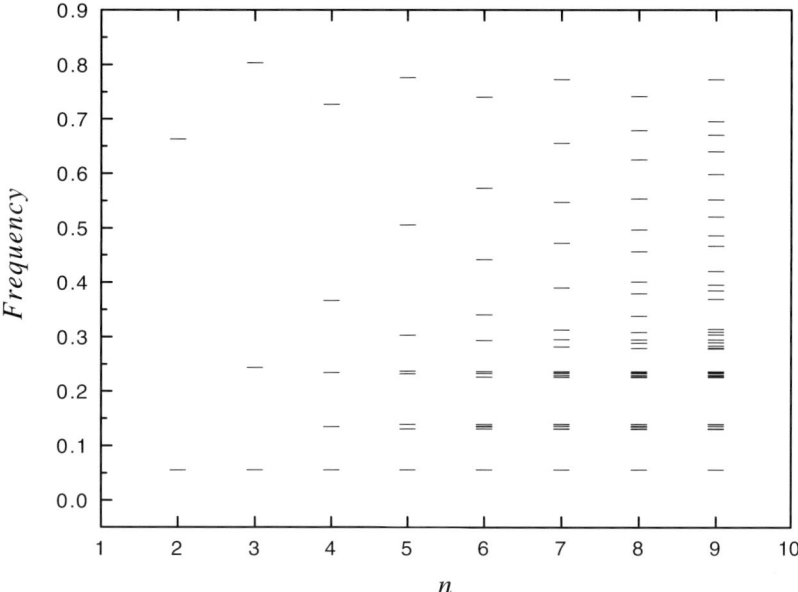

Fig. 6.22. Frequency spectra for the surface plasmon-polariton modes up to the 9th Fibonacci generation number. Here the in-plane dimensionless wavevector $k_x a$ is equal to 0.25 (after Mauriz et al. [61]).

The discrete partition function is now given by

$$Z = \sum_{i=1}^{N} \exp(-\beta \epsilon_i), \qquad (6.46)$$

where N is the number of surface modes of the spectrum, which for the Fibonacci quasiperiodic system is equal to the Fibonacci number F_n. Once we know the partition function, we can calculate the specific heat using the standard Eqs. (6.42) and (6.43), where f_N and g_N are given by

$$f_N = \sum_{i=1}^{N} \epsilon_i^2 \exp(-\beta \epsilon_i) \qquad (6.47)$$

and

$$g_N = \sum_{i=1}^{N} \epsilon_i \exp(-\beta \epsilon_i). \qquad (6.48)$$

The contribution of the surface plasmon-polaritons to the specific heat of the Fibonacci structure, as a function of the temperature, is depicted in Fig. 6.23.

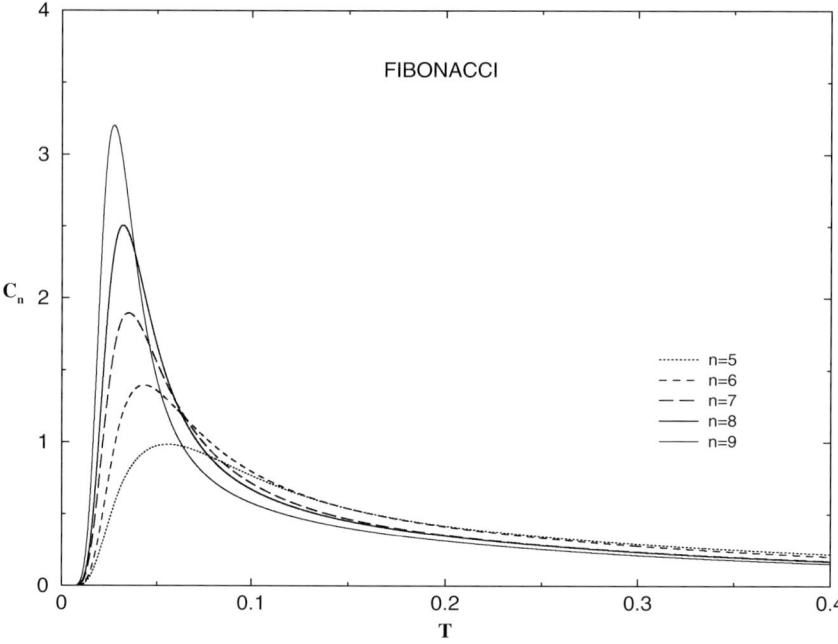

Fig. 6.23. Surface plasmon-polariton specific heat versus temperature for the Fibonacci superlattice. The different lines correspond to various values of the Fibonacci generation number n (after Mauriz et al. [61]).

From there, one can see that, by contrast with the continuum bulk case, the specific heat at low temperature is now a well-behaved function with *no oscillations* in its profile! A possible explanation is that these oscillations may reflect the fractal aspect of the spectra only for a large number of energy levels, which implies a large Fibonacci generation number n. Alternatively, we surmise that if the bulk bands of the plasmon-polaritons were included then they may show this oscillatory behavior.

The polariton specific heat discussed here for a 1D structure may be quite different from that in real 3D structures. Although this limitation adds an extra complication to the experiments in realizing the low dimensionality, it should be possible to circumvent this problem. For instance, for a boson system (e.g. the polaritons case treated here) we may model such a 1D structure by considering the spectra obtained only for a given in-plane wavevector k_x. As one of the most suitable experimental techniques to probe the polariton spectra is inelastic light scattering spectroscopy (either Raman or Brillouin scattering), this in-plane wavevector k_x can be controlled through the incident angle θ of the light in an oblique-scattering geometry [62,63]. There will be a further discussion of the light-scattering formalism in Chapter 11.

References

[1] E.L. Albuquerque and M.G. Cottam, Phys. Rep. 376 (2003) 225.
[2] A.A. Yamaguchi, T. Saiki, T. Tada, T. Nnomiya, K. Misawa, T. Kobayashi, M. Kuwata-Gonokami and T. Yao, Solid State Commun. 75 (1990) 955.
[3] F. Laruelle and B. Etienne, Phys. Rev. B 37 (1988) 4816.
[4] D. Munzar, L. Bocaek, J. Humlicek and K. Ploog, J. Phys.: Condens. Matter 6 (1994) 4107.
[5] T.C. Halsey, M.H. Jensen, L.P. Kadanoff, I. Procaccia and B.I. Shraiman, Phys. Rev. A 33 (1986) 1441.
[6] E.L. Albuquerque and M.G. Cottam, Phys. Rep. 233 (1993) 68.
[7] J. Bellisard and E. Scoppola, Commun. Math. Phys. 85 (1982) 301.
[8] J. Bellisard, B. Iochum, E. Scoppola and D. Testard, Commun. Math. Phys. 125 (1989) 527.
[9] J. Bellisard, A. Bovier and J.-M. Ghez, Commun. Math. Phys. 135 (1991) 379.
[10] M. Kolár and M.K. Ali, Phys. Rev. B 39 (1989) 6538.
[11] M. Kolár and M.K. Ali, J. Phys.: Condens. Matter 1 (1989) 823.
[12] M. Kolár and M.K. Ali, Phys. Rev. B 41 (1990) 7108.
[13] M.S. Vasconcelos and E.L. Albuquerque, Phys. Rev. B 57 (1998) 2826.
[14] M.S. Vasconcelos and E.L. Albuquerque, Physica B 222 (1996) 113.
[15] B. Simon, Adv. Appl. Math. 3 (1982) 463.
[16] B. Simon and B. Soulliard, J. Stat. Phys. 36 (1984) 273.
[17] C.S. Ryu, Y.G. Oh and M.H. Lee, Phys. Rev. B 48 (1993) 132.
[18] A. Süto, J. Stat. Phys. 56 (1989) 525.
[19] P.D. Kirkman and J.B. Pendry, J. Phys. C 17 (1984) 4327.
[20] P. Hawrylak and J.J. Quinn, Phys. Rev. Lett. 57 (1986) 380.
[21] F. Dominguez-Adame, A. Sánchez and E. Diez, Phys. Rev. B 50 (1994) 17736.
[22] M. Widom, D. Bensimon, L.P. Kadanoff and S.J. Shenker, J. Stat. Phys. 32 (1983) 443.
[23] I. Procaccia in: Proceedings of Nobel Symposium on Chaos and Related Problems [Phys. Scr. T9 (1985) 40].
[24] S.H. Strogatz, Nonlinear Dynamics and Chaos, Addison-Wesley, Reading, 1994.
[25] A. Bunde and S. Havlin, Eds., Fractals and Disordered Systems, Springer-Verlag, Heidelberg, 1991.
[26] A.L. Olemskoi and A.Y. Flat, Phys. Uspekhi 36 (1993) 1087.
[27] T.C. Halsey, P. Meakin and I. Procaccia, Phys. Rev. Lett. 56 (1986) 854.
[28] M. Kohmoto, B. Sutherland and C. Tang, Phys. Rev. B 35 (1987) 1020.
[29] L. de Arcangelis, S. Redner and A. Coniglio, Phys. Rev. B 31 (1985) 4725.
[30] T. Vicsek, Fractal Growth Phenomena, World Scientific, Singapore, 1989.
[31] G. Paladin and A. Vulpiani, Phys. Rep. 156 (1987) 148.
[32] J.L. McCauley, Phys. Rep. 189 (1990) 225.
[33] M.S. Vasconcelos, E.L. Albuquerque and E. Nogueira Jr., Physica A 268 (1999) 165.
[34] P. Grassberger, R. Badii and A. Politi, J. Stat. Phys. 51 (1988) 135.
[35] A.B. Chhabra and R.V. Jensen, Phys. Rev. Lett. 62 (1989) 1327.
[36] A.B. Chhabra, C. Meneveau and K.R. Srenivasan, Phys. Rev. A 40 (1989) 5284.
[37] C. Tsallis, J. Stat. Phys. 52 (1988) 479.
[38] M.L. Lyra and C. Tsallis, Phys. Rev. Letters 80 (1998) 53.
[39] M.S. Vasconcelos and E.L. Albuquerque, Solid State Commun. 117 (2001) 495.
[40] N. Raj and D.R. Tilley, in: The Dielectric Function of Condensed Systems, Eds., L.V. Keldysh, D.A. Kirzhnitz and A.A. Maradudin, North-Holland, Amsterdam, 1989.

[41] T. Dumelow, T.J. Parker, S.R.P. Smith and D.R. Tilley, Surf. Sci. Rep. 17 (1993) 151.
[42] C. Tsallis, L.R. da Silva, R.S. Mendes, R.O. Vallejos and A.M. Mariz, Phys. Rev. E 56 (1997) R4922.
[43] R.O. Vallejos, R.S. Mendes, L.R. da Silva and C. Tsallis, Phys. Rev. E 58 (1998) 1346.
[44] R.O. Vallejos and C. Anteneodo, Phys. Rev. E 58 (1998) 4134.
[45] G. Gumbs, G.S. Dubey, A. Salman, B.S. Mahmoud and D. Huang, Phys. Rev. B 52 (1995) 210.
[46] A. Arneodo, E. Bacry, S. Jaffard and J.F. Muzy, J. Stat. Phys. 87 (1997) 179.
[47] A. Erzan and J.-P. Eckmann, Phys. Rev. Lett. 78 (1997) 3245.
[48] Y. Meurice, S. Niermann and G. Ordaz, J. Stat. Phys. 87 (1997) 237.
[49] A. Petri and G. Roucco, Phys. Rev. B 51 (1995) 11399.
[50] P. Carpena, A.V. Coronado and P.B-Galván, Physica A 287 (2000) 37.
[51] J.C. Kimball and H.L. Frisch, J. Stat. Phys. 89 (1977) 453.
[52] H.-O. Peitgen, H. Jurgens and D. Saupe, Chaos and Fractals, Springer-Verlag, Heidelberg, 1992.
[53] E.M.F. Curado and M.A. Rego-Monteiro, Phys. Rev. E 61 (2000) 6255.
[54] P.W. Mauriz, E.L. Albuquerque and M.S. Vasconcelos, Phys. Rev. B 63 (2001) 184203.
[55] P. Carpena, V. Gasparian and M. Ortuño, Phys. Rev. B 51 (1995) 12813.
[56] H. Saleur and D. Sornette, J. Phys. (France) I 6 (1996) 327.
[57] D. Karevsky and L. Turban, J. Phys. A 29 (1996) 3461.
[58] P. Carpena, A.V. Coronado and P.B-Galván, Phys. Rev. E 61 (2000) 2281.
[59] S. Alexander and R. Orbach, J. Phys. (France) Lett. 43 (1982) L625.
[60] R. Rammal and G. Toulouse, J. Phys. (France) Lett. 44 (1983) L13.
[61] P.W. Mauriz, M.S. Vasconcelos and E.L. Albuquerque, Surf. Sci. 482 (2001) 537.
[62] R. Merlin, J.P. Valladares, A. Pinczuk, A.C. Gossard and J.H. English, Solid State Commun. 84 (1992) 87.
[63] E.L. Albuquerque, Solid State Commun. 99 (1996) 311.

Chapter 7

Magnetic Polaritons

In this chapter we cover the basics of excitations in magnetic materials while other, more specialized, aspects are discussed in Chapters 8 and 9. To summarize, we have so far given a brief introduction to the concept of a magnon (or spin wave) in Section 1.7 assuming the dominant interactions between magnetic moments to come from Heisenberg-type exchange coupling. Subsequently in Section 3.4 we showed how to derive the bulk magnetic susceptibility matrix for ferromagnets and antiferromagnets in a regime where exchange interactions were ignored. The magnetic interactions (magnetic dipole–dipole interactions) in this case came about through the application of Maxwell's equations. The susceptibility relations were then employed to discuss magnetic polaritons in bulk materials in Section 3.5, where we included retardation effects in the general analysis.

It is well known that the two main interactions between the different atomic sites in magnetic materials are due to exchange interactions and magnetic dipole–dipole interactions. In general the exchange interactions are of short range, typically providing coupling only to the nearest and next-nearest neighbors, but they are relatively strong and are responsible for the overall scheme of magnetic ordering. On the other hand, the dipole–dipole interactions are usually much weaker, but they are long range in character. As a consequence, the dipole–dipole interactions can become important (and eventually dominate) for the *dynamical* effects if we consider sufficiently small values of $k \equiv |\vec{k}|$, implying that the wavelength is large compared to the lattice parameter, where $\vec{k}$ is the wavevector of the excitation. For a bulk ferromagnet the exchange effects are characterized at small wavevectors by a term proportional to JSa^2k^2, as can be seen from Eq. (1.38). By contrast, the dipole–dipole terms are characterized (as we shall see later) by terms proportional to the static magnetization M, and hence a rough criterion for the exchange terms to be small is $JSa^2k^2 \ll M$. At even smaller values of k than required by this condition, namely when $k \sim \omega/c$, the retardation effects become important in the electromagnetic equations, and this is the regime of the magnetic polaritons. Therefore, in terms of $\vec{k}$, we can summarize the different regions of magnetic behavior as follows (see also Fig. 7.1):

(a) *The exchange region.* The wavevector $\vec{k}$ is sufficiently large that the exchange effects dominate over the dipole–dipole terms in determining the magnon

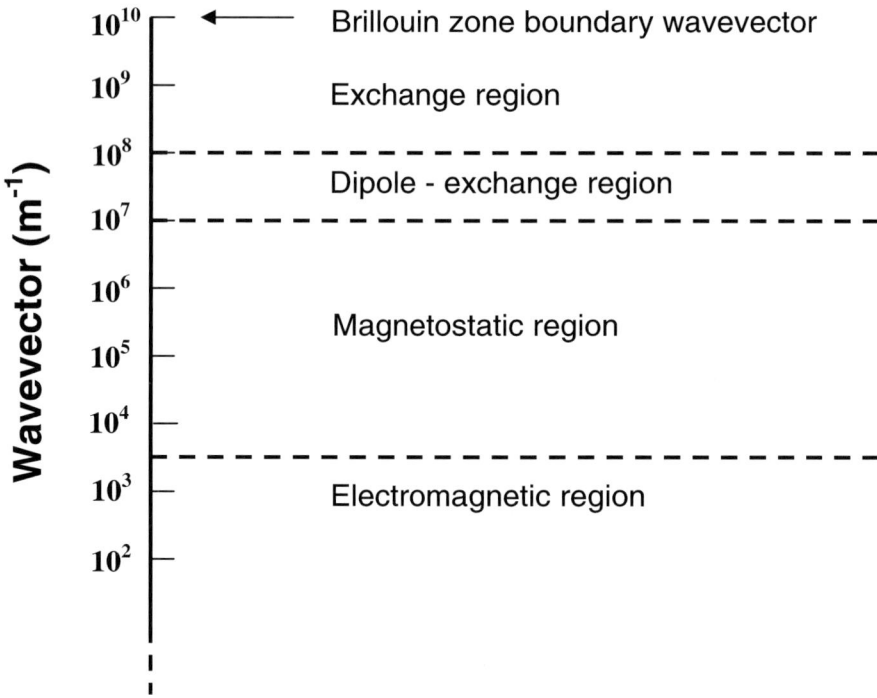

Fig. 7.1. Different regions of magnetic behavior in terms of the wavevector $\vec{k}$ (after Cottam and Tilley [14]).

(or spin-wave) properties. Typically this is the case for $k > 10^8 \, \text{m}^{-1}$. For comparison the Brillouin zone boundary wavevector is of order $10^{10} \, \text{m}^{-1}$.

(b) *The dipole-exchange region.* Here the exchange and dipole–dipole contributions are comparable for the spin-wave dynamics, and neither of them can be ignored. Typically this case corresponds to $10^7 \, \text{m}^{-1} < k < 10^8 \, \text{m}^{-1}$.

(c) *The magnetostatic region.* This is the case in which $\vec{k}$ is sufficiently small that exchange effects can be ignored, while the dipole–dipole terms can be treated within the magnetostatic approximation (i.e. the retardation effects are neglected in Maxwell's equations). Typically this occurs for $3 \times 10^3 \, \text{m}^{-1} < k < 10^7 \, \text{m}^{-1}$.

(d) *The electromagnetic region.* In this case the dipole–dipole terms have to be treated using the full form of Maxwell's equations with retardation (e.g. as in Section 3.5 for bulk materials). Typically this occurs for $k < 3 \times 10^3 \, \text{m}^{-1}$.

Similar information about the spin-wave dynamics is provided by Fig. 3.7, except that here we have chosen also to identify the dipole-exchange region, since

7.1. EXCHANGE SPIN WAVES IN THIN FILMS

it corresponds to some experimental situation (e.g. in the Brillouin scattering of light). The k-values quoted above may vary for different materials. The regions of interest as regards magnetic polaritons are the electromagnetic region and (as a limiting case) the magnetostatic region.

The surface spin waves (or magnons) in thin films are discussed first in Section 7.1 for the exchange region and then in Section 7.2 for the magnetostatic region (including some comments on the dipole-exchange case). It is shown how these results can be generalized to periodic superlattices in Section 7.3. The remaining sections of this chapter then deal with specific examples of magnetic polaritons in layered structures.

Although polaritons in dielectrics and metals were extensively studied, both theoretically and experimentally, they have received less attention in magnetic materials. One reason is that in ferromagnets the frequency of a magnetic polariton ω is typically in the microwave part of the spectrum (e.g. 1–20 GHz). Recalling that we are interested in $k \sim \omega/c$ for polaritons, the corresponding wavelength $2\pi/k$ becomes larger than most practical sample sizes. However, polaritons are of interest for antiferromagnets (and other magnetic materials with larger energy gaps in their spectra) since the antiferromagnetic resonance frequency typically lies in the far infrared (e.g. 200 GHz to a few THz), making sample sizes more realistic. Nevertheless, in both cases, the unretarded limit (the magnetostatic regime) is of considerable interest. In general, it is found that surface magnetic polaritons can often have a more interesting behavior than their dielectric counterparts because the modes depend strongly on the relative orientation of the crystal axes, the external magnetic field $\vec{H}_0$, the surface orientation, and the propagation wavevector (see e.g. Ref. [1]).

7.1 Exchange Spin Waves in Thin Films

In order to introduce the concept of surface spin waves we shall generalize the calculation given in Section 1.7 for an infinite Heisenberg ferromagnet to the case of a thin-film geometry. The spin Hamiltonian is again assumed to be given by Eq. (1.29), consisting of an exchange part and a Zeeman term for an applied magnetic field H_0 in the z-direction. We employ a microscopic model in which the film is assumed to consist of N atomic layers parallel to xy-plane. We label the layers by integer n ($=1, 2, \ldots, N$) with $n=1$ denoting the top surface (at $z=0$) and $n=N$ denoting the lower surface at $z=-(N-1)a$, where a is the lattice constant. As in Section 1.7, the lattice structure is taken to be simple cubic for convenience.

An equation of motion can be written down for the spin operator S_i^+ at any site i using Eq. (1.30), which again yields Eq. (1.32) when the random-phase approximation (RPA) at low temperature $T \ll T_c$ is employed. There is now translational symmetry only in the xy-plane, and so we look for solutions of the form

$$S_i^+ = s_n(\vec{k}_\parallel) \exp(i\vec{k}_\parallel \cdot \vec{r}_{\parallel j}) \exp(i\omega t) \tag{7.1}$$

in accordance with Bloch's theorem. Here $\vec{k}_\| = (k_x, k_y)$ and $\vec{r}_\| = (x, y)$ are 2D vectors. From Eqs. (1.32) and (7.1) it is found that the amplitude coefficients $s_n(\vec{k}_\|)$ satisfy the following set of coupled equations for the different layers:

$$[\omega - g\mu_B H_0 - SJ - 4SJ_S\Lambda(\vec{k}_\|)]s_1 + SJs_2 = 0 \quad (n=1), \tag{7.2}$$

$$[\omega - g\mu_B H_0 - 2SJ - 4SJ\Lambda(\vec{k}_\|)]s_n + SJ(s_{n+1} + s_{n-1}) = 0 \quad (2 \leq n \leq N-1), \tag{7.3}$$

$$[\omega - g\mu_B H_0 - SJ - 4SJ'_S\Lambda(\vec{k}_\|)]s_N + SJs_{N-1} = 0 \quad (n=N), \tag{7.4}$$

where we introduce the notation

$$\Lambda(\vec{k}_\|) = [2 - \cos(k_x a) - \cos(k_y a)]/2. \tag{7.5}$$

As before, J is the nearest-neighbor exchange parameter inside the ferromagnet, while we allow for possible surface modification of the exchange through parameters J_S and J'_S denoting the exchange interaction between a pair of nearest neighbors that are both in the top $(n=1)$ or bottom $(n=N)$ layer, respectively.

Eqs. (7.2)–(7.4) can be solved by a variety of techniques to obtain the frequencies of the various spin-wave modes [2–4]. However, a simple approach is to seek the bulk spin-wave solutions using

$$s_n(\vec{k}_\|) = A(\vec{k}_\|)\exp(ik_z na) + B(\vec{k}_\|)\exp(-ik_z na), \tag{7.6}$$

where k_z is a real wavevector component and the two terms represent propagation in the positive and negative z-direction. On substituting this into Eqs. (7.2)–(7.4), we find that the dispersion relation of the bulk modes is formally the same as in Eq. (1.34), *except* that k_z now has to satisfy

$$\tan(k_z d) = \frac{\text{Im}\{[\Delta - \exp(-ik_z a)][\Delta' - \exp(-ik_z a)]\}}{\text{Re}\{[\Delta - \exp(-ik_z a)][\Delta' - \exp(-ik_z a)]\}}. \tag{7.7}$$

Here $d = (N-1)a$ is the thickness of the film, while Δ is a parameter associated with the upper surface and defined by

$$\Delta = \left[1 + 4(1 - J_S/J)\Lambda(\vec{k}_\|)\right]^{-1}, \tag{7.8}$$

and Δ' for the lower surface is defined in a similar way with J'_S replacing J_S. Eq. (7.7) will be satisfied only by certain discrete values of k_z, so this represents a "quantization" of k_z, and hence of the bulk spin-wave frequency, due to the finite film thickness.

It can also be shown that surface spin waves are predicted. In this case, instead of Eq. (7.6), we seek attenuated solutions for s_n of the form

$$s_n(\vec{k}_\|) = A'(\vec{k}_\|)\exp(-\kappa na) + B'(\vec{k}_\|)\exp(\kappa na), \tag{7.9}$$

where we take $\text{Re}(\kappa) > 0$ as a localization condition. Proceeding as before, we find solutions $\omega = \omega_S(\vec{k}_\parallel)$, where [3]

$$\omega_S(\vec{k}_\parallel) = g\mu_B H_0 + 4SJ\Lambda(\vec{k}_\parallel) + SJ[2 - \exp(\kappa a) - \exp(-\kappa a)] \quad (7.10)$$

and κ satisfies

$$\tanh(\kappa d) = \frac{\{[\Delta - \exp(\kappa a)][\Delta' - \exp(\kappa a)] - [\Delta - \exp(-\kappa a)][\Delta' - \exp(-\kappa a)]\}}{\{[\Delta - \exp(\kappa a)][\Delta' - \exp(\kappa a)] + [\Delta - \exp(-\kappa a)][\Delta' - \exp(-\kappa a)]\}}. \quad (7.11)$$

It is easier to discuss the surface modes first in the case of large film thickness ($d \to \infty$). We then have $\tanh(\kappa d) \to 1$, and Eq. (7.11) has two possible solutions for κ corresponding to $\exp(-\kappa a) = \Delta$ (provided $|\Delta| < 1$) and $\exp(-\kappa a) = \Delta'$ (provided $|\Delta'| < 1$). If the existence conditions are satisfied (and this depends on the ratios J_S/J and J'_S/J), the two modes are essentially as for semi-infinite media, being localized at the upper and lower surfaces, respectively. Taking the upper surface we therefore have

$$\omega_S(\vec{k}_\parallel) = g\mu_B H_0 + 4SJ\Lambda(\vec{k}_\parallel) - SJ(\Delta - 1)^2/\Delta. \quad (7.12)$$

It is easy to show that the localization condition can be satisfied in two ways. Either $J_S < J$, in which case we have an *acoustic surface spin wave*, or $J_S > 5J/4$, in which case we have an *optical surface spin wave*. These are spin waves that occur below or above the bulk modes, respectively. Some numerical examples are shown in Fig. 7.2. The results may be generalized (through redefinitions of Δ and Δ') to include surface anisotropy (or "pinning").

For a finite film thickness the analysis of the bulk and surface modes is more complicated, but it can be shown generally that there are r surface modes (with $r = 0$, 1, or 2 only, depending on the surface parameters) and $N - r$ quantized bulk modes [4,5].

It is worthwhile to conclude this subsection with a few remarks. We have chosen here to follow a microscopic approach, based on the operator equation of motion and the Heisenberg Hamiltonian. It can be generalized to calculate spin-dependent Green functions using Eq. (A.30) in Appendix A (see e.g. Ref. [6] and references therein). Similar calculations can also be made for Heisenberg antiferromagnets by taking account of the two sublattices. Finally, the calculations for ferromagnets and antiferromagnets in the regime of small wavevector can alternatively be carried out using a macroscopic approach based on the torque equation of motion [see Eq. (3.36)]. These topics are reviewed in Ref. [5].

7.2 Magnetostatic Modes in Thin Films

In the magnetostatic region we have wavevectors such that $ka \ll 1$, and so a continuum approach is applicable in which the dipole–dipole terms are treated using

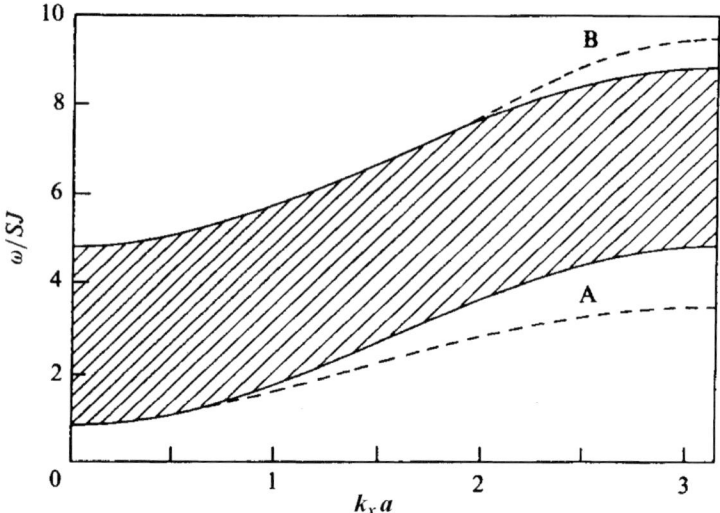

Fig. 7.2. Spin-wave frequency (in units of SJ) versus $k_x a$ in a semi-infinite Heisenberg ferromagnet for propagation vector $\vec{k}_\| = (k_x, 0)$. The bulk spin-wave region is shown shaded, together with two surface spin-wave branches corresponding to: A, $J_S/J = 0.5$; B, $J_S/J = 1.8$. Also $g\mu_B H_0/SJ = 1.0$.

Maxwell's equations. The analysis makes use of the bulk-medium susceptibility relationships, which we derived formally in Section 3.4 for ferromagnets and for two-sublattice uniaxial antiferromagnets. In either case the general result is that the susceptibility matrix has the gyromagnetic form quoted in Eq. (3.51). Here we shall restate the components $\chi_a(\omega)$ and $\chi_b(\omega)$ in a form that is convenient for the present discussion. For a ferromagnet we have

$$\chi_a(\omega) = \omega_M \omega_0/(\omega_0^2 - \omega^2), \tag{7.13}$$

$$\chi_b(\omega) = \omega_M \omega/(\omega_0^2 - \omega^2), \tag{7.14}$$

where we have introduced the following characteristic frequencies associated with the applied magnetic field and the magnetization:

$$\omega_0 = -\gamma H_0, \qquad \omega_M = -\gamma M. \tag{7.15}$$

Thus there are poles (zeros of the denominator) for the frequency $\omega = \pm\omega_0$, where ω_0 represents the ferromagnetic resonance (FMR) frequency. For an antiferromagnet we may write

$$\chi_a(\omega) = \tfrac{1}{2}(\chi^+ + \chi^-), \qquad \chi_b(\omega) = \tfrac{1}{2}(\chi^+ - \chi^-), \tag{7.16}$$

where

$$\chi^\pm = \frac{2\omega_A \omega_M}{\omega_A(2\omega_E + \omega_A) - (\omega \mp \omega_0)^2}. \tag{7.17}$$

7.2. MAGNETOSTATIC MODES IN THIN FILMS

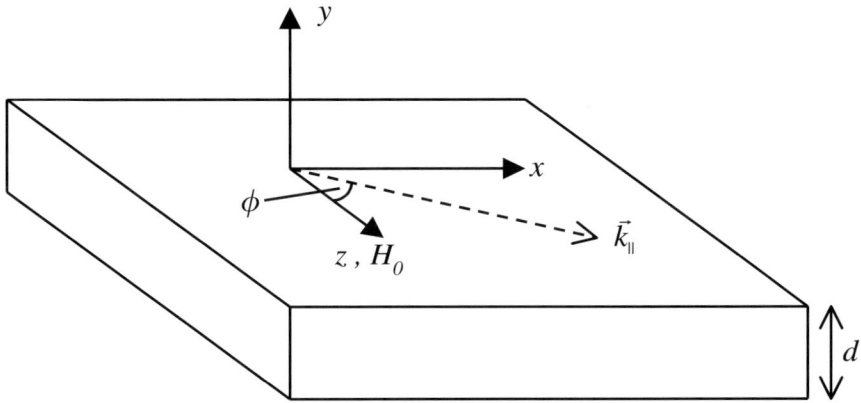

Fig. 7.3. Assumed geometry for calculating the magnetostatic modes in a ferromagnetic film magnetized parallel to its surfaces.

Here the additional frequencies are related to the anisotropy and exchange:

$$\omega_A = -\gamma H_A, \qquad \omega_E = -\gamma H_E. \tag{7.18}$$

We denote by H_A and H_E the anisotropy and intersublattice-exchange fields, respectively; the latter is just of magnitude λM in the notation of Section 3.4. It may be noted that, in the case of zero applied field ($\omega_0 = 0$), we have $\chi^+ = \chi^-$ and so the susceptibility matrix becomes diagonal. In this case the poles are at $\omega = \pm \omega_{AF}$, where ω_{AF} is the antiferromagnetic resonance (AFMR) frequency:

$$\omega_{AF} = [\omega_A(2\omega_E + \omega_A)]^{1/2}. \tag{7.19}$$

The interesting configuration for surface magnetostatic modes in a thin film occurs when the static magnetization $\vec{M}$ (or sublattice magnetization in the case of an antiferromagnet) is parallel to the surfaces, and we consider this first before examining what happens when $\vec{M}$ is in the perpendicular orientation.

7.2.1 Magnetization Parallel to the Film Surfaces

We begin with a ferromagnet, for simplicity, and adopt the geometry in Fig. 7.3 where the in-plane wavevector $\vec{k}_\parallel$ is taken to be at an angle ϕ to the z-axis. The film of thickness d occupies the region between $y = 0$ and $y = -d$. The theory for the mode frequencies was first worked out by Damon and Eshbach [7]. Some review accounts are to be found, e.g. in Refs. [8,9].

The magnetostatic approximation corresponds to neglecting retardation effects (i.e. effectively taking the limit $c \to \infty$), so the relevant Maxwell's equations are

$$\vec{\nabla} \times \vec{h}(\vec{r}) = 0, \qquad \vec{\nabla} \cdot [\vec{h}(\vec{r}) + \vec{m}(\vec{r})] = 0. \tag{7.20}$$

The first of these means that a scalar potential ψ, the *magnetostatic potential*, can be introduced by writing
$$\vec{h} = \vec{\nabla}\psi. \qquad (7.21)$$
Then, using the susceptibility relation in Eq. (3.51) to relate the *fluctuating* magnetization $\vec{m}$ to $\vec{h}$, the divergence equation in Eq. (7.20) gives
$$(1 + \chi_a)\left(\frac{\partial^2 \psi}{\partial x^2} + \frac{\partial^2 \psi}{\partial y^2}\right) + \frac{\partial^2 \psi}{\partial z^2} = 0. \qquad (7.22)$$
This holds inside the ferromagnetic film; for the regions outside (assumed to be non-magnetic) $\chi_a = 0$ and so
$$\nabla^2 \psi = 0. \qquad (7.23)$$
As a consequence of the translational invariance in the x- and z-directions, it follows that $\psi(\vec{r})$ must be of the form $\psi_1(y)\exp(i\vec{k}_\parallel \cdot \vec{r}_\parallel)$, where $\vec{r}_\parallel = (x, z)$. For Eqs. (7.22) and (7.23) to be satisfied in their respective regions and for ψ_1 to vanish at $y = \pm\infty$, the general form of the solution is
$$\psi_1(y) = A_1 \exp(-k_\parallel y), \qquad y > 0, \qquad (7.24)$$
$$\psi_1(y) = A_2 \exp(ik_y y) + A_3 \exp(-ik_y y), \qquad 0 > y > -d, \qquad (7.25)$$
$$\psi_1(y) = A_4 \exp(k_\parallel y), \qquad y < -d, \qquad (7.26)$$
where $k_\parallel = (k_x^2 + k_z^2)^{1/2} > 0$. Using Eq. (7.22) the quantity k_y, which can be real or imaginary, satisfies
$$(1 + \chi_a)(k_y^2 + k_\parallel^2) - \chi_a k_z^2 = 0. \qquad (7.27)$$

The coefficients A_j ($j = 1, 2, 3, 4$) in Eqs. (7.24)–(7.26) can be found by applying the usual electromagnetic boundary conditions at the film surfaces $y = 0$ and $y = -d$. These are equivalent to requiring that ψ must be continuous at each boundary and that $(h_y + m_y)$ inside the ferromagnet at $y = 0$ and $y = -d$ must be equal to h_y just outside. The solvability condition for the four homogeneous equations with the four coefficients leads to
$$k_\parallel^2 + 2k_\parallel k_y (1 + \chi_a)\cot(k_y d) - k_y^2(1 + \chi_a)^2 - k_x^2 \chi_b^2 = 0. \qquad (7.28)$$

If Eq. (7.27) is employed to eliminate k_y from the above equation, we arrive at an implicit dispersion relation for the magnetostatic modes.

It is easiest to examine the solution in the special case when the in-plane propagation is parallel or antiparallel to the x-direction (i.e. $k_z = 0$ or angle $\phi = \pi/2$). This is referred to as the *Voigt configuration*. In this case we note that Eq. (7.27) is always satisfied if $\chi_a(\omega) = -1$, which from Eq. (7.13) means $\omega = \pm\omega_B$, where
$$\omega_B = [\omega_0(\omega_0 + \omega_M)]^{1/2}. \qquad (7.29)$$

7.2. MAGNETOSTATIC MODES IN THIN FILMS

This corresponds to the bulk magnetostatic modes of the film; it is independent of the wavevector components k_x and k_y. The other way in which Eq. (7.27) can be satisfied in the Voigt configuration is when $k_y = \pm i k_\parallel$. These two imaginary values of k_y correspond to surface magnetostatic modes with the frequency

$$\omega_S = \tfrac{1}{2}\left[(2\omega_0 + \omega_M)^2 - \omega_M^2 \exp(-2k_\parallel d)\right]^{1/2}. \qquad (7.30)$$

Despite their common frequency, it is important to note that the two surface solutions, in terms of the A_j coefficients in Eqs. (7.24)–(7.26), are different. If $k_x > 0$ it can be shown that k_y equal to $ik_\parallel$ and $-ik_\parallel$ correspond to surface states localized near the lower ($y = -d$) and upper ($y = 0$) surface, respectively, and vice versa if $k_x < 0$. This property, which is not shared by the surface modes in the exchange regime, is an example of *non-reciprocal propagation*. It has been convincingly confirmed by inelastic light-scattering experiments (see e.g. Refs. [10,11]). We note in this geometry that $\omega_S > \omega_B$, and some limiting surface mode frequencies are $[\omega_0(\omega_0 + \omega_M)]^{1/2}$, when $k_\parallel d \ll 1$, and $(2\omega_0 + \omega_M)/2$, when $k_\parallel d \gg 1$. The k-dependence of the modes ω_B and ω_S is shown schematically in Fig. 7.4.

Using Eqs. (7.24) to (7.28) the magnetostatic modes can also be investigated for other directions of the in-plane wavevector (i.e. when $k_z \neq 0$). For example, if $\vec{k}_\parallel$ is along the z-direction (parallel to H_0), meaning angle $\phi = 0$, it is easy to show that no localized surface mode exists. However, the bulk-mode spectrum in this case consists of a series of discrete (or "quantized") modes. For a thick film ($d \to \infty$) they merge to form a continuum, with the dispersion relation becoming

$$\omega_B = \left[\omega_0^2 + \omega_0 \omega_M \frac{k_y^2}{(k_y^2 + k_\parallel^2)}\right]^{1/2}. \qquad (7.31)$$

It is clear from the above expression that the frequency of these $\phi = 0$ bulk modes decreases with increasing $k_\parallel$, giving them the property that their group velocity parallel to the surface, namely $\partial \omega_B / \partial k_\parallel$, is negative. For this reason they are sometimes called *magnetostatic backward bulk modes*.

For general values of ϕ the behavior is more complicated than described above. However, it can be shown [7] that non-reciprocal surface magnetostatic modes exist for the range $\phi_c < \phi < \pi - \phi_c$, where the critical angle ϕ_c corresponds to

$$\sin(\phi_c) = \left[\omega_0/(\omega_0 + \omega_M)\right]^{1/2}. \qquad (7.32)$$

The mode frequency decreases as ϕ decreases from $\pi/2$ (the Voigt geometry), eventually merging with the bulk region at the critical value defined above.

The magnetostatic theory for a ferromagnetic film with an in-plane magnetization direction has also been generalized to the dipole-exchange region, where exchange effects (still in a long-wavelength continuum approximation) are included. There are two alternative approaches to be found in the literature. One involves supplementing the torque equation of motion, Eq. (3.36), to include dynamical

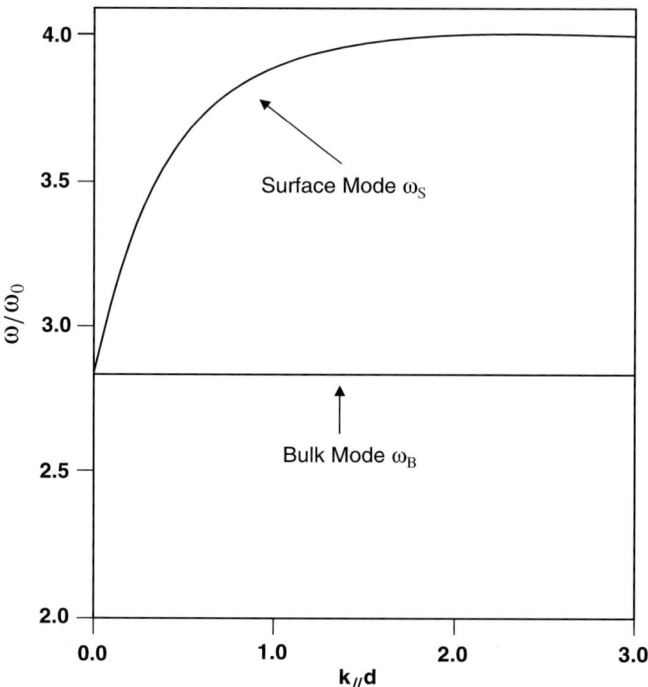

Fig. 7.4. The surface and bulk magnetostatic modes in a ferromagnetic film, magnetized parallel to its surfaces, as a function of $k_\parallel d$ for $k_z = 0$ and $\omega_M/\omega_0 = 6$.

effects of exchange in the effective-field term [2]. The other is a tensorial Green-function method in which the dipole–dipole interactions are introduced perturbatively [12]. To a good approximation, for the Voigt configuration it is found that the spectrum of the film consists of [2]

(i) a magnetostatic-type surface mode with frequency still given to a very good approximation by Eq. (7.30), and

(ii) a series of standing bulk spin waves with real wavevector component k_y and having discrete frequencies.

A simplified treatment gives the k_y values as $n\pi/d$ with $n = 1, 2, 3, \ldots$, and it may then be shown that Eq. (7.29) generalizes to

$$\omega_B = \left\{ \left[\omega_0 + Dk_\parallel^2 + D(n\pi/d)^2\right]\left[\omega_0 + \omega_M + Dk_\parallel^2 + D(n\pi/d)^2\right] \right\}^{1/2}. \quad (7.33)$$

More recently a microscopic theory, valid for general wavevectors and ultrathin films, for the dipole-exchange spin waves in ferromagnets was reported in Ref. [13].

Another extension of the magnetostatic theory, due to Stamps and Camley [14], was to antiferromagnetic films with sublattice magnetizations parallel to the

7.2. MAGNETOSTATIC MODES IN THIN FILMS

surfaces. The calculation of the mode frequencies proceeds in a manner similar to the ferromagnetic case, except that Eqs. (7.16) and (7.17) are employed for the susceptibility. For simplicity, we focus here on the limiting case of a thick film ($d \to \infty$) and the Voigt configuration ($k_z = 0$). When the applied field is zero, there are two bulk-mode frequencies $\omega_B^{\pm}$ given by

$$\omega_B^+ = \left[\omega_A(2\omega_E + \omega_A) + 2\omega_A\omega_M\right]^{1/2}, \tag{7.34}$$

$$\omega_B^- = \left[\omega_A(2\omega_E + \omega_A)\right]^{1/2}. \tag{7.35}$$

We note that the frequency ω_B^- of the lower branch is the same as the AFMR frequency in Eq. (7.19). The two branches are non-degenerate in frequency as a consequence of the dipole–dipole effects (making $\omega_M \neq 0$). The surface-mode spectrum consists of a single branch with frequency

$$\omega_S = \left[\omega_A(2\omega_E + \omega_A) + \omega_A\omega_M\right]^{1/2}, \tag{7.36}$$

so that it lies between the two bulk modes. In the presence of an applied field, it is found that ω_S is increased by an amount equal to ω_0 if $k_x > 0$ and decreased by ω_0 if $k_x < 0$. This is another example of non-reciprocal propagation, i.e. the property that

$$\omega_S(\vec{k}_\parallel) \neq \omega_S(-\vec{k}_\parallel). \tag{7.37}$$

It is different, however, from the non-reciprocity in the ferromagnetic case, where there is no surface mode *at the same surface* when $\vec{k}_\parallel$ is reversed. There will be further examples and a discussion of non-reciprocal modes in antiferromagnets in Chapter 8.

7.2.2 Magnetization Perpendicular to the Film Surfaces

For completeness, we now turn our attention to the case when the magnetization (or sublattice magnetization) is perpendicular to the film surfaces. When the magnetostatic calculation is repeated in this geometry it is found that no surface modes are predicted, essentially because the localization condition cannot be satisfied; this holds for both ferromagnets and uniaxial antiferromagnets.

The quantized bulk modes can, however, be studied in both materials following the same approach as in Subsection 7.2.1. For a ferromagnet it is found that

$$\omega_B = \left[\omega_0^2 - \omega_0\omega_M \frac{k_z^2}{(k_z^2 + k_\parallel^2)}\right]^{1/2}, \tag{7.38}$$

where the z-axis (in the direction of magnetization) is now perpendicular to the surfaces. The group velocity $\partial \omega_B / \partial k_\parallel$ is easily shown to be positive for these modes; they are called *magnetostatic forward bulk modes*, in contrast to the backward bulk modes of Eq. (7.31).

The above modes are found to be either symmetric or antisymmetric with respect to the mid-plane of the film, leading to the expressions

$$k_z \tan(k_z d) = k_\|, \qquad \text{symmetric modes,} \tag{7.39}$$

$$k_z \cot(k_z d) = -k_\|, \qquad \text{antisymmetric modes.} \tag{7.40}$$

These equations determine the series of discrete wavevector components k_z. The above equations can be solved rather easily by graphical methods.

7.3 Spin Waves in Magnetic Superlattices

The results for magnetic films in the previous section can readily be extended to infinite and semi-infinite periodic superlattices, for example by using the transfer-matrix formalism just as described in Chapter 5 for non-magnetic cases. An excellent review of this topic, focusing on the theory, has been given by Barnas [15], while other reviews by Grünberg [11] and Hillebrands [16] emphasize the experimental studies.

In this section we provide an overview of the excitations in periodic magnetic superlattices, treating separately the exchange and magnetostatic regions. We do this relatively briefly and for some simple theoretical models; then later in this chapter, as well as in Chapters 8 and 9, some specific examples and applications will be covered in more detail for both periodic and quasiperiodic magnetic structures.

7.3.1 Exchange Region

The exchange across the interface between two adjacent ferromagnetic materials can sometimes be quite complicated, having long-range oscillatory terms and so-called *biquadratic exchange* contributions as well as the usual Heisenberg-type (or bilinear) exchange [17]. As a consequence, the equilibrium spin configurations near an interface may be modified (providing an example of magnetic *surface reconstruction*), and there can be considerable theoretical challenges just in solving this static problem before tackling the dynamical problem of finding the spin-wave excitations. We shall be discussing examples of this behavior later (see Chapter 9), some of it in the context of giant magnetoresistance (GMR) [17], but for the present introduction we consider just the simplest case. This is the ferromagnetic/ferromagnetic superlattice with ferromagnetic exchange across the interfaces, so that no consideration of surface reconstruction is needed.

Specifically, we consider alternating simple-cubic ferromagnets with the same value of the lattice constant a and with (001) interfaces. The nearest-neighbor exchange parameter within each constituent material A and B is denoted by J_A and J_B respectively, and I denotes the exchange coupling across an interface. The case of an infinite periodic superlattice was analyzed by Albuquerque et al.

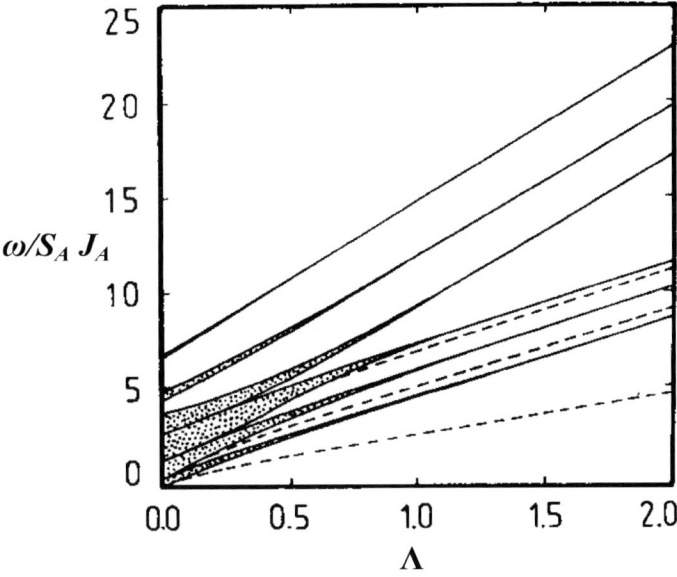

Fig. 7.5. Bulk (shaded regions) and surface (broken lines) spin waves (in units of $S_A J_A$) in a semi-infinite ferromagnetic superlattice in terms of $\Lambda(\vec{k}_\parallel)$ defined in Eq. (7.5). Each period consists of three atomic layers of material A and three of material B. For the spin quantum numbers it is assumed that $S_A = S_B$, while for the exchange parameters $J_B/J_A = 1$, $I/J_A = 1.4$, and $J_{AS}/J_A = 0.5$ (after Barnaś [15]).

[18] and subsequently extended by Barnas [15] to semi-infinite and finite superlattices, allowing the study of the superlattice surface spin waves. In these cases a transfer matrix can be constructed by following closely the methods presented in Section 5.1. The approximate spin equations of motion for the operators S_i^+ are written down for each atomic layer in a superlattice period, by analogy with Eqs. (7.1)–(7.4) for a film. Within each component of the superlattice, the solution is written as a linear combination of waves travelling in the positive and negative z-direction (perpendicular to the interfaces) as in Eq. (7.6). The equations of motion for the spins adjacent to the interfaces relate the amplitudes in the two superlattice constituents. Thus a 2×2 transfer matrix can be built up as before. From this, the bulk and surface modes can be deduced using Eqs. (5.27) and (5.34) [plus the analog of Eq. (5.41)], respectively, for a semi-infinite superlattice.

A numerical example of the resulting dispersion relations for the superlattice bulk spin waves (shaded regions) and surface spin waves (broken lines) is given in Fig. 7.5, where Λ is defined in terms of the in-plane wavevector $\vec{k}_\parallel$ in Eq. (7.5). There are six bulk bands in this case, corresponding to their being a total of six atomic layers assumed in each unit cell (three in each of the ferromagnets A and B). However, the number of surface branches is smaller, depending sensitively on the choice of the exchange parameter J_{AS} at the exterior surface of the structure, assumed to be in material A.

7.3.2 Magnetostatic Region

Here we consider periodic superlattices in which ferromagnetic and non-magnetic layers alternate. The ferromagnetic layers are coupled, even though they are separated by a non-magnetic layer, because the magnetostatic scalar potential extends to the region outside any magnetic layer, having a decay length with distance from the surface like $1/k_\parallel$ [see Eq. (7.24)]. We begin with the case of the magnetization parallel to the interfaces, for which the theory was first developed by Camley et al. [19] and Grünberg and Mika [20].

Choosing coordinate axes as in Fig. 7.3, we outline a calculation for the dispersion relations in a semi-infinite superlattice for the Voigt geometry ($\phi = \pi/2$). We employ a transfer-matrix method as in Section 5.1, but now we frame it in terms of the magnetostatic scalar potential introduced in Subsection 7.2.1. Thus, for the ferromagnetic film in the nth unit cell, i.e. when $-(n-1)L > y > -(n-1)L - a$, we write

$$\psi_1(y) = A_n \exp\left\{ik_y\left[y + (n-1)L\right]\right\} + B_n \exp\left\{-ik_y\left[y + (n-1)L\right]\right\}, \quad (7.41)$$

by analogy with Eq. (5.5), where A_n and B_n are constants. The quantity k_y is defined by Eq. (7.27) provided we put $k_z = 0$ for the Voigt geometry. The magnetic layer (A) has thickness a, while the non-magnetic (or spacer) layer (B) has thickness b. A similar expression to Eq. (7.41), but with different amplitudes and with modified exponential factors, holds in the non-magnetic layer of the nth cell. The boundary conditions at any interface are just the same as quoted in Subsection 7.2.1. It is therefore a straightforward matter to construct the transfer matrix and hence deduce the dispersion relations of the magnetostatic modes. For the bulk modes of the superlattice the result is

$$\cos(QL) = \cosh(k_\parallel a)\cosh(k_\parallel b) + Z(\omega)\sinh(k_\parallel a)\sinh(k_\parallel b), \quad (7.42)$$

where Q, as before, denotes the real Bloch wavevector, $L = a + b$ is the size of the superlattice unit cell, and the factor Z is defined in terms of the susceptibility components given in Eqs. (7.13) and (7.14):

$$Z(\omega) = \frac{1}{2}[1 + \chi_a(\omega)] - \frac{\chi_b(\omega)[1 + \chi_b(\omega)]}{[1 + \chi_a(\omega)]}. \quad (7.43)$$

The solutions of Eqs. (7.42) and (7.43) correspond to a band of frequencies bounded by the curves for $QL = 0$ and $QL = \pi$.

Any surface mode of the semi-infinite superlattice must correspond to solutions of Eq. (7.42) for values of $Q = i\beta$ that are complex with the correct sign of Re(β) for the envelope function to decay with distance from the surface (i.e. as $y \to -\infty$). It is also necessary to satisfy the magnetostatic boundary conditions at the surface $y = 0$. If the exterior medium is non-magnetic, the implicit dispersion relation for a surface mode is

$$2 + \chi_a(\omega) + \text{sgn}(k_x)\chi_b(\omega) = 0 \quad (7.44)$$

provided $a > b$, meaning the magnetic films are thicker than the spacer layers. It is remarkable that Eq. (7.44) does not depend on the layer thicknesses, except through the stated inequality. Thus the surface-mode frequency is the same as that for a semi-infinite ferromagnet; also the dependence on the sign of k_x is another indication of non-reciprocal propagation of the surface mode. If $a < b$ there is no superlattice surface mode, because we cannot then satisfy the localization condition.

The above predictions have been very clearly demonstrated by Brillouin light scattering experiments (see e.g. the reviews in Refs. [5,11]).

Calculations for semi-infinite ferromagnetic/non-magnetic superlattices in the case of the magnetization being perpendicular to the interfaces have also been reported [21]. The expressions for the susceptibility components in Eqs. (7.13) and (7.14) can again be employed provided the applied field H_0 is replaced by the static internal field $H_i = H_0 - N_z M$. This reduction is caused by the demagnetizing factor in the film; the z-axis is chosen to be perpendicular to the interfaces (as in Subsection 7.2.2) and so $N_z = 1$. The calculation for the dispersion relations follows straightforwardly using the transfer-matrix method, as detailed in [21]. If the thickness of the outermost ferromagnetic layer (i.e. the one at the superlattice surface) is allowed to be different from that in the other cells, then it can be shown that surface superlattice modes are predicted. This is noteworthy, because no surface modes exist in a single magnetic film in this perpendicular geometry. Some extensions of these calculations, including also the antiferromagnetic/non-magnetic superlattices will be discussed for the case of quasiperiodic structures later in this chapter.

7.4 Rare-Earth Superlattices

For a first example of magnetic superlattices involving applications to specific materials, we consider the rare-earth metals. Their magnetic properties in bulk samples were the subject of intensive studies in the 1960s (for major reviews, see Refs. [22,23]). Much of the interest at that time had been stimulated by the fascinating variety of magnetic ordering found in the various heavy rare-earth metals (such as Tb, Dy, Ho, Er, and Tm) on cooling them down from the paramagnetic region. In particular, the neutron diffraction technique allowed a determination of the various complex orderings [24,25]. For each of the rare-earth metals a transition at a Néel temperature T_N was observed to an oscillatory antiferromagnetic configuration of a helical or linear-oscillatory type. At a lower Curie temperature T_C, further transitions to simple ferromagnetic, ferromagnetic spiral, or anti-phase domain-type configurations were observed.

Rare-earth materials display, apart from the simple ferromagnetic alignment, several magnetic ordering schemes. To determine these magnetic states (as a necessary preliminary before solving for the magnetic excitations), it is usually adequate to treat the localized moments by using molecular-field (or mean-field) theory [26–28], which ignores the spin fluctuations. Among these magnetic orderings, two of the most important are the so-called spiral and cone states. The materials all crystallize in the hexagonal-close-packed (hcp) structure.

The principal mechanism responsible for spiral and cone states is the competition between ferromagnetic nearest-neighbor and antiferromagnetic next-nearest-neighbor exchange terms. The basic feature of these states is the rotation of the magnetization direction from one close-packed plane to the next one by an angle ϕ. Furthermore, in the cone state the direction of the static magnetization makes an angle θ with the normal to the close-packed plane, while in the spiral state the direction of the magnetization lies in the plane ($\theta = \pi/2$). The exchange derives from the so-called Ruderman–Kittel–Kasuya–Yosida (RKKY) interaction, which is a long-range oscillatory process via the polarization of the conduction electrons [29–31]. Another contribution to the total energy comes from the anisotropy energy of the unstrained lattice resulting from interaction of each rare-earth ion with the crystalline electric field due to the other charged rare-earth ions in the hcp lattice [32,33]. However this is a minor contribution compared with the RKKY interaction, and so it will not be considered here.

For example, spiral states occur in dysprosium (Dy) in the temperature interval between T_N and T_C, namely 179 K $> T >$ 85 K. The spiral turn angle between the moments of successive hexagonal layers varies with temperature, decreasing from about 43° at the Néel temperature T_N to about 26° at the Curie temperature T_C. Cone states occur, for example, in holmium (Ho) for $T < T_C$ (= 20 K). A good account of the RKKY Hamiltonian, which successfully describes these states, is to be found in Ref. [34].

Following the success of modern crystal growth techniques in producing rare-earth superlattices with high-quality interfaces, the interest in these materials is currently undergoing a fast revival, with exciting new perspectives for the future [35–37]. In this section we develop a theory of magnetic polaritons, which propagate in superlattices made up of a rare-earth magnet intercalated by a non-magnetic material. We consider both the spiral and the cone states in the rare-earth constituent of the superlattice, and we include the magnetostatic (or unretarded) limit as part of our discussions.

We have shown in the previous chapters on non-magnetic polaritons in superlattices, as well as in Section 7.3 on magnetic superlattices, that the bulk and surface modes can be conveniently described by a transfer-matrix formalism. In these cases the transfer matrix T provides relationships between the field amplitudes as one traverses across one or more periods of the superlattice. Another approach, which we shall introduce here, is known as the *effective-medium method*, and it is very suitable for describing excitations in the far-infrared region, corresponding to the long-wavelength regime of $\lambda \gg L$, where λ is the wavelength of the excitation and L is the superlattice period. In this regime, which is the case for many polaritons, it might be anticipated that the equations for optical propagation reduce to those of an effective bulk uniaxial medium. The macroscopic symmetry of the superlattice is obviously uniaxial (with the growth direction being the uniaxis) because of its layered structure. In cases where λ is not large compared to the superlattice period, the details of transmission across the interfaces within a periodicity length become important, and the transfer-matrix method is then needed.

7.4. RARE-EARTH SUPERLATTICES

The effective-medium method for superlattices was established by Agranovich and Kravstov in 1985 [38], and since then it has been successfully employed for various materials [39–42]. We justify the use of this technique in the present context because many rare-earth resonance frequencies lie in the infrared frequency range of 0.1–10 THz. Therefore, it should be feasible for the far-infrared studies of semiconductor superlattices, as pursued by several experimental groups (see e.g. Refs. [43–45]), to be extended to these magnetic systems.

We consider the superlattice to be made up of alternating layers of a rare-earth magnet and a non-magnetic material, in such a way that the coordinate z-axis is chosen parallel to the easy axis of the magnetic films. The thickness of the rare-earth magnetic (non-magnetic) layer is a (b), and therefore the size of the superlattice unit cell is $L = a + b$. The system fills the half-space $y \leq 0$, with its surface parallel to the xz-plane. In the regime $y > 0$ we have vacuum. The geometry is therefore analogous to that depicted in Fig. 7.3. Surface polariton propagation is taken to be along the x-axis, normal to the cone axis, and parallel to the surface (i.e. it corresponds to the Voigt geometry).

The non-magnetic layers will be described by a dielectric constant ϵ_2 and magnetic permeability μ_2, both assumed to be independent of frequency. On the other hand, the rare-earth magnet is described by the dielectric function given by [46]

$$\epsilon_1(\omega) = 1 - \frac{\Omega_1^2}{\omega(\omega + i\gamma_1)} - \frac{\Omega_2^2}{\omega(\omega + i\gamma_2)}, \qquad (7.45)$$

where Ω_1 and Ω_2 are two plasma frequencies associated with two collision time constants γ_1^{-1} and γ_2^{-1}. Also its gyrotropic magnetic permeability tensor can be expressed as

$$\mu_1 = \begin{pmatrix} 1 + \chi_a(\omega) & i\chi_b(\omega) & 0 \\ -i\chi_b(\omega) & 1 + \chi_a(\omega) & 0 \\ 0 & 0 & 1 \end{pmatrix}, \qquad (7.46)$$

where, by an extension of the theory presented in Section 3.4, the susceptibility elements are

$$\chi_{a,b}(\omega) = \frac{1}{2}\left(\frac{A_+}{D_+} \pm \frac{A_-}{D_-}\right). \qquad (7.47)$$

The definitions for χ_a and χ_b correspond to the upper and lower signs on the right-hand side, respectively, while

$$A_\pm/(\gamma\mu N) = -S\left(K_2 + 6K_4 S^2 \cos^2\theta\right)\sin^2\theta - \left[f(k_0)\cos\theta - f(2k_0)\cos\theta \pm \omega\right]\cos\theta, \qquad (7.48)$$

where we denote $f(k) = 2J_1 S \cos(kd_c/2) - 2J_2 S \cos(kd_c)$, and

$$D_\pm = \left[f(0)\cos\theta - f(k_0)\cos\theta \pm \omega\right]\left[f(k_0)\cos\theta - f(2k_0)\cos\theta \pm \omega\right]$$

$$-2S\left[K_2 + 6K_4 S^2 \cos^2\theta\right]\left[f(k_0) - (1/2)f(0) - (1/2)f(2k_0)\right]\sin^2\theta. \qquad (7.49)$$

Here J_1 and J_2 describe the ferromagnetic nearest-neighbor and the antiferromagnetic next-nearest-neighbor exchange, respectively, while S is the total spin quantum number, N is the number of layers per unit volume, μ is the magnetic moment per layer, and K_2 and K_4 are the coefficients of the anisotropy fields. The equilibrium condition in mean-field theory gives the relation [34]

$$4K_4 S^3 \cos^3\theta + 2S(K_2 + 2J_2 - J_1 + J_1^2/8J_2)\cos\theta - \gamma H_0 = 0. \tag{7.50}$$

The external magnetic field H_0 acts in the z-direction, parallel to the easy axis of the magnetic films; its value is assumed to be below that for the spin-flop phase transition. Finally, k_0 is defined as the wavevector of the equilibrium configuration, so that $k_0 d_c = 2\phi$, with d_c denoting the unit cell length along the crystallographic c-axis (hence $d_c/2$ is the distance between the close-packed planes).

Now, applying the standard results from the effective-medium method, in which one deduces relationships between the spatially averaged electric and magnetic fields [38,47,48], we can write down expressions for the effective dielectric tensor ϵ_{eff} and the effective magnetic permeability tensor μ_{eff} as

$$\epsilon_{eff} = \begin{pmatrix} f_1\epsilon_1 + f_2\epsilon_2 & 0 & 0 \\ 0 & f_1\epsilon_1^{-1} + f_2\epsilon_2^{-1} & 0 \\ 0 & 0 & f_1\epsilon_1 + f_2\epsilon_2 \end{pmatrix}, \tag{7.51}$$

$$\mu_{eff} = \begin{pmatrix} \mu_{xx} & i\mu_{xy} & 0 \\ -i\mu_{xy} & \mu_{yy} & 0 \\ 0 & 0 & 1 \end{pmatrix}, \tag{7.52}$$

where

$$\mu_{xx} = \left\{\left(f_1^2 + f_2^2\right)\mu_a(\omega) + f_1 f_2\left[1 + \mu_a^2(\omega) - \chi_b^2(\omega)\right]\right\}/D, \tag{7.53}$$

$$\mu_{yy} = \left[\mu_a(\omega)/f_1\chi_b(\omega)\right]\mu_{xy} = \mu_a(\omega)/D, \tag{7.54}$$

$$D = \left[f_1 + f_2\mu_a(\omega)\right]. \tag{7.55}$$

Here $\mu_a(\omega) = 1 + \chi_a(\omega)$, and the superlattice fractions are $f_j = j/L$ ($j = a, b$).

The dispersion relation for the magnetic polaritons in the unretarded (or magnetostatic) limit can now be found through Maxwell's equations as in Subsection 7.2.1. Thus we introduce the magnetostatic potential ψ as in Eq. (7.21), while Eq. (7.22) is generalized to

$$\mu_{xx}\frac{\partial^2\psi}{\partial x^2} + \mu_{yy}\frac{\partial^2\psi}{\partial y^2} + \frac{\partial^2\psi}{\partial z^2} = 0. \tag{7.56}$$

With the minor difference that $\mu_{xx} \neq \mu_{yy}$ in the present case, the calculation of the bulk and surface modes is analogous to that in Subsection 7.2.1 (taking the thick-film limit $d \to \infty$ and the Voigt geometry). The bulk modes satisfy

$$\mu_{xx}k_x^2 + \mu_{yy}k_y^2 = 0. \tag{7.57}$$

7.4. RARE-EARTH SUPERLATTICES

Unlike the previous case of Eq. (7.29), the solutions for ω_B depend on the wavevector components, so a band of excitations is predicted.

For the magnetostatic surface modes, we obtain the following implicit dispersion relation:

$$(1 + \mu_{xy})^2 - \mu_{xx}\mu_{yy} = 0 \tag{7.58}$$

with the constraint (as a localization condition) that

$$1 + \mu_{xy} < 0. \tag{7.59}$$

This dispersion relation leads to the analog of the well-known Damon–Eshbach magnetostatic mode (see Subsection 7.2.1).

It is now of interest to turn to the retarded modes in order to deal with the infrared region. In this case, we cannot use the magnetostatic scalar potential; it is necessary to employ the complete form of Maxwell's equation in the form

$$\vec{\nabla} \times \vec{H} = \frac{\epsilon_{\mathit{eff}}}{c} \frac{\partial \vec{E}}{\partial t}, \tag{7.60}$$

where c is the velocity of light in vacuum. The calculation now proceeds like in Section 4.2 for the single-interface polaritons in anisotropic media. The difference is that for the effective medium representing the superlattice we have both a dielectric tensor and a magnetic permeability tensor of the forms already quoted. From the determinantal condition obtained from the linear equations for coefficients of the magnetic field components, we find that the dispersion relation for the bulk modes is given by

$$\mu_{xx}k_x^2 + \mu_{yy}k_y^2 = \epsilon_{zz}(\omega/c)^2(\mu_{xx}\mu_{yy} - \mu_{xy}^2) \tag{7.61}$$

with ϵ_{zz} denoting the zz-component of the tensor ϵ_{eff} defined in Eq. (7.51).

The retarded surface modes are obtained by writing down localized solutions for the magnetic field in the regions $y < 0$ and $y > 0$, and then proceeding as before (in Subsection 7.2.1). After some straightforward but tedious algebra, the desired surface dispersion relation is obtained as

$$\alpha_0^2(\mu_{xx}\mu_{yy} - \mu_{xy}^2) - 2k_x\alpha_0\mu_{xy} = (k_x^2 - \epsilon_{zz}\mu_{yy}\omega^2/c^2), \tag{7.62}$$

provided the constraint

$$\alpha_0(\mu_{xx}\mu_{yy} - \mu_{xy}^2) - k_x\mu_{xy} < 0 \tag{7.63}$$

is satisfied. Here we have defined $\alpha_0 = [k_x^2 - (\omega/c)^2]^{1/2}$.

Examples of the dispersion curves given by Eqs. (7.61) and (7.62) calculated for a superlattice formed from cone-state Ho alternating with ZnF$_2$ are shown in Fig. 7.6 [49]. The physical parameters used here for Ho ($\theta = 4\pi/9$, $\phi = \pi/6$) are those given in Ref. [23]. For the non-magnetic material (ZnF$_2$), we have taken $\epsilon_2 = 8$ and $\mu_2 = 1$. Although the asymmetry between the positive-k_x and negative-k_x sides is not very noticeable, even in the absence of the magnetic field (Fig. 7.1a) the surface-polariton spectrum exhibits non-reciprocity in the sense of Eq. (7.37).

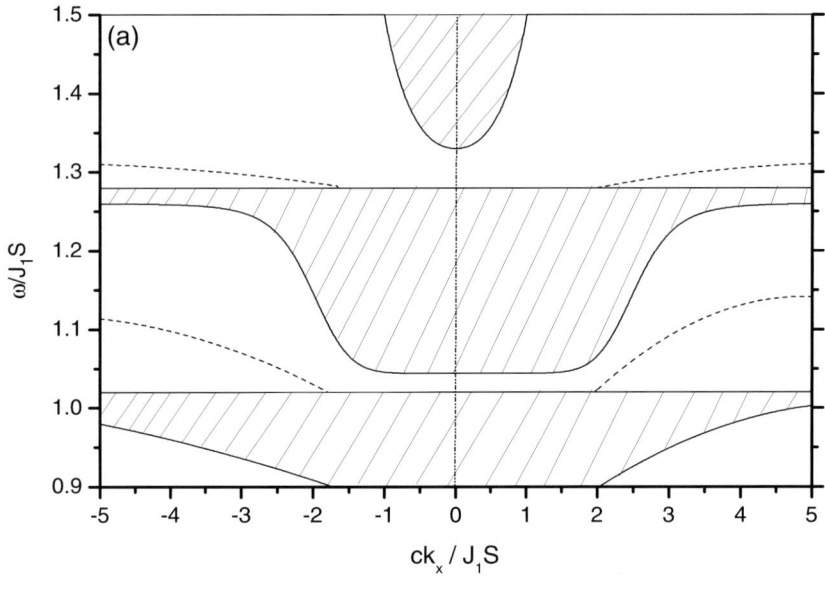

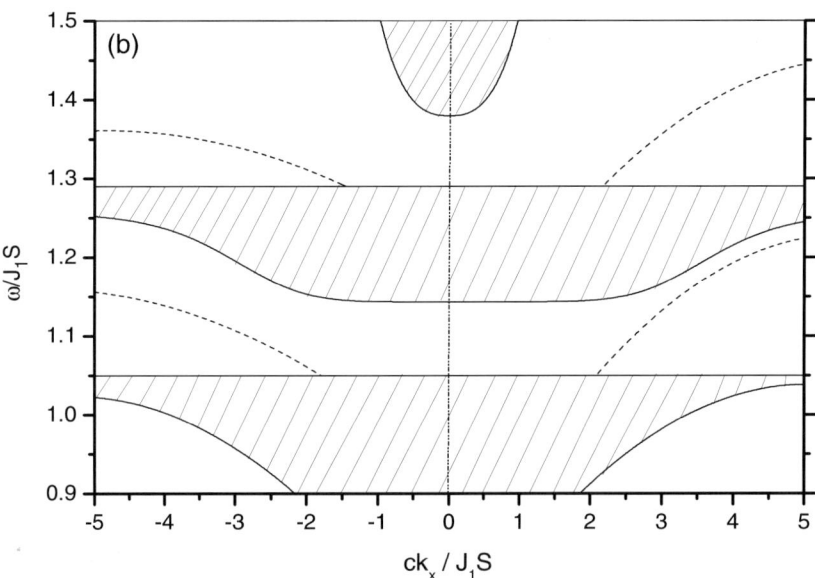

Fig. 7.6. Magnetic polariton spectrum for the cone-state case. Here the bulk modes are shown shaded, while the surface modes are represented by dashed lines. The geometry of the superlattice is such that $f_1 = 0.25$ and $f_2 = 0.75$. The physical parameters used here, which are appropriate to Ho, are described in the text. The values of the external magnetic field are (a) $H_0 = 0$ and (b) $H_0 = 200$ G (after Albuquerque and Da Silva [49]).

This non-reciprocal propagation appears here due to the non-zero off-diagonal term μ_{xy} in the permeability tensor given by Eq. (7.52). The three bulk bands, shown shaded here, are a consequence of the two poles in the susceptibility [see Eq. (7.47)], and are given only for the cone state. The surface modes (dotted lines), which are sensitive to the value of the external magnetic field, lie, as usual, in the regions between the bulk bands. Even with only a fairly modest applied magnetic field, there is a striking increase in the non-reciprocity in both the upper and lower surface branches. Fig. 7.1b describes this situation with $H_0 = 200$ G. The bulk bands are also modified by the application of H_0, although not to the same extent.

For completeness we also present in Fig. 7.7 the spectrum for the case of spiral-state Dy as the rare-earth component of the superlattice, using $\theta = \pi/2$ and $\phi = \pi/6$ for the physical parameters [23]. Although the calculation for this case is just the limiting situation where $\theta = \pi/2$ in the cone-state case, some interesting physical properties come from this structure as well. First we note that, because the off-diagonal term μ_{xy} of the magnetic permeability tensor [see Eq. (7.54)] vanishes for $H_0 = 0$, the spectrum is now reciprocal in the absence of the magnetic field as in Fig. 7.7a. Furthermore, this structure behaves just like a superlattice with alternating layers of a simple antiferromagnet and a non-magnetic material. This can be seen by taking the limiting case for the susceptibility, and it results in only two bulk bands in the absence of an external magnetic field. However, if an external magnetic field is applied, the upper bulk continuum is split into two bulk bands, as shown in Fig. 7.7b. Again, a remarkable non-reciprocity is apparent in the lower-frequency surface modes, but the reciprocal aspect of the bulk bands is unaffected due to the symmetry of the bulk dispersion relation in spiral rare-earths. We note the appearance of a new surface mode, which does not exist in the magnetostatic limit, appearing on the $-k_x$ side of the spectrum. It is a consequence of the antiferromagnetic character of the spiral rare-earth mode, and its properties are analogous to those found in other antiferromagnetic/non-magnetic superlattices in the polariton region [50].

7.5 Metamagnetic Thin Films

Mixed magnetic systems, including those in which layers of two different materials are juxtaposed in either a finite structure or a superlattice, are of particular interest theoretically and experimentally because of the competing interactions and interface properties (see e.g. Refs. [8,17]). For example, superlattices composed of alternating layers of a ferromagnet and an antiferromagnet will involve competing exchange interactions, while more generally other mixed magnetic systems with competing exchange terms present exciting challenges from both the theoretical and experimental points of view [51–53]. For instance, strongly competing interactions can lead the mixed magnet to exhibit frustration effects and often into spin-glass behavior [54–56].

Another example involving mixed (or competing) exchange is provided by *metamagnetic* materials. Basically, these have a layered magnetic structure in which ferromagnetically coupled spins in a layer are also coupled by a weak

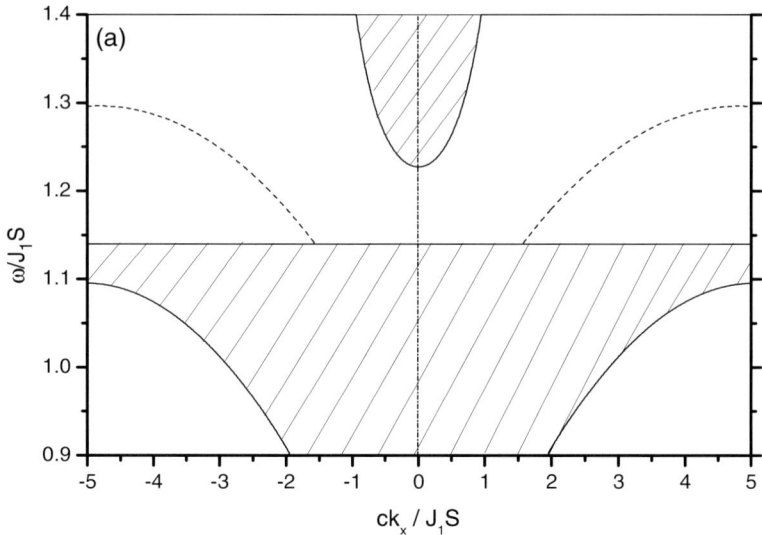

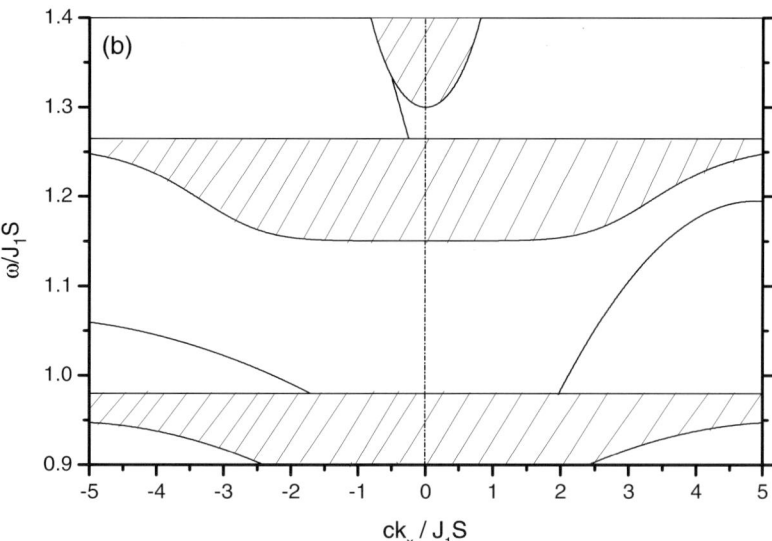

Fig. 7.7. The same as in Fig. 7.6, but for the spiral state, assuming Dy (after Albuquerque and Da Silva [49]).

7.5. METAMAGNETIC THIN FILMS

interlayer antiferromagnet exchange to the adjacent layers in the material. Such systems, including $FeBr_2$ and $FeCl_2$, have been extensively studied as regards their bulk properties for many years (see e.g. Refs. [57,58]). Typically, the ferromagnetically ordered layers have magnetization directions that alternate up and down as one moves from layer to layer; thus the total spontaneous magnetization adds up to zero. However, if a magnetic field H_0 is applied perpendicular to the layers, it may be sufficiently strong to overcome the weak antiferromagnetic exchange, and the overall ordering is then ferromagnetic with a non-zero total magnetization. Another characteristic of these magnetic materials is that the application of H_0 causes them to undergo a phase transition directly from the antiferromagnetic (AFM) phase, occurring at zero or small H_0, to the ferromagnetic (FM) phase at larger H_0. By contrast with typical antiferromagnets, this occurs without an intermediate spin-flop phase, and it is a consequence of the large single-ion anisotropy characteristic of the metamagnets [57].

Since metamagnets are layered structures, they are natural candidates in which to study the surface and bulk spin waves. In recent theoretical work on surface spin waves in metamagnets in both the AFM and FM phases, a rich spectrum was predicted for the metamagnets $FeBr_2$ and $FeCl_2$. As mentioned, they both have a large magnetic anisotropy, which is comparable to the ferromagnetic intralayer exchange but much larger than the weak antiferromagnetic interlayer coupling. The two materials have the same hexagonal structure of magnetic ions within a layer, but the stacking of the layers is different (see Fig. 7.8). Thus they are interesting cases for comparison, and it has been found that the structural effects lead to a difference in the surface spin-wave dispersion relations and existence conditions [59–61]. Their bulk samples have been studied experimentally by techniques like neutron scattering [58] and light scattering [62], mainly in the AFM phase. Theoretical studies were also made, leading to satisfactory agreement with experiment [63]. They typically have a ferromagnetic resonance below 1 cm^{-1} in which all sublattices (layers) precess in phase and, in addition, one or more higher-frequency exchange resonances in which some of the sublattices (layers) are not in phase.

In this section we shall explore the nature of the spectrum of surface magnetic polaritons propagating in metamagnetic thin films. As mentioned earlier, magnetic polaritons, or the coupled electromagnetic and spin-wave modes, have been discussed by many authors in different ferromagnetic and antiferromagnetic arrangements and are a topic of continuing interest; some additional references are Refs. [64–68]. In thin films, one finds both surface polaritons, where the excitation is localized near the surface, and guided modes, where the excitation has a standing-wave-like character. This is analogous to the situation with non-magnetic polaritons (see Chapter 4).

We consider a geometry in which the film surfaces are in the xy-plane, perpendicular to the magnetization and the external magnetic field directed along the z-direction. The surface polaritons are taken to propagate along the x-axis (see Fig. 7.8). The thin film with thickness L occupies the region $0 < z < L$, and it is assumed that the medium bounding the film is vacuum (with $\epsilon = \mu = 1$).

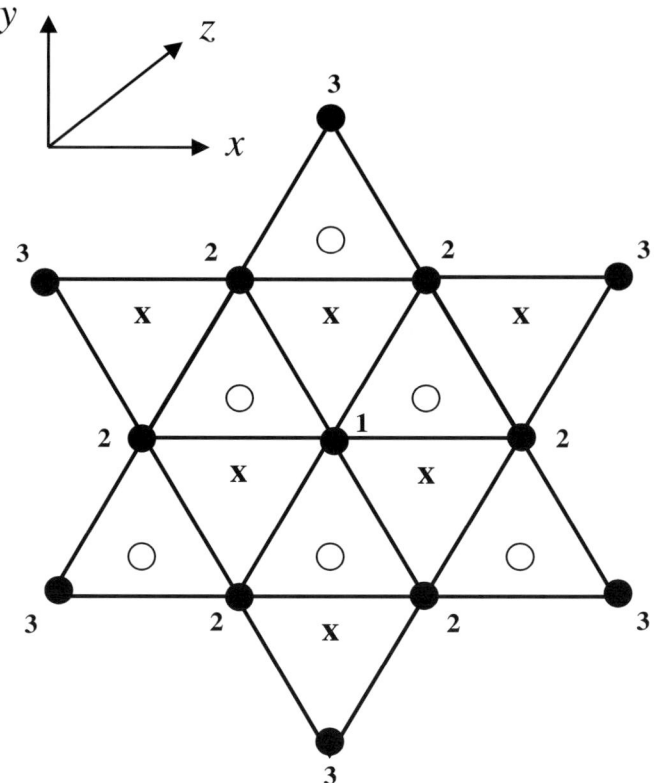

Fig. 7.8. Planar view of the hexagonal arrangement of Fe^{2+} ions (solid circles) in the ferromagnetically ordered layers of the metamagnets $FeBr_2$ and $FeCl_2$. The nearest- and next-nearest neighbors to the ion labelled 1 are those labelled 2 and 3, respectively. In $FeBr_2$ the ions in adjacent layers are directly above and below the solid circles. In $FeCl_2$ the layers are staggered with respect to one another; the circles and crosses represent the positions of the ions in adjacent layers above and below.

The spin Hamiltonian for uniaxial metamagnets can be expressed as [59,60]

$$H = \sum_{i,j} J_{ij} \vec{S}_i \cdot \vec{S}_j - (\tfrac{1}{2}) \sum_{i,i'} J_{ii'} \vec{S}_i \cdot \vec{S}_{i'} - (\tfrac{1}{2}) \sum_{j,j'} J_{jj'} \vec{S}_j \cdot \vec{S}_{j'} + H_Z + H_{anis},$$

(7.64)

where the Zeeman term due to the applied field is

$$H_Z = -\gamma H_0 \left[\sum_i S_i^z + \sum_j S_j^z \right],$$

(7.65)

7.5. METAMAGNETIC THIN FILMS

and the anisotropy term is approximated by

$$H_{anis} = -D\left[\sum_i (S_i^z)^2 + \sum_j (S_j^z)^2\right]. \tag{7.66}$$

Here i and i' denote sites on one sublattice (i.e. one type of layer in the metamagnetic crystal), while j and j' denote sites on the other sublattice (the set of adjacent layers). Also J_{ij} is the interlayer antiferromagnetic exchange interaction and $J_{ii'}$, $J_{jj'}$ are the intralayer ferromagnetic exchange terms. The uniaxial anisotropy coefficient D is related to the effective anisotropy field H_A by the relationship $H_A = (2S - 1)D$, valid at low temperatures. The other notation is as before.

The operator equation of motion for $S_i^+ = S_i^x + iS_i^y$ is now employed as in Eq. (1.30) but with the full form of the Hamiltonian H defined above. The RPA at low temperatures can then be applied as in Section 1.7. It is relatively straightforward to show that the spin-wave dispersion relations in a bulk metamagnetic material are as follows (see e.g. Ref. [69] for more details):

(a) FM phase (one branch):

$$\cos(ka) = [-1 + 3(J_1 + J_2)/J_3] - (J_1/J_3)\varphi_M(\vec{k}_\parallel) + (\Delta_M^+ - \hbar\omega)/2J_3S, \tag{7.67}$$

where

$$\varphi_M(\vec{k}_\parallel) = \cos(k_x a) + 2\cos(k_x a/2)\cos(k_y a\sqrt{3}/2)$$
$$+ 2(J_2/J_1)\left[\cos(k_y a\sqrt{3}) + 2\cos(k_x a3/2)\cos(k_y a\sqrt{3}/2)\right] \tag{7.68}$$

and $\Delta_M^\pm = \gamma[H_0 \pm H_A]$. Here J_1 and J_2 are the nearest-neighbor and next-nearest-neighbor exchange terms, respectively; J_3 is the weak antiferromagnetic interlayer exchange term.

(b) AFM phase (two branches):

$$\cos(ka) = \pm(1/2SJ_3)\left\{\left[\lambda + (\Delta_M^+ - \hbar\omega)\right]\left[\lambda - (\Delta_M^- - \hbar\omega)\right]\right\}^{1/2}, \tag{7.69}$$

where

$$\lambda = 2J_1 S\left[3(1 + J_2/J_1) + (J_3/J_1) - \varphi_M(\vec{k}_\parallel)\right]. \tag{7.70}$$

For these results we have assumed the FeBr$_2$ structure and k denotes the wavevector component in the z-direction. There are analogous expressions for the FeCl$_2$ case. We shall not quote the surface spin-wave dispersion relation; these are given in the references mentioned earlier.

We now turn to our main objective of calculating the polariton dispersion relations in metamagnetic films. Before doing this, we need expressions for the dynamic response of a metamagnet, i.e. its magnetic susceptibility as a function of the frequency ω for the assumed geometry. We have already shown in Section 3.4 how such a calculation is carried out, and we obtained expressions applicable to simple ferromagnets and antiferromagnets. The main steps involved in generalizing those earlier results to metamagnets (in both the FM and AFM phases) are outlined in the following paragraphs.

We suppose that the total magnetizations are equal to $\vec{M}_i$ and $\vec{M}_j$ for the two sublattices. Their classical equation of motion takes the same form as in Eq. (3.36):

$$d\vec{M}_p/dt = \gamma(\vec{M}_p \times \vec{H}^{\text{eff}}), \quad p = i, j. \tag{7.71}$$

Here p keeps track of the type of site (i or j) and the total effective field is

$$\vec{H}^{\text{eff}} = \vec{H}_0 + \vec{H}_A - \vec{H}_E + (\tfrac{1}{3})(\vec{M}_i + \vec{M}_j) + \vec{h}(t), \tag{7.72}$$

where $\vec{h}$ is the fluctuating driving field at frequency ω. The $\vec{H}^{\text{eff}}$ field implicitly depends on p because the direction of $\vec{H}_A$ is in the positive (negative) z-direction for $p = i$ (j) in the AFM phase. The term $\vec{H}_E$ describes the effective field of the interlayer exchange coupling, and $(\tfrac{1}{3})(\vec{M}_i + \vec{M}_j)$ is the Lorentz term describing local field corrections. The latter is important here, since the contribution it makes is neither vanishing (as in a ferromagnet) nor negligible (as in an antiferromagnet).

The equation of motion Eq. (7.71) is now linearized to relate the fluctuating magnetizations $\vec{m}_p$ ($p = i, j$) to the driving term $\vec{h}$. As in Eqs. (3.38) and (3.39), it is again convenient to use the rotating wave representation to obtain for the coupled equations of motion

$$(\omega + \omega_{ij})m_i^+ + \omega_{ii}m_j^+ = \omega_{Mii}h^+, \tag{7.73}$$

$$\omega_{jj}m_i^+ - (\omega - \omega_{ji})m_j^+ = \omega_{Mjj}h^+. \tag{7.74}$$

These are analogous to Eqs. (3.40) and (3.41), but now we have introduced the frequencies

$$\omega_{ij} = \gamma[H_E - H_A - H_0 - (\tfrac{1}{3})M_j^z], \tag{7.75}$$

$$\omega_{ji} = \gamma[-H_E - H_A + H_0 + (\tfrac{1}{3})M_i^z], \tag{7.76}$$

$$\omega_{\mu\mu} = \mp\gamma[H_E - (\tfrac{1}{3})M_\mu^z], \tag{7.77}$$

$$\omega_{M\mu\mu} = \mp\gamma M_\mu^z. \tag{7.78}$$

In Eqs. (7.77) and (7.78) μ can be i or j, with the upper and lower signs referring to i and j, respectively. The remaining part of the calculation now becomes formally similar to the antiferromagnetic case discussed in Section 3.4. The results in the FM phase can likewise be deduced as before.

7.5. METAMAGNETIC THIN FILMS

Once we have found the permeability tensor appropriate to the metamagnetic material from the above analysis, the calculation of the polariton spectrum is straightforward and proceeds by analogy with the non-magnetic cases in previous chapters. We use Maxwell's curl equations in the magnetic film. After eliminating the electric-field variable $\vec{E}$, we obtain the following wave equation for the magnetic field:

$$\nabla^2 \vec{H} - \vec{\nabla}(\vec{\nabla} \cdot \vec{H}) - (\epsilon\bar{\mu}/c^2)\partial^2 \vec{H}/\partial t^2 = 0, \tag{7.79}$$

while the divergence condition gives

$$\vec{\nabla} \cdot (\bar{\mu}\vec{H}) = 0. \tag{7.80}$$

For the vacuum regions the electromagnetic wave equation is simply

$$c^2 \nabla^2 \vec{H} - \partial^2 \vec{H}/\partial t^2 = 0. \tag{7.81}$$

The solutions of Eqs. (7.79)–(7.81) are of the form

$$\vec{H} = \vec{H}_1 \exp(-\beta z) \exp(ik_\| x) \exp(-i\omega t) \quad \text{for} \quad z > L, \tag{7.82}$$

$$\vec{H} = [\vec{A}\exp(ik_z z) + \vec{B}\exp(-ik_z z)] \exp(ik_\| x) \exp(-i\omega t) \text{ for } 0 < z < L, \tag{7.83}$$

$$\vec{H} = \vec{H}_3 \exp(\beta z) \exp(ik_\| x) \exp(-i\omega t) \quad \text{for} \quad z < 0, \tag{7.84}$$

where k_z is real for guided modes and imaginary (namely, $k_z = i\alpha$, with α real and positive) for surface modes. As a condition for a bounded excitation, β is real and positive. The above expressions are non-trivial solutions provided

$$\alpha^2 = (\mu_{xx}/\mu_{zz})k_\|^2 - (\omega/c)^2 \epsilon_\perp \mu_V, \tag{7.85}$$

$$\beta^2 = k_\|^2 - \omega^2/c^2. \tag{7.86}$$

The quantity μ_V is called the Voigt permeability and is defined by

$$\mu_V = \mu_{xx} - \mu_{xz}^2/\mu_{zz}. \tag{7.87}$$

The determination of the dispersion relation now requires the application of the usual electromagnetic boundary conditions at $z=0$ and $z=L$, namely, the continuity of the tangential components of the magnetic and electric fields. After a bit of algebra, we obtain [70]

$$\frac{g_+ - \beta(\mu_{xx}\mu_{zz} - \mu_{xz}^2)}{g_- - \beta(\mu_{xx}\mu_{zz} - \mu_{xz}^2)} \exp(-k_z L) = \frac{g_+ + \beta(\mu_{xx}\mu_{zz} - \mu_{xz}^2)}{g_- + \beta(\mu_{xx}\mu_{zz} - \mu_{xz}^2)} \exp(k_z L), \tag{7.88}$$

where we define the factors $g_\pm = \mu_{xz} k_\| \pm \mu_{zz} k_z$. Eq. (7.88) is the main analytical expression describing the propagation of the surface magnetic polaritons. It can be used to reproduce results previously reported for a single interface by taking $L \to \infty$, i.e. the case of a semi-infinite metamagnetic medium [71,72].

In the magnetostatic limit, where $k_\parallel \gg \omega/c$, the results simplify. The decay constant in the vacuum regions reduces to $\beta = k_\parallel$, and

$$k_z = (\mu_{xx}/\mu_{zz})^{1/2} k_\parallel \qquad (7.89)$$

inside the film. Eq. (7.88) then becomes

$$(\mu_{xx}\mu_{zz} - \mu_{xz}^2 + 1)\tanh\left[(\mu_{xx}/\mu_{zz})^{1/2}|k_\parallel|L\right] + 2\mu_{zz}(\mu_{xx}/\mu_{zz})^{1/2} = 0. \qquad (7.90)$$

As might be expected for this geometry, the dispersion relation in this limit is reciprocal.

We now discuss our analytical results in more detail by making specific applications to the metamagnets $FeBr_2$ and $FeCl_2$ in their FM and AFM phases, respectively. As discussed already, these materials have different crystallographic arrangements of the magnetic ions, which will lead to differences in the polariton spectra. Although they have the same trigonal arrangement of the magnetic ion Fe^{2+} within each ferromagnetically ordered layer, the two materials differ in the stacking arrangement of the layers (see Fig. 7.8). In $FeBr_2$, the Fe^{2+} ions in one layer are directly above and below those in the adjacent layers. On the other hand, for $FeCl_2$, the stacking of the layers is staggered, leading to three nearest neighbors in each of the adjacent layers.

Fig. 7.9 shows the magnetic polariton dispersion relations for bulk and surface modes considering a metamagnetic thin film of thickness $L = 200\,\mu m$ for $FeBr_2$ in its FM phase [73]. This material has spin $S = 1$ and critical temperature $T_c = 14.2\,K$. The approximate values of the exchange and other relevant parameters are known from Raman scattering experiments [58,62]: $J_1 = 5.07\,cm^{-1}$, $J_2 = -1.2\,cm^{-1}$, $J_3 = 1.45\,cm^{-1}$, $H_A = 7.34\,cm^{-1}$, and $M_i^z = M_j^z = 1.45\,kG$. We have plotted the reduced frequency ω/ω_{FM} against the dimensionless in-plane wave vector $k_x L$, taking $\vec{k}_\parallel = (k_x, 0)$. The resonance frequency ω_{FM} is associated with a precession of the total magnetization $\vec{M} = \vec{M}_i + \vec{M}_j$ about the effective static field:

$$\omega_{FM} = (1/2)(\omega_{ji} - \omega_{ij}) + (\tfrac{1}{2})[(\omega_{ij} + \omega_{ji})^2 - 4\omega_{ii}\omega_{jj}]^{1/2}. \qquad (7.91)$$

From the definitions it may be shown that this quantity increases linearly with the external applied field $\vec{H}_0$.

The nearly vertical dashed lines in this figure are the light lines $\omega = ck_x$ in vacuum. We have considered a free-standing metamagnetic film, assuming that the bounding media are both the same (namely vacuum). The generalization to the case when the media are different, as would occur for a film on a substrate, is straightforward. Because of the finite film thickness, the bulk modes are actually quantized, corresponding to different standing waves in the film. However, for simplicity, we just show the bulk-mode regions shaded. There are two bulk bands, and in the gap regions four surface modes can propagate (shown as full lines identified by the letter S). As $|k_x L|$ increases, they asymptotically approach a well-defined value for their frequencies. The spectra are now non-reciprocal, i.e.

7.5. METAMAGNETIC THIN FILMS

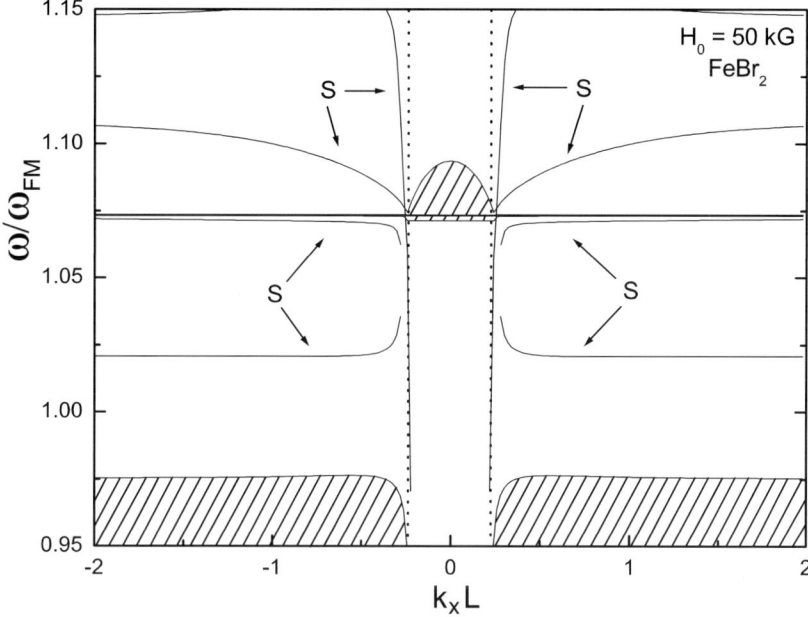

Fig. 7.9. Dispersion relation of bulk (shaded areas) and surface/guided magnetic polariton modes for the metamagnet FeBr$_2$ in its FM phase as a function of the in-plane dimensionless wave vector $k_x L$. We have considered a free-standing film and an external applied magnetic field $H_0 = 50$ kG, which is sufficient to overcome the weak antiferromagnetic interlayer coupling (after Guimarães and Albuquerque [73]).

Eq. (7.37) applies. The non-reciprocity becomes more pronounced in the supported film case, and is due to the non-vanishing off-diagonal term $\mu_{xz} = -\mu_{zx}$ in the magnetic permeability tensor. The magnetic field profile for the $+k_x L$ mode is biased towards one surface whereas the field profile for the $-k_x L$ mode is biased towards the other surface.

For completeness, we also show in Fig. 7.10 the magnetic polariton modes for the case of FeCl$_2$ in its AFM phase [73]. We have considered the film thickness equal to 200 μm, and have plotted the reduced frequency ω/ω_{AFM} against the dimensionless in-plane wavevector $k_x L$. Here, ω_{AFM} is associated with a precession about the exchange field $\vec{H}_E$ of the AFM phase; it is given by an expression similar to Eq. (7.91), provided we replace the plus sign before the square root term by a minus sign. The other physical parameters, in accordance with data from neutron and Raman scattering [63], are $J_1 = 5.5$ cm^{-1}, $J_2 = -1.2$ cm^{-1}, $J_3 = 0.28$ cm^{-1}, $H_A = 9.6$ cm^{-1}, and $M_i^z = -M_j^z = 1.2$ kG. There are three bulk regions (shown shaded), and in the gap regions five surface modes can propagate (the full lines identified by the letter S). It is possible to distinguish two different types of surface modes: those that persist as $|k_x L| \to \infty$, and the other one (the high-frequency one) that merges into a bulk region at a finite value of $|k_x L|$. The former are real

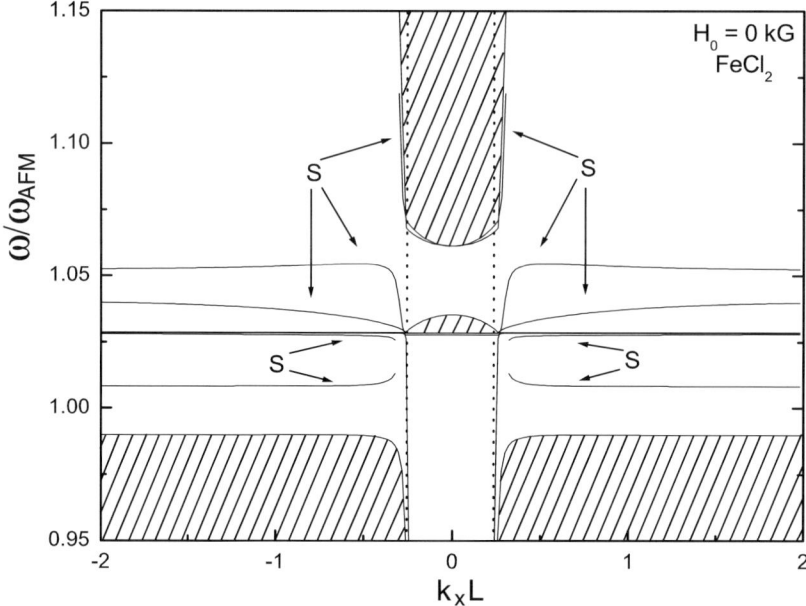

Fig. 7.10. Same as in Fig. 7.9, but for the metamagnet FeCl$_2$ in its AFM phase, in the absence of an external magnetic field (after Guimarães and Albuquerque [73]).

surface modes and stop abruptly near the light lines. The bulk bands are broader than those in the FM phase (see Fig. 7.9), essentially because the polaritons are more localized within the film as a consequence of the small J_3 value and the different stacking of layers.

7.6 Quasiperiodic Structures

In this final section we shall consider the polariton spectra in quasiperiodic structures whose building blocks are composed of a magnetic layer (building block A) and a non-magnetic material (building block B) stacked alternately following either a Fibonacci or a Double-period sequence. Our model is based on a transfer-matrix formalism to simplify the algebra, which is otherwise quite involved. The localization and scaling properties of the spectra are also presented and discussed. We are therefore extending the calculations for superlattices described in Subsection 7.3.2 in two ways: the retardation effects are included to obtain the polaritons, and quasiperiodicity effects are introduced.

We consider first, for simplicity, a periodic structure in which the coordinate y-axis is chosen parallel to the easy axis of the magnetic layers. The thickness of the magnetic (non-magnetic) layer is a (b), and therefore the size of the periodic superlattice unit cell is $L = a + b$. The system fills the half-space $z \leq 0$, with its surface parallel to the xy-plane, and in the region $z > 0$ we have vacuum. Surface

7.6. QUASIPERIODIC STRUCTURES

polariton propagation is taken to be along the x-axis, parallel to the surface (the Voigt geometry). Assuming s-polarization the electric and magnetic fields have the overall form

$$\vec{H}_j(x,z,t) = (H_{xj}, 0, H_{zj})\exp(ik_x x - i\omega t), \tag{7.92}$$

$$\vec{E}_j(x,z,t) = (0, E_{yj}, 0)\exp(ik_x x - i\omega t). \tag{7.93}$$

For the magnetic layer A the above field components are

$$H_{xA}(z) = A_{1A}^n \exp(-k_A z) + A_{2A}^n \exp(k_A z), \tag{7.94}$$

$$H_{zA}(z) = (-i/\mu_0)\left[\mu_{eff}^- A_{1A}^n \exp(-k_A z) + \mu_{eff}^+ A_{2A}^n \exp(k_A z)\right], \tag{7.95}$$

$$E_{yA}(z) = (-i/k_x)\left[\mu^- A_{1A}^n \exp(-k_A z) + \mu^+ A_{2A}^n \exp(k_A z)\right] \tag{7.96}$$

with the notation that

$$\mu_0 = -k_x^2 + \mu_1 \epsilon_A \omega^2/c^2, \tag{7.97}$$

$$\mu_{eff}^\pm = \pm k_x k_A + \mu_2 \epsilon_A \omega^2/c^2, \tag{7.98}$$

$$\mu^\pm = \mu_1 - \mu_2 \mu_{eff}^\pm/\mu_0, \tag{7.99}$$

$$k_A^2 = k_x^2 - \mu_V \epsilon_A \omega^2/c^2. \tag{7.100}$$

Here the combination $\mu_V = \mu_1 - \mu_2^2/\mu_1$ is called the Voigt permeability. The dielectric constant of medium A is ϵ_A. The basic permeability terms μ_1 and μ_2 appearing in Eqs. (7.97)–(7.100) are simply related to the susceptibility components derived earlier by

$$\mu_1(\omega) = 1 + \chi_a(\omega), \qquad \mu_2(\omega) = \chi_b(\omega). \tag{7.101}$$

The explicit expressions for χ_a and χ_b are given in Eqs. (7.13) and (7.14) for ferromagnets and in Eqs. (7.16) and (7.17) for two-sublattice uniaxial antiferromagnets.

On the other hand, for the non-magnetic layer we have

$$H_{xB}(z) = A_{1B}^n \exp(-k_B z) + A_{2B}^n \exp(k_B z), \tag{7.102}$$

$$H_{zB}(z) = (ik_x/k_B)[A_{1B}^n \exp(-k_B z) - A_{2B}^n \exp(k_B z)], \tag{7.103}$$

$$E_{yB}(z) = (-i\omega/ck_B)[A_{1B}^n \exp(-k_B z) - A_{2B}^n \exp(k_B z)], \tag{7.104}$$

where ϵ_B is the dielectric constant of medium B and

$$k_B = [k_x^2 - \epsilon_B \omega^2/c^2]^{1/2}. \tag{7.105}$$

Next, using the electromagnetic boundary conditions at the interfaces $z = nL + a$ and $z = (n+1)L$, we find the following equations for the amplitudes of the electromagnetic fields:

$$A^n_{1A} f_A + A^n_{2A} \bar{f}_A = A^n_{1B} + A^n_{2B}, \qquad (7.106)$$

$$\mu^-_{\text{eff}} A^n_{1A} f_A + \mu^+_{\text{eff}} A^n_{2A} \bar{f}_A = -\mu_0 (k_x/k_B)(A^n_{1B} - A^n_{2B}), \qquad (7.107)$$

$$A^n_{1B} f_B + A^n_{2B} \bar{f}_B = A^{n+1}_{1A} + A^{n+1}_{2A}, \qquad (7.108)$$

$$-\mu_0 (k_x/k_B)(A^n_{1B} f_B - A^n_{2B} \bar{f}_B) = \mu^-_{\text{eff}} A^{n+1}_{1A} + \mu^+_{\text{eff}} A^{n+1}_{2A}, \qquad (7.109)$$

where f_j and $\bar{f}_j$ are defined in Eq. (5.15). By introducing for each medium the two-column vector

$$|A^n_j> = \begin{bmatrix} A^n_{1j} \\ A^n_{2j} \end{bmatrix} \qquad (7.110)$$

and following the standard procedure to determine the transfer matrix $\bar{T}$ (see Section 5.1), we find that the magnetic polariton bulk dispersion relation is given formally by Eq. (5.27) with $\bar{T} = \bar{N}_A^{-1} \bar{M}_A \bar{N}_B^{-1} \bar{M}_B$, where (for $j = A, B$)

$$\bar{M}_j = \begin{pmatrix} f_j & \bar{f}_j \\ \mu^-_{\text{eff}} f_j & \mu^+_{\text{eff}} \bar{f}_j \end{pmatrix}, \qquad (7.111)$$

$$\bar{N}_j = \begin{pmatrix} 1 & 1 \\ k_x/k_B & -k_x/k_B \end{pmatrix}. \qquad (7.112)$$

We now consider a semi-infinite superlattice, occupying the region $z \leq 0$. The theory is again closely analogous to that in Section 5.1. The dispersion relation for the surface modes is given formally by Eqs. (5.34) and (5.41), where the Bloch wavevector Q has been replaced by $\beta = iQ$ in such a way that $\text{Re}(\beta) > 0$, and λ is redefined as

$$\lambda = (k_x - \mu^+_{\text{eff}} k_v)/(\mu^-_{\text{eff}} k_v - k_x) \qquad (7.113)$$

with $k_v = [k_x^2 - \omega^2/c^2]^{1/2}$.

This method can now be extended to superlattices with a periodic structure formed in a more complex manner. In the present context, we have quasiperiodic structures in mind, and we have already shown for non-magnetic polaritons in Chapter 6 how the calculations are extended using higher-order generations of a chosen quasiperiodic sequence to define a new unit cell. It is then a straightforward matter, in principle, to obtain the transfer matrix $\bar{T}$ for any structure under consideration. For the Fibonacci and Double-period cases the transfer matrices are generated formally using Eqs. (6.4) and (6.15), respectively.

We now discuss some numerical examples by making applications to specific magnetic materials. To study the structural dependence we define $R = b/a$ as the ratio between the thicknesses of the non-magnetic and the magnetic layers, respectively. Taking the magnetic material A to be the ferromagnet EuS, the

7.6. QUASIPERIODIC STRUCTURES

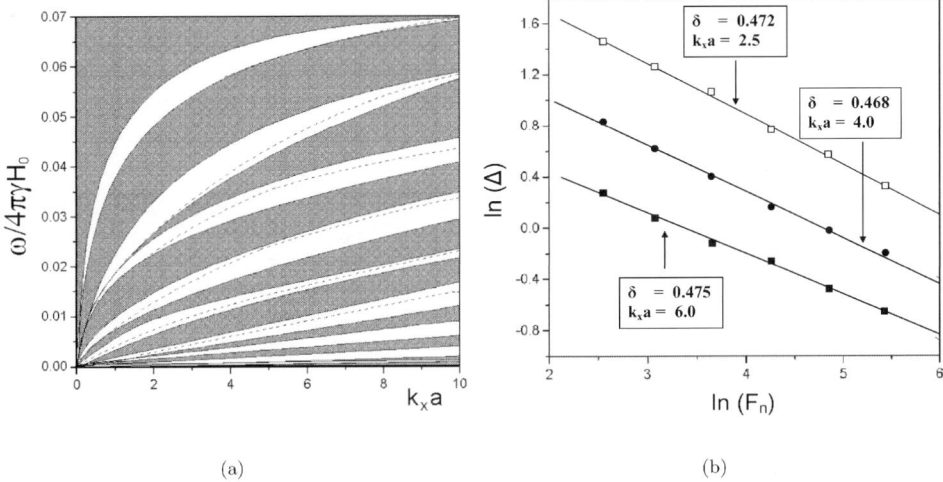

Fig. 7.11. (a) Ferromagnetic polariton spectrum showing the reduced frequency $\omega/4\pi\gamma H_0$ versus $k_x a$ for a 6th sequence, Fibonacci quasiperiodic superlattice. The physical parameters are given in the text. (b) Log–log plot of the total width of the allowed regions Δ against the Fibonacci number F_n, for several generations of the Fibonacci sequence (after Albuquerque and Guimarães [74]).

polariton spectrum in terms of $\omega/4\pi\gamma H_0$ plotted versus $k_x a$ for the 6th generation of the Fibonacci sequence is shown in Fig. 7.11a [74]. The physical parameters used here are $H_0 = 13.5$ kG, $M = 1.0$ kG, $\epsilon_A = 1.5$, and $R = 0.5$. The non-magnetic material, considered to be ZnF$_2$, has $\epsilon_B = 8$. Here the surface modes are represented by the dashed lines, while the bulk bands are characterized by the shaded areas, which are bounded by the curves for $QL = 0$ and $QL = \pi$ in an alternating fashion. For the surface modes, whenever the generation number n is greater than 2, we have a natural broken symmetry for the structure. Thus we do not need to vary the thickness of the magnetic (or non-magnetic) materials in order to obtain surface modes, as was the case in other structures [21]. This effect was evident for all the sequences with $n > 2$. We emphasize the importance of imposing the condition $\text{Re}(\beta) > 0$, because otherwise we might obtain non-physical surface modes. Such modes are characterized by having an exponentially *increasing* amplitude, and must therefore be excluded.

The total width of the allowed regions in energy for the ferromagnetic Fibonacci sequence decreases with n as the power law $\Delta \sim F_n^{-\delta}$, where F_n is the Fibonacci number. The log–log plot of the bandwidths Δ against the generation number is shown in Fig. 7.11b for several values of the in-plane wavevector $k_x a$. As we can infer, the exponent δ depends only slightly on the common in-plane wavevector $k_x a$.

Finally in Fig. 7.12a we show the spectrum for the antiferromagnet MnF$_2$ for the 3rd generation of the Double-period sequence [74]. Here the vertical axis represents the reduced frequency $(\omega - \omega_{AF})/\omega_{AF}$, where ω_{AF} is the antiferromagnetic

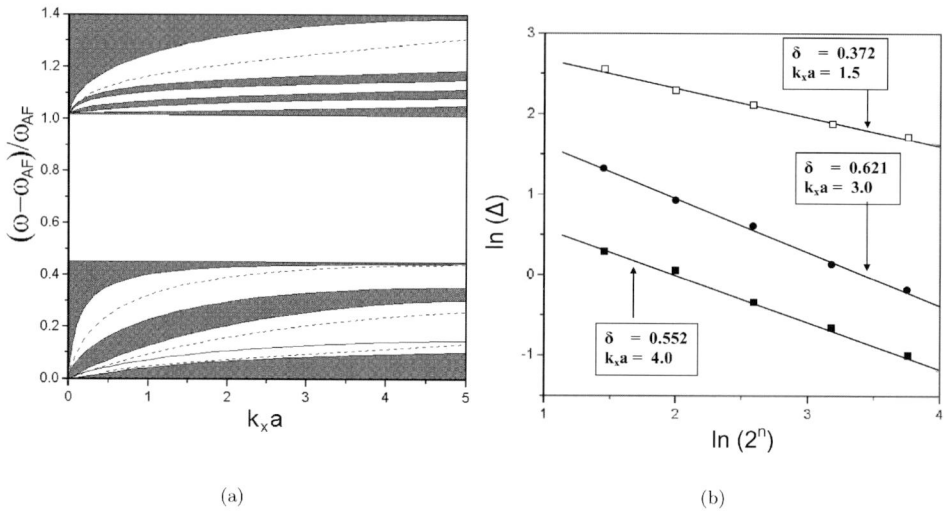

Fig. 7.12. (a) Antiferromagnetic polariton spectrum for the reduced frequency $(\omega - \omega_{AF})/\omega_{AF}$ versus $k_x a$ for a 3rd sequence, Double-period superlattice. (b) Log–log plot of the total width of the allowed regions Δ against 2^n, for several generations of the Double-period sequence (after Albuquerque and Guimarães [74]).

resonance frequency [see Eq. (7.19)]. The physical parameters, appropriate to MnF$_2$, are $H_0 = 550$ kG, $H_A = 7.87$ kG, $M = 1.0$ kG, and $\epsilon_A = 5.5$. The number of bulk bands increases as 2^n, with n being the Double-period generation number. In this case the application of an external field H_0 gives rise to a gap between two regions of solutions for the modes (with $\mu_1 < 0$). If we set the external field equal to zero, the two regions collapse into one, still showing the same self-similar behavior as in the previous case. Also, the modes are sensitive to the value of the anisotropy field, having their relative position governed by the magnitude of H_A. The scaling behavior of the allowed bandwidth Δ is $\Delta \sim (2^n)^{-\delta}$, where here, as in the Fibonacci case, the exponent δ only depends weakly on the common wavevector $k_x a$, as can be seen in Fig. 7.12b.

For both of the structures described here, a systematic study was made of the effects of the varying the physical parameters. It is found that the size of the bulk bands is quite sensitive to the ratio R. It is noteworthy that surface solutions are obtained in the quasiperiodic cases even when $R > 1$ (i.e. the non-magnetic spacer thickness is greater than that of the magnetic material). This contrasts with the behavior discussed in Subsection 7.3.2 for the corresponding periodic superlattice.

On the experimental side, the technique of Brillouin light scattering spectroscopy has proved to be important in probing the excitations discussed in this chapter and testing the theoretical predictions [11,16,75]. A detailed analysis of light scattering from the magnetic polaritons and magnetostatic modes requires evaluating the linear response of the system to an external source, such as an electromagnetic wave. The analysis can be done through the use of the Green-function techniques. There is further discussion of this in Chapter 11 and in Appendix A. Possible applications of the modes studied here include spin valves and other multilayered devices.

References

[1] D.R. Tilley, in: Linear and Nonlinear Spin Waves in Magnetic Films and Superlattices, Ed. M.G. Cottam, World Scientific, Singapore, 1994.
[2] T. Wolfram and R.E. Dewames, Prog. Surf. Sci. 2 (1972) 233.
[3] M.G. Cottam, J. Phys. C 9 (1976) 2121.
[4] H. Puszkarskii, Prog. Surf. Sci 9 (1979) 191.
[5] M.G. Cottam and D.R. Tilley, Introduction to Surface and Superlattice Excitations, 2nd. ed., Institute of Physics Publishing, Bristol, 2004.
[6] M.G. Cottam and A.A. Maradudin, in: Surface Excitations, Eds., V.M. Agranovich and R. Loudon, North-Holland, Amsterdam, 1984.
[7] R.W. Damon and J.R. Eshbach, J. Phys. Chem. Solids 19 (1961) 308.
[8] Linear and Nonlinear Spin Waves in Magnetic Films and Superlattices, Ed. M. G. Cottam, World Scientific, Singapore, 1994.
[9] P. Kabos, in: High Frequency Processes in Magnetic Materials, Eds. G. Srinivasan and A.N. Slavin, World Scientific, Singapore, 1995.
[10] P. Grünberg, Prog. Surf. Sci. 18 (1985) 1.
[11] P. Grünberg, in: Light Scattering in Solids V, Eds. M. Cardona and G. Güntherodt, Springer, Berlin, 1989.
[12] B.A. Kalinikos and A.N. Slavin, J. Phys. C 19 (1986) 7013.
[13] R.N. Costa Filho, M.G. Cottam and G.A. Farias, Phys. Rev. B 62 (2000) 6545.
[14] R.L. Stamps and R.E. Camley, J. Appl. Phys. 56 (1984) 3497.
[15] J. Barnaś, in: Linear and Nonlinear Spin Waves in Magnetic Films and Superlattices, Ed. M.G. Cottam, World Scientific, Singapore, 1994.
[16] B. Hillebrands, in: Light Scattering in Solids VII, Eds. M. Cardona and G. Güntherodt, Springer, Berlin, 2000.
[17] Ultrathin Magnetic Structures, Vols. I and II, Eds., J.A.C. Bland and B. Heinrich, Springer-Verlag, Berlin, 1994.
[18] E.L. Albuquerque, P. Fulco, E.F. Sarmento and D.R. Tilley, Solid State Commun. 58 (1986) 61.
[19] R.E. Camley, T.S. Rahman and D.L. Mills, Phys. Rev. B 27 (1983) 261.
[20] P. Grünberg and K. Mika, Phys. Rev. B 27 (1983) 2955.
[21] R.E. Camley and M.G. Cottam, Phys. Rev. B 35 (1987) 189.
[22] R.J. Elliott, in: Magnetism, Vol. IIA, Eds., G.T. Rado and H. Suhl, Academic Press, New York, 1965.
[23] B.R. Cooper, in: Solid State Physics, Vol. 21, Eds., F. Seitz, D. Turnball and H. Ehrenreich, Academic Press, New York, 1968.
[24] W.C. Koehler, E.O. Wollan, M.K. Wilkinson and J.W. Cable, in: Rare Earth Research, Ed., E.V. Kleber, Macmillan, New York, 1961.
[25] W.C. Koehler, J. Appl. Phys. 36 (1965) 1078.
[26] R.J. Elliott, Phys. Rev. 124 (1961) 346.
[27] H. Miwa and K. Yosida, Progr. Theoret. Phys. (Kyoto) 26 (1961) 693.
[28] T.A. Kaplan, Phys. Rev. 124 (1961) 329.

[29] M.A. Ruderman and C. Kittel, Phys. Rev. 96 (1954) 99.
[30] T. Kasuya, Progr. Theoret. Phys. (Kyoto) 16 (1956) 45.
[31] K. Yosida, Phys. Rev. 106 (1957) 893.
[32] K.W.H. Stevens, Proc. Phys. Soc. (London) A65 (1952) 209.
[33] R.J. Elliot and K.W.H. Stevens, Proc. Roy. Soc. A219 (1953) 387.
[34] N.S. Almeida and D.R. Tilley, Phys. Rev. B 43 (1991) 11145.
[35] C.F. Majkrzak, J. Kwo, M. Hong, Y. Yafet, D. Gibbs, C.L. Chien and J. Bohr, Adv. Phys. 40 (1991) 99.
[36] T.R. Thurston, G. Helgesen, D. Gibbs, J.P. Hill, B.D. Gaulin and G. Shirane, Phys. Rev. Lett. 70 (1993) 3151.
[37] P.M. Gehring, K. Hirota, C.F. Majkrzak and G. Shirane, Phys. Rev. Lett. 71 (1993) 1087.
[38] V.M. Agranovich and V.E. Kravstov, Solid State Commun. 55 (1985) 85.
[39] N.S. Almeida and D.L. Mills, Phys. Rev. B 38 (1988) 6698.
[40] N.S. Almeida and D. R. Tilley, Solid State Commun. 73 (1990) 23.
[41] E.L. Albuquerque and P. Fulco, Phys. Stat. Sol. (b) 182 (1994) 357.
[42] E.L. Albuquerque and P. Fulco, Z. Phys. B 100 (1996) 289.
[43] S. Perkowitz, R. Sudharsanan, K.A. Harris, J.W. Cook Jr., J.F. Schetzina and N.J. Schulman, Phys. Rev. B 36 (1987) 9290.
[44] E. Jahne, A. Roselar and K. Ploog, Superlattice and Microstructure 9 (1991) 219.
[45] B. Samson, T. Dumelow, A.A. Hamilton, T.J. Parker, S.R.P. Smith, D.R. Tilley, C.T.B. Foxon, D. Hilton and K.J. Moore, Phys. Rev. B 46 (1992) 2375.
[46] Y.V. Knyazev and M.M. Noskov, Phys. Stat. Sol. (b) 80 (1972) 11.
[47] N. Raj and D. R. Tilley, Solid State Commun. 55 (1985) 373.
[48] W.M. Liu, G. Eliasson and J.J. Quinn, Solid State Commun. 55 (1985) 533.
[49] E.L. Albuquerque and L.R. da Silva, Phys. Lett. A 218 (1996) 333.
[50] M.C. Oliveros, N.S. Almeida, D.R. Tilley, J. Thomas and R.E. Camley, J. Phys.: Condens. Matter 4 (1992) 8497.
[51] K. Held, M. Ulmke, N. Blumer and D. Vollhardt, Phys. Rev. B 56 (1997) 14469.
[52] S. Galam, C.S.O. Yokoi and S.R. Salinas, Phys. Rev. B 57 (1998) 8370.
[53] G.C. DeFotis et al, Phys. Rev. B 62 (2000) 6421.
[54] J.A. Mydosh, Spin Glass: An Experimental Introduction, Taylor & Francis, London, 1993.
[55] Z. Neda, Phys. Rev. B 50 (1994) 3011.
[56] K. Zenmyo and H. Kubo, J. Phys. Soc. Jpn. 65 (1996) 4045.
[57] F. Keffer, Handbuch der Physik 18 (1966) 1.
[58] W.B. Yelon and C. Vettier, J. Phys. C 8 (1975) 2760.
[59] J.H. Baskey and M.G. Cottam, Phys. Rev. B 42 (1990) 4304.
[60] D.H.A.L. Anselmo, E.L. Albuquerque and M.G. Cottam, J. Appl. Phys. 83 (1998) 6955.
[61] E. Meloche and M.G. Cottam, Phys. Stat. Solidi (a) 196 (2003) 165.
[62] G.C. Psaltakis, G. Mischler, D.J. Lockwood, M.G. Cottam, A. Zwick and S. Legrand, J. Phys. C 17 (1984) 1735.
[63] G.C. Psaltakis and M.G. Cottam, J. Phys. C 15 (1982) 4847.
[64] J. Barnaś, J. Phys.: Condens. Matter 2 (1990) 7173.
[65] K. Abraha, D.E. Brown, T. Dumelow, T.J. Parker and D.R. Tilley, Phys. Rev. B 50 (1994) 6808.
[66] M.R.F. Jensen, S.A. Feivent, T.J. Parker and R.E. Camley, J. Phys.: Condens. Matter 9 (1997) 7233.

[67] V.V. Tarakanov, V.I. Khizhnyi, A.P. Korolyuk and M.B. Strugatsky, Physica B 284 (2000) 1452.
[68] R.T. Tagiyeva and M. Saglam, Solid State Commun. 122 (2002) 413.
[69] J.M. de Souza, E.L. Albuquerque, D.H.A.L. Anselmo and G.A. Farias, Physica B, 324 (2003) 218.
[70] E.S. Guimarães and E.L. Albuquerque, Solid State Commun. 122 (2002) 413.
[71] N. Raj and D.R. Tilley, Phys. Rev. B 35 (1987) 189.
[72] E.S. Guimarães and E.L. Albuquerque, J. Appl. Phys. 89 (2001) 7401.
[73] E.S. Guimarães and E.L. Albuquerque, Phys. Lett. A 307 (2003) 172.
[74] E.L. Albuquerque and E.S. Guimarães, Physica A 277 (2000) 399.
[75] A.S. Borovik-Ramanov and N.M. Kreines, Phys. Rep. 81 (1982) 351.

Chapter 8

Magnetic Polaritons in Spin-Canted Systems

The application of a magnetic field *perpendicular* to the easy axis of an antiferromagnet can give rise to interesting effects in the excitation spectrum. At high magnetic fields, a significant magnetization component normal to the easy axis will be induced. We may thus expect to find a corresponding magnetic-field-induced surface spin wave that propagates parallel to the easy axis, with possible non-reciprocal propagation characteristics [1]. In this respect, the behavior would be similar to the Damon–Eshbach surface wave (see Subsection 7.2.1). Furthermore, in this physical situation the spins are canted due to the applied field, as depicted in Fig. 8.1.

Spin canting has been the subject of a great number of investigations in the past decades. In some cases it occurs spontaneously (i.e. even in the absence of an applied magnetic field) due to a lowering of symmetry produced by certain types of single-ion anisotropy (such as in NiF_2) or due to an antisymmetric-exchange mechanism (as in $FeBO_3$). These materials have a weak ferromagnetic moment due to the canting, and this makes them of particular interest for light scattering from the spin waves; a review is given in Ref. [2]. By contrast, in other cases the canting is produced by the application of an external field, as will be considered in Section 8.1.

The effects of spin canting can be observed in a number of natural and synthetic ferrimagnetic iron oxides [3], Tb_3Rh and TDAE-C-60 single crystals [4,5], as well as in some manganese compounds of the type XMn_2, where X can be Er (erbium, with a spin-canted magnetic structure at about 15 K and a spin reorientation at about 10 K [6]), Dy (dysprosium, with a spin-canted magnetic structure at about 40 K and a spin reorientation observed at 36 K [7,8]), and Ho (holmium, with a spin-canted magnetic structure at about 31 K and a spin reorientation found at the Curie point of 25 K [9]). Also, for YFe_2 compounds with $Y = Nb_{1-x}Zr_x$, neutron diffraction measurements in the composition range $0.18 \leq x \leq 0.22$ provide evidence of a low-temperature spin-canted structure in which the antiferromagnetic magnetic arrangement is modified by the appearance of a basal-plane ferromagnetic component [10]. Further, in ferromagnetic materials with an external field applied parallel to the surface, but having a strong single-ion surface anisotropy with easy axis normal to the surface, the magnetization in the bulk of the crystal will be parallel to the surface but there can be spin canting near the surface (see e.g. Ref. [11]). Moreover, in some manganese oxide films, the so-called manganites,

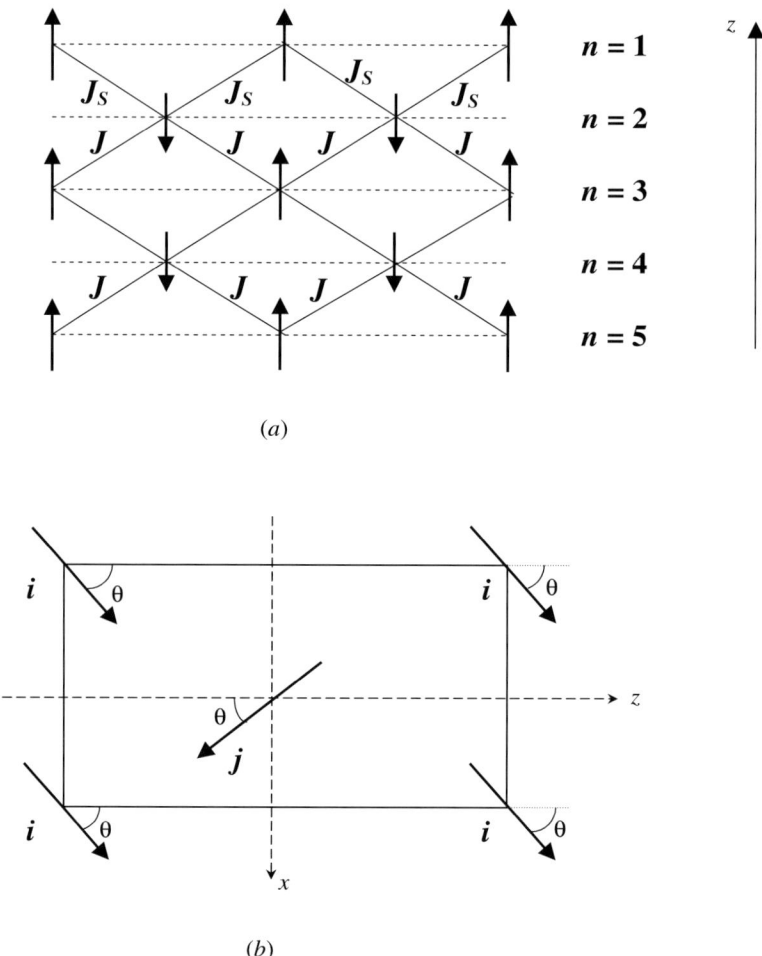

Fig. 8.1. Schematic representation of a canted antiferromagnetic crystal with body-centered tetragonal structure. We take the z-direction as the easy axis and the x-direction as the direction of the applied field. (a) Side view of a (001) surface in the absence of an applied field; (b) orientations of the canted spins for the sublattices i and j when a magnetic field is applied.

with an intrinsic antiferromagnetic spin structure, there have been observations of a huge magnetoresistance that arises from the spin-independent electron scattering due to spin canting [12,13]. In manganite compounds like $R_{1-x}D_x MnO_3$, with R being a metal or rare-earth ion and D a divalent ion, the metallic antiferromagnetism is compatible with a double-exchange theory of the canted spin arrangement; an important aspect of this theory is a self-stabilizing mechanism with the anisotropic antiferromagnetic ordering acting as the main symmetry breaker [14]. Experimental data reveal that the magnetic order parameter in low-doped manganite is only weakly coupled to lattice strains [15]. Neutron scattering techniques used to study the structure and spin dynamics of the compound $La_{0.85}Sr_{0.15}MnO_3$ show anomalies in the Bragg peak intensities consistent with the onset of a spin-canted phase below the Curie temperature, which appears to be associated with a small gap in the spin-wave spectrum at the wavevector $\vec{k} = (0, 0, 0.5k_{ZB})$, where k_{ZB} is the Brillouin zone-boundary wavevector component [16].

Artificial structures, such as magnetic multilayers, have led to an increase of interest in recent years regarding their spin-canted states. Multilayers composed of two antiferromagnets such as FeF_2/CoF_2 [17] and NiO/CoO [18] have been constructed and studied by neutron diffraction and thermal measurements. In the ferromagnetic/antiferromagnetic bilayer system, such as $Fe/KMnF_3$, interesting interfacial exchange properties can be inferred due to the magnetic transition from the antiferromagnetic to the weakly ferromagnetic spin-canted state [19].

The spin dynamics in such canted systems is of special interest in this chapter. We shall consider the propagation of magnetic polaritons (and their magnetostatic counterparts) in antiferromagnets of the type XF_2 (where X = Mn, Fe, Co, etc.), which have a body-centered tetragonal (or rutile) structure. A (001) surface layer, for example, is occupied by spins of just one sublattice type, and the surface therefore has the effect of removing any equivalence that would otherwise exist between the two sublattices (see Fig. 8.1a).

The first experimental evidence for the existence of a magnetic polariton in such an antiferromagnetic structure was found in the late 1970s [20] by a high-resolution far-infrared study of its antiferromagnetic resonance spectra. Investigations of interface and guided magnetic polaritons for a three-layer system composed of semi-infinite yttrium iron garnet (YIG), which behaves essentially like a ferromagnet, and a semi-infinite antiferromagnet (FeF_2) separated by a non-magnetic dielectric layer of finite thickness followed soon afterwards [21,22]. Theoretical treatments were developed by Camley [23–25] and Barnaś [26], among others. For these materials the exchange energy term is substantial, and the maximum canting angle realized in a typical laboratory experiment is modest, being of the order of a few degrees in external magnetic fields of the order of 100 kG. Reversing the direction of propagation or reversing the direction of the applied magnetic field leads to a mode of a different frequency. The degree of this non-reciprocity can be controlled by varying the strength of the applied field. Time-reversal and space-inversion symmetry arguments require the propagation of the surface wave along the magnetic field direction to retain its full left–right symmetry; hence these cases provide interesting examples of the influence of symmetry on the non-reciprocity.

From a mathematical point of view, the non-reciprocal modes appear when there are non-zero off-diagonal terms in the magnetic susceptibility tensor (see Section 8.2), just as in some of the other cases studied in Chapter 7. For instance, experimental observation and interpretation of high-resolution far-infrared reflectivity and attenuated total reflection (ATR) spectra allowed the mapping out of dispersion curves for the antiferromagnet FeF_2. These dispersion curves clearly display the relationship between the frequency and wavevector of magnetic polaritons, and show the position of the bulk and surface modes and surface resonances in zero, low, and relatively strong applied fields [27]. The spectra also show huge non-reciprocity in infrared reflectivity at low fields, a feature unique to magnetic systems [28]. Besides, as we will prove later using the stability condition for magnetic equilibrium, the convergence to the infinite-bulk canting angle θ in the surface region can be over several magnetic layers for MnF_2, which has a relatively low anisotropy field and a low resonance frequency. There is a contrasting behavior near the surface of materials with a larger anisotropy (e.g. CoF_2), as we shall see later.

8.1 The Magnetic Hamiltonian

To gain a better understanding of the magnetic polariton propagation in these body-centered tetragonal antiferromagnets, we consider first the Heisenberg Hamiltonian describing the magnetic excitations. This can be expressed as

$$H = \sum_{i,j} J_{ij} \vec{S}_i \cdot \vec{S}_j - g\mu_B H_0 \left(\sum_i S_i^x + \sum_j S_j^x \right) + H_{anis}, \quad (8.1)$$

where the sums in the first term are taken over nearest-neighbor exchange-coupled sites i and j on opposite sublattices, and H_0 is a static external magnetic field perpendicular to the easy axis and pointing in the x-direction. Other notations are as before, while H_{anis} represents the uniaxial anisotropy (along the z-direction) given by

$$H_{anis} = -g\mu_B H_A \left(\sum_i S_i^z - \sum_j S_j^z \right) \quad (8.2)$$

with H_A being the effective anisotropy field.

The overall geometry of the canted spins was indicated in Fig. 8.1b. The first step in analyzing these systems is to transform to new sets of axes, as represented by rotations about the y-axis in Fig. 8.2, that are related to the equilibrium orientations of the spins. We therefore make the axis transformations $zx \to z_1 x_1$ ($z_2 x_2$) for each sublattice, representing rotations through the canting angle θ, in such a way that the z_1-direction (z_2-direction) is the direction of the spin alignment in the i-sublattice (j-sublattice). The total Hamiltonian [Eq. (8.1)] in a bulk sample now becomes

$$H = \sum_{i,j} J_{ij} \left[(c_0^2/2)(S_i^+ S_j^+ + S_i^- S_j^-) - (s_0^2/2)(S_i^+ S_j^- + S_i^- S_j^+) + (c_0^2 - s_0^2) S_i^{z_1} S_j^{z_2} \right]$$

$$- (\omega_A c_0 + \omega_0 s_0) \left(\sum_i S_i^{z_1} - \sum_j S_j^{z_2} \right), \quad (8.3)$$

where we use the new axes to define $S_i^\pm = S_i^{x_1} \pm i S_i^y$, and there are similar expressions for $S_j^\pm$. Also we write $c_0 = \cos\theta$ and $s_0 = \sin\theta$ as a shorthand notation. The frequencies $\omega_0 = g\mu_B H_0$ and $\omega_A = g\mu_B H_A$ relate to the applied field

8.1. THE MAGNETIC HAMILTONIAN

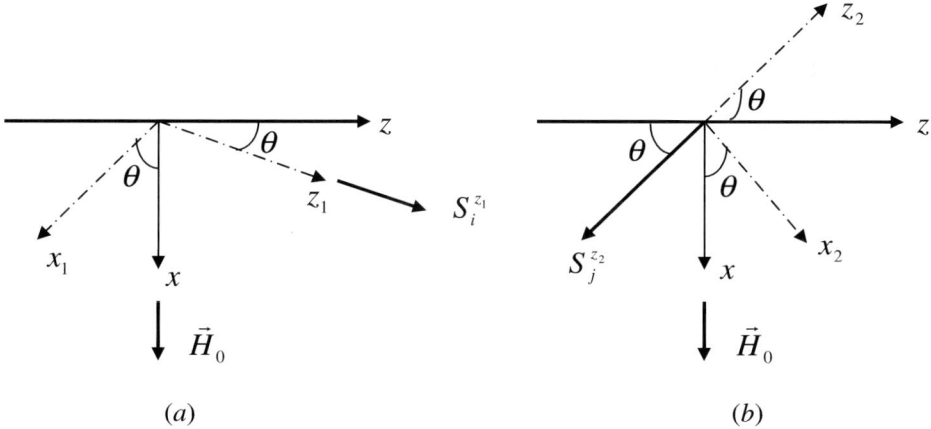

Fig. 8.2. Geometry representing rotations through the canting angle θ of the coordinate axes for each sublattice to a new set of equilibrium-related axes: (a) $zx \to z_1 x_1$; (b) $zx \to z_2 x_2$.

and anisotropy field as before. The stability condition defines the canting angle θ through the transcendental equation

$$-\omega_A \tan\theta + \omega_0 - 2\omega_E \sin\theta = 0 \tag{8.4}$$

where the RPA approximation (see Section 1.7) has been used to replace $S_i^{z_1}$ and $S_j^{z_2}$ by S at low temperatures $T \ll T_N$, and we have defined $\omega_E = SJ(0)$. Note that for a bcc cubic structure, with lattice parameter equal to a, we have

$$J(\vec{k}) = 8J \cos(k_x a/2) \cos(k_y a/2) \cos(k_z a/2). \tag{8.5}$$

This yields $J(0) = 8J$, with J being the nearest-neighbor exchange term.

Before introducing the surface effects, it is instructive to derive the bulk-magnon dispersion relation in an infinite medium for this canted structure. Using the Heisenberg equation of motion for the spin operators $S^\pm$ and taking into account the transformed Hamiltonian [Eq. (8.3)], we find after a bit of algebra

$$\left[\omega - \omega_A c_0 - \omega_0 s_0 - (c_0^2 - s_0^2)\omega_E\right] S_{1\vec{k}}^+ = \left(c_0^2 S_{2\vec{k}}^+ - s_0^2 S_{2\vec{k}}^-\right) SJ(\vec{k}), \tag{8.6}$$

$$\left[\omega + \omega_A c_0 + \omega_0 s_0 + (c_0^2 - s_0^2)\omega_E\right] S_{2\vec{k}}^+ = \left(-c_0^2 S_{1\vec{k}}^+ + s_0^2 S_{1\vec{k}}^-\right) SJ(\vec{k}), \tag{8.7}$$

$$\left[\omega + \omega_A c_0 + \omega_0 s_0 + (c_0^2 - s_0^2)\omega_E\right] S_{1\vec{k}}^- = \left(-c_0^2 S_{2\vec{k}}^- + s_0^2 S_{2\vec{k}}^+\right) SJ(\vec{k}), \tag{8.8}$$

$$\left[\omega - \omega_A c_0 - \omega_0 s_0 - (c_0^2 - s_0^2)\omega_E\right] S_{2\vec{k}}^- = \left(c_0^2 S_{1\vec{k}}^- - s_0^2 S_{1\vec{k}}^+\right) SJ(\vec{k}), \tag{8.9}$$

where wavevector Fourier transforms have been used for the spin operators $S_i^\pm$ and $S_j^\pm$ by analogy with Eq. (1.33). The bulk dispersion relation for the canted system is then found by writing down the determinant of the coefficients of the spin operators in Eqs. (8.6)–(8.9), and then setting this equal to zero for a non-trivial solution. The result is

$$\omega^2 = (a_1 \pm b_1)^2 - b_2^2, \tag{8.10}$$

where we have defined

$$a_1 = -\omega_A c_0 - \omega_0 s_0 - \omega_E \cos(2\theta), \tag{8.11}$$

$$b_1 = s_0^2 SJ(\vec{k}), \qquad b_2 = c_0^2 SJ(\vec{k}). \tag{8.12}$$

From the stability condition we find that the canting angle θ (using small angle approximations) is

$$\theta = \omega_0/(\omega_A + 2\omega_E). \tag{8.13}$$

Note that, for $\theta = 0$, Eq. (8.10) leads to the well-known antiferromagnetic resonance frequency with $\omega_{AF}^2 = \omega_A^2 + 2\omega_A\omega_E$ in the case of $\vec{k} = 0$.

We next generalize the above calculation to the more interesting problem of the surface and bulk magnons in semi-infinite Heisenberg canted antiferromagnets. We consider the antiferromagnet to be occupying the half-space $z < 0$ and the directions of average spin alignment to be parallel and antiparallel to the z-axis for sublattices i and j, respectively. As we can see from Fig. 8.1a, for body-centered tetragonal antiferromagnets each layer (labelled with index n) parallel to the surface contains magnetic sites of one sublattice type only, and the nearest-neighbor exchange coupling is to sites in the *adjacent* layers only. Equations analogous to Eqs. (8.6)–(8.9) can be written down for each value of n, just as in Section 7.1 for the simpler calculation presented there. In doing this, we assume wave-like solutions in the xy-plane for the spin operators as

$$S_i^+ = t_n(\vec{k}_\|) \exp(i\vec{k}_\| \cdot \vec{\rho}_i) \exp(-i\omega t), \tag{8.14}$$

$$S_i^- = v_n(\vec{k}_\|) \exp(i\vec{k}_\| \cdot \vec{\rho}_i) \exp(-i\omega t). \tag{8.15}$$

Here we are considering a Fourier component for the spin operators with angular frequency ω, and a 2D wavevector $\vec{k}_\| = (k_x, k_y)$ parallel to the surface. The factor $\exp(i\vec{k}_\| \cdot \vec{\rho}_i)$, where $\vec{\rho} = (x, y)$, is in accordance with Bloch's theorem and the translational symmetry of the system in the xy-plane. Because of the surface, there is no such factor for the z-direction, and the terms $t_n(\vec{k}_\|)$ and $v_n(\vec{k}_\|)$ depend on z through the layer index n.

The infinite series of coupled equations for $t_n(\vec{k}_\|)$ and $v_n(\vec{k}_\|)$ can now be solved for the bulk and surface modes by following the approach used in Section 7.1, i.e. we seek solutions that are either wave-like or exponentially decaying in the z-direction. For example, the coupled equations are solved for surface modes using the ansatz

$$t_n(\vec{k}_\|) = C(\vec{k}_\|) \exp(-n\kappa c) \tag{8.16}$$

8.1. THE MAGNETIC HAMILTONIAN

with a similar expression for $v_n(\vec{k}_\parallel)$. Here c is the lattice parameter corresponding to the z-direction in the body-centered tetragonal lattice, and κ is the attenuation factor of any surface magnon. We note that $\mathrm{Re}(\kappa) > 0$ for localization. The bulk modes are found using an ansatz analogous to Eq. (7.6), and are in fact just the same as found in the infinite bulk material.

Although the calculation for the surface mode is relatively straightforward, we do not intend to present here the explicit form of the dispersion relation. Instead, we shall present numerical results to illustrate the spectra. However, we point out that care has to be taken in dealing with the canting angle, which can vary with distance from the surface. It is easily shown that the stability condition is now given by the set of finite-difference relations [1]

$$4SJ\left[\sin(\theta_n + \theta_{n-1}) + \sin(\theta_n + \theta_{n+1})\right] = \omega_0 \cos\theta_n - \omega_A \sin\theta_n, \qquad (8.17)$$

where, for $n=1$, the θ_{n-1} term has to be omitted. Numerical results show that the convergence of θ_n to the infinite-bulk canting angle θ (to within, say, 10%) can be over several layers (e.g. 13 layers for MnF_2) or almost localized in the surface layer (e.g. two layers for FeF_2 and one layer for CoF_2). These differences are mainly due to differences in ω_A (which is small in MnF_2) for these materials.

Fig. 8.3 shows the calculated spin-wave spectrum for the case of CoF_2 [1]. The assumed physical parameters are $\omega_A = 241\,\mathrm{kG}$, $\omega_0 = 20\,\mathrm{kG}$, and $\omega_E = SJ(0) = 326\,\mathrm{kG}$. The degenerate bulk modes that would occur for $\theta = 0$ are now split into two well-defined bulk regions in the spectrum. The high-frequency branch is broad and it is shown shaded. The low-frequency one is represented by the thin line in the spectrum. Between these branches one can see the acoustic surface magnon branch. We have shown the surface branch for three different values of the attenuation factor, namely $\kappa c = 0.5$ (chain-dotted line), 1.0 (dotted line), and 10 (full line). Observe that there are small differences in the surface lines for the wavevector range $0 < k_x a/\pi < 0.8$. For larger wavevectors up to the limit of the Brillouin zone they essentially are merged into one line.

Fig. 8.4 shows the calculated spin-wave spectrum for FeF_2 [1]. The physical parameters are $\omega_A = 215\,\mathrm{kG}$, $\omega_0 = 20\,\mathrm{kG}$, and $\omega_E = SJ(0) = 626\,\mathrm{kG}$. Again, the doubly degenerate bulk mode that occurs for $\theta = 0$ (in zero applied magnetic field) is now split due to the canting into two well-defined bulk regions in the spectrum. Although the lower branch is still narrow, it is broader than in the CoF_2 case discussed previously. The surface branch in this case is more sensitive to the attenuation factor. Indeed, for $\kappa c = 0.5$ (chain-dotted line) and 1.0 (dotted line), the surface branch emerges from the high-frequency bulk band at different wavevector values corresponding to $k_x a/\pi = 0.17$ and 0.2, respectively, whereas for $\kappa c = 10$ (full line) it starts at $k_x a/\pi = 0$. The surface branches essentially merge into one single line for $k_x a/\pi = 0.9$ and above.

The application of the external magnetic field also causes *non-reciprocal* propagation. However, for the data used here, the differences between $\omega(k_x)$ and $\omega(-k_x)$ are too close to be resolved for the scale used in Figs. 8.3 and 8.4. This would not be the case for MnF_2 and/or for larger applied fields (making the canting angle larger).

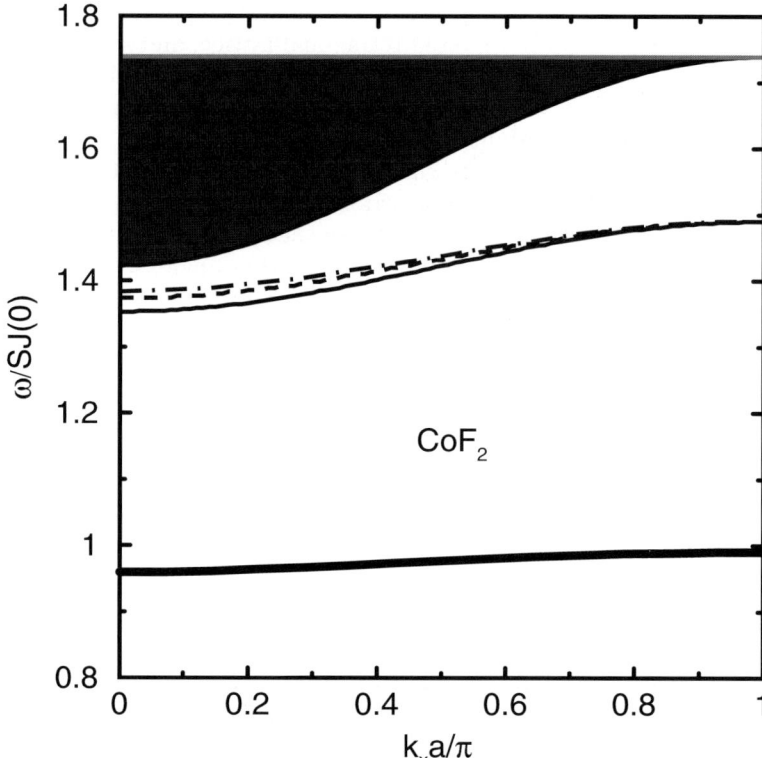

Fig. 8.3. Spin-wave spectra for the reduced frequency $\omega/SJ(0)$ versus $k_x a/\pi$ for CoF$_2$. The physical parameters are given in the main text. The shaded area is the high-frequency bulk mode, and the single lines are the surface modes for several values of the attenuation factor κc. There is also a low-frequency bulk mode (the lower thin line) (after Albuquerque et al. [1]).

Although there is extensive theoretical literature for spin waves at the surfaces of Heisenberg ferromagnets or antiferromagnets, the experimental data are comparatively sparse to date in comparison to the data for the dipole-exchange and magnetostatic regions. It would be of interest to have more experimental results for surface spin waves in the exchange-dominated region to test our theoretical results described here. Techniques involving spin-polarized particle scattering [29,30] and recent advances in inelastic neutron scattering [31] are perhaps the most appropriate and promising for experiments on the modes presented here.

8.2 Magnetic Polaritons in Canted Antiferromagnets

We have already pointed out in the introduction to Chapter 7 that antiferromagnetic materials are of particular interest because their resonance frequencies usually lie in the infrared region with typical values ranging from 250 GHz to a few

8.2. MAGNETIC POLARITONS IN CANTED ANTIFERROMAGNETS

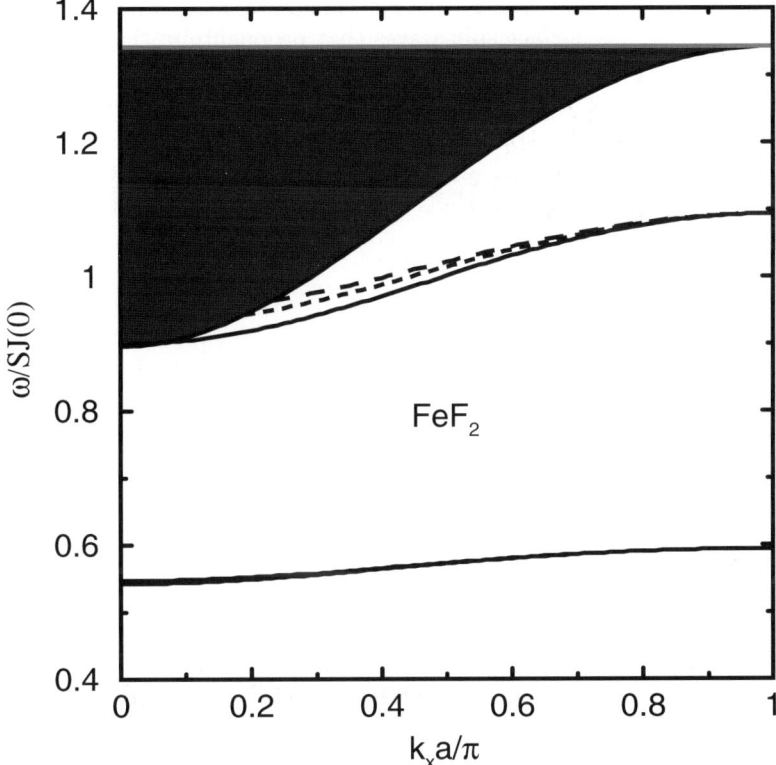

Fig. 8.4. The same as in Fig. 8.3, but for FeF$_2$ (after Albuquerque et al. [1]).

THz. Thus, the wavelengths (and penetration depths) involved range from millimeters to a few hundred micrometers, making these materials excellent options for probing the polariton modes and for interesting research [32]. Further, in contrast to the phonon-polariton case, the two-sublattice antiferromagnet with its easy axis (and sublattice magnetizations) parallel to the surface provides another interesting illustration of the influence of symmetry on the non-reciprocity. We recall that this relates to the property of certain surface magnetic waves in which their dispersion relation is different, depending on whether they propagate from right to left, or vice versa, along a given direction in the surface. The most frequently cited example is that of the Damon–Eshbach surface wave in a ferromagnetic film (see Section 7.2); a more general review can be found in Ref. [33]. The non-reciprocal behavior can be realized when the propagation direction of the spin wave makes a non-zero angle with the magnetization, assumed parallel to the surface and to an external magnetic field applied to the material. The two-sublattice antiferromagnet, with its easy direction and sublattice magnetization parallel to the surface, provides an interesting illustration of this case, as we mentioned briefly at the end of Subsection 7.2.1. In zero external magnetic field, where the net magnetization is zero, one has surface spin waves that propagate in the direction normal to the easy axis. No non-reciprocity occurs in this case. Application of a magnetic field parallel to the easy axis induces the non-reciprocal behavior [34].

On the other hand, the application of a magnetic field perpendicular, rather than parallel, to the easy axis of an antiferromagnet also gives rise to an interesting physical situation, namely it leads to a geometry in which the spins on each sublattice become canted relative to the easy axis, as we discussed in terms of the Heisenberg model in the previous section.

In this section it is appropriate to generalize the earlier calculations to consider the polariton modes when spin canting occurs. Specifically we investigate the magnetic polaritons in antiferromagnets which have a body-centered tetragonal (or rutile) structure, i.e. crystals of the type XF_2, with X = Mn and Fe in this case. The (001) surface layer is occupied by spins of one sublattice type, and the surface therefore has the effect of removing any equivalence between the two sublattices, as noted before. We shall derive the polariton dispersion relation taking into account not only the unretarded (magnetostatic) limit, but also the retarded modes. The latter case must be included to explain experimental observations of the excitation in the far-infrared region. Experimental light-scattering techniques of the Raman and Brillouin types are the most appropriate tools to probe these modes.

The Hamiltonian of the spin system was already studied in the previous section. It is defined by Eq. (8.1) relative to the crystallographic axes and can be re-expressed as Eq. (8.3) relative to new axes taken along the equilibrium direction of the canted spins (see Fig. 8.2). The first step in determining the polariton spectrum is to deduce the form of the permeability tensor for the canted antiferromagnet. This part of the calculation is analogous to that described in Section 7.5 for metamagnets, provided we use the appropriate Hamiltonian. Thus we consider, for each sublattice, the magnetizations denoted by $\vec{M}_i$ and $\vec{M}_j$, respectively. Their classical equation of motion is formally the same as in Eq. (7.71), where now the total sublattice magnetizations and effective field are expressed as

$$\vec{M}_i = S\hat{z}_1 + \vec{m}_i \exp(-i\omega t), \tag{8.18}$$

$$\vec{M}_j = -S\hat{z}_2 + \vec{m}_j \exp(-i\omega t), \tag{8.19}$$

$$\vec{H}_i^{eff} = H_0\hat{x} + H_A\hat{x}_1 - H_E\hat{z}_2 + \vec{h}(t), \tag{8.20}$$

$$\vec{H}_j^{eff} = H_0\hat{x} + H_A\hat{x}_2 - H_E\hat{z}_1 + \vec{h}(t). \tag{8.21}$$

Here $\vec{m}_p$ (with $p=i,j$) and $\vec{h}$ are the fluctuating magnetizations and fluctuating driving fields at frequency ω. On substituting Eqs. (8.18)–(8.21) into Eq. (7.71), and making the usual linear spin-wave approximation of neglecting small terms that are of second order in the fluctuations, we obtain, after some algebra, a linear

8.2. MAGNETIC POLARITONS IN CANTED ANTIFERROMAGNETS

relationship between $\vec{m} \equiv \vec{m}_i + \vec{m}_j$ and $\vec{h}$. This has the form

$$m_\alpha = m_\alpha^0 + \sum_\beta \chi_{\alpha\beta}(\omega) h_\beta \qquad (\alpha, \beta = x, y, z), \tag{8.22}$$

where $\vec{m}^0$ is the static magnetization in the canted system induced by the external field $\vec{H}_0$. The quantity $\chi_{\alpha\beta}$ in the other term is the susceptibility tensor; it is related to the required permeability tensor by $\bar{\mu} = \bar{I} + \bar{\chi}$, where $\bar{I}$ is the identity (unit) tensor. The permeability tensor is found to be [35]

$$\bar{\mu} = \begin{pmatrix} \mu_{xx} & 0 & -i\mu_{xz} \\ 0 & \mu_{yy} & 0 \\ i\mu_{xz} & 0 & \mu_{zz} \end{pmatrix}, \tag{8.23}$$

where the elements are given by

$$\mu_{xx} = 1 + \omega_M \omega_A \left[(\omega_\perp^2 - \omega_+^2)^{-1} + (\omega_\perp^2 - \omega_-^2)^{-1} \right], \tag{8.24}$$

$$\mu_{yy} = 1 + \omega_M \omega_A \cos^2\theta \left[(\omega_\parallel^2 - \omega_+^2)^{-1} + (\omega_\parallel^2 - \omega_-^2)^{-1} \right], \tag{8.25}$$

$$\mu_{zz} = 1 + \omega_M (\omega_0 \sin\theta + \omega_A \cos 2\theta) \left[(\omega_\perp^2 - \omega_+^2)^{-1} + (\omega_\perp^2 - \omega_-^2)^{-1} \right], \tag{8.26}$$

$$\mu_{xz} = \omega_M \omega_A \cos\theta \left[(\omega_\perp^2 - \omega_+^2)^{-1} - (\omega_\perp^2 - \omega_-^2)^{-1} \right] \tag{8.27}$$

with $\omega_\pm = \omega \pm \omega_0$. The two resonance frequencies appearing in the above expressions are

$$\omega_\parallel^2 = \omega_{AF}^2 \cos^2\theta, \tag{8.28}$$

$$\omega_\perp^2 = \omega_\parallel^2 + 2\omega_0 \omega_E \sin\theta, \tag{8.29}$$

where ω_{AF} is the well-known antiferromagnetic resonance frequency given by Eq. (7.19).

It is helpful to comment on the properties of the so-called Voigt permeability $\mu_V = \mu_{xx} - \mu_{xz}^2/\mu_{zz}$, which was defined following Eq. (7.100), near the magnetic resonance frequencies $\omega_{AF} \pm \omega_0$. Using the expressions for the permeability components as given above, we find that μ_V has two poles and two zeros. The poles correspond to the appearance of bulk continua where bulk polariton modes may exist, and in the frequency gaps (where $\mu_V < 0$) between these bands we may expect to find surface polaritons.

Now that the dynamical magnetic susceptibility tensor for the spin-canted body-centered tetragonal structure has been determined, we can find the dispersion relations for the retarded and unretarded polariton modes that can propagate in this structure. First, if we ignore retardation effects, then the relevant Maxwell's equations are those quoted in Eq. (7.20); they allow the introduction of the magnetostatic scalar potential ψ defined by Eq. (7.21). Inside the canted antiferromagnet ψ satisfies

$$\mu_{xx}\, \partial^2\psi/\partial x^2 + \mu_{yy}\, \partial^2\psi/\partial y^2 + \mu_{zz}\, \partial^2\psi/\partial z^2 = 0, \tag{8.30}$$

which is just a generalization of Eq. (7.22). It leads directly to the *magnetostatic bulk mode dispersion relation*

$$\mu_{xx}k_x^2 + \mu_{yy}k_y^2 + \mu_{zz}k_z^2 = 0, \tag{8.31}$$

as can be seen by seeking a wave-like solution for ψ with wavevector $\vec{k} = (k_x, k_y, k_z)$.

For surface modes in the magnetostatic limit, we suppose that the crystal is truncated at $z=0$, with vacuum outside in the region $z>0$. We also adopt the Voigt geometry (i.e. in-plane propagation perpendicular to the applied field direction), which occurs when $k_x = 0$ for our assumed surface geometry, which is different from that in Section 7.2. The magnetostatic scalar potentials are then given by

$$\psi = A \exp(ik_y y - i\omega t) \exp(\alpha z) \tag{8.32}$$

for $z < 0$ (inside the magnetic material), and

$$\psi = B \exp(ik_y y - i\omega t) \exp(-k_\| z) \tag{8.33}$$

for $z > 0$ (in the vacuum region), denoting $k_\| \equiv |\vec{k}_\| | = |k_y|$. Here $\alpha > 0$ is a decay term that ensures the localized behavior expected for a surface mode. Application of the usual electromagnetic boundary conditions at the crystal/vacuum interface would give the desired dispersion relation (by analogy with Subsection 7.2.1).

For the *retarded modes*, instead of using Eq. (7.20), we must employ the full form of the relevant Maxwell's equation as quoted in Eq. (7.60). This involves c as the velocity of light in the vacuum and $\epsilon_{\it eff} = \eta^2$, where η is the refractive index of the medium. From a determinantal condition, obtained from the linear homogeneous equations for the coefficients of the magnetic field components, we find that the dispersion relation for the bulk modes in the Voigt geometry ($k_x = 0$) is given by

$$\mu_{zz}k_z^2 + \mu_{yy}k_y^2 = (\eta\omega/c)^2(\mu_{zz}\mu_{yy} - \mu_{yz}^2). \tag{8.34}$$

To study the surface modes, we consider localized solutions of $\vec{H}$ with respect to the z-coordinate. It is straightforward, although tedious, to set up a system of equations involving the Cartesian components of $\vec{H}$, whose solvability condition then yields the desired surface dispersion relation. We quote the final result as

$$\alpha_0^2(\mu_{zz}\mu_{yy} - \mu_{yz}^2) - 2k_x\alpha_0\mu_{yz} = k_x^2 - \mu_{yy}(\eta\omega/c)^2 \tag{8.35}$$

provided that the constraint (a localization condition)

$$\alpha_0(\mu_{zz}\mu_{yy} - \mu_{yz}^2) - k_x\mu_{yz} < 0 \tag{8.36}$$

is satisfied.

8.2. MAGNETIC POLARITONS IN CANTED ANTIFERROMAGNETS

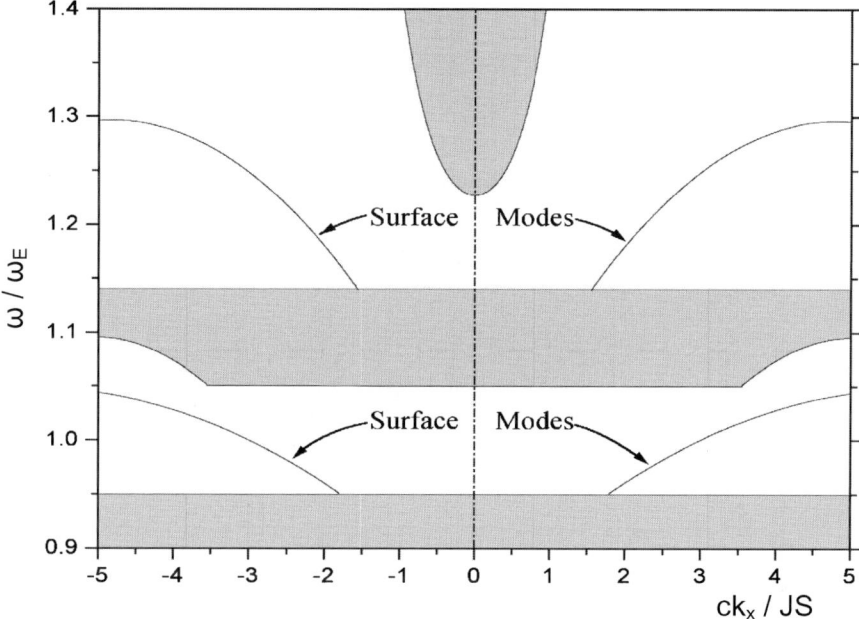

Fig. 8.5. Magnetic polariton spectra (retarded modes) for the reduced frequency ω/ω_E versus ck_x/JS for FeF$_2$. The physical parameters are given in the main text. For completeness we also show the bulk modes (shaded areas). The single lines are the surface modes. Observe that there is negligible non-reciprocity in the propagation of the surface modes (after Guimarães and Albuquerque [35]).

An example of the dispersion curves corresponding to Eqs. (8.34) and (8.35) calculated for the antiferromagnetic FeF$_2$ crystal is shown in Fig. 8.5 [35]. The physical parameters used here are $\omega_E = 626\,\text{kG}$, $\omega_A = 215\,\text{kG}$, $\omega_0 = 20\,\text{kG}$, and $\omega_M = 0.6\,\text{kG}$ [36]. For simplicity, we have not considered a damping term in the susceptibilities, which are defined by Eqs. (8.24)–(8.27), because we want to stress the effects due to the canting angle in the polariton spectra. For a derivation of the susceptibility tensor, with damping and non-linear terms included, we refer the reader to Refs. [37,38]. Of course, when damping is present in the magnetic system, the dispersion equations Eqs. (8.34) and (8.35) no longer have *real* wavevector and frequency solutions. They can be satisfied, however, for real frequencies and complex wavevectors, whose solutions represent dissipative waves that have finite path lengths.

Because the exchange field is substantially greater than the other effective fields in the system, the bulk canting angle that one realizes in typical laboratory experiments is modest; its value is 0.0136 rad for the data used here. In the first layer ($z = 0$) of the material we have a surface value of $\theta_S = 0.0197$ rad. The convergence to the bulk canted angle θ_B, for the perturbation to be less than 10%, is achieved in the third magnetic layer. Although the asymmetry of the bands is not noticeable, the polariton spectra in principle have a small non-reciprocity, in

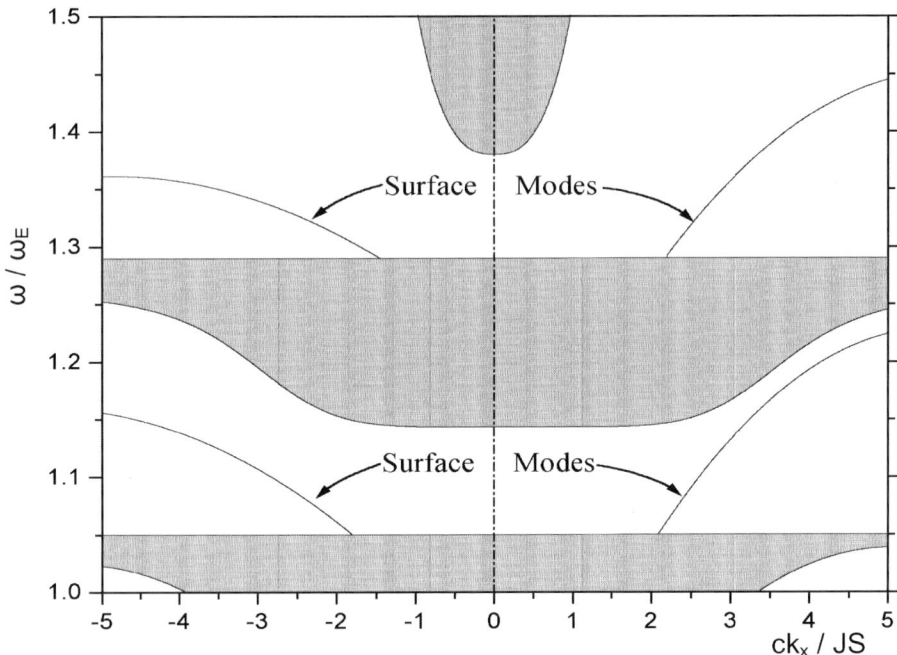

Fig. 8.6. The same as in Fig. 8.5, but for MnF$_2$. Observe the striking non-reciprocal effects for this material, even for a small external magnetic field (after Guimarães and Albuquerque [35]).

the sense of Eq. (7.37). In the present case, this non-reciprocal propagation appears due to the non-vanishing off-diagonal term μ_{xz} and μ_{zx} in the permeability tensor. The three bulk bands, shown shaded here, are due to the two poles in the permeability. The surface modes (full lines), which are very sensitive to the presence of the external magnetic field, lie as usual in the regions between the bulk bands.

This situation changes quite dramatically for the case of MnF$_2$, even with a modest external magnetic field (taken as 20 kG). This can be seen from Fig. 8.6, where there is now a clear appearance of non-reciprocity for the upper- and lower-frequency surface modes [35]. The bulk bands are also affected, although not to the same extent. The physical parameters used are $\omega_E = 550$ kG, $\omega_A = 7.9$ kG, and $\omega_M = 0.56$ kG [36]. Convergence to the bulk canting angle θ_B, for the perturbation to be less than 10%, is now achieved in the 14th magnetic layer.

It would be of interest to consider systems in which the exchange coupling between the antiferromagnetic moments is weaker than those shown here. In this way we expect that even a modest field (like the one considered here) can induce larger canting angles, allowing one to enter a regime where the surface modes split off further from the bulk manifold. A good candidate could be Fe double layers, with an ultrathin layer of Cr between each film. A semi-infinite periodic or quasiperiodic array of such antiferromagnetically coupled films could be an interesting geometry for application of the theory presented here, in the appropriate wavelength regime.

8.3 Magnetic Polaritons in Spin-Canted Thin Films

As indicated in Section 7.1, the excitations in magnetic thin films have been extensively investigated; some review accounts covering both experiment and theory can be found in Refs. [30,39–41]. In thin films, one obtains both surface polaritons, in which the excitation is localized near the surface, and guided modes, where the excitation has a standing-wave-like character within the film [42]. Early theoretical works on surface polaritons on antiferromagnets [24] have now been verified experimentally for FeF_2, both through infrared reflectivity [43,44] and through infrared ATR measurement [45]. They have given a renewed impetus to extend theoretical calculations for more complex and realistic structures. In magnetic superlattices, early studies make use of the so-called *effective-medium* theory [46], in which both the dielectric tensor and the permeability tensor components are given as spatial averages of the dielectric and permeability constants of the constituent magnetic layers [47,48]. Such a description is valid when the characteristic wavelengths of the excitation are much longer than the superlattice period. We have already made use of an effective-medium approximation in Section 7.4 in the discussion of rare-earth periodic superlattices.

In this section we consider the propagation of magnetic polaritons in unsupported spin-canted thin films of the rutile-structure antiferromagnets XF_2 (with X = Mn, Fe). While this is obviously an extension of the previous section to include two surfaces or interfaces, it is more than just a complicated geometrical extension of the previous works since new effects, like perturbed surface modes (or guided modes) and spatial-quantization features in the polariton spectra, are introduced. Experimental studies have been reported in MnF_2, for instance, by Lui et al. [49] in which, at low resonance frequency, a field sweep at fixed frequency can be employed. Alternatively, conventional far-infrared resonance (FIR) spectroscopy can be utilized to produce frequency-sweep spectra at a fixed field [50].

We again use the Voigt configuration, where the direction of propagation is perpendicular to the magnetic moments and to the applied external magnetic field. We employ formal analytical equations for the magnetic polariton dispersion relations, as derived, for example, in Chapter 7 for a film geometry, taking both the retarded region [see Eq. (7.88)] and the magnetostatic regime [Eq. (7.90)]. It is necessary only to substitute the appropriate permeability components for the present application to canted antiferromagnets.

The magnetic film is assumed to occupy the region $0 < z < L$, with L being its thickness, and to be bounded on each side by the same medium (a symmetrical geometry), taken as vacuum ($\epsilon = \mu = 1$). For numerical purposes, the magnetic material is considered to be either FeF_2 or MnF_2, and it will be of interest to compare the two cases. The thickness of the magnetic film is equal to 200 μm, and

210   CHAPTER 8. MAGNETIC POLARITONS IN SPIN-CANTED SYSTEMS

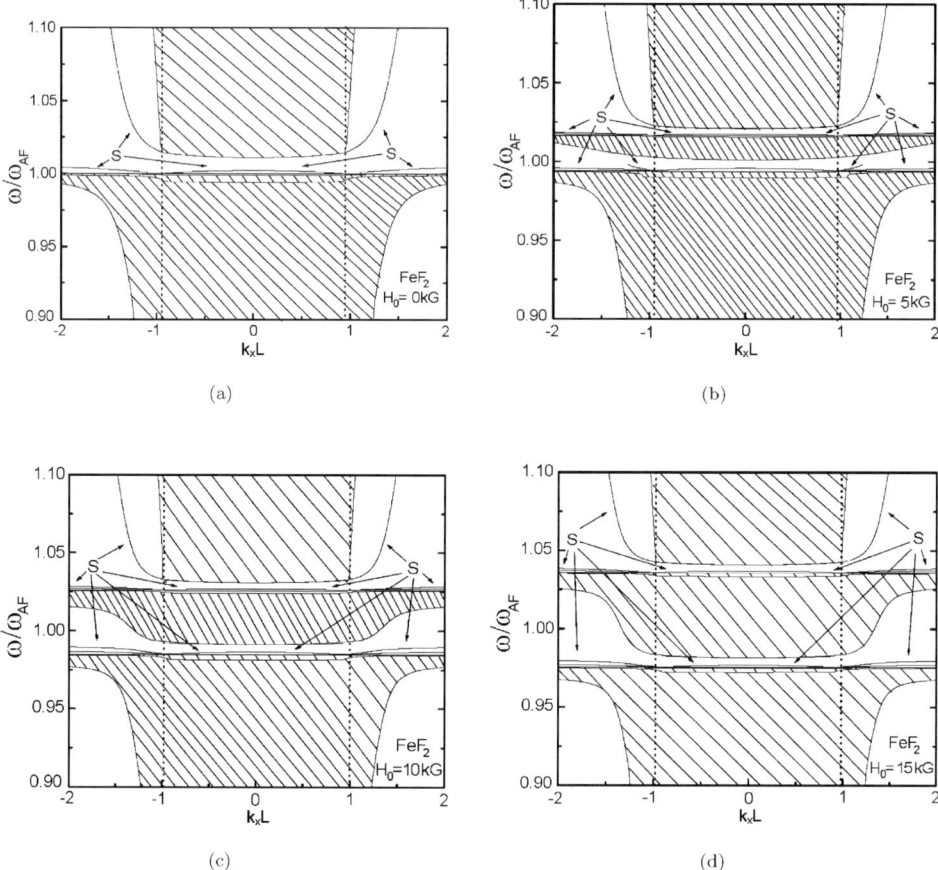

Fig. 8.7. (a) Dispersion relation of bulk (shaded areas) and surface/guided spin waves for the FeF$_2$ free-standing magnetic film, measured by the reduced frequency ω/ω_{AF} as a function of the in-plane dimensionless wavevector $k_x L$, in the absence of an external applied magnetic field. (b) Same as in (a), but now for an external applied magnetic field $H_0 = 5$ kG. (c) Same as in (a), but now for an external applied magnetic field $H_0 = 10$ kG. (d) Same as in (a), but now for an external applied magnetic field $H_0 = 15$ kG (after Guimarães and Albuquerque [51]).

several values of the external magnetic field $\vec{H}_0$ are considered in order to vary the amount of canting.

In Fig. 8.7 we present the magnetic polariton spectra in the form of plots of the reduced frequency ω/ω_{AF} against the dimensionless wavevector $k_x L$, where we take for simplicity $\vec{k}_\parallel = (k_x, 0)$ [51]. The nearly vertical dashed lines are the light lines $\omega = ck_x$ in vacuum. We use the physical parameters and permeabilities for FeF$_2$ [52]; these are the same as quoted in the previous section. FeF$_2$ has a resonant frequency of about 1.5 THz, and it is very suitable for studies by FIR spectroscopy in either grating or Fourier-transform form. An important difference between the film and the semi-infinite case [1] is that k_z (inside the film) can now be either purely real or purely imaginary, whereas for the semi-infinite medium it can only be imaginary. For real k_z the magnetic field is oscillatory within the film; as in the analogous case of a dielectric film, the mode is then called a guided wave. On the other hand, for imaginary k_z, the fields are a combination of decaying exponentials at each surface, so that in effect one is dealing with a surface polariton at one surface whose properties are modified by the presence of the other surface. This again has an analog in dielectric films, as well as for the surface spin waves in Heisenberg ferromagnetic films (see Section 7.1). Here we have to consider free-standing magnetic film on the assumption that the media in $z > L$ and $z < 0$ are both the same (vacuum). The generalization to the case when the media are different, as would occur for a film on a substrate, is straightforward.

Because of the finite film thickness, the bulk modes are now quantized, corresponding to different standing waves in the film, and the bulk bands become broken up into individual discrete modes. However, for the sake of simplicity, we just show the upper and lower limits of the bulk modes in Fig. 8.7, considering the region inside as a band, as in the case of the semi-infinite medium. In the absence of an external magnetic field, the polariton spectra have three bulk bands (characterized by the hatched regions) and three surface and guided modes (identified by the letter S). The $+k_x$ and $-k_x$ modes are essentially mirror images of each other when $H_0 = 0$, with the localization exactly reversed by a reversal of the direction of propagation. This is true for all modes in this symmetric structure, and it is expected since a reversal of propagation is essentially equivalent to rotating the film through 180° about the external magnetic field. The situation changes when the symmetry is broken due to the presence of an external magnetic field. The evolution of the dispersion curves as the external magnetic field H_0 increases is shown by a comparison of Figs. 8.7b–d. First the spectra become non-reciprocal, although this reciprocity could be better seen in the supported film case. The magnetic field profile for the $+k_x$ mode is now biassed towards one interface ($z = L$) whereas the field profile for the $-k_x$ mode is biassed towards the other interface ($z = 0$). Observe that for $H_0 = 5$ kG (Fig. 8.7b) a new bulk band appears at $1.01 < \omega/\omega_{AF} < 1.02$. This bulk band becomes broader as the external magnetic field increases (see Figs. 8.7c and d). New surface modes appear between this bulk band and the bulk bands that were already present in zero external magnetic field, making the spectra more complex.

In Fig. 8.8 we present the magnetic polariton spectra using the physical parameters and permeabilities for MnF$_2$ [53]. The values used here are $\omega_E = 550$ kG, $\omega_A = 7.78$ kG, and $\omega_M = 0.6$ kG. MnF$_2$ has a resonant frequency of about 240 GHz, and it is more suitable for the ferromagnetic resonance of a magnetic field sweep at fixed frequency. The spectra are quite different from the previous case, although they keep some of the general properties like the $+k_x$ and $-k_x$ mirror images of each other in the absence of the external magnetic field. Non-reciprocal propagation of the surface modes is then seen in Figs. 8.8b–d for magnetic fields equal to 0.2, 0.4, and 0.9 kG, respectively. The surface mode curves are now restricted (because of localization conditions) to certain values of $|k_x|L$, and most of

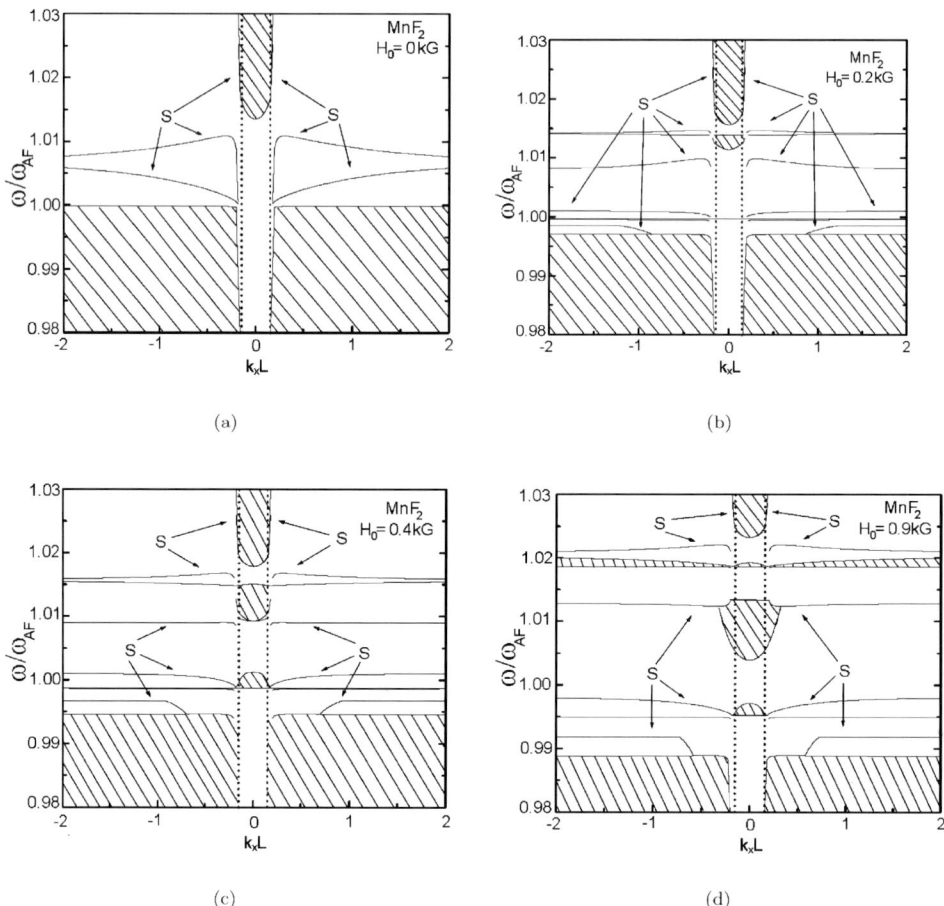

Fig. 8.8. Same as in Fig. 8.7 but for the MnF$_2$ free-standing magnetic film, taking the following values for the external applied magnetic field: (a) $H_0 = 0$ kG; (b) $H_0 = 0.2$ kG; (c) $H_0 = 0.4$ kG; (d) $H_0 = 0.9$ kG (after Guimarães and Albuquerque [51]).

them start at the intersection between the light lines (shown dashed) and the bulk bands (the light line now marks the boundary of the bounded modes). Observe that for $H_0 = 0.2$ kG the high-frequency bulk band splits into two bands, with a surface mode at $\omega/\omega_{AF} \simeq 1.015$ between them. The evolution of this bulk band is quite interesting. For an external magnetic field equal to 0.4 kG it becomes broader (see Fig. 8.8c), and it splits into two new bulk bands at a higher magnetic field (0.9 kG). Both bands expand outside the light line (see Fig. 8.8d). There is also the appearance of a new bulk band at $\omega/\omega_{AF} \simeq 1.0$ (thin solid line) and a new surface mode close to it (see Fig. 8.8b). This band evolves towards the central region and its profile resembles the other bulk band located at $1.018 < \omega/\omega_{AF} < 1.02$.

Finally, we remark that polaritons and their related properties in magnetic systems have acquired greater relevance in the light of recent developments in the application of high-resolution Fourier-transform spectroscopy to magnetism (for a review see Ref. [54]). The accessible frequency range is from 200 GHz upwards, so that this technique should be applicable to the materials considered here, whose resonant frequencies are in this range. One complication arises because the permeability tensor $\mu(\omega)$ depends sensitively on the direction of the applied magnetic field $\vec{H}_0$ with respect to the crystal axis. However, as explained, for the geometry considered here a careful analysis of the corresponding canting angle and its influence on the dynamic permeability tensor, as was done in obtaining Eqs. (8.24)–(8.27), can overcome this problem.

References

[1] E.L. Albuquerque, R.N. Costa Filho and M.G. Cottam, J. Appl. Phys. 87 (2000) 5938.
[2] M.G. Cottam and D.J. Lockwood, Light Scattering in Magnetic Solids, Wiley, New York, 1986.
[3] Q.A. Pankhurst and R.J. Pollar, Phys. Rev. Lett. 67 (1991) 248.
[4] E. Talik, W. Witas, J. Kusz, A. Winiarski, T. Mydlarz, M. Neumann and H. Bohm, Physica B 293 (2000) 75.
[5] R. Blinc, P. Cevc, D. Arcon, A. Omerzu, M. Mehring, S. Knorr, A. Grupp, A.L. Barra and G. Chouteau, Phys. Rev. B 58 (1998) 14416.
[6] E. Talik, M. Kulpa, T. Mydlarz, J. Kusz and H. Bohm, J. Alloy Compd. 348 (2003) 12.
[7] E. Talik, M. Kulpa, T. Mydlarz, J. Kusz and H. Bohm, J. Alloy Compd. 308 (2000) 30.
[8] C. Ritter, R. Cywinski, S.H. Kilcoyne and S. Mondal, J. Phys.: Condens. Matter 4 (1992) 1559.
[9] C. Ritter, S.H. Kilcoyne and R. Cywinski, J. Phys.: Condens. Matter 3 (1991) 727.
[10] M.R. Crook, B.R. Coles, C. Ritter and R. Cywinski, J. Phys.: Condens. Matter 8 (1996) 7785.
[11] D.L. Mills, Phys. Rev. B 39 (1989) 12306.
[12] K. Chahara, T. Ohno, M. Kasai, and Y. Kozono, Appl. Phys. Lett. 63 (1993) 1990.
[13] Z. Jirak, J. Hejtmanek, K. Knizek, M. Marysko, E. Pollert, M. Dlouha, S. Vratislav, R. Kuzel and M. Hervieu, J. Mag. Mag. Mat. 250 (2002) 275.
[14] I.V. Solovyev and K. Terakura, Phys. Rev. B 63 (2001) 174425.
[15] D. Maurer, R. Heichele, N. Lingg, V. Muller and A.M. Balbashov, Eur. Phys. J. B 19 (2001) 425.
[16] L. Vasiliu-Doloc, J.W. Lynn, A.H. Moudden, A.M. de Leon-Guevara and A. Revcolevschi, Phys. Rev. B 58 (1998) 14913.
[17] C.A. Ramos, D. Lederman, A.R. King and V. Jaccarino, Phys. Rev. Lett. 65 (1990) 2913.
[18] J.A. Borchers, M.J. Carey, R.W. Erwin, C.F. Majkrzak and A.E. Berkowitz, Phys. Rev. Lett. 70 (1993) 1878.
[19] Z. Celinski, D. Lucic, N. Cramer, R.E. Camley, R.B. Goldfarb and D. Skrzypek, J. Mag. Mag. Mat. 202 (1999) 480.
[20] R.W. Sanders, V. Jaccarino and S.M. Rezende, Solid State Commun. 28 (1978) 907.
[21] C. Shu and A. Caillé, Solid State Commun. 42 (1982) 233.
[22] C. Thibaudeau and A. Caillé, Phys. Rev. B 32 (1985) 5907.
[23] R.E. Camley and D.L. Mills, Phys. Rev. B 26 (1982) 1280.
[24] R.L. Stamps and R.E. Camley, Phys. Rev. B 40 (1989) 596.

[25] R.L. Stamps and R.E. Camley, Phys. Rev. B 40 (1989) 609.
[26] J. Barnaś, J. Phys.: Condens. Matter 2 (1990) 7173.
[27] M.R.F. Jensen, S.A. Feiven, T.J. Parker and R.E. Camley, J. Phys.: Condens. Matter, 9 (1997) 7233.
[28] M.R.F. Jensen, S.A. Feiven, T.J. Parker and R.E. Camley, Phys. Rev. B 55 (1997) 2745.
[29] S.F. Alvarado, E. Kisker and M. Campagna in: Magnetic Properties of Low-Dimensional Systems, Eds., L.M. Falicov and J.L. Moran-Lopez, Springer, Heidelberg, 1986.
[30] Ultrathin Magnetic Structures, Vols. I and II, Eds., J.A.C. Bland and B. Heinrich, Springer-Verlag, Berlin, 1994.
[31] A. Schreyer, T. Schmitte, R. Siebrecht, P. Bödecker, H. Zabel, S.H. Lee, R.W. Erwin, C.F. Majkrzak, J. Kwo and M. Hong, J. Appl. Phys. 87 (2000) 5443.
[32] C. Thibaudeau and A. Caillé, Solid State Commun. 87 (1993) 643.
[33] R.E. Camley, Surf. Sci. Rep. 7 (1987) 103.
[34] B. Luthi, R.E. Camley and D.L. Mills, Phys. Rev. B 28 (1983) 1475.
[35] E.S. Guimarães and E.L. Albuquerque, J. Appl. Phys. 89 (2001) 7401.
[36] R.E. Camley and M.G. Cottam, Phys. Rev. B 35 (1987) 189.
[37] S.-C. Lim, J. Osman and D.R. Tilley, J. Phys.: Condens. Matter, 10 (1998) 1891.
[38] S.-C. Lim, J. Osman and D.R. Tilley, J. Phys. D, 33 (2000) 2899.
[39] Linear and Nonlinear Spin Waves in Magnetic Films and Superlattices, Ed. M. G. Cottam, World Scientific, Singapore, 1994.
[40] P. Grünberg, in: Light Scattering in Solids V, Eds. M. Cardona and G. Güntherodt, Springer, Berlin, 1989.
[41] B. Hillebrands, in: Light Scattering in Solids VII, Eds. M. Cardona and G. Güntherodt, Springer, Berlin, 2000.
[42] D.R. Tilley in: Electromagnetic Surface Excitations, Eds., R.F. Wallis and G.I. Stegeman, Springer-Verlag, Berlin, 1986.
[43] L. Remer, B. Luthi, H. Sauer, R. Geick and R.E. Camley, Phys. Rev. Lett. 56 (1986) 2752.
[44] D.E. Brown, T. Dumelow, T.J. Parker, K. Abraha and D.R. Tilley, Phys. Rev. B 49 (1994) 12266.
[45] M.R.F. Jensen, T.J. Parker, K. Abraha and D.R. Tilley, Phys. Rev. Lett. 75 (1995) 3756.
[46] V.M. Agranovich and V.E. Kravtsov, Solid State Commun. 55 (1985) 85.
[47] N.S. Almeida and D.L. Mills, Phys. Rev. B 38 (1988) 6698.
[48] F.G. Elmzughi and R.E. Camley, J. Phys.: Condens. Matter 9 (1997) 1039.
[49] M. Lui, C.A. Ramos, A.R. King and V. Jaccarino, J. Appl. Phys. 67 (1990) 5518.
[50] K. Abraha, D.E. Brown, T. Dumelow, T.J. Parker, and D.R. Tilley, Phys. Rev. B 50 (1994) 6808.
[51] E.S. Guimarães and E.L. Albuquerque, Solid State Commun. 122 (2002) 413.
[52] M.C. Oliveros, N.S. Almeida, D.R. Tilley, J. Thomas and R.E. Camley, J. Phys.: Condens. Matter 4 (1992) 8497.
[53] D.H.A.L. Anselmo, M.G. Cottam and E.L. Albuquerque, J. Phys.: Condens. Matter 12 (2000) 1041.
[54] K. Abraha and D.R. Tilley, Surf. Sci. Rep. 24 (1996) 125.

Chapter 9

Metallic Magnetic Multilayers

The study of metallic magnetic multilayers has been one of the most active fields in the last decade (for a review see the articles contained in Ref. [1]). It presents exciting challenges from both the theoretical and experimental points of view. In a pioneering piece of work, Grünberg et al. [2] reported evidence of an antiferromagnetic Heisenberg (or bilinear) exchange coupling between ultrathin magnetic films. The system consisted of ferromagnetic films (Fe) with interfilm exchange coupling via an intervening non-ferromagnetic layer (Cr), giving the trilayer Fe/Cr/Fe structure. In simple metals such as Cu(001) and Ag(001) the bilinear exchange coupling exhibits strong oscillations with a short wavelength (of the order of two atomic monolayers). The exchange coupling through non-ferromagnetic interlayers in this case is found to be strongly affected by the interface roughness. Realistic interfaces often consist of finite terraces, along with other inhomogeneities, which result in variations of the interlayer thickness [3].

Shortly afterwards Baibich et al. [4] noticed a sudden fall in the electrical resistance of Fe/Cr magnetic multilayers when an external magnetic field was applied. The effect was so striking that it was called *giant magnetoresistance* (GMR), and recently it has been widely considered for applications in information storage technology [5–7]. Through magnetoresistance measurements, Parkin and collaborators [8] observed an oscillatory (or alternating ferromagnetic–antiferromagnetic) behavior of the exchange coupling in magnetic metallic multilayers as a function of the non-magnetic spacer thickness. This work was seminal to a number of experimental studies on Fe/Cr/Fe structures with different non-magnetic spacer thickness. Accounts of the physical origin of the long-range exchange, including the so-called RKKY mechanisms involving the conduction electrons, are given in Ref. [1].

Later Rührig et al. [9] showed evidence of a preferred non-collinear alignment (at 90°) of the magnetization directions between the ferromagnetic layers (Fe) for values of the non-magnetic spacer thickness where the bilinear exchange term was small. This behavior could not be explained by considering only the usual Heisenberg-type bilinear exchange term in the free magnetic energy of the system. It was necessary, in addition, to take into account biquadratic exchange coupling (BEC) to achieve the stabilization of the non-collinear alignments [10]. Slonczewski [11] gave a phenomenological interpretation for this biquadratic magnetic coupling, considering spatial fluctuations of the chromium layer thickness amounting to around one atomic monolayer. The corresponding fluctuations of the short-period oscillatory term of the usual Heisenberg-type exchange coupling induce static waves of magnetization whose energy has the observed biquadratic form. Experimental results [12] show that the Fe/Cr/Fe sample enhances the effect of biquadratic coupling which aligns the magnetization of the Fe layers at 90° to each other. The bilinear and biquadratic coupling strengths can be extracted by fitting the experimental results with a theory that treats the static and dynamic responses on an equal footing. However, in order to obtain reliable fits between theory and experiment, it is necessary to study the variation of some spin-wave property [13–15].

It is helpful at this point to make clear the distinction between bilinear and biquadratic exchange. If we consider two interface spins, denoted by $\vec{S}_1$ and $\vec{S}_2$, separated by a spacer layer, the bilinear coupling is just the usual Heisenberg interaction of the form $-J_{bl}\vec{S}_1 \cdot \vec{S}_2$. This favors parallel or antiparallel alignment, depending on whether the exchange term J_{bl} is positive or negative. By contrast, the biquadratic coupling is described by $J_{bq}(\vec{S}_1 \cdot \vec{S}_2)^2$. If $J_{bq} > 0$, which is the practical case, then the minimum energy is when the spin directions are perpendicular. If both types of exchange occur, then there is a competing effect, as we shall discuss later.

Parallel to the achievements obtained using the trilayer Fe/Cr/Fe structures, other non-magnetic metallic spacers were successfully investigated for similar effects. These included, among others, Cu [16] (through Brillouin scattering) and Au [17] (by using the confocal magneto-optical Kerr effect to determine the bilinear exchange coupling). Ferromagnetic resonance (FMR) and surface magneto-optical Kerr effect (SMOKE) studies have shown that the interfaces in bcc Fe/Cu/Fe(001) trilayers grown on a Ag(001) single-crystal substrate can be significantly improved by choosing an appropriate growth procedure [18]. The interpretation of magnetization loops for Fe/Cu/Fe trilayers requires the simultaneous presence of bilinear and biquadratic exchange coupling between the magnetic layers such that the strength of the biquadratic exchange coupling increases with increasing terrace width. Considering Al as the spacer, Edwards et al. [19], using a simple model, found that intrinsic biquadratic exchange is associated with higher harmonics in the oscillatory exchange coupling. Consequently it is much smaller than the bilinear exchange and also falls off more rapidly with increasing temperature. Further, the magnetization loops for Fe/Ag/Fe trilayers can be explained in a satisfactory way only by including the simultaneous presence of bilinear and biquadratic exchange coupling [20].

Until recently, BEC was always found to be small compared to the bilinear exchange term. However, Rezende and collaborators [21–23] published a series of experimental results which showed that, in Fe/Cr/Fe samples, BEC was comparable to the bilinear exchange coupling. The existence of bilinear and biquadratic exchange coupling in epitaxial Fe/Cr/Fe/Ag/GaAs(100) structures was also reported [24]. The longitudinal and polar magneto-optical Kerr effect (MOKE) and Brillouin light scattering (BLS) measurements have been combined to determine values for the bilinear and biquadratic coupling strengths and, while the phase and period of the oscillations in the interlayer coupling agree well with those reported by other researchers, the magnitude of the coupling strength was found to be reduced. The dependence of biquadratic coupling strength upon Cr thickness was well described by a simple power law and may provide a useful test of extrinsic biquadratic coupling models. Therefore, BEC can play a remarkable role in the properties of magnetic multilayers.

In this chapter we intend to go beyond the study of magnetic polaritons in magnetic structures, as discussed in the last two chapters, focusing our attention on other magnetic properties (e.g. magnetoresistance, magnetization and ferromagnetic resonance spectra) of ultrathin metallic magnetic structures of the Fe/Cr/Fe type grown following quasiperiodic sequences. We use a phenomenological model that includes the Zeeman energy and cubic anisotropy energy, together with the bilinear and biquadratic exchange energies [25–27]. The assumed physical parameters are based on experimental data recently reported [28] for samples where BEC is comparable in magnitude to the bilinear exchange. When BEC is sufficiently large, a striking self-similar pattern emerges in the magnetoresistance curves, as discussed below.

9.1 Magnetoresistance Self-Similar Spectra

It is well known that GMR occurs also in non-periodic granular systems, like Cu–Co alloys consisting of ultrafine Co-rich precipitate particles in a Cu-rich matrix [29]. Due to the fact that precipitate particles of these heterogeneous alloys have an average diameter and an average spacing similar to magnetic film thicknesses, the origin of GMR in granular systems is similar to that in magnetic films [30]. Further, quasiperiodic systems that show magnetoresistive properties can be a first step towards a better understanding of magnetoresistance in granular systems. On the other hand, from a technological point of view (as we will show later) the combination of BEC and quasiperiodicity permits us to control the magnetic field regions where magnetoresistance remains almost constant before saturation.

In what follows we investigate the influence of quasiperiodicity on the magnetoresistance properties of ultrathin metallic magnetic films. In particular, we shall be interested in Fe/Cr(100) multilayer structures that follow a Fibonacci sequence. The well-known trilayer Fe/Cr/Fe is a magnetic counterpart of the 3rd Fibonacci generation (ABA). We remark that only the odd Fibonacci generations have a magnetic counterpart because they start and finish with an A (Fe) building block. Therefore we only take into account this kind of generation of sequences, which implies an even number of Fe layers to ensure a real magnetic counterpart. However, we also avoid the intriguing behavior when even and odd numbers of Fe layers are considered, as discussed in Ref. [31].

Fig. 9.1 shows schematically the 3rd and 5th Fibonacci generations and their magnetic counterparts, where $t(d)$ is the thickness of a single Fe(Cr) layer. It is important to note the occurrence of a double Fe layer whose thickness is $2t$ in the 5th generation. It is easy to show that quasiperiodic magnetic films, for any Fibonacci generation, will be composed of single Cr layers, single Fe layers, and double Fe layers. The number of Fe and Cr layers for a given generation n is

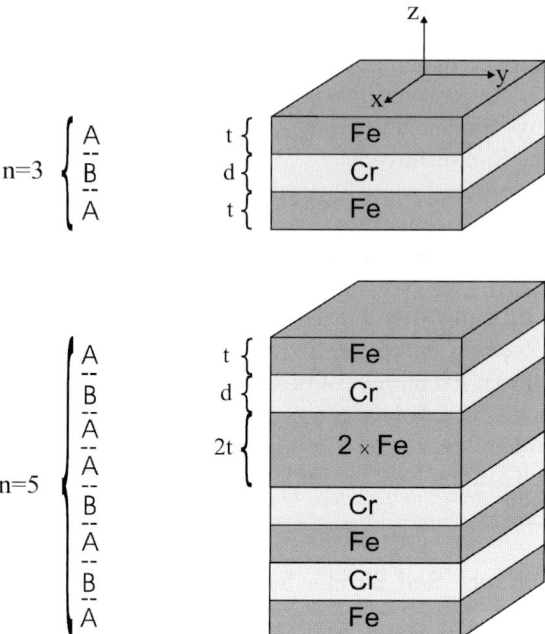

Fig. 9.1. The 3rd and 5th Fibonacci generations together with their magnetic counterparts.

related to the Fibonacci number as follows. The number of single Cr layers is F_{n-2}, the number of single Fe layers is $2F_{n-2} - F_{n-1} + 2$, and the number of double Fe layers is $F_{n-1} - F_{n-2} - 1$. These values can be deduced from properties given in Section 2.3.

We start by considering ferromagnetic films to have their magnetization in the xy-plane and we take the z-axis as the growth direction. The strong demagnetizing field in the z-direction will usually suppress any tendency for the magnetization to tilt out of plane. The global behavior of the system is well described by a simple macroscopic theory in terms of the magnetic energy per unit area [28]. We write this as

$$E_T = E_Z + E_{bl} + E_{bq} + E_a, \qquad (9.1)$$

where E_Z is the Zeeman energy, E_{bl} is the bilinear energy, E_{bq} is the biquadratic energy, and E_a is the four-fold magnetocrystalline anisotropy energy. More explicitly, for a system of m magnetic films we have

$$E_T = \sum_{i=1}^{m} \left[-t_i M_i H_0 \cos(\theta_i - \theta_H) + \tfrac{1}{4} t_i K_a \sin^2(2\theta_i) \right]$$
$$+ \sum_{i=1}^{m-1} \left[-J_{bl} \cos(\theta_i - \theta_{i+1}) + J_{bq} \cos^2(\theta_i - \theta_{i+1}) \right]. \qquad (9.2)$$

9.1. MAGNETORESISTANCE SELF-SIMILAR SPECTRA

Here J_{bl} is the usual bilinear exchange coupling that favors antiferromagnetic (ferromagnetic) alignment if it is negative (positive). We are concerned here with $J_{bl} < 0$ because magnetoresistive effects occur only for this case. J_{bq} is the BEC that favors a 90° alignment between magnetizations in adjacent films, and is experimentally found to be positive [9]. Also K_a is the cubic anisotropy constant that renders the (100) direction an easy direction, H_0 is the external in-plane magnetic field, and θ_H specifies its angular orientation. From now on we consider $\theta_H = 0$, which means that the magnetic field is applied along the easy axis. The thickness, angular orientation (relative to the easy axis), and magnetization of the ith Fe layer are denoted by t_i, θ_i, and M_i, respectively. It is usual to write the total magnetic energy in terms of experimental parameters (having the dimensions of effective fields) for each interaction, namely $H_a = 2K_a/M_S$, $H_{bl} = J_{bl}/tM_S$, and $H_{bq} = J_{bq}/tM_S$, where M_S is the total saturation magnetization, and we assume $M_i = M_S$. Therefore from Eq. (9.2) we have

$$\frac{E_T}{tM_S} = \sum_{i=1}^{m} (t_i/t) \left[-H_0 \cos(\theta_i) + \tfrac{1}{8} H_a \sin^2(2\theta_i) \right]$$

$$+ \sum_{i=1}^{m-1} \left[-H_{bl} \cos(\theta_i - \theta_{i+1}) + H_{bq} \cos^2(\theta_i - \theta_{i+1}) \right]. \quad (9.3)$$

The equilibrium orientations θ_i of the magnetizations are calculated numerically by minimizing the magnetic energy given by Eq. (9.3). We remark that it has proved difficult to generate accurately the configurations for larger structures, especially when BEC is strong [32]. Nevertheless, we have results for sufficiently large generations to infer important information about the effect of the quasiperiodicity.

As regards the underlying theory, the spin-dependent scattering mechanism is generally accepted as being responsible for the GMR effect [30]. It has been shown that GMR varies linearly with $\cos(\Delta\theta)$ when the electrons behave as a free-electrons gas, i.e. there are no barriers between adjacent films [33]. Here $\Delta\theta$ is the angular difference between adjacent magnetization vectors. In metallic systems like Fe/Cr this angular dependence is valid. Once the set $\{\theta_i\}$ is found, we obtain normalized values for the magnetoresistance and magnetization component in the field direction, i.e.

$$\frac{R(H_0)}{R(0)} = \sum_{i=1}^{m-1} \frac{1 - \cos(\theta_i - \theta_{i+1})}{2(m-1)} \quad (9.4)$$

and

$$\frac{M(H_0)}{M_S} = \sum_{i=1}^{m} t_i M_i \cos(\theta_i) \Big/ \sum_{i=1}^{m} t_i M_i. \quad (9.5)$$

The quantity $R(0)$ in Eq. (9.4) denotes the resistance at zero field. When $m=2$ we recover the usual expression for the trilayer Fe/Cr/Fe [33].

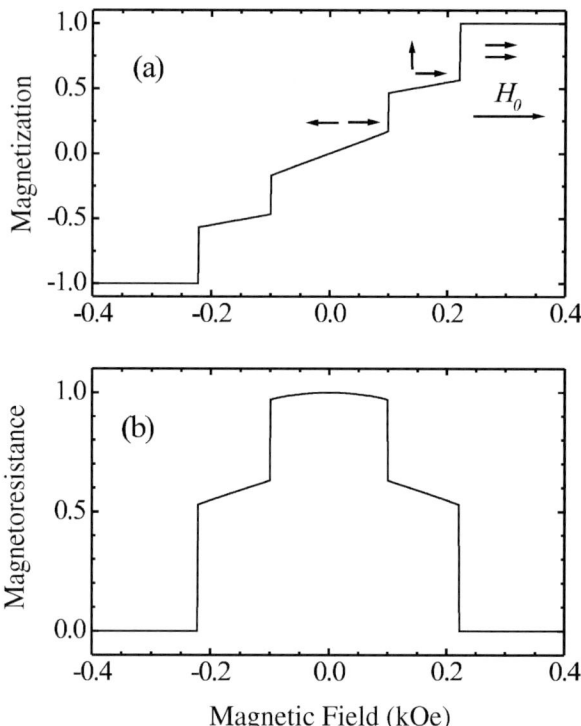

Fig. 9.2. (a) Magnetization and (b) magnetoresistance versus magnetic field for the 3rd Fibonacci generation with $H_{bl} = -150$ Oe and $H_{bq} = 50$ Oe. In (a) the arrows indicate the relative positions of the magnetizations in each phase (after Bezerra et al. [25]).

Now we present numerical calculations of the magnetization and magnetoresistance for Fibonacci ultrathin magnetic films. Previous studies on phase diagrams either did not take into account the biquadratic interaction [34,35] or did not consider properly the so-called 90° phase [36], where the magnetizations in the two magnetic films are nearly perpendicular to each other. We assume the cubic anisotropy field to be $H_a = 0.5$ kOe, corresponding to Fe(100) with $t > 30$ Å grown by sputtering. We choose values of the bilinear and biquadratic fields, H_{bl} and H_{bq}, such that they lie in three regions of interest:

(a) close to the region of the first antiferromagnetic–ferromagnetic transition where H_{bl} is moderate [9];

(b) near the maximum antiferromagnetic peak where H_{bl} reaches its maximum value [4];

(c) in the second antiferromagnetic peak where H_{bl} is small and equal to H_{bq} [22].

In Fig. 9.2 we show the curves for the normalized magnetization and magnetoresistance versus the magnetic field (applied along the easy axis) for the 3rd

9.1. MAGNETORESISTANCE SELF-SIMILAR SPECTRA

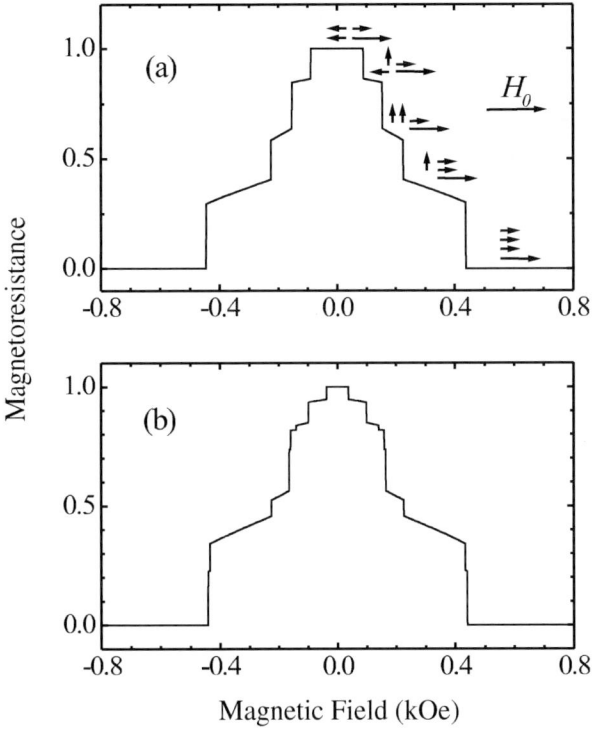

Fig. 9.3. Magnetoresistance versus magnetic field for the (a) 5th and (b) 7th Fibonacci generations with the same parameters as in Fig. 9.2. In (a) the arrows indicate the relative positions of the magnetizations in each phase (after Bezerra et al. [25]).

Fibonacci generation, i.e. corresponding to the Fe/Cr/Fe trilayer [25]. We assume $H_{bl} = -0.15\,\text{kOe}$ and $H_{bq} = 0.05\,\text{kOe}$ (so that $|H_{bl}| > H_{bq}$). These parameters correspond to a realistic sample with Cr thickness equal to 15 Å. From there one can identify two first-order phase transitions at $H_1 \sim 100\,\text{Oe}$ and $H_2 \sim 220\,\text{Oe}$. There are three magnetic phases:

(a) an antiferromagnetic phase occurs in the low-field region ($H_0 < 100\,\text{Oe}$);

(b) a 90° phase prevails in the midfield region ($100\,\text{Oe} < H_0 < 220\,\text{Oe}$);

(c) a saturated phase exists for $H_0 > 220\,\text{Oe}$. These numerical calculations indicate that a first-order phase transition occurs when $H_a > 2(|H_{bl}| + 2H_{bq})$.

Next we focus the discussion on the magnetoresistance curves, looking for their self-similarity pattern, and we leave the discussion of the magnetization curves for the next section. Fig. 9.3 shows the normalized magnetoresistance curves for the 5th and 7th Fibonacci generations with the same experimental parameters considered in Fig. 9.2 [25]. For the 5th generation the angular orientations of

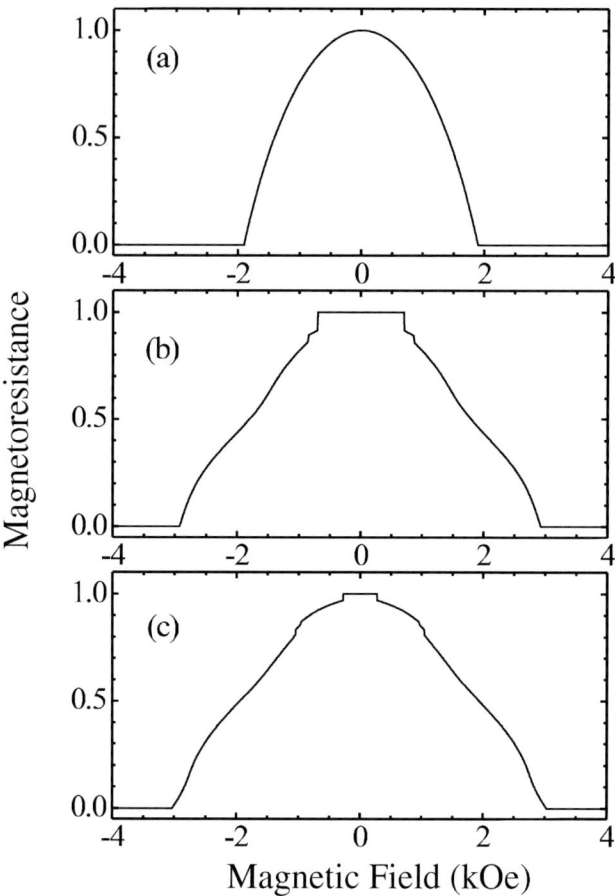

Fig. 9.4. Magnetoresistance versus magnetic field for the (a) 3rd, (b) 5th, and (c) 7th Fibonacci generations with $H_{bl} = -1.0\,\text{kOe}$ and $H_{bq} = 0.1\,\text{kOe}$ (after Bezerra et al. [25]).

magnetizations are shown by the arrows, where the Fe double layer is indicated by the bigger arrow. We can identify four first-order phase transitions, where each one is due to a 90° jump of magnetization. For the 7th generation there are eight first-order phase transitions, and nine magnetic phases ranging from the antiferromagnetic phase ($H_0 < 38\,\text{Oe}$) to the saturated one ($H_0 > 440\,\text{Oe}$) are present. We note a clear *self-similar pattern* of the magnetoresistance curves on comparing Figs. 9.2b and 9.3a and b. In other words, the pattern of the trilayer Fe/Cr/Fe is always present in the next generation.

By contrast, when $H_{bl} = -1.0\,\text{kOe}$ and $H_{bq} = 0.1\,\text{kOe}$ (the case of $|H_{bl}| \gg H_{bq}$), corresponding to a sample with Cr thickness equal to 10 Å, the self-similarity is *not* observed (see Fig. 9.4). For this set of parameters, the majority of phase transitions are of second order and we have found numerically that this occurs when $H_a < 2(|H_{bl}| + 2H_{bq})$.

9.1. MAGNETORESISTANCE SELF-SIMILAR SPECTRA

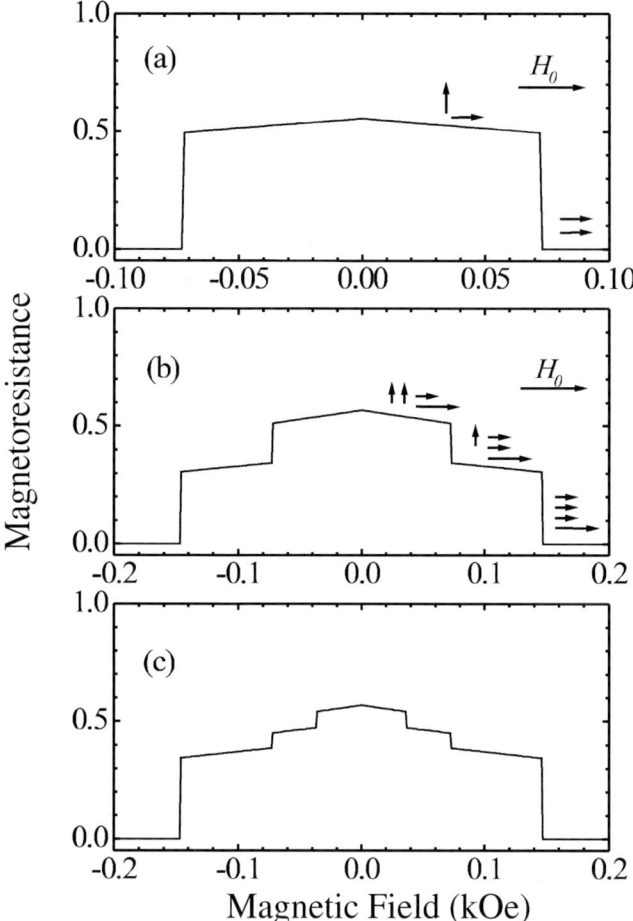

Fig. 9.5. Same as Fig. 9.4 but with $|H_{bl}| = H_{bq} = 35$ Oe, which corresponds to a sample with Cr thickness equal to 25 Å. Note a striking self-similar pattern. In (a) and (b) the arrows indicate the relative positions of the magnetizations in each phase (after Bezerra et al. [25]).

However, when H_{bq} is increased (taking the case of $|H_{bl}| = H_{bq} = 35$ Oe), we observe again a striking *self-similar pattern* (see Fig. 9.5). Each new transition occurs for a value of the magnetic field that is about one half of the previous one. For this set of parameters the magnetoresistance is approximately half of its value at zero magnetic field, because the magnetizations of the adjacent Fe films are nearly perpendicular to each other due to the strong biquadratic field. For the 3rd generation (see Fig. 9.5a) there is only one transition at $H_1 \sim 70$ Oe and two magnetic phases: a 90° phase at $H_0 < 70$ Oe and a saturated phase at $H_0 > 70$ Oe [22]. In the 5th generation (see Fig. 9.5b) there are two transitions at $H_1 \sim 70$ Oe and $H_2 \sim 140$ Oe. For the 7th generation (see Fig. 9.5c) there are three transitions at $H_1 \sim 35$ Oe, $H_2 \sim 70$ Oe, and $H_3 \sim 140$ Oe.

The above numerical results point to a dependence of the self-similar magnetoresistance on first-order phase transitions and on the strength of BEC. A possible explanation is that BEC reinforces the long-range quasiperiodic order which is responsible for the self-similarity in quasiperiodic systems. This was an unexpected effect of this exchange coupling and, as far as we know, this is the first system that exhibits magnetoresistance with fractal properties. From a technological point of view, magnetoresistance with defined flat regions (as in Figs. 9.3 and 9.5) opens new perspectives in information storage technology with the possibility of a recording system having more than two states. Fibonacci ultrathin magnetic films can be realized experimentally following the procedures of Refs. [37,38] to grow the samples. The above argument is reinforced by previous works on the correlation lengths of magnetic systems with BEC; see, for example, Ref. [39].

9.2 Magnetization Profiles

The effects of quasiperiodicity on magnetization curves in metallic magnetic multilayers will be investigated next. We are interested in new magnetic phases and alignments that are due only to the quasiperiodicity of the system. We discuss Fe/Cr(100) structures that follow Fibonacci and Double-period quasiperiodic sequences.

We consider that magnetic films are uniformly magnetized and behave as single domains. The interfilm exchange coupling between adjacent ferromagnetic films is weak compared to the strong exchange coupling between spins within a given film. Therefore we can represent ferromagnetic films in terms of classical magnetization vectors $\vec{M}$ composed of real spins within films that are strongly coupled by intrafilm exchange. These classical magnetizations interact through interfilm exchange coupling and they can also have some anisotropy depending on the structure studied. It should be noted that this system is isomorphous to a one-dimensional chain of classical spins.

The global behavior of this system is well described by a phenomenological theory in terms of the magnetic free energy expressed in Eqs. (9.1)–(9.3). Once the magnetic free energy is determined, we can calculate the equilibrium configuration for specific values of the experimental parameters as a function of the external applied field. In simple situations the equilibrium configuration can be obtained analytically by equating to zero the derivatives of the magnetic energy with respect to the orientation angle in each film. However, in most cases this leads to transcendental equations which cannot be analytically solved. In general, we want to find the global minimum of the energy function

$$E_T = E_T(\theta_1, \theta_2, \ldots, \theta_p). \tag{9.6}$$

Here θ_p can assume values in the range $[0, 2\pi]$ and the set defines a p-dimensional space. When the dimension of this space is high, the E_T function has a rough surface, i.e. there are many local minima that make it difficult to find the global minimum. There are many numerical methods to solve this type of problem [40].

9.2. MAGNETIZATION PROFILES

Here we describe two methods that were successfully used, namely the simulated annealing method and the gradient method.

The *simulated annealing method* (or SA method) was introduced by Kirkpatrick et al. [41]. It comes from the fact that the process of heating (annealing) and slowly cooling a metal brings it into a more uniform crystalline state, which is believed to be the state where the free energy of bulk matter takes its global minimum. The role played by the temperature is to allow the configurations to reach higher energy states with a probability given by Boltzmann's exponential law. Thus they can overcome energy barriers that would otherwise force them to remain in local minima. In general, a simulated annealing technique can be described as follows:

(a) Choose an initial point in the parameter space, corresponding to an initial configuration $\{\theta_j\}$, and calculate the associated energy E_j.

(b) Choose a second point in the parameter space, corresponding to a second configuration $\{\theta_{j+1}\}$, and calculate its associated energy E_{j+1}.

(c) If $\Delta E = E_{j+1} - E_j < 0$, then $\{\theta_{j+1}\}$ is adopted as the new configuration of the system.

(d) If $\Delta E \geq 0$, we define the probability $p = \exp(-\Delta E/k_B T)$ and choose a random number x such that $0 \leq x \leq 1$. If $x \geq p$, then $\{\theta_{j+1}\}$ is the new configuration of the system. Otherwise, $\{\theta_j\}$ is maintained as the configuration of the system.

(e) This procedure is executed again and again until equilibrium is reached.

The second method is the *gradient method* [42], which is based on the directional derivative of the E_T function in a search of its global minimum. We calculate the gradient of E_T with relation to the set $\{\theta_i\}$,

$$\nabla E_T = \sum_{i=1}^{p} \frac{\partial E_T}{\partial \theta_i} \hat{\theta}_i, \qquad (9.7)$$

where $\hat{\theta}_i$ is a unit vector. Then we execute the following algorithm to find the equilibrium configuration:

(a) We generate a configuration in the parameter space $\{\theta_j\}$ from which we calculate the associated energy E_j and its gradient.

(b) A second (displaced) point in the parameter space is generated by $\{\theta_{j+1}\} = \{\theta_j\} - \alpha \nabla E_T$, where α controls the size of the displacement.

(c) The energy of the second point is calculated and, if $E_{j+1} > E_j$, the parameter α is divided by two and we go back to step (b). Otherwise, we generate a new configuration from $\{\theta_{j+1}\}$ and the whole process is repeated until there is essentially no change.

In the last step the reduction of α is limited by the required precision. This limit is reached when $|\alpha \nabla E_T| < \epsilon$, where the value for ϵ can be chosen.

We now present results in which both of the above methods are used to obtain the equilibrium positions of the magnetizations. Each method was applied for each value of the applied magnetic field and for each set of experimental parameters. We adopt the configuration with the lowest energy furnished by both methods as the equilibrium configuration. Some numerical results applied to quasiperiodic magnetic multilayers are now described. In all cases we consider the cubic anisotropy effective field $H_{ca} = 0.5\,\mathrm{kOe}$ which corresponds to Fe(100) with $t > 30\,\text{Å}$. We use two sets of experimental values for the bilinear and biquadratic exchange coupling. The first has $H_{bl} = -1.0\,\mathrm{kOe}$ and $H_{bq} = 0.1\,\mathrm{kOe}$, which lies in the region of the first antiferromagnetic peak of bilinear exchange coupling, corresponding to a sample whose Cr thickness is about $10\,\text{Å}$. The second set has $H_{bl} = -0.035\,\mathrm{kOe}$ and $H_{bq} = 0.035\,\mathrm{kOe}$, which is in the region of the second antiferromagnetic peak of bilinear exchange coupling, corresponding to a sample whose Cr thickness is about $25\,\text{Å}$.

The magnetization curves for the first set of parameters of the Fibonacci magnetic multilayers are shown in Fig. 9.6 [26]. For the 3rd generation (which corresponds to the Fe/Cr/Fe trilayer) in the low-field region, the magnetizations are antiparallel. As the field increases, they continuously rotate toward the field direction (indicating a second-order phase transition) and saturation is reached when the external magnetic field $H_0 \sim 1.91\,\mathrm{kOe}$. For the 5th generation there are two first-order phase transitions at $H_1 \sim 0.71\,\mathrm{kOe}$ and $H_2 \sim 0.87\,\mathrm{kOe}$. Saturation is reached at $H_0 \sim 2.93\,\mathrm{kOe}$. For the 7th generation there are three first-order phase transitions at $H_1 \sim 0.28\,\mathrm{kOe}$, $H_2 \sim 0.96\,\mathrm{kOe}$, and $H_3 \sim 1.06\,\mathrm{kOe}$. Saturation is reached at $H_0 \sim 3.03\,\mathrm{kOe}$. For this set of parameters the majority of the transitions are of second order. We note that, due to the different Fe layer thickness for the 5th and 7th generations, the magnetization is non-zero in this case even when $H_0 = 0$.

In Fig. 9.7 we show the results calculated using the second set of parameters [26]. For the 3rd generation, due to the strong biquadratic field, there is no antiparallel phase in the low-field region. Two magnetic phases occur in this case: $90°$ ($H_0 < 72\,\mathrm{Oe}$) and saturated ($H_0 > 72\,\mathrm{Oe}$). The 5th generation presents three magnetic phases:

(a) $90°$ ($H_0 < 72\,\mathrm{Oe}$);

(b) almost saturated ($72\,\mathrm{Oe} < H_0 < 0.14\,\mathrm{kOe}$);

(c) saturated ($H_0 > 0.14\,\mathrm{kOe}$).

The 7th generation presents four magnetic phases from $90°$ ($H_0 < 36\,\mathrm{Oe}$) to the saturated regime ($H_0 > 0.14\,\mathrm{kOe}$). All transitions are of first order. Note the striking self-similar pattern shown by the magnetization profile.

Figs. 9.8a and b show the results calculated for the Double-period magnetic multilayers, considering the first set of parameters [26]. For the 2nd generation, due

9.2. MAGNETIZATION PROFILES

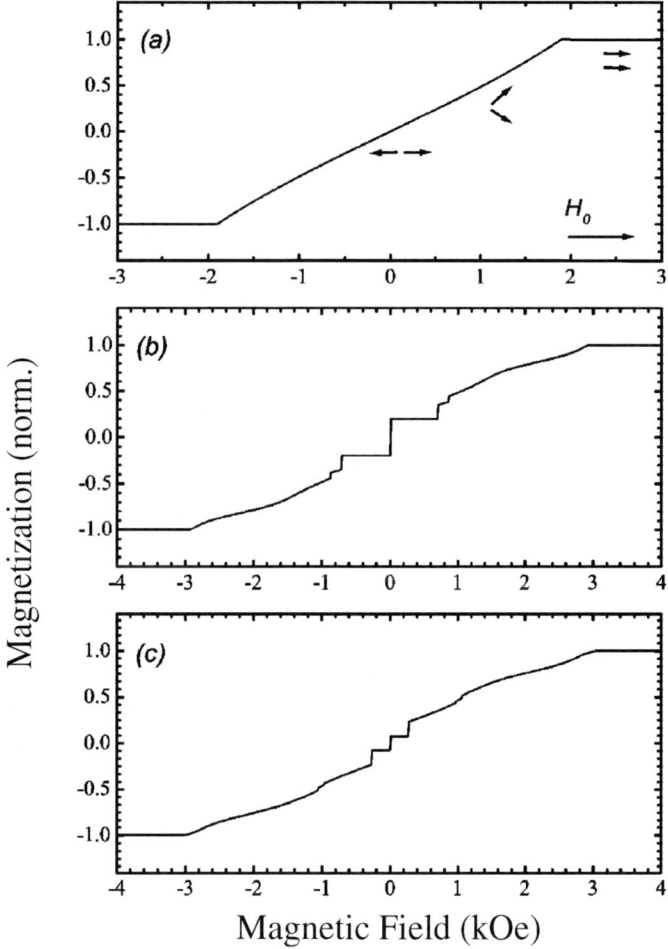

Fig. 9.6. Magnetization versus applied magnetic field for the (a) 3rd, (b) 5th, and (c) 7th Fibonacci generations with $|H_{bq}|/|H_{bl}| = 0.10$, corresponding to a realistic sample whose Cr thickness is about 10 Å. We have considered the cubic anisotropy effective field $H_{ca} = 0.5$ kOe, which corresponds to Fe(100) with $t > 30$ Å (after Bezerra et al. [26]).

to the double Fe layer, the magnetization has about one-third of its saturation value for zero magnetic field. There is a first-order phase transition from an antiparallel to an asymmetric phase at $H_0 \sim 0.69$ kOe. In this latter phase, the magnetizations are asymmetrically oriented along the magnetic field. When $H_0 \sim 1.34$ kOe the saturated phase emerges. For the 4th generation, for zero magnetic field, the magnetization has about 10% of its saturation value due to the different thickness of the Fe layers. There is a first-order phase transition at $H_0 \sim 0.29$ kOe and saturation is reached at $H_0 \sim 3.27$ kOe. All other phase transitions are of second order.

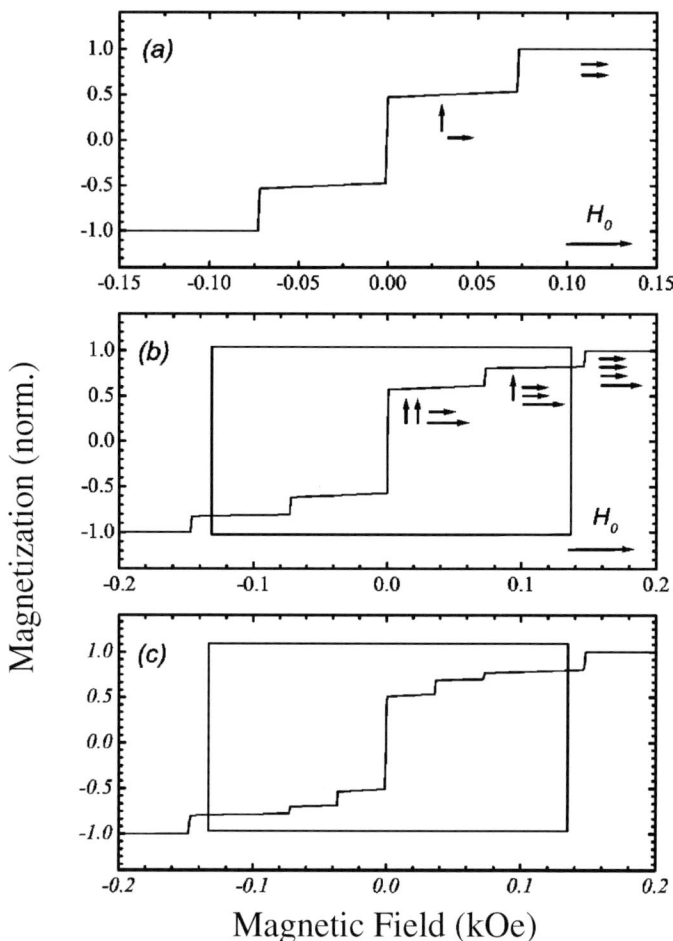

Fig. 9.7. Same as Fig. 9.6, but with $|H_{bq}|/|H_{bl}| = 1.0$, corresponding to a realistic sample whose Cr thickness is about 25 Å (after Bezerra et al. [26]).

By contrast, for the second set of parameters (see Fig. 9.9) all transitions are of first order. There is no antiparallel phase in the low-field region, due to the strong biquadratic field. For the 2nd generation, the magnetization is about two-thirds of its saturation value when $H_0 = 0$. There are two magnetic phases: (i) 90° ($0 < H_0 < 72$ Oe) and (ii) saturated ($H_0 > 72$ Oe). For the 4th generation, the magnetization for zero external magnetic field is about half of its saturation value. Four magnetic phases are present, from 90° ($0 < H_0 < 38$ Oe) to the saturated phase ($H_0 > 0.14$ kOe). As in the Fibonacci case, a self-similar pattern is present in the magnetization curves.

To conclude this section, we would now like to emphasize two aspects: (i) the effect of the different thickness of Fe layers, and (ii) the effect of BEC. First, the

9.2. MAGNETIZATION PROFILES

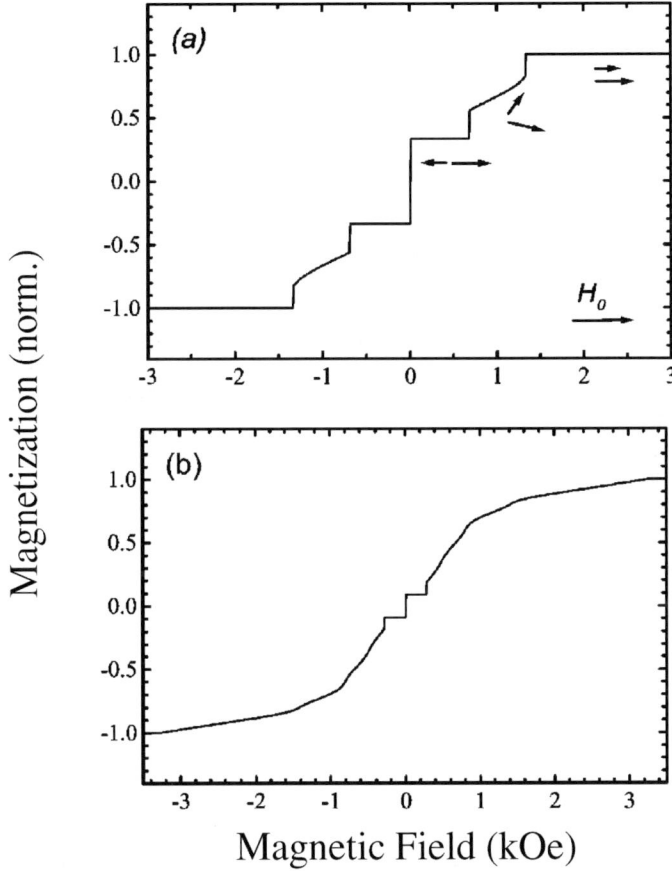

Fig. 9.8. Magnetization versus applied magnetic field for the (a) 2nd and (b) 4th Double-period generations with $|H_{bq}|/|H_{bl}| = 0.10$. The cubic anisotropy effective field is $H_{ca} = 0.5$ kOe. The Cr thickness is 10 Å (after Bezerra et al. [26]).

effect of different Fe-layer thickness is evident in the low-field region. In that case, there is a net magnetization even if the alignment is antiparallel and the external magnetic field is zero. Besides, the nature of the phase transitions is changed by the different thickness (e.g. Fig. 9.6a shows only second-order phase transitions, while Fig. 9.8a shows an additional first-order phase transition). These results suggest that, by varying the thickness of Fe layers, it is possible to tailor magnetic multilayers to exhibit desired specific phase transitions and critical fields. However, as the thickness of Fe layers increases, the crystalline anisotropy of Fe(100) films on Cr(100) also increases. Fortunately, as a characteristic of the quasiperiodic multilayers considered here, the maximum size of multiple Fe layers is just two (for the Fibonacci case) and three (for the Double-period case), no matter what the value of their generation numbers. Also, for a thickness greater than 40 Å, crystalline anisotropy reaches saturation [28].

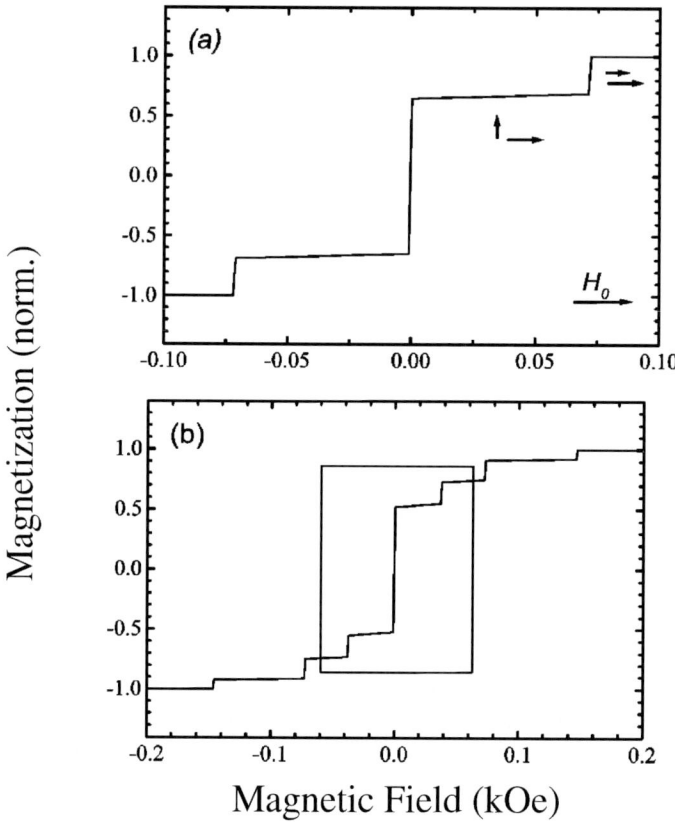

Fig. 9.9. Same as Fig. 9.8, but with $|H_{bq}|/|H_{bl}| = 1.0$, corresponding to a realistic sample, whose Cr thickness is about 25 Å (after Bezerra et al. [26]).

Second, BEC plays a remarkable role in determining the features of magnetization curves. For example, when the bilinear exchange coupling prevails, the majority of the transitions are of second order (see Figs. 9.6 and 9.8). However, when BEC is comparable to bilinear exchange, in the presence of a stronger crystalline anisotropy the transitions are characterized by discontinuous jumps in the magnetization, indicative of first-order phase transitions. This can be considered as the basic signature of BEC (see Figs. 9.7 and 9.9), although for the case where there is no biquadratic term, a first-order phase transition appears due to the anisotropy [43]. Furthermore, as can be seen from these figures, the magnetization curves of higher generations reproduce the magnetization curves of lower generations. This self-similar behavior is a general characteristic of quasiperiodic systems, although it is not present when the bilinear exchange prevails (see Figs. 9.6 and 9.8). A possible explanation for these different types of behavior is that the BEC induces long-range correlations that emphasize the quasiperiodicity of the system. These long-range correlations enable the whole structure to respond to (or *sense*) its quasiperiodicity, which is reflected in the magnetization curves. This argument is reinforced by previous works on the correlation lengths of magnetic systems with BEC (see, for example, Sørensen and Young [39]).

The most appropriate experimental property for studying the magnetization curves of magnetic films is the magneto-optical Kerr effect (or MOKE) [22]. However, because MOKE measurements provide surface sensitivity only on the scale of the optical penetration depth ($\sim 10\,\text{Å}$), it is necessary also to make use of superconducting quantum interface device (SQUID) magnetometry [31]. These two techniques prove to be complementary in understanding the switching behavior of the multilayer films, as far as the magnetization curves are concerned. We hope that these theoretical results will serve to stimulate more experimental studies of these structures.

9.3 Ferromagnetic Resonance Curves

Here we provide calculations for the FMR curves of Fe/Cr(100) thin films and multilayers that follow the Fibonacci sequence. Our approach is based on the equation of motion for the small-signal magnetization deviation from the equilibrium directions, which can be deduced from the total magnetic energy expression. An analogous approach has already been employed for other magnetic systems in Chapters 7 and 8.

Spin-wave studies of magnetic systems with both bilinear and biquadratic exchange interactions are quite recent [44–46]; they are of interest because they give important information about the coupling parameters, particularly for noncollinear configurations. Bilinear and biquadratic exchange coupling were also used to describe compensated, partially compensated, and uncompensated interfaces between a ferromagnet and an antiferromagnet [47]. Based on the exchange-coupled resonance theory, which we describe later in this section, the angular dependence of FMR measurements was recently studied [48] on several series of symmetric and asymmetric Co/Ru/Co structures. However, only the bilinear exchange coupling was observed in these systems. The biquadratic (BEC) contribution was more than two orders of magnitude smaller.

As noted earlier, BEC in real systems usually favors perpendicular alignment of the film magnetization, whereas the bilinear one favors either parallel or antiparallel configurations, depending on its sign. The inclusion of both of these exchange terms leads to interesting physical properties. We choose the ratio between the biquadratic and bilinear exchange terms according to experimental data. Although the exchange coupling parameters may oscillate as functions of the non-magnetic (Cr) layer thickness, we can justify the use of the fixed physical parameters in our calculations, because for the quasiperiodic Fibonacci sequence considered here there are only single Cr layers, no matter which generation is considered.

The spin-wave dispersion relations for trilayer structures such as Fe/Cr/Fe have been calculated by several authors using various models. The earlier calculations [49,50], which were made before the interlayer exchange coupling was discovered, considered only the effect of magnetic dipolar interaction on surface and bulk magnetostatic modes in coupled films. Subsequently, solutions that included the BEC terms, as well as the usual exchange and dipolar contributions, were worked out by Hillebrands [51] for an arbitrary number of magnetic layers. For the case of the trilayer structure (Fe/Cr/Fe), his dispersion relation was obtained from a system of 16 linear coupled equations. This required the use of appropriate numerical tools which complicated the interpretation of the observed spectra and encouraged several other authors to develop alternative calculations [52–55].

We now present a derivation of the spin-wave dispersion relation based on the equations of motion for the small-signal magnetization deviations from the equilibrium directions, similar to those derived by Cochran and co-workers [56,57]. The starting point is the torque equation of motion in the absence of damping, as expressed in Eq. (3.36), where i is now a label for each magnetic film (and takes values from 1 to N). The equation relates the magnetization $\vec{M}_i$ in film i to the effective magnetic field $\vec{H}_{eff}$ in that layer. For each magnetic film we employ a Cartesian coordinate system, henceforth referred to as the local axes system, in which the transformed z-axis coincides with the equilibrium direction of the magnetization (see Fig. 9.10). For each film, the magnetization can then be expressed in terms of static and fluctuating parts as

$$\vec{M}_i = M_{iz_i}\hat{e}_{z_i} + m_{iy}\hat{e}_y + m_{ix_i}\hat{e}_{x_i}. \tag{9.8}$$

Here we assume that $|m_{ix_i}| \ll M_{iz_i}$ and $|m_{iy}| \ll M_{iz_i}$, and we shall make linear spin-wave approximations as usual. Note that the y-axis of the local axes system coincides with the crystalline y-axis (so we do not need the index i for the y-component in the local axes system). The transformation to connect the components of the magnetization to the original variables is given by (see Fig. 9.10)

$$M_{ix} = M_{iz_i}\sin\theta_i + m_{ix_i}\cos\theta_i,$$
$$M_{iy} = m_{iy}, \tag{9.9}$$
$$M_{iz} = M_{iz_i}\cos\theta_i - m_{ix_i}\sin\theta_i.$$

Likewise, the effective magnetic field in layer i is expressed as

$$\vec{H}_{eff} = H_{iz_i}\hat{e}_{z_i} + h_{iy}\hat{e}_y + h_{ix_i}\hat{e}_{x_i}. \tag{9.10}$$

Thus, substituting the magnetization given by Eq. (9.8) and the effective field, Eq. (9.10), into the torque equation, Eq. (3.36), and linearizing, we obtain for the non-zero components

$$dm_{iy}/dt = -\gamma_i(M_{iz_i}h_{ix_i} - m_{ix_i}H_{iz_i}), \tag{9.11}$$

$$dm_{ix_i}/dt = -\gamma_i(m_{iy}H_{iz_i} - M_{iz_i}h_{iy}). \tag{9.12}$$

The effective field is related to the total energy by the expression

$$\vec{H}_{eff} = -\nabla_M E_T, \tag{9.13}$$

9.3. FERROMAGNETIC RESONANCE CURVES

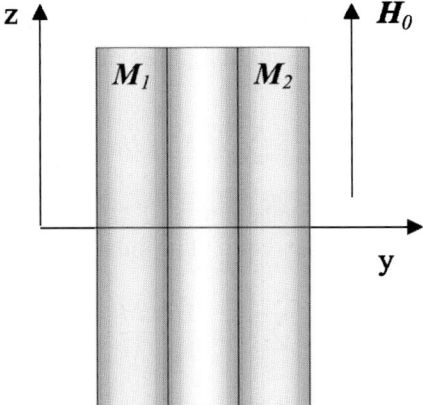

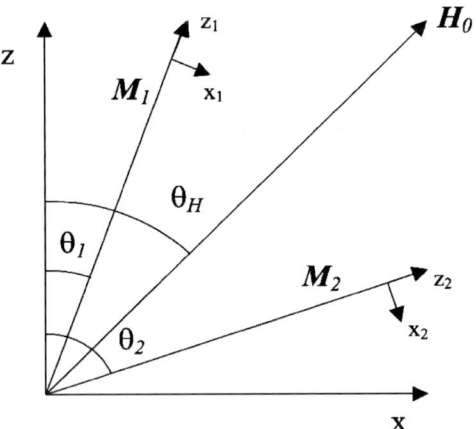

Fig. 9.10. Illustration of the local axes system for the FMR calculation. The transformed z-directions (z_1 and z_2) are chosen to be aligned with the equilibrium orientations of the layer magnetization.

and the total magnetic energy is given here by

$$E_T = E_Z + E_{bl} + E_{bq} + E_d + E_{ca} + E_{sa}, \tag{9.14}$$

where E_Z, E_{bl}, E_{bq} have already been defined for the macroscopic case in Section 9.1, and E_d, E_{ca}, E_{sa} are, respectively, the dipolar, cubic anisotropy, and surface anisotropy energies. We now define all components of the total energy individually in a form appropriate to our present calculation.

First, the Zeeman interaction energy is found from the scalar product of the magnetization with the external applied magnetic field. In the case of N magnetic films it is given by

$$E_Z = -\sum_{i=1}^{N} d_i \vec{M}_i \cdot \vec{H}_0, \tag{9.15}$$

where d_i is the width of ferromagnetic layer i, and $\vec{M}_i$ and $\vec{H}_0$ are the magnetization and external magnetic field, respectively. The bilinear and biquadratic energies are defined as

$$E_{bl} = -\sum_{i=1}^{N-1} J_{bl} \frac{\vec{M}_i \cdot \vec{M}_{i+1}}{|\vec{M}_i||\vec{M}_{i+1}|} \tag{9.16}$$

and

$$E_{bq} = \sum_{i=1}^{N-1} J_{bq} \left(\frac{\vec{M}_i \cdot \vec{M}_{i+1}}{|\vec{M}_i||\vec{M}_{i+1}|} \right)^2, \tag{9.17}$$

where J_{bl} and J_{bq} are the bilinear and biquadratic exchange constants, as before. The dipolar (or demagnetizing) energy is represented here by the approximate expression

$$E_d = 2\pi \sum_{i=1}^{N} d_i (\vec{M}_i \cdot \hat{e}_y)^2. \tag{9.18}$$

Finally, the cubic and surface anisotropy energies can be written as

$$E_{ca} = \sum_{i=1}^{N} \frac{d_i K_{ca}}{|\vec{M}_i|^4} \left(M_{ix}^2 M_{iy}^2 + M_{ix}^2 M_{iz}^2 + M_{iy}^2 M_{iz}^2 \right) \tag{9.19}$$

and

$$E_{sa} = -\sum_{i=1}^{N} \frac{K_s}{|\vec{M}_i|^2} (\vec{M}_i \cdot \hat{e}_y)^2, \tag{9.20}$$

where K_s is the surface anisotropy constant.

The required effective fields h_{ix_i}, h_{iy}, and H_{iz_i} can then be calculated by making use of the above energy terms, together with Eq. (9.13):

$$h_{ix_i} = -\frac{\partial E_T}{\partial m_{ix_i}}, \quad h_{iy} = -\frac{\partial E_T}{\partial m_{iy}}, \quad H_{iz_i} = -\frac{\partial E_T}{\partial M_{iz_i}}. \tag{9.21}$$

Writing explicitly the resulting equations of motion for the N magnetic films, it is necessary to solve a system of $2N$ coupled equations given by

$$\frac{1}{\gamma} \sum_{i=1}^{N} \frac{dm_{iy}}{dt} = M_{iz_i} \left(\frac{\partial E_T}{\partial m_{ix_i}} \right) - m_{ix_i} \left(\frac{\partial E_T}{\partial M_{iz_i}} \right),$$

$$\frac{1}{\gamma} \sum_{i=1}^{N} \frac{dm_{ix_i}}{dt} = m_{iy} \left(\frac{\partial E_T}{\partial M_{iz_i}} \right) - M_{iz_i} \left(\frac{\partial E_T}{\partial m_{iy}} \right). \tag{9.22}$$

9.3. FERROMAGNETIC RESONANCE CURVES

For example, in the case of the trilayer Fe/Cr/Fe we have $N=2$, corresponding to the 3rd Fibonacci sequence ($n=3$). For the case of seven layers, Fe/Cr/2Fe/Cr/Fe/Cr/Fe, we have $N=4$, corresponding to the 5th Fibonacci generation ($n=5$), and so on.

Now, using Eq. (9.14) to evaluate the derivatives in Eq. (9.22) and considering normal-mode solutions of the type

$$m_{ix_i} = m^0_{ix_i} \exp(-i\omega t), \quad m_{iy} = m^0_{iy} \exp(-i\omega t) \tag{9.23}$$

corresponding to a frequency ω, we eventually find in the simplest case of $N=2$ that

$$\begin{bmatrix} -\frac{i\omega}{\gamma} & H_1 & 0 & H_2 \\ -H_3 & -\frac{i\omega}{\gamma} & H_4 & 0 \\ 0 & H_1 & -\frac{i\omega}{\gamma} & H_2 \\ H_3 & 0 & -H_4 & -\frac{i\omega}{\gamma} \end{bmatrix} \begin{bmatrix} m_{1x_1} \\ m_{1y} \\ m_{2x_2} \\ m_{2y} \end{bmatrix} = 0. \tag{9.24}$$

Here we define the quantities H_1 to H_4 by

$$H_1 = H_0 \cos(\theta_1 - \theta_H) + H_{bl} \cos(\theta_1 - \theta_2) - (H_{ca}/2)[\sin^2(2\theta_1) - 2]$$
$$- 2H_{bq} \cos^2(\theta_1 - \theta_2) + 4\pi M_s - H_{sa}, \tag{9.25}$$

$$H_2 = -H_{bl} + 2H_{bq}\cos(\theta_1 - \theta_2), \tag{9.26}$$

$$H_3 = H_0 \cos(\theta_1 - \theta_H) + H_{bl}\cos(\theta_1 - \theta_2) - 2H_{bq}\cos(\theta_1 - \theta_2)$$
$$-(H_{ca}/2)[4\sin^2(2\theta_1) - 2], \tag{9.27}$$

$$H_4 = -H_{bl}\cos(\theta_1 - \theta_2) + 2H_{bq}\cos[2(\theta_1 - \theta_2)]. \tag{9.28}$$

The vanishing of the 4×4 secular determinant found from the matrix in Eq. (9.24) provides the condition for non-trivial solutions. In terms of ω the result is

$$(\omega/\gamma)^4 + \alpha_0(\omega/\gamma)^2 + \alpha_1 = 0, \tag{9.29}$$

which may be solved to yield the dispersion relation for the trilayer case as

$$(\omega/\gamma)^2 = -(\alpha_0/2) \pm \sqrt{(\alpha_0/2)^2 - \alpha_1}. \tag{9.30}$$

The quantities α_0 and α_1 appearing in the above expressions are given by

$$\alpha_0 = -H_2H_4 + H_1H_4 - H_1H_3 + H_2H_3, \tag{9.31}$$

$$\alpha_1 = -H_1 H_4 G_1 G_4 + H_1 H_3 G_1 G_3 - H_2 H_3 G_2 G_3 + H_2 H_4 G_2 G_4. \tag{9.32}$$

Finally the G_j coefficients are obtained from H_j by interchanging the layer index i labels 1 and 2 in Eqs. (9.25)–(9.28).

The case of four magnetic films can be treated using the same method. When $N=4$ we find that Eq. (9.24) is replaced by a similar matrix equation of higher dimension:

$$\begin{bmatrix} -i\omega/\gamma & H_2 & 0 & H_3 & 0 & 0 & 0 & 0 \\ -H_5 & -i\omega/\gamma & H_6 & 0 & 0 & 0 & 0 & 0 \\ 0 & G_1 & -i\omega/\gamma & G_2 & 0 & G_3 & 0 & 0 \\ G_4 & 0 & -G_5 & -i\omega/\gamma & G_6 & 0 & 0 & 0 \\ 0 & 0 & 0 & I_1 & -i\omega/\gamma & I_2 & 0 & I_3 \\ 0 & 0 & I_4 & 0 & -I_5 & -i\omega/\gamma & I_6 & 0 \\ 0 & 0 & 0 & 0 & 0 & J_1 & -i\omega/\gamma & J_2 \\ 0 & 0 & 0 & 0 & J_4 & 0 & -J_5 & -i\omega/\gamma \end{bmatrix} \begin{bmatrix} m_{1x_1} \\ m_{1y} \\ m_{2x_2} \\ m_{2y} \\ m_{3x_3} \\ m_{3y} \\ m_{4x_4} \\ m_{4y} \end{bmatrix} = 0. \tag{9.33}$$

The coefficients H_i, G_i, I_i, and J_i are specified by

$$X_1 = H_0 \cos(\theta_i - \theta_H) + H_{bl} \cos(\theta_i - \theta_{i+1}) - (H_{ca}/2)[\sin^2(2\theta_i) - 2]$$

$$-2H_{bq} \cos^2(\theta_i - \theta_{i+1}) + 4\pi M_s - H_{sa}, \tag{9.34}$$

$$X_2 = -H_{bl} + 2H_{bq} \cos(\theta_i - \theta_{i+1}), \tag{9.35}$$

$$X_3 = H_0 \cos(\theta_i - \theta_H) + H_{bl} \cos(\theta_i - \theta_{i+1}) - 2H_{bq} \cos[2(\theta_i - \theta_{i+1})]$$

$$-(H_{ca}/2)[4\sin^2(2\theta_i) - 2], \tag{9.36}$$

$$X_4 = H_{bl} \cos(\theta_i - \theta_{i+1}) - 2H_{bq} \cos[2(\theta_i - \theta_{i+1})], \tag{9.37}$$

$$Y_1 = H_0 \cos(\theta_{i+1} - \theta_H) + H_{bl}\Big[\cos(\theta_i - \theta_{i+1}) + \cos(\theta_{i+1} - \theta_{i+2})\Big]$$

$$-(H_{ca}/2)\Big[\sin^2(2\theta_{i+1}) - 2\Big] - 2H_{bq}\Big[\cos^2(\theta_i - \theta_{i+1}) + \cos^2(\theta_{i+1} - \theta_{i+2})\Big]$$

$$+4\pi M_s - H_{sa}, \tag{9.38}$$

$$Y_2 = -H_{bl} + 2H_{bq} \cos(\theta_{i+1} - \theta_{i+2}), \tag{9.39}$$

9.3. FERROMAGNETIC RESONANCE CURVES

$$Y_3 = H_0 \cos(\theta_{i+1} - \theta_H) + H_{bl}\left[\cos(\theta_i - \theta_{i+1}) + \cos(\theta_{i+1} - \theta_{i+2})\right]$$

$$-2H_{bq}\left\{\cos[2(\theta_i - \theta_{i+1})] + \cos[2(\theta_{i+1} - \theta_{i+2})]\right\}$$

$$-(H_{ca}/2)\left[4\sin^2(2\theta_{i+1}) - 2\right], \tag{9.40}$$

$$Y_4 = H_{bl}\cos(\theta_{i+1} - \theta_{i+2}) - 2H_{bq}\cos[2(\theta_{i+1} - \theta_{i+2})], \tag{9.41}$$

together with the additional relations

$$\begin{aligned}
X_1 &= H_2, & X_2 &= H_3, & X_3 &= H_5, & X_4 &= H_6 & \text{for } i = 1, \\
Y_1 &= G_2, & Y_2 &= G_3, & Y_3 &= G_5, & Y_4 &= G_6 & \text{for } i = 1, \\
Y_1 &= I_2, & Y_2 &= I_3, & Y_3 &= I_5, & Y_4 &= I_6 & \text{for } i = 2, \\
X_1 &= J_2, & X_2 &= J_1, & X_3 &= J_5, & X_4 &= J_4 & \text{for } i = 3,
\end{aligned} \tag{9.42}$$

and finally $G_1 = H_3$, $G_3 = I_1$, $G_4 = H_6$, and $G_6 = I_4$. Although the definition of all the coefficients is rather tedious, the resulting 8×8 determinantal condition is rather easily applied to give the desired dispersion relation as the solutions of

$$\left(\frac{\omega}{\gamma}\right)^8 + \alpha_0\left(\frac{\omega}{\gamma}\right)^6 + \alpha_1\left(\frac{\omega}{\gamma}\right)^4 + \alpha_2\left(\frac{\omega}{\gamma}\right)^2 + \alpha_3 = 0, \tag{9.43}$$

where the α coefficients are defined elsewhere [27] and will not be quoted here.

It is a straightforward matter in principle to generalize the preceding theory to larger values of N. It is found quite generally that the spin-wave dispersion relation for a system with N magnetic films is given by an Nth-order polynomial in $(\omega/\gamma)^2$.

We now present some numerical results for the spin-wave dispersion relations in the case of magnetic thin films following a quasiperiodic Fibonacci sequence. Our results are divided into three groups, depending on the physical parameters used. The first parameter set, which is near the first antiferromagnetic peak (strong bilinear exchange coupling), has $H_{bq} = -0.1 H_{bl} = 0.1\,\text{kOe}$. A second parameter set, near the first ferromagnetic–antiferromagnetic transition (moderate bilinear exchange coupling), has $H_{bq} = -\frac{1}{3}H_{bl} = 0.05\,\text{kOe}$. Finally a third set, near the second antiferromagnetic peak (weak bilinear exchange coupling comparable to the biquadratic exchange coupling), has $H_{bq} = -H_{bl} = 0.035\,\text{kOe}$. In all cases we have assumed the same values for the cubic anisotropy field, $H_{ca} = 0.5\,\text{kOe}$, and the surface anisotropy field, $H_{sa} = 2\,\text{kOe}$.

Figs. 9.11–9.13 show the spin-wave spectra for the case of seven layers, i.e. four magnetic thin films separated by three non-magnetic spacers (see Fig. 9.1) [27]. We consider the three sets of parameters specified above and plot the frequency shift against the external magnetic field. Here dashed lines represent the acoustic

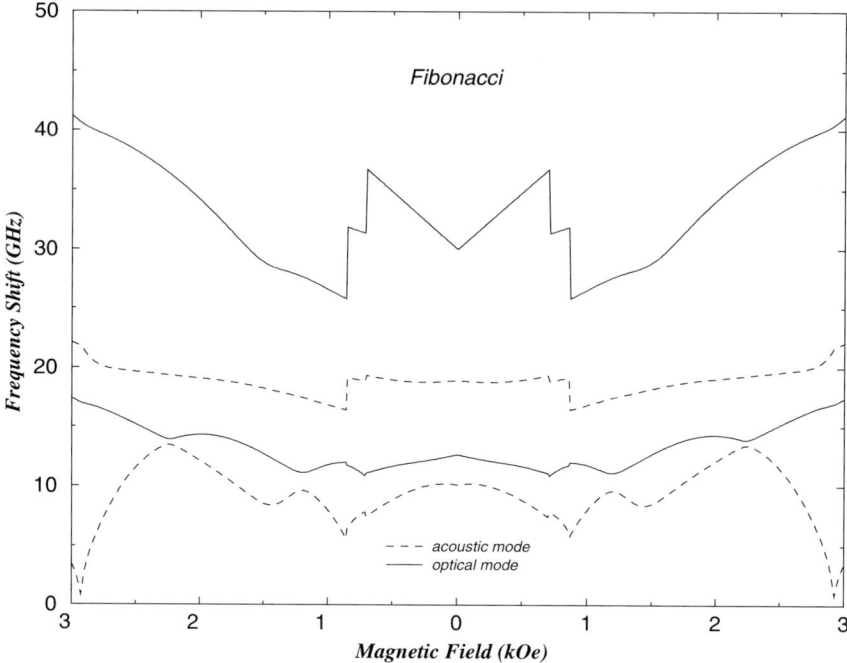

Fig. 9.11. Spin-wave dispersion relations for the seven-layer case taking the ratio of the biquadratic to the bilinear term equal to 0.1. We have plotted the frequency shift (in GHz) against the external magnetic field (in kOe). The acoustic (optical) modes are shown by the dashed (full) lines (after Mauriz et al. [27]).

modes, while solid lines represent the optical modes. Altogether we have four dispersion curves because there are four magnetic layers. The most pronounced effect of BEC, as seen by comparing the three figures, is to shift the frequency of the optical mode downwards in the AF (antiferromagnetic) and FM (ferromagnetic) phases and upwards in the central region (which corresponds to the 90° magnetic phase).

Similarly, we show in Figs. 9.14–9.16 the spin-wave spectra for the more robust (and complex) case of 17 layers, i.e. nine magnetic thin films separated by eight non-magnetic spacers [27]. There are now five acoustic modes and four optical modes in the spectra. In Figs. 9.14 and 9.15 the first four curves (in ascending frequency) represent a total superposition (or degeneracy) of the acoustic and optical modes (solid lines), and the fifth curve represents the optical mode (dashed line). In Fig. 9.16 we can see five curves, where the first one is a partial superimposition of the acoustic mode (dashed line) and the optical one (dotted line). The second, third, and fourth curves are complete superpositions of the acoustic and optical modes (solid lines). The fifth and last curve is a single acoustic mode (dashed line). There are some important overall conclusions that we can infer from this case:

9.4. THERMODYNAMIC PROPERTIES

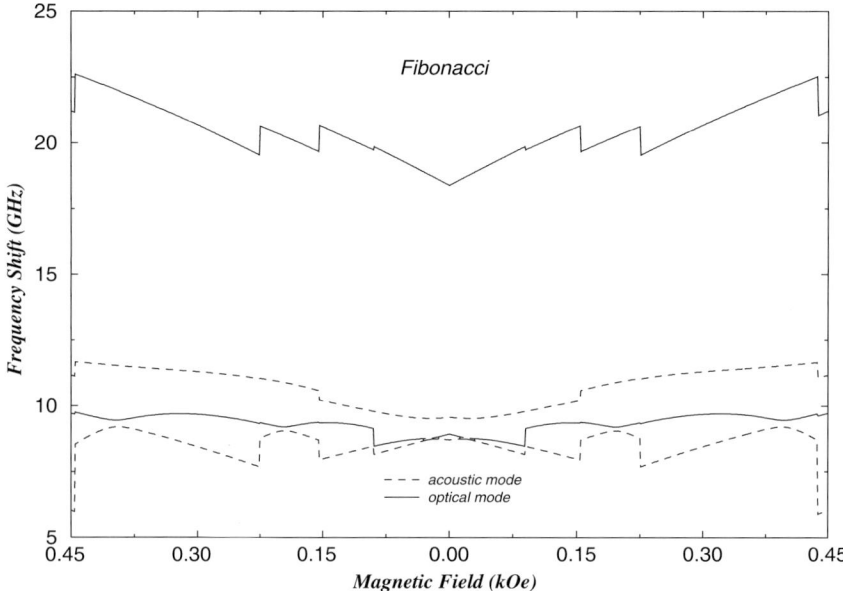

Fig. 9.12. Same as Fig. 9.11 but taking the ratio of the biquadratic to the bilinear term equal to $\frac{1}{3}$ (after Mauriz et al. [27]).

(a) the reduced values of the biquadratic term (see Figs. 9.14 and 9.15) are related only to the transition between the 90° magnetic phase and the saturated region, and therefore it is not possible to determine the two coupling fields accurately;

(b) surprisingly, the same behavior was observed in Fig. 9.16 in the region around the second AF peak;

(c) apart from these regions, the magnetization curves show qualitatively similar behavior.

9.4 Thermodynamic Properties

We now will study the spin-wave contribution to the specific heat for Fibonacci quasiperiodic structures composed of ferromagnetic films, each described by the Heisenberg model, with biquadratic and bilinear coupling between them. The general formalism for calculating specific heat has been described in Chapter 6, where some applications to non-magnetic systems were given. Although some of the previous properties, which were described in Section 6.5 for the plasmon-polariton contribution of the specific heat, will also occur here, we shall see that new features appear in this case. The most important of these is an interesting broken symmetry related to the interlayer biquadratic term [58,59], which has no analog in the previous case.

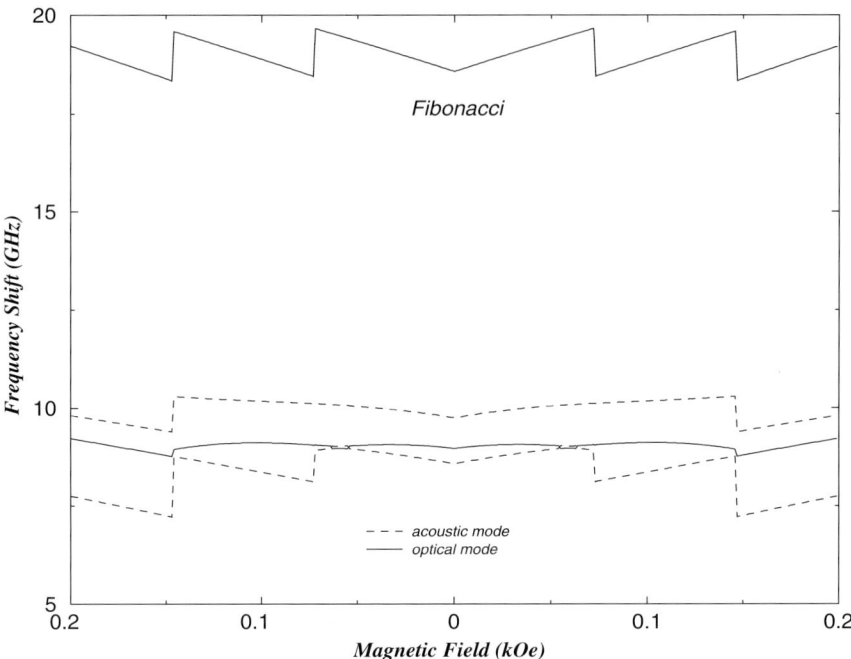

Fig. 9.13. Same as Fig. 9.11 but taking the ratio of the biquadratic to the bilinear term equal to 1 (after Mauriz et al. [27]).

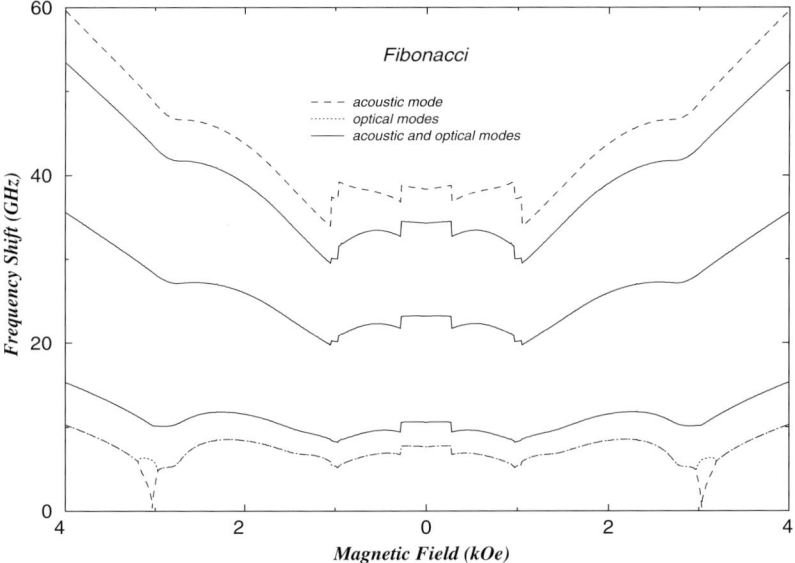

Fig. 9.14. Spin-wave dispersion relations for the 17-layer case taking the ratio of the biquadratic to the bilinear term equal to 0.1. We have plotted the frequency shift (in GHz) against the external magnetic field (in units of kOe) (after Mauriz et al. [27]).

9.4. THERMODYNAMIC PROPERTIES

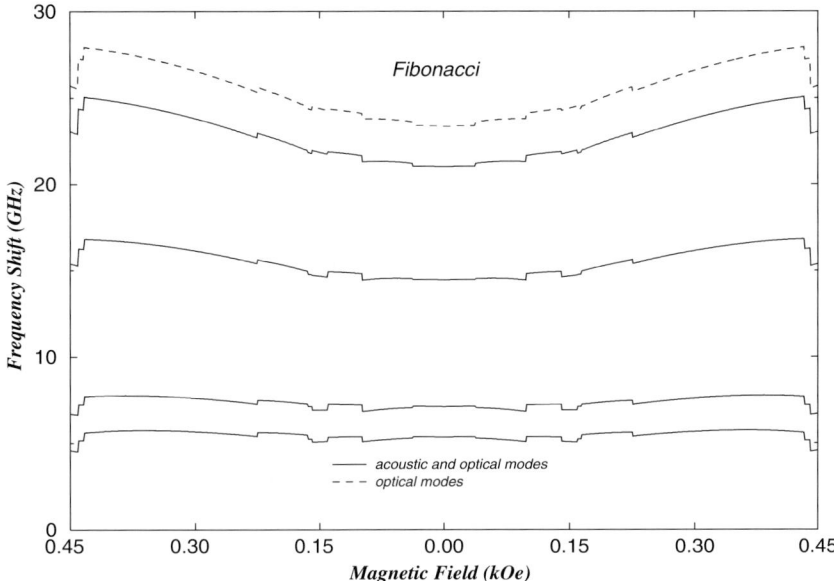

Fig. 9.15. Same as Fig. 9.14 but taking the ratio of the biquadratic to the bilinear term equal to $\frac{1}{3}$ (after Mauriz et al. [27]).

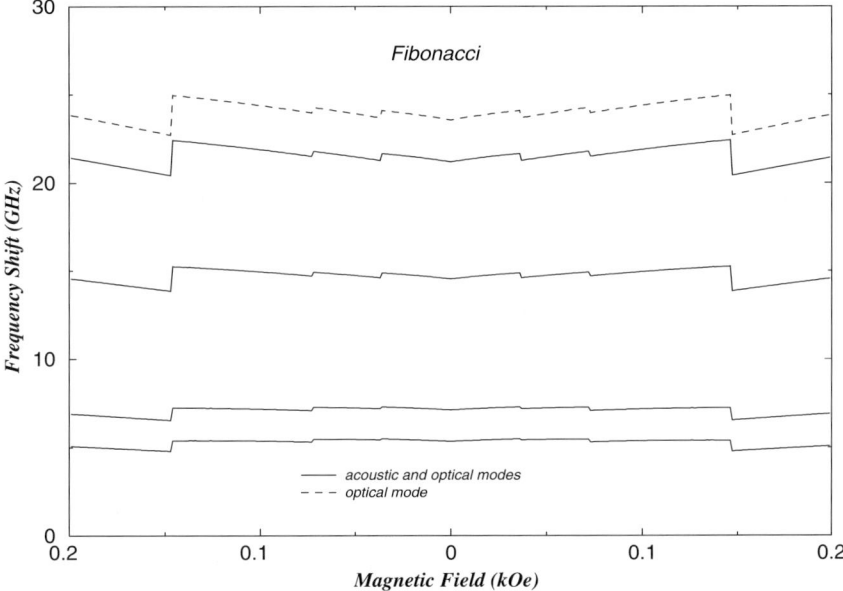

Fig. 9.16. Same as Fig. 9.14 but taking the ratio of the biquadratic to the bilinear term equal to 1 (after Mauriz et al. [27]).

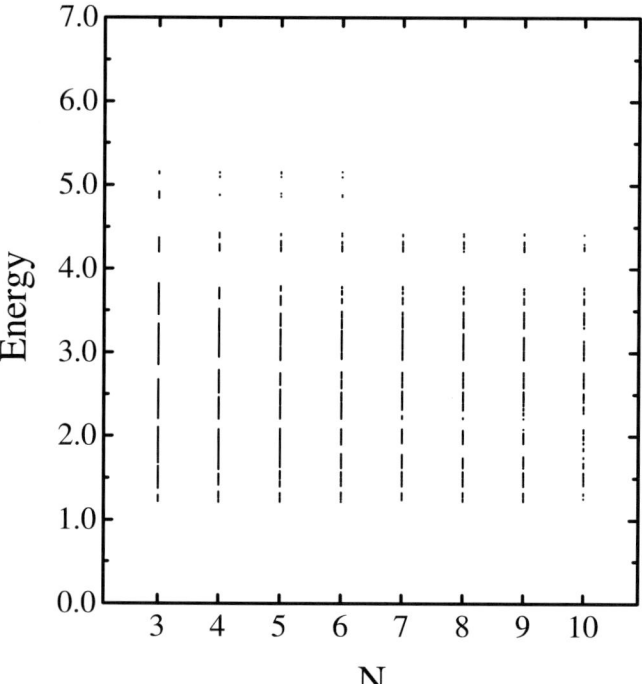

Fig. 9.17. Energy spectra of spin waves for the quasiperiodic Fibonacci structure versus the generation number when a biquadratic interlayer exchange term is present (see the text). Here we consider the dimensionless in-plane wavevector $k_x a = 2.0$ (after Bezerra et al. [59]).

The spin-wave fractal spectra for the Fibonacci superlattices with BEC is shown in Fig. 9.17 for a fixed value of the in-plane dimensionless wavevector, namely $k_x a = 2.0$ [59]. We can clearly see the forbidden and allowed energies as the Fibonacci generation number n is increased, going up to the 10th generation (where the unit cell is composed of $F_9 = 55$ A and $F_8 = 34$ B building blocks). The number of allowed bands is equal to three times the Fibonacci number F_n of the corresponding generation. For the above calculations we have assumed the spin quantum numbers $S_A = 1$ and $S_B = \frac{3}{2}$, and have taken values of the ratio between the interlayer biquadratic and bilinear exchange terms, defined as

$$R = J_{bq}/J_{bl}, \tag{9.44}$$

to be in accordance with recent experimental data [21,22,28].

Fig. 9.18 shows the log–log plot of the spin-wave specific heat as a function of temperature in the low-temperature regime and in the absence of the BEC term ($R = 0$) [58]. We see that there is an interesting harmonic oscillation of the specific heat, which is a *log-periodic* function of the temperature, i.e. $C_n(T) = \alpha C_n(aT)$, where α is a constant and a is any number (see Fig. 9.19). These log-periodic

9.4. THERMODYNAMIC PROPERTIES

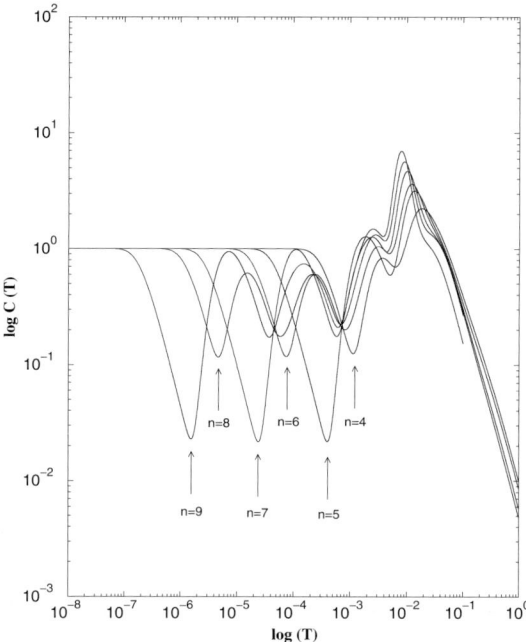

Fig. 9.18. Log–log plot of the spin-wave specific heat versus temperature for the generation numbers of the Fibonacci quasiperiodic sequence in the absence of the biquadratic term. Observe a different behavior for the even ($n=4,6,8$) and odd ($n=5,7,9$) generation numbers (after Bezerra et al. [58]).

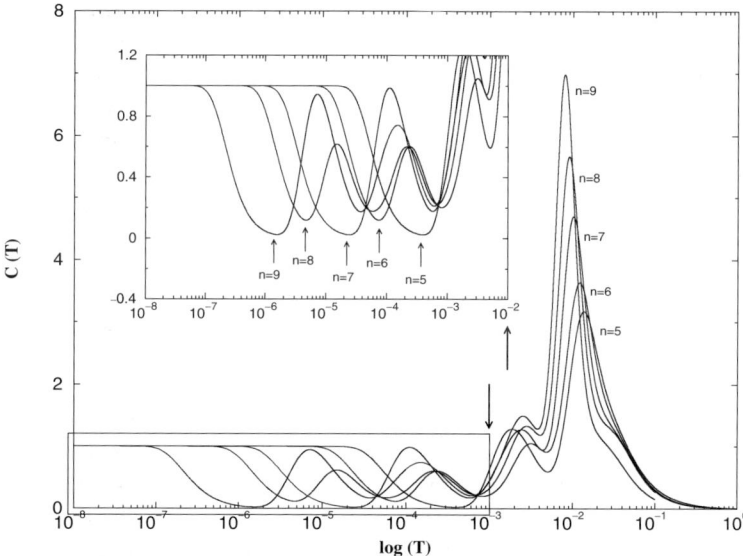

Fig. 9.19. Log-periodic profile of the spin-wave specific heat versus temperature for several values of the generation numbers of the Fibonacci sequence (after Bezerra et al. [58]).

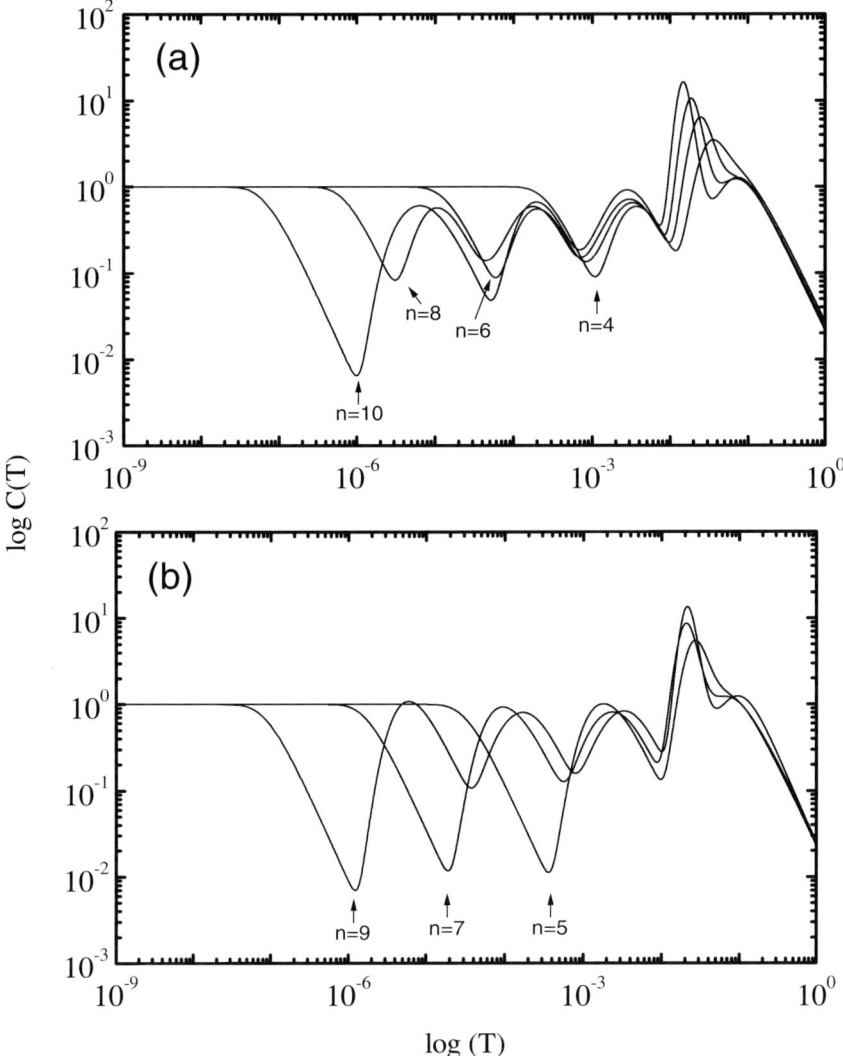

Fig. 9.20. Same as Fig. 9.17, but for the ratio between the interlayer biquadratic and bilinear exchange terms $R = 0.2$. From this value upwards the even symmetry starts to be broken (after Bezerra et al. [59]).

oscillations can be traced back to the log-periodicity of the density of state's spectral staircase. They resemble very much the shape of the devil's staircase obtained from idealized Cantor sets. The number of oscillations depends on the generation number of the Fibonacci sequence: the bigger the system, the greater the number of oscillations. An important feature, however, is a characteristic of the Fibonacci structure already found for other excitations, namely, the well-defined even- and odd-parity spectra related to the generation number of the Fibonacci structure.

9.4. THERMODYNAMIC PROPERTIES

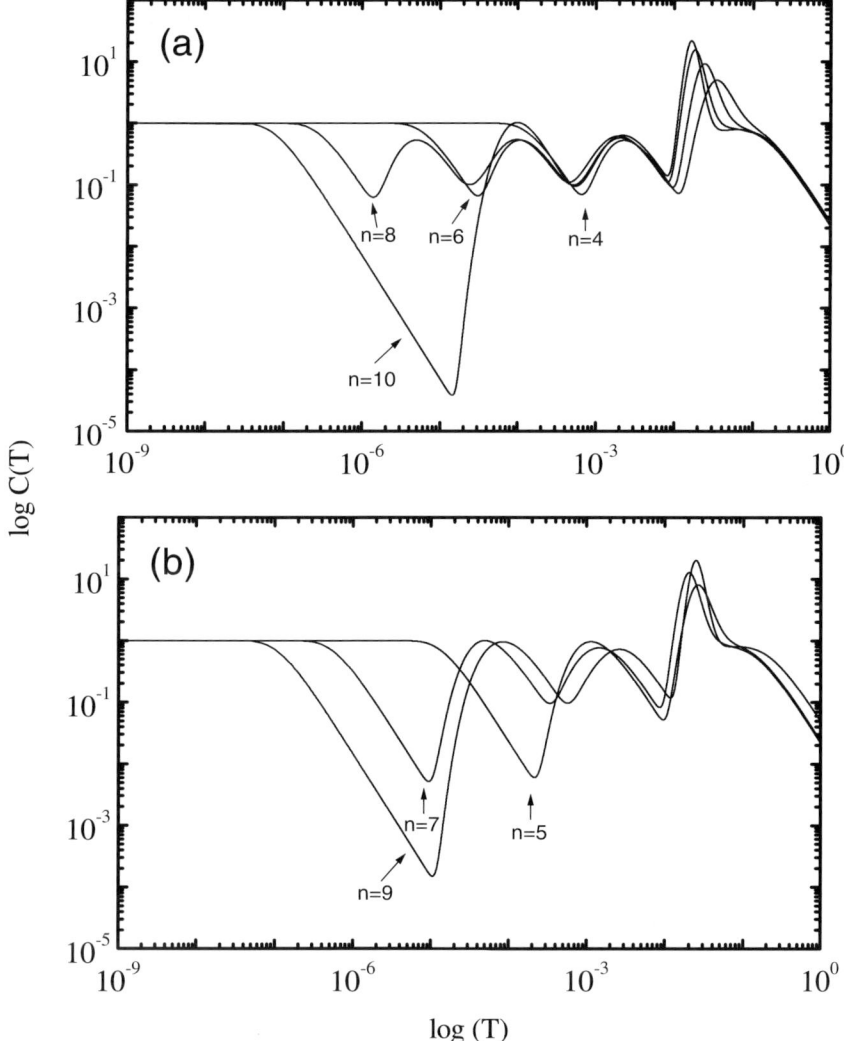

Fig. 9.21. Same as Fig. 9.17, but for the ratio between the interlayer biquadratic and bilinear exchange terms $R = 0.4$. From this value upwards the odd symmetry also starts to be broken (after Bezerra et al. [59]).

These harmonic oscillations can be identified as the *signature* of the Fibonacci structure, and there is no counterpart either in the idealized triadic Cantor set or in the other quasiperiodic structures discussed in this book.

However, it turns out that when BEC is present ($R \neq 0$) an important modification takes place. By contrast with the spectra found in earlier work [58], the two different symmetrical profiles of oscillations are broken for different ratios between the biquadratic and bilinear terms. To explain and reinforce this intriguing behavior, we show in Figs. 9.20 and 9.21 the evolution of this broken-symmetry property for $R = 0.2$ (where the even-mode symmetry starts to be broken) and 0.4 (where the odd-mode symmetry also starts to be broken), respectively [59]. A possible explanation is that these different types of behavior arise because the biquadratic exchange term reduces the effective coupling between the adjacent magnetic layers at the interface (see Refs. [46,60]). Also BEC induces long-range correlations (as we noted earlier) that emphasize the quasiperiodicity of the system. In other words, it looks like these effects make the whole structure *see better* its quasiperiodicity, in effect increasing its degree of disorder as R increases! It turns out that the spin-wave energy band structure is more disordered and, as a consequence, this is reflected in the specific heat curves, which do not have the well-defined standard profiles found in the absence of BEC. This argument is reinforced by previous works on the correlation lengths of magnetic systems that exhibit BEC (see e.g. Ref. [39]).

References

[1] Ultrathin Magnetic Structures, Vols. I and II, Eds., J.A.C. Bland and B. Heinrich, Springer-Verlag, Berlin, 1994.

[2] P. Grünberg, R. Schreiber, Y. Pang, M.O. Brodsky and H. Sowers, Phys. Rev. Lett. 57 (1986) 2442.

[3] Z. Celinski, B. Heinrich and J.F. Cochran, J. Magn. Magn. Mater. 145 (1995) L1.

[4] M.N. Baibich, J.M. Broto, A. Fert, F. Nguyen Van Dau, F. Petroff, P. Etienne, G. Creuzet, A. Friederich and J. Chazelas, Phys. Rev. Lett. 61 (1988) 2472.

[5] W.J. Gallagher, S.S.P. Parkin, Y. Lu, X. Bian, A. Marley, K.P. Roche, R.A. Altman, S.A. Rishton, C. Jahnes, T.M. Shaw and G. Xiao, J. Appl. Phys. 81 (1997) 3741.

[6] J.M. Daughton, J. Appl. Phys. 81 (1997) 3758.

[7] B.A. Gurney, V.S. Speriosu, D.R. Wilhoit, H. Lafakis, R.E. Fontana Jr., D.E. Heim and M. Dovek, J. Appl. Phys. 81 (1997) 3998.

[8] S.S.P. Parkin, N. More and K.P. Roche, Phys. Rev. Lett. 64 (1990) 2304.

[9] M. Rührig, R. Schafer, A. Hubert, R. Mosler, J.A. Wolf, S. Demokritov and P. Grünberg, Phys. Stat. Sol. (a) 125 (1991) 635.

[10] U. Köbler, K. Wagner, R. Wiechers, A. Fuss and W. Zinn, J. Magn. Magn. Mater. 103 (1992) 236.

[11] J.C. Slonczewski, Phys. Rev. Lett. 67 (1991) 3172.

[12] P. Vavassori, M. Grimsditch and E.E. Fullerton, J. Magn. Magn. Mater. 223 (2001) 284.

[13] B. Heinrich, J.F. Cochran, M. Kowalewski, J. Kirschner, Z. Celinski, A.S. Arrott and K. Myrtle, Phys. Rev. B 44 (1991) 9348.

[14] S.M. Rezende, F.M. de Aguiar, A. Azevedo, M.A. Lucena, C. Chesman, P. Kabos and C.E. Patton, Phys. Rev. B 55 (1997) 8071.

[15] M.A. Lucena, F.M. de Aguiar, S.M. Rezende, A. Azevedo, C. Chesman and S.S.P. Parkin, J. Appl. Phys. 81 (1997) 4770.

[16] G. Gubbiotti, L. Albini, G. Carlotti, A. Montecchiari, M. De Crescenzi, R. Zivieri, L. Giovannini and F. Nizzoli, Surface Science 482 (2001) 970.

[17] J. Unguris, R.J. Celotta and D.T. Pierce, Phys. Rev. Lett. 79 (1997) 2734.

[18] B. Heinrich, Z. Celinski, J.F. Cochran, A.S. Arrott, K. Myrtle and S.T. Purcell, Phys. Rev. B 47 (1993) 5077.

[19] D.M. Edwards, J.M. Ward and J. Mathon, J. Magn. Magn. Mater. 126 (1993) 380.

[20] Z. Celinski, B. Heinrich and J.F. Cochran, J. Appl. Phys. 73 (1993) 5966.
[21] A. Azevedo, C. Chesman, S.M. Rezende, F.M. de Aguiar, X. Bian and S.S.P. Parkin, Phys. Rev. Lett. 76 (1996) 4837.
[22] C. Chesman, A. Azevedo, S.M. Rezende, F.M. de Aguiar, X. Bian and S.S.P. Parkin, J. Appl. Phys. 81 (1997) 3791.
[23] C. Chesman, M.A. Lucena, M.C. de Moura, A. Azevedo, F.M. de Aguiar and S.M. Rezende, Phys. Rev. B 58 (1998) 101.
[24] R.J. Hicken, C. Daboo, M. Gester, A.J.R. Ives, S.J. Gray and J.A.C. Bland, Thin Solid Films 275 (1996) 199.
[25] C.G. Bezerra, J.M. de Araujo, C. Chesman and E.L. Albuquerque, Phys. Rev. B 60 (1999) 9264.
[26] C.G. Bezerra, J.M. de Araujo, C. Chesman and E.L. Albuquerque, J. Appl. Phys. 89 (2001) 2286.
[27] P.W. Mauriz, E.L. Albuquerque and C.G. Bezerra, J. Phys.: Condens. Matter 14 (2002) 1785.
[28] S.M. Rezende, C. Chesman, M.A. Lucena, A. Azevedo, F.M. Aguiar and S.S.P. Parkin, J. Appl. Phys. 84 (1998) 958.
[29] A.E. Berkowitz, J.R. Mitchell, M.J. Carey, A.P. Young, S. Zhang, F.E. Spada, F.T. Parker, A. Hutten and G. Thomas, Phys. Rev. Lett. 68 (1992) 3745.
[30] S. Zhang and P.M. Levy, J. Appl. Phys. 73 (1993) 5315.
[31] R.W. Wang, D.L. Mills, E.E. Fullerton, J.E. Matson and S.D. Bader, Phys. Rev. Lett. 72 (1994) 920.
[32] N.S. Almeida and D.L. Mills, Phys. Rev. B 52 (1995) 13504.
[33] A. Vedyayev, B. Dieny, N. Ryzhanova, J.B. Genin and C. Cowache, Europhys. Lett. 25 (1994) 465.
[34] B. Dieny and J.P. Gavigan, J. Phys.: Condens. Matter 2 (1990) 187.
[35] W. Folkerts, J. Magn. Magn. Mater. 94 (1991) 302.
[36] H.J. Elmers, G. Liu, H. Fritsche and U. Gradmann, Phys. Rev. B 52 (1995) R696.
[37] E.E. Fullerton, M.J. Conover, J.E. Mattson, C.H. Sowers and S.D. Bader, Phys. Rev. B 48 (1993) 15755.
[38] X. Bian, H.T. Hardner and S.S.P. Parkin, J. Appl. Phys. 79 (1996) 4980.
[39] E.S. Sørensen and A.P. Young, Phys. Rev. B 42 (1990) 754.
[40] R. Horst and P.M. Darlos, Eds., Handbook of Global Optimization, Kluwer, Dordrecht, 1995.
[41] S. Kirkpatrick, C.D. Geddat Jr. and M.P. Vecchi, Science 220 (1983) 671.
[42] W.H. Press, S.A. Teukolski, W.T. Vetterling and B.P. Flanery, Numerical Recipes, Cambridge Univ. Press, Cambridge, 1998.
[43] W. Folkerts and S.T. Purcell, J. Magn. Magn. Mater. 111 (1992) 306.
[44] M. Maccio, M.G. Pini, P. Politi and A. Rettori, Phys. Rev. B 49 (1994) 3283.
[45] J. Barnaś, Phys. Stat. Sol. (b) 203 (1997) 221.
[46] C.G. Bezerra and M.G. Cottam, Phys. Rev. B 41 (2002) 530.
[47] Z. Zhang, L. Zhou, P.E. Wigen and K. Ounadjela, Phys. Rev. B 50 (1994) 6094.
[48] R.L. Stamps, Phys. Rev. B 61 (2000) 12174.
[49] P. Grünberg, J. Appl. Phys. 51 (1980) 4338; 52 (1981) 6824.
[50] R.E. Camley, T.S. Rahman and D.L. Mills, Phys. Rev. B 23 (1981) 1226.
[51] B. Hillebrands, Phys. Rev. B 41 (1990) 530.
[52] M. Vohl, J. Barnaś and P. Grünberg, Phys. Rev. B 39 (1989) 12003.

[53] B. Heinrich, S.T. Purcell, J.R. Dutcher, K.B. Urquhart, J.F. Cochran and A.S. Arrott, Phys. Rev. B 38 (1988) 12879.

[54] J.J. Krebs, P. Lubitz, A. Chaiken and G.A. Prinz, Phys. Rev. Lett. 63 (1989) 1645.

[55] P.E. Wigen and Z. Zhang, Braz. J. Phys. 22 (1992) 267.

[56] J.F. Cochran, J. Rudd, W.B. Muir, B. Heinrich and Z. Celinski, Phys. Rev. B 42 (1990) 508.

[57] B. Heinrich and J.F. Cochran, Adv. Phys. 42 (1993) 523.

[58] C.G. Bezerra, E.L. Albuquerque, A.M. Mariz, L.R. da Silva and C. Tsallis, Physica A 294 (2001) 415.

[59] C.G. Bezerra, E.L. Albuquerque and M.G. Cottam, Physica A 301 (2001) 372.

[60] E.L. Albuquerque, C.G. Bezerra and C. Chesman, Surf. Sci. 532 (2003) 47.

Chapter 10

Exciton-Polaritons

The concept of an exciton as a bound electron–hole pair in semiconductors was introduced in Section 1.6, along with its relevance to the optical characterization of semiconductor materials (in the bulk case). Then in Section 3.3 we considered the bulk exciton-polaritons, deriving their dispersion relationship and stressing the important role of spatial dispersion. We now wish to explore exciton-polaritons in greater depth, particularly in multilayer samples where surface and interface effects will dominate.

Although the electronic properties of semiconductor superlattices and multi-quantum wells (MQWs) have attracted enormous attention in the last two decades, considerably less work has been done on the optical characterization of these artificial specimens, and this is where exciton-polaritons have an important role. Exact equations for optical propagation in a layered medium were first discussed by Rytov [1], while later a more complete formalism was developed by Yariv and co-workers [2,3]. All of this was applicable to periodic systems, usually characterized by having just two alternating media A and B, which may have different thicknesses and dielectric functions.

The study of exciton-polaritons as coupled-mode excitations made up from dipole-active excitons interacting with photons in the so-called excitonic region (see Section 3.3) began with the famous papers of Pekar [4], Hopfield [5], and Hopfield and Thomas [6], among others. The main feature of these papers is that the finite wavevector dependence in the dispersion equation for the total energy of the exciton could play a decisive role in the optical properties of a crystal in the exciton energy region. Indeed the description of exciton-polaritons requires the inclusion of a spatial-dispersion term Dk^2 in the definition of the dielectric function, e.g. see Eqs. (3.21) and (3.22). We shall here assume the general form with phonon and plasma terms:

$$\epsilon(k,\omega) = \epsilon_\infty \left(\frac{\omega_L^2 - \omega^2 + Dk^2}{\omega_T^2 - \omega^2 + Dk^2 - i\omega\Gamma} \right) + \epsilon_\infty \left(1 - \frac{\omega_p^2}{\omega^2} \right). \quad (10.1)$$

As before, ϵ_∞ is the background dielectric constant, ω_L and ω_T are the longitudinal and transverse (LO and TO) phonon frequencies, respectively, ω_p is the plasma frequency, and Γ is an empirical damping constant.

In this chapter we present a comprehensive account of the main properties of the exciton-polaritons (bulk and surface modes) propagating in thin films and periodic superlattices. Particular attention needs to be given to the problem of choosing additional boundary conditions (ABCs), as mentioned in Section 3.3. Some applications related to the experimental technique of attenuated total reflection (ATR) are briefly mentioned (but see Section 11.5 for a discussion about it). To date, the most useful experimental method for probing exciton-polaritons is resonant Brillouin scattering (RBS), and this is discussed in Section 11.4 of the next chapter (along with other light-scattering techniques).

10.1 Thin Films

Exciton-polaritons in direct band-gap semiconductors were extensively studied in the 1970s and 1980s (for a review see Ref. [7]). However, there have only recently been investigations of their optical properties (photoluminescence, reflectance spectra, emission spectra, etc.) for supported films (with sapphire as the substrate) as well as for unsupported films made up of wide-band-gap semiconductors, like gallium nitride (GaN). Today this topic has developed into an active research field. For example, when a nitride semiconductor is located on a substrate, dramatic changes are produced in the intensity of the optical response, the energy of the optical transitions, the radiative lifetimes, and the emission patterns [8–12].

There has recently been a general rekindling of interest in group III nitride semiconductor compounds. These compounds display noticeable crystalline robustness as well as good chemical and thermal stability, and have a direct band gap at the zone center. Advances have been made in the production of several novel optoelectronic devices (e.g. green and blue light-emitting diodes) [13] and a promising breakthrough in the area of UV detectors, heterojunctions bipolar transitions, etc. [14]. Overall these structures provide excellent options for technological applications and are attractive objects of research.

In this section we study the spectrum of the more energetic exciton coupled with a photon (s- and p-polarized), namely the A coupled exciton mode. This is done for a GaN buffer thin film at low temperature, grown with strain on the C-plane sapphire substrate. Our main aim is the investigation of the modified optical properties of the exciton-polariton bulk modes, in both the frequency and time domains, to gain a better understanding of the dynamics of the excitation in confined systems. Furthermore, we study the confinement and localization of the surface and guided modes outside the bulk-mode boundaries.

The bulk exciton-polariton dispersion relation is found by solving Maxwell's equation with the use of the dielectric function quoted earlier. The result is well known, as discussed in Section 3.3, namely

$$\epsilon(k,\omega)\left[k^2 - \epsilon(k,\omega)\omega^2/c^2\right] = 0, \qquad (10.2)$$

where c is the speed of light in vacuum. The solution of this equation gives a longitudinal mode if $\epsilon(k,\omega) = 0$ [see Eq. (3.26)], whereas the transverse mode corresponds to $\epsilon(k,\omega) = c^2k^2/\omega^2$ [see Eq. (3.25)].

10.1. THIN FILMS

We make the extension now to obtain the exciton-polariton spectra in supported GaN films. Suppose that a *p*-polarized light is incident from the vacuum (we call it region I) on the surface (in the xy-plane) of the excitonic medium (region II), which is supported by the sapphire substrate defining region III. The excitonic medium is taken to occupy the region $-d/2 < z < d/2$, with d being the thickness of the film. Two transverse modes with wavevectors $\vec{k}_1$ and $\vec{k}_2$ and one longitudinal mode with wavevector $\vec{k}_L$ can propagate in the medium. In order to study polariton propagation in these media, we will apply the usual electromagnetic boundary conditions at the interfaces. However, these provide us with only two boundary conditions, so an ABC is required. This was discussed in Section 3.3, where we mentioned some possible forms for the ABCs. As summarized by Halevi [15], the most frequently adopted view is that just below the surface of the excitonic medium, there is an exciton-free (or dead) layer of thickness a in which the excitonic field $\vec{P}$ is zero, whereas on the other side of it either $\vec{P}$ or its derivative $d\vec{P}/dz$ is equal to zero. While the dead layer means that there is an unknown input parameter for macroscopic models based on ABCs, a microscopic theory can be used to determine the dead-layer thickness. On the other hand, if a *non-linear* optical-response field is being taken into account, the ABC theory for linear response has to be extended in such a way that the non-local effect appears in the spectra. This is the case even when the system size is much smaller than the relevant light wavelengths [16].

To summarize, the ABCs proposed up to now can be cast into the form [17]

$$\gamma \vec{P} + \beta \, d\vec{P}/dz = 0, \tag{10.3}$$

to be applied at the interfaces of the excitonic media. The extreme cases of the above are $\gamma = 1, \beta = 0$ (which leads to $\vec{P} = 0$; we subsequently call it ABC1 as a shorthand) and $\gamma = 0, \beta = 1$ (meaning $d\vec{P}/dz = 0$; we call it ABC2). Another case that is sometimes of interest is $\gamma = \beta = 1$ (meaning $\vec{P} + d\vec{P}/dz = 0$; we call it ABC3).

Following the well-known Born–Huang treatment for dielectric solids, the excitonic polarization field $\vec{P}$ satisfies Eq. (3.19) in the so-called dielectric approximation, where the macroscopic electric field is given by

$$\nabla^2 \vec{E} - (\epsilon_\infty/c^2) \partial^2 \vec{E}/\partial t^2 = (1/\epsilon_0 c^2) \partial^2 \vec{P}/\partial t^2 \tag{10.4}$$

from Maxwell's equations. Assuming plane-wave solutions $\exp[i(\vec{k}\cdot\vec{r} - \omega t)]$ for all the fields, it is now easy to find the dispersion relations for the transverse and longitudinal modes.

For the case of *p*-polarized light waves, the electric field is in the xz-plane, while $\vec{H}$ is in the y-direction. From Maxwell's equation $\nabla \times \vec{E} = -\partial \vec{B}/\partial t$, we can find the following differential equation for the electric field in the non-excitonic media:

$$d^2\vec{E}/dz^2 - \alpha_J^2 \vec{E} = 0, \tag{10.5}$$

where $\alpha_J^2 = k_x^2 - \epsilon_J \omega^2/c^2 \equiv -k_{Jz}^2$ with $J = 1$ and 3, corresponding to regions I and III. This equation has the following solution for the x-component of the

electric field:
$$E_{xJ} = E_1 \exp(-\alpha_J z) + E_2 \exp(\alpha_J z). \tag{10.6}$$

Next, using Maxwell's equation for the divergence of $\vec{E}$, we can find E_{zJ} in regions I and III as
$$E_{zJ} = (ik_x/\alpha_J)\Big[E_1 \exp(-\alpha_J z) - E_2 \exp(\alpha_J z)\Big]. \tag{10.7}$$

Therefore, the electric field can be written in region I ($z > d/2$) as
$$\vec{E} = (1, 0, ik_x/\alpha_1) E_1 \exp[-\alpha_1(z - d/2)]. \tag{10.8}$$

Likewise, for region III ($z < -d/2$), we have
$$\vec{E} = (1, 0, -ik_x/\alpha_3) E_3 \exp[\alpha_3(z + d/2)]. \tag{10.9}$$

On the other hand, in region II ($-d/2 < z < d/2$), we have the two transverse solutions in the excitonic medium given by
$$\vec{E}_T = \sum_{j=1,2} \Big\{ \Big[A_j \exp(-k_j^z z) + B_j \exp(k_j^z z)\Big], 0,$$
$$(ik_x/k_j^z)\Big[A_j \exp(-k_j^z z) - B_j \exp(k_j^z z)\Big] \Big\}, \tag{10.10}$$

with $k_j^{z\,2} = k_x^2 - k_j^2$ (and $j = 1, 2$, corresponding to the two transverse modes), while for the longitudinal mode we have
$$\vec{E}_L = \Big\{ \Big[A_L \exp(-k_L^z z) + B_L \exp(k_L^z z)\Big], 0,$$
$$(ik_L^z/k_x)\Big[A_L \exp(-k_L^z z) - B_L \exp(k_L^z z)\Big] \Big\}, \tag{10.11}$$

with $k_L^{z\,2} = k_x^2 - k_L^2$. Here k_L is the wavevector for the longitudinal mode. Therefore, in region II, we have for the total electric field,
$$\vec{E} = \vec{E}_T + \vec{E}_L. \tag{10.12}$$

The excitonic polarization field is given by
$$\vec{P} = \epsilon_0 \chi_L \vec{E}_L + \sum_{j=1}^{2} \epsilon_0 \chi_j \vec{E}_T. \tag{10.13}$$

The dispersion relation can then be found by applying the electromagnetic boundary conditions (continuity of E_x and D_z) together with the chosen form of ABC (for P_x and P_z) at $z = \pm d/2$. The eight coupled linear equations can be rewritten in matrix form as
$$\bar{X}_p \bar{u}_p = 0, \tag{10.14}$$

10.1. THIN FILMS

where $\bar{u}_p$ is the column matrix formed by the above eight unknown coefficients related to the electric fields defined by Eqs. (10.8)–(10.11), namely $(E_1, E_3, A_1, A_2, A_L, B_1, B_2, B_L)$, and $\bar{X}_p$ is an 8×8 matrix. Using ABC1, for example, and the electromagnetic boundary conditions at $z = \pm d/2$, together with Eqs. (10.8), (10.9) and (10.12), we find for the matrix [18]

$$\bar{X}_p = \begin{bmatrix} -1 & 0 & \bar{f}_1 & \bar{f}_2 & f_1 & f_2 & \bar{f}_L & -f_L \\ -\frac{\epsilon_1 k_x}{\alpha_1} & 0 & \frac{\epsilon_2(k_1)k_x}{k_1^z}\bar{f}_1 & \frac{\epsilon_2(k_2)k_x}{k_2^z}\bar{f}_2 & -\frac{\epsilon_2(k_1)k_x}{k_1^z}f_1 & -\frac{\epsilon_2(k_2)k_x}{k_2^z}f_2 & 0 & 0 \\ 0 & 0 & \chi_1\bar{f}_1 & \chi_2\bar{f}_2 & \chi_1 f_1 & \chi_2 f_2 & -\epsilon_\infty \bar{f}_L & -\epsilon_\infty f_L \\ 0 & 0 & \frac{\chi_1 k_x}{k_1^z}\bar{f}_1 & \frac{\chi_2 k_x}{k_2^z}\bar{f}_2 & -\frac{\chi_1 k_x}{k_1^z}f_1 & -\frac{\chi_2 k_x}{k_2^z}f_2 & \frac{\epsilon_\infty k_L^z}{k_x}\bar{f}_L & \frac{\epsilon_\infty k_L^z}{k_x}f_L \\ 0 & -1 & f_1 & f_2 & \bar{f}_1 & \bar{f}_2 & f_L & -\bar{f}_L \\ 0 & \frac{\epsilon_2 k_x}{\alpha_3} & \frac{\epsilon_2(k_1)k_x}{k_1^z}f_1 & \frac{\epsilon_2(k_2)k_x}{k_2^z}f_2 & -\frac{\epsilon_2(k_1)k_x}{k_1^z}\bar{f}_1 & -\frac{\epsilon_2(k_2)k_x}{k_2^z}\bar{f}_2 & 0 & 0 \\ 0 & 0 & \chi_1 f_1 & \chi_2 f_2 & \chi_1 \bar{f}_1 & \chi_2 \bar{f}_2 & -\epsilon_\infty f_L & -\epsilon_\infty \bar{f}_L \\ 0 & 0 & \frac{\chi_1 k_x}{k_1^z}f_1 & \frac{\chi_2 k_x}{k_2^z}f_2 & -\frac{\chi_1 k_x}{k_1^z}\bar{f}_1 & -\frac{\chi_2 k_x}{k_2^z}\bar{f}_2 & \frac{\epsilon_\infty k_L^z}{k_x}f_L & \frac{\epsilon_\infty k_L^z}{k_x}\bar{f}_L \end{bmatrix}.$$

(10.15)

Here the electrical susceptibility is,

$$\chi_j = S/\epsilon_0(Dk_j^2 - \Omega^2), \qquad (10.16)$$

where $\Omega^2 = \omega^2 - \omega_0^2 + i\omega\Gamma$, $f_j = \exp(k_j^z d/2)$, and $\bar{f}_j = 1/f_j$, with $j = 1, 2$ (for the two transverse modes) and $j = L$ (for the longitudinal mode). The required dispersion relations are finally obtained from the condition for $\bar{X}_p$ to have a vanishing determinant (det $\bar{X}_p = 0$).

Fig. 10.1 shows the p-polarized exciton-polariton spectrum calculated for a supported thin film of GaN on a sapphire substrate and in the absence of damping [18]. The physical parameters used here, suitable for the A exciton in bulk GaN, are [8,14] $\epsilon_\infty = 8.75$, $\hbar\omega_0 = 3.487$ eV, $4\pi\alpha_0 = 15 \times 10^{-3}$, and $M = 1.3 m_0$ (see Section 3.3 for the notation). We have plotted the reduced frequency ω/ω_0 as a function of the dimensionless wavevector $k_x d$, with k_x being the in-plane wavevector, taking the film thickness d equal to 50 nm. As we can see, the bulk-mode continuum is now replaced by several undamped normal modes that have mixed surface and guided character in the spectrum (the full lines in the figure). Their energies correspond to the discretized values of the exciton-polariton wavevector along the z-axis. For actual structures, the real parts of the energies correspond to the splitting between the exciton energies in different quantum wells. In our case, for small d (having the magnitude of the Bohr radius in GaN, which is of the order of 10 nm) the size of this effect is 0.057 meV, in the region of the resonance (i.e. where $\omega/\omega_0 \simeq 1$). This is consistent with the spacing of lines in the spectrum depicted in Fig. 10.1. For large d (much bigger than the Bohr radius), the discrete levels evolve into the

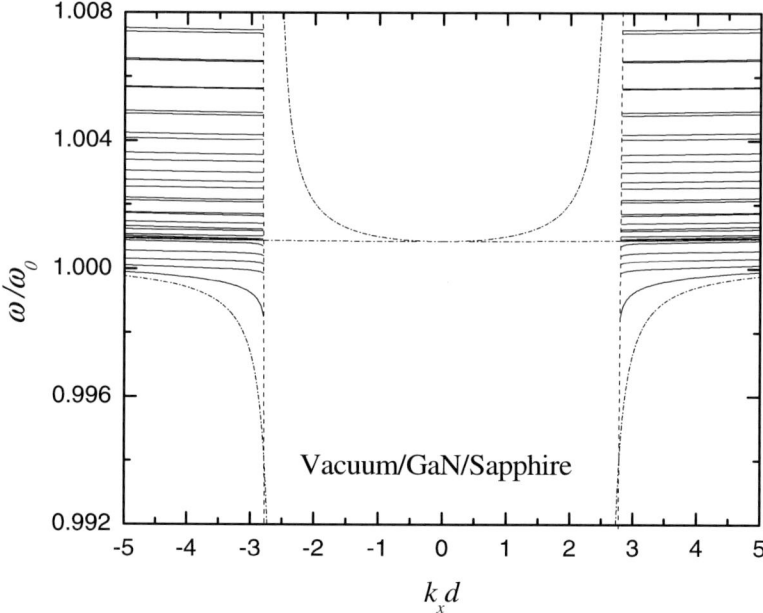

Fig. 10.1. Exciton-polariton spectrum in the case of p-polarized modes, for a 50 nm GaN film on a sapphire substrate, with ABC chosen as $\vec{P}=0$ and zero damping. Here the almost vertical dashed lines are the light lines, while the chain-dotted lines correspond to the two transverse modes and one longitudinal bulk mode. The full lines represent the calculated surface and guided modes (after Vasconcelos et al. [18]).

superlattice bands along the z-direction. These are described in Ref. [19] (but are not shown here).

In principle, the imaginary part of the energies would yield the radiative width of the excitons. Their behavior is similar to that calculated for the crossover from 2D excitons to bulk polaritons in molecular layers [20]. A similar behavior was also found in multiple quantum wells [21]. However, the modes depicted in Fig. 10.1 are undamped and therefore have zero radiative widths.

The chain-dotted lines in Fig. 10.1 describe the propagation of the bulk transverse polariton modes corresponding to the transverse wavevectors k_j ($j=1,2$) and the longitudinal wavevector k_L. We remark that the polariton spectrum is highly directional, since it comes from non-thermalized excitons. This means that, for application to real thin films, the energy differences between excitons localized in different quantum wells (due to the well-width fluctuations) provide a dephasing mechanism. Consequently, they tend to disappear for the region of $k_x d$ between the light lines in the spectrum shown in Fig. 10.1. We do not expect to see this effect experimentally, unless high-quality materials are used; unfortunately it is not easy to fabricate them [22,23]. Finally, we note that the polariton energy levels become less dense below the resonance energy equal to $\hbar\omega_0$.

The numerical calculations presented here were carried out assuming ABC1. We have also made a similar numerical analysis using ABC2; the differences compared with Fig. 10.1 were found to be minor in this example.

For completeness, we now present the exciton-polariton dispersion relation for the case of s-polarization, where the magnetic field is in the xz-plane and $\vec{E}$ is in the y-direction. There is no propagation of the longitudinal mode k_L. Following a similar calculation as done above for the p-polarization case, the dispersion relation can be found as the solution of a matrix equation as in Eq. (10.14). However, the column matrix $\bar{u}_s$ is now formed from the six coefficients $(E_1, E_3, A_1, A_2, B_1, B_2)$, and the 6×6 matrix $\bar{X}_s$ is

$$\bar{X}_s = \begin{bmatrix} -1 & 0 & \overline{f}_1 & \overline{f}_2 & f_1 & f_2 \\ -\alpha_1 & 0 & k_1^z \overline{f}_1 & k_2^z \overline{f}_2 & -k_1^z f_1 & -k_2^z f_2 \\ 0 & 0 & k_1^z \chi_1 \overline{f}_1 & k_2^z \chi_2 \overline{f}_2 & -k_1^z \chi_1 f_1 & -k_2^z \chi_2 f_2 \\ 0 & -1 & f_1 & f_2 & \overline{f}_1 & \overline{f}_2 \\ 0 & \alpha_3 & k_1^z f_1 & k_2^z f_2 & -k_1^z \overline{f}_1 & -k_2^z \overline{f}_2 \\ 0 & 0 & k_1^z \chi_1 f_1 & k_2^z \chi_2 f_2 & -k_1^z \chi_1 \overline{f}_1 & -k_2^z \chi_2 \overline{f}_2 \end{bmatrix}. \quad (10.17)$$

The lower dimensionality of the matrix, compared with Eq. (10.15), is due to the absence of the longitudinal mode. Again the condition $\det \bar{X}_s = 0$ gives the dispersion relation. The s-polarized exciton-polariton spectrum (not shown here) is found to be qualitatively similar to the p-polarization case; the minor differences are mainly related to the number of surface and guided modes outside the bulk mode lines.

Among the experiments to probe these modes, the most relevant are the photoreflectance (PR) and photoluminescence (PL) studies for GaN thin films (on sapphire) carried out by Gil et al. [24]. The 2K photoluminescence spectrum was taken using a 325 nm line of the HeCd laser at an excitation density of $1\,\mathrm{W\,cm^{-2}}$, which is enough for quantization of the exciton transverse modes as discussed here. The transition energies obtained from the PR spectrum agree with the energies of the exciton-polariton branches shown in Fig. 10.1. Polarized PR and PL measurements were also used to study exciton-polariton structures in a high-quality bulk ZnO single crystal. The energies of the PR resonances corresponded to those of the upper and lower exciton-polariton branches, where A, B, and C excitons couple simultaneously to an electromagnetic wave. Longitudinal–transverse splitting of ground-state exciton-polaritons and resonances due to the first excited states of these excitons were observed due to the large oscillator strength [25].

Although the theory provides a good fit to data for films, it should also provide some insights into the exciton-polariton spectra in quantum wells and microcavities. In all these cases, due to the confinement of the electronic wave functions, the excitons have markedly different characteristics from those in bulk semiconductors. For application to quantum wells and microcavities a microscopic analysis is generally required. Some significant contributions have come from studying the exciton–light coupling in quantum wells within a semi-classical model, taking the exciton resonance frequency to be described by a Gaussian distribution [26] or by the light coupling with ensembles of localized quantum dot-like exciton states [27]. Similar calculations were performed by Citrin [28], who employed a Green-function theory for the coupled exciton-electromagnetic modes of the structure.

Wannier–Mott excitons (see Section 1.6) in the wurzite-type semiconductor material ZnO are stable at room temperature, have an extremely large oscillator strength, and emit blue light. This makes ZnO an excellent candidate for the fabrication of room-temperature lasers where coherent light amplification depends on Bose condensation of the exciton polaritons. Indeed, a model ZnO-based microcavity structure was recently proposed as a structure for the observation of a polariton laser effect [29].

Polariton lasers are devices producing monochromatic and coherent light spontaneously emitted by Bose-condensed exciton-polaritons. They do not require inversion of population and, theoretically, have no threshold. Therefore there are great advantages in using cavity polaritons with respect to excitons to achieve a bosonic phase transition [30]. On the other hand, a pronounced motional-narrowing effect of exciton-polariton modes was found in the time-resolved reflection of GaN/AlGaN single and multiquantum wells [31]. The critical temperature for Bose condensation of exciton-polaritons in the AlGaN microcavity containing nine GaN quantum wells was calculated to be 460 K, suggesting that GaN microcavities may be excellent candidates for realization of room-temperature polariton lasers [32].

The condensation of microcavity exciton-polaritons was confirmed through observation of a phase transition from a classical thermal mixed state to a quantum-mechanical pure state of exciton-polaritons in a GaAs multiquantum-well microcavity, as signaled by a decrease of the second-order coherence function [33]. It has also been suggested that an organic semiconductor microcavity operating in the strong-coupling regime may be a candidate for a polariton laser [34].

10.2 Superlattice Modes

We now extend the formalism of the previous section to calculate the dispersion relations of exciton-polaritons in a superlattice. Specifically, we consider a semi-infinite superlattice structure composed of alternating layers of materials A and B, where at least one of them is excitonic. Here we choose medium A to be excitonic, while for simplicity medium B is not. The layers have thicknesses denoted by a and b, and dielectric functions denoted by $\epsilon_A(k,\omega)$ and $\epsilon_B(\omega)$, respectively, where $\epsilon_A(k,\omega)$ is given by Eq. (10.1) and $\epsilon_B(\omega)$ is given by Eq. (3.13). Both A and B may have a volume density of charge, implying non-zero plasma frequencies. The Cartesian axes are chosen in such a way that the z-axis is normal to the plane of the layers. As before, the structure is terminated at the plane $z = 0$, with the half-space $z < 0$ filled with a material C that has frequency-independent dielectric function ϵ_C.

10.2. SUPERLATTICE MODES

It is assumed that a p-polarized (TM) electromagnetic wave of frequency ω_I and wavevector $\vec{k}_I$ is incident from the external medium C into the superlattice at an angle of incidence θ_I. The use of Maxwell's boundary conditions together with the exciton ABC, having the general form quoted in Eq. (10.3), allows the determination of the electric field in multiple layers. For the present we assume that there is no external magnetic field. The full calculation is complicated to carry out, but an alternative (and simpler) theoretical method is based on the so-called *effective-medium approach* that we employed in earlier chapters. We recall that the idea behind this method is to replace the layered structure of the superlattice by an "effective" single medium that has a modified dielectric function (which will no longer be isotropic). It has been successfully employed in previous works [35–37]. Within this model, in the absence of any excitonic effect, it is well known that we can treat the superlattice as a uniaxial crystal, where the dielectric function is given by [38]

$$\epsilon_{xx}(\omega) = f_a \epsilon_A(\omega) + f_b \epsilon_B(\omega), \tag{10.18}$$

$$\epsilon_{zz}^{-1}(\omega) = f_a \epsilon_A^{-1}(\omega) + f_b \epsilon_B^{-1}(\omega). \tag{10.19}$$

Here we have defined the fractions

$$f_j = j/(a+b) = j/L \quad (j=a,b), \tag{10.20}$$

with L being the size of the superlattice unit cell.

If layer A is now an excitonic medium, a tedious (but straightforward) calculation yields the generalized result that [39]

$$\epsilon_{xx}(k,\omega) = f_a \epsilon_A(\omega,k) + f_b \epsilon_B(\omega) - 2\alpha\epsilon_\infty(\omega_{LA}^2 - \omega_{TA}^2)/\delta D_{xx}, \tag{10.21}$$

$$\epsilon_{zz}^{-1}(k,\omega) = f_a \epsilon_A^{-1}(\omega,k) + f_b \epsilon_B^{-1}(\omega) + 2\alpha(\omega_{LA}^2 - \omega_{TA}^2)/\epsilon_\infty \delta D_{zz}. \tag{10.22}$$

Here we define

$$D_{xx} = Q_{xx}^3 \left[Q_{xx} + \alpha \coth(Q_{xx}a/2) \right], \tag{10.23}$$

$$\delta = \hbar\omega_{TA}/M, \quad \alpha = \beta/\gamma, \tag{10.24}$$

with

$$Q_{xx}^2 = \omega_{LA}^2 - \omega^2 + \delta k^2. \tag{10.25}$$

The quantity D_{zz} is given by an expression similar to D_{xx} provided we replace Q_{xx} by

$$Q_{zz} = (\omega_{TA}^2 - \omega^2 + \delta k^2)^{1/2}. \tag{10.26}$$

Finally, the formal expression for the surface polariton dispersion relation in the effective medium approach is well known from elsewhere [40,41]; in the case of $\epsilon_C = 1$, its expression is

$$(ck_x/\omega)^{1/2} = (\epsilon_{xx} - 1)/(\epsilon_{xx} - \epsilon_{zz}^{-1}). \tag{10.27}$$

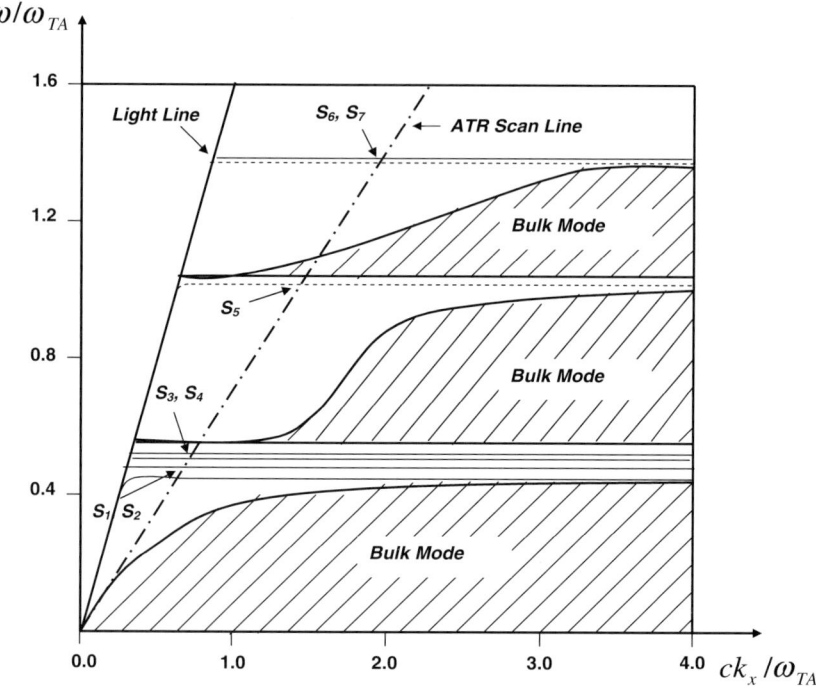

Fig. 10.2. Surface exciton-polariton spectrum in p-polarization, calculated within the effective-medium approach. The physical parameters used here, as well as the labelling of the modes, are described in the main text (after Albuquerque and Fulco [41]).

Now we present some numerical examples to illustrate the bulk and surface polariton spectra. In what follows we assume physical parameters typical of electron concentrations in GaAs/AlGaAs superlattices, and we assume that the medium outside the superlattice is vacuum. We take thicknesses corresponding to $a = 40$ nm and $b/a = 2$. The following parameters are also used in these calculations: $\epsilon_{\infty A} = 10.9$, $\epsilon_{\infty B} = 10.22$, $\omega_{LA} = 5.496$, $\omega_{TA} = 5.057$, $\omega_{LB} = 6.979$, $\omega_{TB} = 6.731$, and $\omega_{pA} = \omega_{pB} = 2.54$ (all the frequencies are in units of 10^{13} rad/s). The damping is taken to be zero in both media. Fig. 10.2 shows the surface polariton dispersion curves together with a constant-angle ATR scan line (see Section 11.5 on ATR for an explanation and discussions). We have plotted a reduced frequency ω/ω_{TA} against a dimensionless wavevector ck_x/ω_{TA}. Here, the chain-dotted line is the ATR scan line and the vacuum light line (also shown) is defined by $\omega = ck_x$. For completeness the bulk bands are included in Fig. 10.2; their dispersion relation is [41]

$$k_z^2 \epsilon_{xx}^{-1} + k_x^2 \epsilon_{zz}^{-1} = (\omega/c)^2. \qquad (10.28)$$

Here k_z, denoting the z-component of the wavevector $\vec{k}$ in the effective medium, is given by

$$k_z^2 = (\omega/c)^2 \epsilon_{xx} - (\epsilon_{xx}/\epsilon_{zz})k_x^2. \tag{10.29}$$

In this example we use ABC1 (corresponding to $\vec{P}=0$), which implies $\alpha=0$. The points with the labels S_1 to S_7 are identified as being the crossing of the ATR scan line with the many branches of the surface polariton dispersion curve. We observe that the pair of modes (S_1, S_2) as well as (S_3, S_4) and (S_6, S_7) are so close together that it is difficult to differentiate one from the other. It is possible to distinguish two different types of surface polaritons. Those that are characterized by having both ϵ_{xx} and ϵ_{zz} negative represent the *real* surface modes; they are represented by solid curves in Fig. 10.2. On the other hand, the broken curves represent the case when ϵ_{xx} is negative, although ϵ_{zz} is positive. These modes are known as *virtual* surface modes. They may also be present in cases where there is no excitonic medium. The polariton branches obtained with and without spatial dispersion are similar. However, if Fig. 10.2 is compared with the corresponding result obtained by the dispersionless model, we may conclude that the spatial dispersion in layer A increases the number of polariton branches in the spectrum. Furthermore, the additional polariton modes can have a more substantial wavevector dependence than those present in the dispersionless model.

Although we have restricted our theoretical analysis to Pekar's ABC ($\vec{P}=0$, or ABC1), it is illustrative of how the calculations would proceed for other choices of ABCs, such as those mentioned earlier. The use of other ABCs, with α different from zero, can lead to a qualitative change in the behavior of the dielectric function Eqs. (10.21) and (10.22), since their poles are shifted to a different frequency range [39].

10.3 Superlattice Modes in the Presence of a Magnetic Field

Here we extend the calculations of the previous section by including an external magnetic field $\vec{H}_0$ applied to the superlattice parallel to the interfaces. We chose the field to be along the y-direction (see Fig. 10.3).

We again employ the theoretical model based on the modified effective-medium approach. The inclusion of an applied magnetic field generalizes the permittivity, as we shall describe shortly. It now takes on the form that we call magneto-exciton (or, more loosely, magnetoplasmon); we prefer the former name here because it emphasizes the SD property. Theoretical results (e.g. Ref. [42]) suggest that in a bulk semiconductor containing minority as well as majority carriers, the magneto-exciton reflectivity can in principle be used to determine the densities and effective masses of both species of carrier, so this is a motivation for this section. Like the permeability tensor of a magnetic medium, the magneto-exciton permittivity is not diagonal (see Section 5.5), so that expressions are more complicated than those used in the absence of an external magnetic field.

Fig. 10.3 shows the geometry used here. As before, the superlattice fills the half-space $z > 0$, with the layers and interfaces in the xy-plane. The semi-infinite

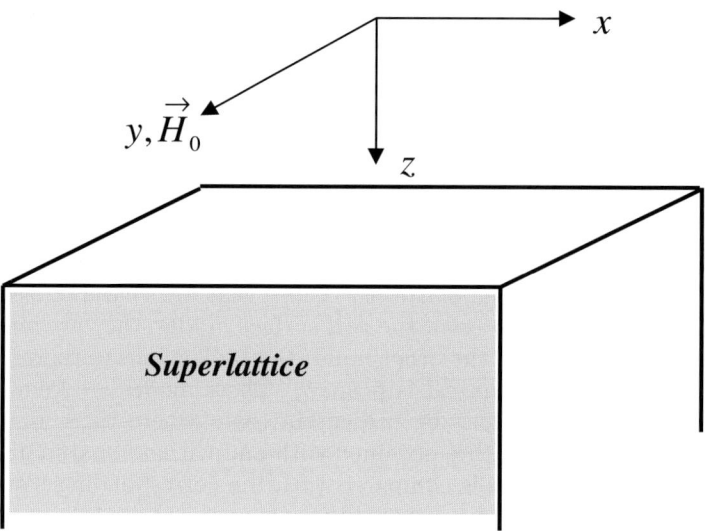

Fig. 10.3. Geometry used in calculations for the exciton-polaritons in the presence of an external magnetic field. The layers of the superlattice are parallel to the free surface (the xy-plane at $z=0$) and the external magnetic field is pointing in the y-direction.

superlattice structure is composed of alternating layers of materials A and B, one of which (medium A) is excitonic. The structure is terminated at the plane $z=0$, with the half-space $z<0$ filled with a material that has a frequency-independent dielectric function ϵ_C.

If an external magnetic field $\vec{H}_0$ is now applied along the y-axis, the frequency-dependent dielectric tensor $\bar{\epsilon}_A$ of layer A can be written in the same form as Eq. (5.106), namely

$$\bar{\epsilon}_A = \begin{bmatrix} \epsilon_{1A} & 0 & -i\epsilon_{2A} \\ 0 & \epsilon_{3A} & 0 \\ i\epsilon_{2A} & 0 & \epsilon_{1A} \end{bmatrix}, \qquad (10.30)$$

but the definitions of the matrix elements are generalized to include dispersion. Their full expressions are

$$\epsilon_{1A} = \epsilon_{\infty A}\left(1 + \frac{\omega_{LA}^2 - \omega^2 + Dk^2}{\omega_{TA}^2 - \omega^2 + Dk^2 - i\omega\Gamma_A} + \frac{\omega_{pA}^2}{\omega_{cA}^2 - \omega(\omega - i\Gamma_A)}\right), \quad (10.31)$$

$$\epsilon_{2A} = \epsilon_{\infty A}\frac{\omega_{pA}^2 \omega_{cA}}{\omega(\omega_{cA}^2 - \omega^2)}, \qquad (10.32)$$

$$\epsilon_{3A} = \epsilon_{\infty A}\left(1 + \frac{\omega_{LA}^2 - \omega^2 + Dk^2}{\omega_{TA}^2 - \omega^2 + Dk^2 - i\omega\Gamma_A} - \frac{\omega_{pA}^2}{\omega(\omega - i\Gamma_A)}\right). \qquad (10.33)$$

10.3. SUPERLATTICE MODES IN THE PRESENCE OF A MAGNETIC FIELD

Here ω_{pA} and ω_{cA} are, respectively, the plasma frequency and the cyclotron frequency of medium A, which is considered to be excitonic. Eqs. (10.30)–(10.33) are also applicable to medium B, provided the spatial dispersion parameter D is set equal to zero and index A is replaced by index B.

Following Refs. [38,39], the effective dielectric tensor for the superlattice can be obtained from the relation of the averaged displacement vector $\langle \vec{D} \rangle$ to the averaged electric field $\langle \vec{E} \rangle$, where the average is taken over the unit cell of the superlattice. These fields are given explicitly by

$$\langle D_x \rangle = f_a D_x^A + f_b D_x^B + (4\pi/L) \int_0^a P_x(z)\, dz, \tag{10.34}$$

$$\langle D_z \rangle = D_z^A = D_z^B, \tag{10.35}$$

$$\langle E_x \rangle = E_x^A = E_x^B, \tag{10.36}$$

$$\langle E_z \rangle = f_a E_z^A + f_b E_z^B - (4\pi/\epsilon_\infty L) \int_0^a P_z(z)\, dz. \tag{10.37}$$

Here, $\vec{P}(z)$ denotes the excitonic polarization vector (in layer A).

Using the property that E_x and D_z are independent of z inside layer A, and making use of the standard electromagnetic boundary conditions together with the general form of ABC [see Eq. (10.3)], we can eventually find the following expression for the effective dielectric tensor:

$$\langle \bar{\epsilon} \rangle = \begin{bmatrix} \epsilon_{xx} & 0 & -i\epsilon_{xz} \\ 0 & \epsilon_{yy} & 0 \\ i\epsilon_{xz} & 0 & \epsilon_{zz} \end{bmatrix}, \tag{10.38}$$

where

$$\epsilon_{xx}(k,\omega) = \epsilon_{xz}(k,\omega) + \sum_{i=a}^{b} f_i \epsilon_{1i} - \sum_{i=a}^{b} f_i(\epsilon_{2i}^2/\epsilon_{1i}) - J_{xx}, \tag{10.39}$$

$$\epsilon_{yy}(k,\omega) = \sum_{i=a}^{b} f_i \epsilon_{3i}, \tag{10.40}$$

$$\epsilon_{zz}^{-1}(\omega,k) = \sum_{i=a}^{b} f_i \epsilon_{1i}^{-1} + J_{zz}, \tag{10.41}$$

$$\epsilon_{xz}(k,\omega) = \left[\sum_{i=a}^{b} f_i(\epsilon_{2i}^2/\epsilon_{1i})\right]^2 \left[\sum_{i=a}^{b} f_i \epsilon_{1i}^{-1}\right]^{-1}. \tag{10.42}$$

In the above, we have introduced the additional notation

$$J_{xx} = 2\alpha \epsilon_{\infty A}(\omega_{LA}^2 - \omega_{TA}^2)/\delta D_{xx}, \tag{10.43}$$

$$J_{zz} = 2\alpha(\omega_{LA}^2 - \omega_{TA}^2)/\epsilon_{\infty A}\delta D_{zz}. \tag{10.44}$$

This result for the effective dielectric tensor is algebraically complicated, but it is of rather general applicability; we note that it reduces to the result in Section 10.2 when $H_0 = 0$. Two special cases can now be considered in terms of Eqs. (10.43) and (10.44):

(a) $\alpha = 0$ (ABC1), which implies $J_{xx} = J_{zz} = 0$ (this was already discussed in the previous section for the special case of zero magnetic field);

(b) $\alpha = \infty$ (ABC2), which implies

$$J_{xx} = 2(\omega_{LA}^2 - \omega_{TA}^2)\epsilon_{\infty A}/LQ_{xx}^3 \coth(Q_{xx}a), \tag{10.45}$$

with a similar expression for J_{zz}.

Now we turn to the calculation of the dispersion relations. For this we assume that a p-polarized electromagnetic wave is incident on the superlattice from the outside medium. Using the dielectric approximation described in Section 3.3, in particular Eq. (3.22), we can deduce equations for the coupling of the excitonic polarization field $\vec{P}(z)$ to the macroscopic electromagnetic fields $\vec{E}$ and $\vec{H}$. They are as follows [43]:

(a) the electric field is

$$\vec{E} = (1, 0, -ik_x/\alpha_0)E_0 \exp(\alpha_0 z) \quad \text{for } z < 0, \tag{10.46}$$

$$\vec{E} = \sum_{j=L,1,2} (1, 0, \lambda_j) E_j \exp(-k_j^z z) \quad \text{for } z > 0; \tag{10.47}$$

(b) the magnetic field is

$$\vec{H} = \left[0, (i\omega\epsilon_0\epsilon_c/\alpha_0)E_0 \exp(\alpha_0 z), 0\right] \quad \text{for } z < 0, \tag{10.48}$$

$$\vec{H} = \left[0, (-\epsilon_0\omega/k_x) \sum_{j=L,1,2} (i\epsilon_{xzj} + \lambda_j \epsilon_{xxj}) E_j \exp(-k_j^z z), 0\right] \quad \text{for } z > 0; \tag{10.49}$$

(c) the polarization field is

$$\vec{P} = \epsilon_0 \sum_{j=L,1,2} (1, 0, \lambda_j) \chi_j E_j \exp(-k_j^z z) \quad \text{for } z > 0. \tag{10.50}$$

In Eqs. (10.46)–(10.50) we have defined $\alpha_0 = (k_x^2 - \epsilon_C \omega^2/c^2)^{1/2}$ and (for $j = L$, 1, or 2)

$$k_j^z = (k_x^2 - \epsilon_j^{eff} \omega^2/c^2)^{1/2}, \tag{10.51}$$

$$\epsilon_j^{eff} = \epsilon_{xxj} - (\epsilon_{xzj}^2/\epsilon_{xxj}), \tag{10.52}$$

10.3. SUPERLATTICE MODES IN THE PRESENCE OF A MAGNETIC FIELD

$$\lambda_j = i(k_x^2 - \epsilon_{xxj}\omega^2/c^2)^{-1}(k_x k_j^z + \epsilon_{xzj}\omega^2/c^2), \tag{10.53}$$

$$\chi_j = \epsilon_\infty(\omega_L^2 - \omega_T^2)(\omega_T^2 - \omega^2 - i\omega^2\Gamma + Dk_j^2)^{-1}. \tag{10.54}$$

The label $j = L$, 1, or 2 refers to the longitudinal mode (L) and the two transverse modes (1, 2), with the wavevector $\vec{k}_j$ being in the form $(k_x, 0, k_j^z)$.

The surface polaritons are free oscillations of the electromagnetic fields at an interface, i.e. they occur even in the limit of zero incident field in the preceding calculation. Viewed in terms of the reflectivity, the polaritons must therefore correspond to a pole in the reflectivity coefficients [44,45]. To explore this further, we note that the reflectivity is given, in general, by the well-known formula [46]

$$R_p = \left| \frac{Z_{non\text{-}local} - Z_{local}}{Z_{non\text{-}local} + Z_{local}} \right|^2, \tag{10.55}$$

where Z, the surface impedance of the wave, is defined for the two regions $z<0$ and $z>0$ by

$$Z_{local} = E_x(0^-)/H_y(0^-) \quad \text{for} \quad z < 0, \tag{10.56}$$

$$Z_{non\text{-}local} = E_x(0^+)/H_y(0^+) \quad \text{for} \quad z > 0. \tag{10.57}$$

Therefore, the surface magneto-exciton-polariton dispersion relation (corresponding to the poles of R_p) is given by

$$Z_{local} + Z_{non\text{-}local} = 0. \tag{10.58}$$

This dispersion relation is rather complicated when written out in full, and it is difficult to draw general conclusions from it. However, some physical insight can be obtained if we examine the limit of small non-local effects. This, in turn, has some bearing on the choice of ABCs. Somewhat fortunately, for the case of ABC2 (namely $d\vec{P}/dz = 0$), it can be proved (see e.g. Ref. [47]) that the lowest-order non-local correction vanishes identically. It is then easy to show that in this approximation the surface polaritons propagating perpendicular to the magnetic field H_0 have their formal dispersion relation as [48]

$$\left(\frac{\epsilon_{xx}}{\epsilon_{zz}} k_x^2 - \frac{\omega^2}{c^2}\epsilon^{eff} \right)^{1/2} + \epsilon^{eff}\left(k_x^2 - \frac{\omega^2}{c^2} \right) = k_x \frac{\epsilon_{xz}}{\epsilon_{zz}}, \tag{10.59}$$

which is actually the same form that applies for the dispersionless case when the gyrotropic dielectric tensor is k-independent (e.g. as in the magneto-phonon case), but now the tensor elements are given by Eqs. (10.39)–(10.42), and ϵ^{eff} is defined in Eq. (10.52).

Now we present some numerical examples to illustrate the spectra for the bulk and surface polaritons. In what follows, we assume the same physical parameters as employed for the case in Section 10.2 in the absence of an external magnetic field. Here the external magnetic field is specified through the cyclotron frequency

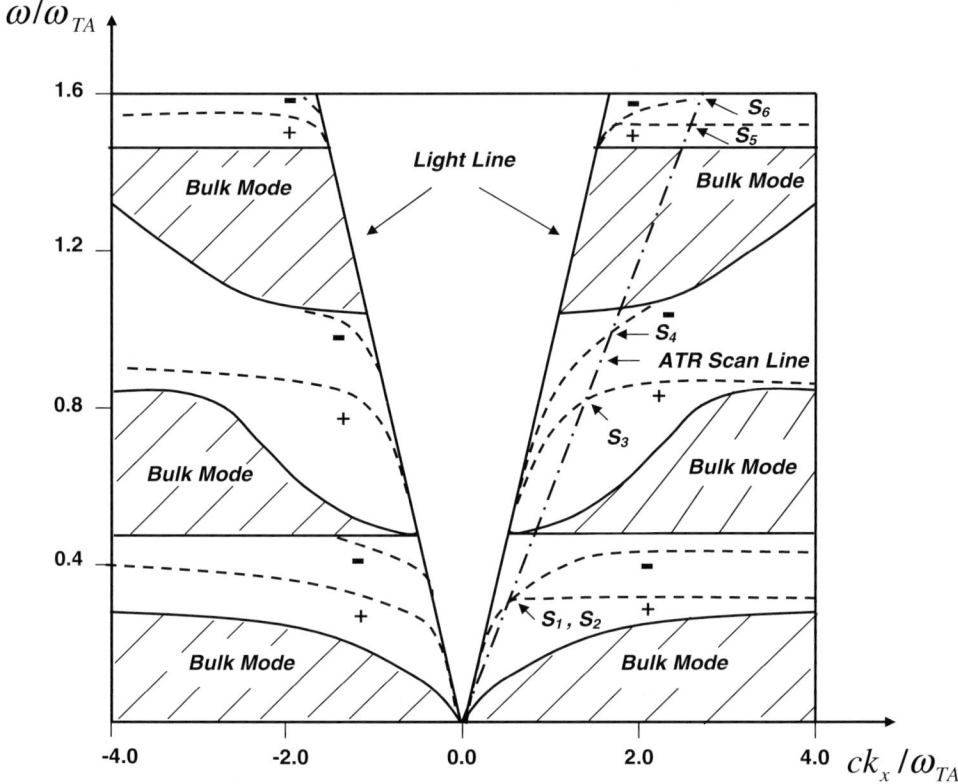

Fig. 10.4. Surface exciton-polariton spectra in p-polarization, calculated within the effective-medium approach in the presence of an external magnetic field. The physical parameters used here, as well as the labelling of the modes, are described in the text (after Albuquerque and Fulco [43]).

ω_c (see Section 5.5), which is assumed to be half the value of the plasma frequency. The parameter D, responsible for the spatial dispersion in medium A, is equal to 2.055 (in units of 10^{13} Hz); the damping factor is taken to be 4.0×10^{-4} ω_{pA} in both media. Fig. 10.4 shows the surface polariton dispersion curves corresponding to Eq. (10.59). As mentioned, it is appropriate for the ABC2 case in the limit of small non-local effects, together with a constant-angle ATR scan line. We have plotted a reduced frequency ω/ω_{TA} against a dimensionless wavevector ck_x/ω_{TA}. Here the chain-dotted line means the ATR scan line. For completeness we have also included the bulk bands, which occupy the continuum region shown shaded in Fig. 10.4. These bulk modes are reciprocal, that is, they are the same for $+k_x$ as for $-k_x$.

On the other hand, the surface polaritons, which are represented by dotted lines in Fig. 10.4, are, as expected, non-reciprocal (just as in the case of surface magnetostatic modes in Section 7.1). The points with the labels S_1 to S_6 represent the crossing of the ATR scan line with the multiple surface polariton branches. All the modes start at their low-frequency end on the vacuum light line. It is noticeable that the degeneracy of the surface modes that occurred in Fig. 10.2 (in the absence of the external magnetic field) is now removed by the applied magnetic field. This leads to the modes labelled by the plus (+) and minus (−) signs in Fig. 10.4. They correspond to propagation in opposite directions for the in-plane component k_x of the wavevector $\vec{k}$. We remark that when there is a magnetic field, a convenient alternative approach experimentally is to use a fixed frequency while the magnetic field is scanned rather than scanning the frequency. Apart from the introduction of non-reciprocal propagation, the external magnetic field produces substantial shifts in the boundaries of the bulk bands and the removal of the degeneracy of the surface modes.

References

[1] S.M. Rytov, Soviet Phys. - J. Exper. Theor. Phys. 2 (1956) 466.

[2] A. Yariv and P. Yeh, J. Opt. Soc. Am. 67 (1977) 438.

[3] P. Yeh, A. Yariv and C.S. Hong, J. Opt. Soc. Am. 67 (1977) 423.

[4] S.I. Pekar, Zh. Eksp. Teor. Fiz. 33 (1957) 1022 [Sov. Phys. JETP 6 (1957) 785].

[5] J.J. Hopfield, Phys. Rev. 112 (1958) 1555.

[6] J.J. Hopfield and D.J. Thomas, Phys. Rev. 132 (1963) 563.

[7] J.L. Birman, in: Excitons, Eds. E.I. Rashba and M.D. Sturge, North-Holland, Amsterdam, 1982.

[8] M. Tchounkeu, O. Briot, B. Gil, J.P. Alexis and R.-L. Aulombard, J. Appl. Phys. 80 (1996) 5352.

[9] P. Stepniewski, K.P. Korona, A. Wysmolek, J.M. Baranowski, K. Pakula, M. Potemski, I. Gregory and S. Porowski, Phys. Rev. B 56 (1997) 15151.

[10] M. Yamaguchi, T. Yagi, T. Sota, T. Deguchi, K. Shimada and S. Nakamura, J. Appl. Phys. 85 (1999) 8502.

[11] S.F. Chichibu, K. Torii, T. Deguchi, T. Sota, A. Setoguchi, H. Nakanishi, T. Azuhata and S. Nakamura, Appl. Phys. Lett. 76 (2000) 1576.

[12] K. Torii, M. Ono, T. Sota, T. Azuhata, S. Chichibu and S. Nakamura, Phys. Rev. B 62 (2000) 4723.

[13] S. Nakamura, T. Mukai and M. Senoh, Appl. Phys. Lett. 64 (1994) 1687.

[14] J.Y. Duboz and M.A. Khan, in: Physics and Applications of Nitride Compounds Semiconductors, Ed. B. Gil, Oxford University Press, 1997.

[15] P. Halevi in: Excitons in Confined Systems, Eds. R. Del Sole, A. D'Andrea and A. Lapiccirella, Springer Verlag, Berlin, 1988.

[16] H. Ishihara and K. Cho, Phys. Rev. B 48 (1993) 7960.

[17] E.L. Albuquerque and C.E.T. Gonçalves da Silva, J. Phys. C 18 (1985) 665.

[18] M.S. Vasconcelos, E.L. Albuquerque, G.A. Farias and V.N. Freire, Solid State Commun. 124 (2002) 109.

[19] R.N. Philp and D.R. Tilley, Phys. Rev. B 44 (1991) 8170.

[20] J. Knoester, Phys. Rev. Lett. 68 (1992) 654.

[21] L.C. Andreani, Phys. Status Solidi (b) 188 (1995) 29.

[22] H. Morkoç, S. Strite, G.B. Gao, M.E. Lin, B. Sversodlov and M. Burns, J. Appl. Phys. 73 (1994) 1363.

[23] H. Morkoç, in: Semiconductor Hetero-epitaxy, Eds. B. Gil and R.L. Aulombard, World Scientific, Singapore, 1995.

[24] B. Gil, S. Clur and O. Briot, Solid State Commun. 104 (1997) 267.

[25] S.F. Chichibu, T. Sota, G. Cantwell, D.B. Eason and C.W. Litton, J. Appl. Phys. 93 (2003) 756.

[26] L.C. Andreani G. Panzarini, A.V. Kakovin and M.R. Vladimirova, Phys. Rev. B 57 (1998) 4670.

[27] A.V. Kavokin, G. Malpuech and W. Langbein, Solid State Commun. 120 (2001) 259.

[28] D.S. Citrin, Phys. Rev. B 49 (1994) 1943; 50 (1994) 5497.

[29] M. Zamfirescu, A. Kavokin, B. Gil, G. Malpuech and M. Kaliteevski, Phys. Rev. B 65 (2002) 161205.

[30] A. Kavokin, G. Malpuech and F.P. Laussy, Phys. Lett. A 306 (2003) 187.

[31] G. Malpuech and A. Kavokin, Appl. Phys. Lett. 76 (2000) 3049.

[32] G. Malpuech, A. Di Carlo, A. Kavokin, J.J. Baumberg, M. Zamfirescu and P. Lugli, Appl. Phys. Lett. 81 (2002) 412.

[33] H. Deng, G. Weihs, C. Santori, J. Bloch and Y. Yamamoto, Science 298 (2002) 199.

[34] D.G. Lidzey, D.D.C. Bradley, M.S. Skolnick, T. Virgili, S. Walker and D.M. Whittaker, Nature 395 (1998) 53.

[35] N.S. Almeida and D.L. Mills, Phys. Rev. B 38 (1988) 6998.

[36] N.S. Almeida and D.R. Tilley, Solid State Commun. 73 (1990) 23.

[37] F.G. Elmznghi, N.C. Constantinou and D.R. Tilley, J. Phys.: Condens. Matter 7 (1995) 315.

[38] N. Raj and D.R. Tilley, Solid State Commun. 55 (1985) 373.

[39] V.M. Agranovich and V.E. Kravstov, Solid State Commun. 55 (1985) 85.

[40] N. Raj, D.R. Tilley and R.E. Camley, J. Phys. C 20 (1987) 5203.

[41] E.L. Albuquerque and P. Fulco, Phys. Status Solidi (b) 182 (1994) 357.

[42] B.L. Johnson, R.E. Camley and D.R. Tilley, Phys. Rev. B 44 (1991) 8837.

[43] E.L. Albuquerque and P. Fulco, Z. Phys. B 100 (1996) 289.

[44] M. Cardona, Am. J. Phys. 39 (1977) 1277.

[45] B.B. Dasgupta and A. Baggchi, Phys. Rev. B 19 (1979) 4935.

[46] K.L. Kliewer and R. Fuchs, Phys. Rev. 172 (1968) 607.

[47] P. Halevi and R. Fuchs, J. Phys. C 17 (1984) 3869; 3889.

[48] M.C. Oliveros, N.S. Almeida and D.R. Tilley, Semicond. Sci. Technol. 8 (1993) 441.

Chapter 11

Experimental Techniques

In this chapter we discuss some of the experimental techniques that can be used to study polariton excitations in periodic and quasiperiodic structures. The purpose is not so much to provide details of the techniques themselves (since there are many other references that cover this topic), but to explore how the results in the earlier chapters can be applied to an experimental analysis. Thus specific numerical applications, based on the theoretical results obtained in earlier chapters, are made.

We start with inelastic light-scattering techniques, which include Raman scattering (which so far is the most useful experimental tool to probe many of the modes covered in this book), Brillouin scattering (which is useful for probing magnetic polaritons, among others), and resonant Brillouin scattering (RBS) (which is appropriate for exciton-polaritons). Then we present the method of attenuated total reflection (ATR), which was originally developed by Otto [1,2] for the study of surface polaritons on metal surfaces. Finally, several other relevant spectroscopic techniques are more briefly described.

11.1 Raman Scattering in Periodic Structures

Raman scattering is the inelastic scattering of light using a grating spectrometer for the detection system. Details of the technique are to be found, for example, in the books by Hayes and Loudon [3] and Cottam and Lockwood [4]. Typical frequency shifts of the light correspond to the range 5–4000 cm^{-1} when expressed in wavenumber units (or about 0.6–500 meV in energy units), making the technique useful for a wide range of excitations in solids.

The first experiments to demonstrate clearly the application of Raman scattering to plasmons in periodic superlattices were due to Olego et al. [5]. These authors made their measurements on doped GaAs/Al$_x$Ga$_{1-x}$As superlattices. The modes correspond fairly well to the model with 2D charge sheets described in Section 5.2, where the charge sheets are here approximated by the GaAs quantum wells. Subsequently, further Raman light-scattering measurements were made on similar superlattices and on finite multilayers by, among others, Sooryakumar et al. [6].

On the theoretical side and at about the same time, considerable efforts were being made to develop a complete theory of Raman scattering, incorporating characteristics such as the spectral width and shape of the Raman peak, integrated

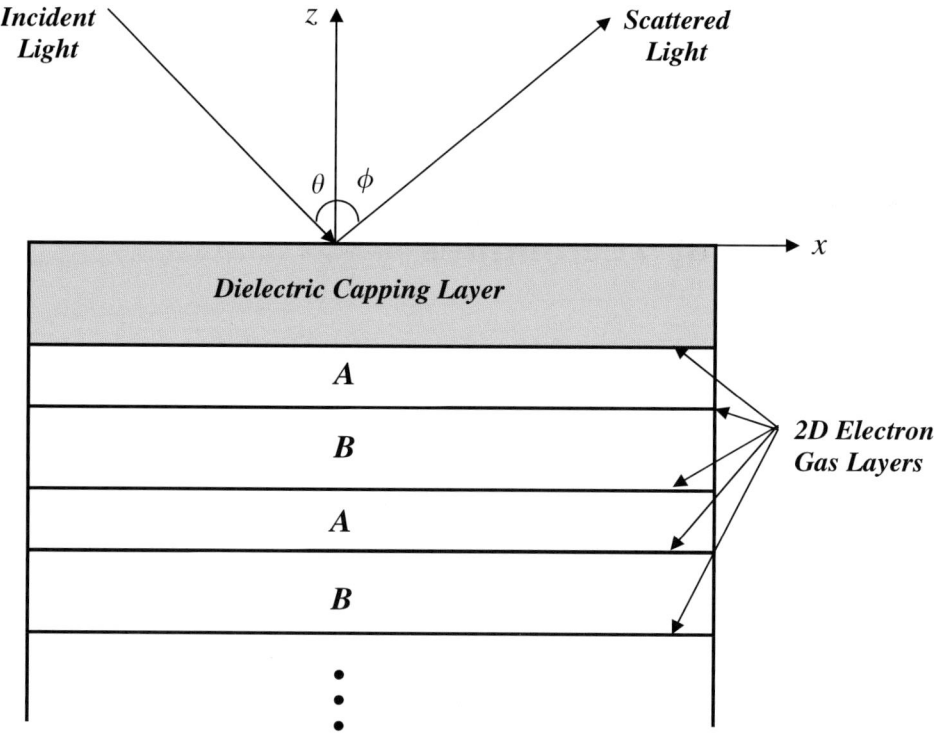

Fig. 11.1. Assumed light-scattering geometry, showing the incident and scattered light beams. The superlattice sample is drawn here as for a charge-sheets superlattice with a capping layer.

intensity, and polarization selection rules, as well as dispersion relations. Using an approach based on the density–density correlation function, a dynamical theory was developed and results were found for the spectral linewidth and shape of the bulk- and surface-plasmon resonances [7–9]. An alternative approach, based on linear-response techniques (see Section A.3 of Appendix A), was adopted by Babiker et al. [10,11] and this proved to be very effective for describing how the plasmon spectral lineshapes and integrated intensities depend on the polarization directions and on the scattering geometry. A typical experimental geometry for polarized Raman scattering by superlattice excitations is depicted in Fig. 11.1. It shows the incident and scattering light beams in the same vertical plane (taken to be the xz-plane) at angles θ and ϕ, respectively, to the z-axis. It is often convenient experimentally to adopt either 180° backscattering (when $\phi = -\theta$) or 90° scattering (when $\theta + \phi = 90°$). The two-component periodic superlattice is characterized by the layers A and B, as in previous chapters. A dielectric capping layer is considered between the outermost layer A and the free surface at $z = 0$.

The general form of the differential cross section, for scattering into an elementary solid angle $d\Omega$ with frequency between ω_s and $\omega_s + d\omega_s$, is (see e.g.

11.1. RAMAN SCATTERING IN PERIODIC STRUCTURES

Refs. [3,4])

$$\frac{d^2\sigma}{d\Omega\,d\omega_s} = \frac{\omega_I \omega_s \bar{S} \cos^2\phi}{4\pi^2 c^2} \frac{\langle |\varepsilon_s|^2 \rangle_\omega}{|\varepsilon_I|^2}, \tag{11.1}$$

where ω_I is the frequency of incident light and $\bar{S}$ is the area of surface through which the scattered beam emerges. The essential part of Eq. (11.1) is that it involves the ratio of the scattered intensity (in terms of the electric-field correlation function $\langle |\varepsilon_s|^2 \rangle_\omega$) to the incident intensity $|\varepsilon_I|^2$. In the limit of scattering from a thick sample, the above result can be rewritten explicitly as [10,11]

$$\frac{d^2\sigma}{d\Omega\,d\omega_s} = \bar{S}[n(\omega)+1]\cos^2\phi \ \text{Im} \sum_{\mu,\nu,\nu',\delta,\delta'} \int dQ \int dQ' \frac{\langle\langle \chi^*_{\delta\nu}(Q); \chi_{\delta'\nu'}(Q') \rangle\rangle_\omega}{(Q+k_z)(Q'+k_z^*)}$$

$$\times (e_s^\mu)^2 e_I^\nu e_I^{\nu'} (f_\nu g_{\mu\delta})^* f_{\nu'} g_{\mu\delta'}. \tag{11.2}$$

Here $n(\omega)$ is the Bose–Einstein thermal factor, $\hat{e}_I$ and $\hat{e}_S$ are the unit polarization vectors of the incident and scattering light, and the summations are over Cartesian components. The coefficients f and g are standard factors describing, respectively, the transmission of the incident and scattered light beams in and out through the sample surface.

The integrations over Q and Q' refer to real wavevector components perpendicular to the surface, and k_z is the component of the light-scattering wavevector in this direction. Assuming ω_I to be approximately equal to ω_S, we have

$$k_z = (2\pi/\lambda)\left[(\epsilon - \sin^2\theta)^{1/2} + (\epsilon - \sin^2\phi)^{1/2}\right], \tag{11.3}$$

where λ is the wavelength of incident light in vacuum and ϵ is the complex dielectric function in the sample. In general, k_z is complex, corresponding to the attenuation of light as it penetrates into the sample. The light-scattering wavevector component k_x parallel to the surface is real and given by

$$k_x = (2\pi/\lambda)(\sin\theta). \tag{11.4}$$

Finally, Eq. (11.2) contains Green functions involving Cartesian components of the dynamic susceptibility χ. For plasmon-polaritons we have [3]

$$\chi_{\delta\nu}(Q) = \sum_\gamma b_{\delta\nu\gamma} E^*_\gamma(Q), \tag{11.5}$$

where $\vec{E}(Q)$ denotes the electric field associated with the excitation of wavenumber Q, and $b_{\delta\nu\gamma}$ is the electro-optic coefficient. We shall assume (for simplicity) that the superlattice constituents have cubic symmetry. Then the components of the electro-optic tensor are non-zero if the subscripts δ, ν, and γ are all different (giving a coefficient b, say), and zero otherwise [12]. Thus to calculate the scattering cross section it is necessary, using Eq. (11.5) for the susceptibility components, to determine electric-field Green functions of the general form $\langle\langle E_\mu(Q); E^*_\nu(Q') \rangle\rangle_{k_x,\omega}$

at in-plane wavevector k_x and frequency ω (see Appendix A for definitions). In applications to *infinitely extended* superlattices, it is convenient to express these Green functions in terms of other quantities, $G_{\mu\nu}(k_x, Q)$, where [13]

$$\langle\langle E_\mu(Q); E_\nu^*(Q')\rangle\rangle_{k_x,\omega} = [8\pi^3/\epsilon_0 V]\delta(Q-Q')G_{\mu\nu}(k_x, Q), \tag{11.6}$$

and V is the (macroscopically large) sample volume. We shall present explicit results for $G_{\mu\nu}$ later, as well as discuss the case of semi-infinite superlattices for which Eq. (11.6) does not apply.

Before proceeding it is important to note that Eq. (11.2) is applicable when the scattering is non-resonant. Thus it will need to be generalized for the case of RBS (i.e. when ω_I is close to a phonon frequency); this is the topic of Section 11.4, and we shall later demonstrate how this is done.

The incident and scattering beams in Fig. 11.1 are in the xz-plane, which is assumed to be perpendicular to an axis of cubic symmetry of the crystal. The superlattice is taken to be effectively infinite; also the incident light is p-polarized, with polarization vector $\hat{e}_I = (\cos\theta, 0, \sin\theta)$, while the scattered light is s-polarized, with polarization vector $\hat{e}_S = (0, 1, 0)$. Then Eq. (11.2) can be used to show that the scattering cross section is proportional to [14]

$$(|b|^2/\text{Im } k_z)[n(\omega)+1]\bar{S}\cos\phi|g_{yy}|^2\text{Im}[R(\theta,\phi)], \tag{11.7}$$

where

$$R(\theta,\phi) = |f_x|^2\cos^2\theta G_{zz}(k_x, -k_z^*) + |f_z|^2\sin^2\theta G_{xx}(k_x, -k_z^*)$$
$$+ \sin\theta\cos\theta\Big[(f_x)^*f_z G_{xz}(k_x, -k_z^*) + (f_z)^*f_x G_{zx}(k_x, -k_z^*)\Big]. \tag{11.8}$$

Here Im k_z represents the reciprocal penetration depth of light into the sample. Also the f and g factors are defined in the present case by

$$f_x = 2(\epsilon - \sin^2\theta)^{1/2}\Big[\epsilon\cos\theta + (\epsilon - \sin^2\theta)^{1/2}\Big]^{-1}, \tag{11.9}$$

$$f_z = 2\cos\theta\Big[\epsilon\cos\theta + (\epsilon - \sin^2\theta)^{1/2}\Big]^{-1}, \tag{11.10}$$

$$g_{yy} = \Big[\cos\phi + (\epsilon - \sin^2\phi)^{1/2}\Big]^{-1}, \tag{11.11}$$

where in Eq. (11.11) a small difference between the incident and scattered light frequencies has been neglected. As we shall see later, the Green functions have poles from which we obtain directly the dispersion relations (this is a general property of Green functions, as mentioned in Section A.3 of Appendix A). In fact, we can write

$$G_{\mu\nu} = U_{\mu\nu} + (V_{\mu\nu}/\Delta), \tag{11.12}$$

so that $\Delta = 0$ yields the dispersion relation of the superlattice plasmon-polaritons.

11.1. RAMAN SCATTERING IN PERIODIC STRUCTURES

The integrated light-scattering intensity for any plasmon-polariton peak (at frequency denoted by ω_0) can next be deduced from Eq. (11.7) by considering the limit in which optical absorption becomes negligible. This corresponds formally to taking the limit in which Im $\epsilon \to 0$. The resonance peak at ω_0 is then represented by a delta-function spike, and its strength is just the integrated intensity $I(\omega_0)$, namely [14]

$$I(\omega_0) = |b^2| \Big[n(\omega_0) + 1\Big] \bar{S} \cos\phi |g_{yy}|^2 (d\Delta/d\omega)^{-1}$$
$$\times \Big[|f_x|^2 \cos^2\theta V_{zz} + |f_z|^2 \sin^2\theta V_{xx}$$
$$+ \sin\theta \cos\theta (f_x^* f_z + f_z^* f_x) V_{xz}\Big], \quad (11.13)$$

where the right-hand side is evaluated at $\omega = \omega_0$ and $Q = k_z$. It only remains for us to evaluate the Green function to complete the determination of the scattering cross section [Eq. (11.7)] and the integrated intensity [Eq. (11.13)].

In general, $G_{\mu\nu}(k_x, Q)$ can be conveniently obtained using the linear-response theory described in Appendix A. The procedure in this case involves solving Maxwell's equations for the electric field $\vec{E}$ in the presence of an externally applied harmonic polarization $\vec{P}_{ext}$ given by

$$\vec{P}_{ext} = \vec{P}_0 \exp[i(Qz + k_x x - \omega t)]. \quad (11.14)$$

From Maxwell's equations, the electric field inside each layer $J = A$ or B of the periodic superlattice should satisfy the following differential equation due to the presence of this external polarization:

$$\nabla^2 \vec{E} + (\epsilon_J \omega^2 / c^2) \vec{E} = -(\omega^2 / \epsilon_0 c^2) \vec{P}_{ext} - (1/\epsilon_0 \epsilon_J) \nabla(\nabla \cdot \vec{P}_{ext}). \quad (11.15)$$

We may solve for the electric field in each layer by proceeding in the same manner as described in Chapter 5 (see Sections 5.1 and 5.2) for the homogeneous problem, i.e. in the absence of $\vec{P}_{ext}$ as the driving term. This includes applying the appropriate electromagnetic boundary conditions at any interface and using Bloch's theorem, Eq. (5.22), associated with the superlattice periodicity. The Green functions are obtained from the linear response of the electric-field components to the components of $\vec{P}_{ext}$. The details are in Refs. [10,13], but this is basically a direct application of Eq. (A.31) in Appendix A: $\vec{P}_{ext}$ is the stimulus, and the response produced in $\vec{E}$ is calculated. Here we consider specific applications of the theory to two different superlattice structures, namely two-component superlattices with $2D$ electron-gas layers [14] and *nipi* superlattices [15].

11.1.1 Two-Component Superlattices with 2D Charge Sheets

We start by considering the two-component ... $ABABABAB$ ... superlattice with building blocks as shown schematically in Fig. 5.1, where the 2D electron-gas layers have an areal carrier concentration $n_A = n_B = n$. The solution of Eq. (11.15) within medium A at the mth unit cell yields

$$E_{xA}(z) = A_{1A}f_A + A_{2A}\bar{f}_A + V_{xA}\exp(iQz), \tag{11.16}$$

$$E_{zA}(z) = (k_x/Q)[-A_{1A}f_A + A_{2A}\bar{f}_A] + V_{zA}\exp(iQz), \tag{11.17}$$

$$H_{yA}(z) = (\omega\epsilon_0\epsilon_A/Q)[A_{1A}f_A - A_{2A}\bar{f}_A] - (\omega\epsilon_0\epsilon_A V_{zA}/k_x)\exp(iQz). \tag{11.18}$$

Here $f_A = \exp(-\alpha_A a)$, $\bar{f}_A = 1/f_A$, with a being the thickness of layer A, and α_A is defined by Eq. (5.8) or (5.9). Also

$$\vec{V}_A = \left[(\epsilon_A\omega^2/c^2)\vec{P}_0 - \vec{k}(\vec{k}\cdot\vec{P}_0)\right]/\left[(Q^2 + \alpha_A^2)\epsilon_0\epsilon_A\right], \tag{11.19}$$

where $\vec{k} = (k_x, 0, Q)$. Similar expressions can be written for electromagnetic fields in medium B.

The standard electromagnetic boundary conditions, i.e. that the x-component of the electric field is continuous across an interface while the y-component of the magnetic field is discontinuous in the presence of a current density at an interface (see e.g. Section 4.3), can now be applied at $z = mL + a$ and $z = (m+1)L$, where $L = a + b$ is the size of the superlattice unit cell. The periodicity property is taken into account through Bloch's theorem. This leads to a set of four linear inhomogeneous equations in A_{jA} and A_{jB} ($j = 1, 2$), which can be expressed in matrix notation as

$$\bar{M}\bar{A} = \bar{a}, \tag{11.20}$$

where $\bar{M}$ is a 4×4 matrix:

$$\bar{M} = \begin{bmatrix} f_A & \bar{f}_A & -1 & -1 \\ -i\bar{f}_A\xi_A & i\bar{f}_A\xi_A & (i\xi_B - \sigma) & -(i\xi_B + \sigma) \\ g & g & -f_B & -\bar{f}_B \\ -g(i\xi_A - \sigma) & g(i\xi_A + \sigma) & if_B\xi_b & -i\bar{f}_B\xi_B \end{bmatrix}. \tag{11.21}$$

The column matrices are

$$\bar{A} = \begin{bmatrix} A_{1A} \\ A_{2A} \\ A_{1B} \\ A_{2B} \end{bmatrix}, \quad \bar{a} = \begin{bmatrix} h_A(V_{xB} - V_{xA}) \\ h_A(c/\omega)^2[(V_{xB} - V_{xA})Q - (V_{zB} - V_{zA})k_x] + h_A\sigma V_{xB} \\ h_A h_B(V_{xB} - V_{xA}) \\ g(c/\omega)^2[(V_{xB} - V_{xA})Q - (V_{zB} - V_{zA})k_x] - g\sigma V_{xA} \end{bmatrix}, \tag{11.22}$$

where $g = \exp(iQL)$ is the Bloch factor, and

$$h_j = \exp(iQj), \quad \xi_j = \epsilon_j/\alpha_j, \quad j = a \text{ or } b. \tag{11.23}$$

11.1. RAMAN SCATTERING IN PERIODIC STRUCTURES

The non-zero Green function $G_{\mu\nu}$ can now be found, making use of Eq. (11.6) together with Eqs. (11.20)–(11.22), through the linear-response approach [10]. Here we quote only the results for the particular case where $\xi_A = \xi_B = \epsilon/\alpha$ [14]:

$$G_{xx}(k_x, Q) = \frac{-\alpha^2}{\epsilon(Q^2 + \alpha^2)} \left(1 - \frac{2\Omega^2 \alpha^2 \Upsilon}{\omega^2(Q^2 + \alpha^2)\Lambda}\right), \tag{11.24}$$

$$G_{xz}(k_x, Q) = G_{zx}(k_x, Q) = (k_x Q/\alpha^2) G_{xx}(k_x, Q), \tag{11.25}$$

$$G_{zz}(k_x, Q) = \frac{-\alpha^2}{\epsilon(Q^2 + \alpha^2)} \left(\frac{\epsilon\omega^2}{c^2} - Q^2 + \frac{2\Omega^2 k_x^2 Q^2 \Upsilon}{\omega^2(Q^2 + \alpha^2)\Lambda}\right), \tag{11.26}$$

$$G_{yy}(k_x, Q) = \frac{\omega^2}{c^2(Q^2 + \alpha^2)} \left(1 + \frac{2\Omega^2 \epsilon \Upsilon'}{c^2(Q^2 + \alpha^2)\Lambda'}\right), \tag{11.27}$$

where $\Omega = (ne^2/\epsilon_0 \epsilon m L)^{1/2}$ is a characteristic frequency similar (but not identical) to that introduced in Section 5.2, and

$$\Upsilon = \cosh(\alpha L) - \cos(QL) + s_1(\alpha L \Omega^2/\omega^2)\Big[\sinh(\alpha b)\cos(Qa)$$

$$+ \sinh(\alpha a)\cos(Qb) - \sinh(\alpha L)\Big], \tag{11.28}$$

$$\Lambda = \cos(QL) - \cosh(\alpha L) + (\alpha L \Omega^2/\omega^2)\Big[s_3 \sinh(\alpha L)$$

$$- s_2(\alpha L \Omega^2/\omega^2)\sinh(\alpha a)\sinh(\alpha b)\Big]. \tag{11.29}$$

Here we have factors $s_1 = s_2 = \frac{1}{2}$ and $s_3 = 1$. The quantities Υ' and Λ' are also defined by Eqs. (11.28) and (11.29), respectively, provided we put $s_1 = \frac{1}{2}$, $s_2 = 0$, and $s_3 = \frac{1}{2}\epsilon(\omega/kc)^2$. We note that the Green functions have poles at either $\Lambda' = 0$ (for G_{yy}) or $\Lambda = 0$ (for the other Green functions). These conditions are just the dispersion relations for s- and p-polarized modes, respectively, in the superlattice; they are consistent with the results found in Section 5.2.

It is easy to substitute the above Green-function results into Eq. (11.7) in order to predict the spectra for Raman scattering from plasmon-polaritons in binary superlattices with a 2D electron gas at the interfaces. Some results are shown in Fig. 11.2, where we have plotted the scattering cross section (in arbitrary units) versus the frequency shift (in meV). The lower-frequency peak at 3.5 meV is just the analog of the peak observed experimentally [5] in non-alternating samples (i.e. where $a = b$). The higher-frequency peak at 5.1 meV is due to an additional branch in the plasmon-polariton spectrum occurring when $a \neq b$. The parameters used here are $\theta = 20°$, $\phi = 70°$, $\lambda \simeq 780$ nm and $T \simeq 10$ K. The complex refractive index was taken to be $\eta = 3.6 + 0.07i$, which corresponds to the value for GaAs at this wavelength. Also the thicknesses of the layers are $a = 40$ nm and $b = 20$ nm. The integrated Raman intensities for the two plasmon-polariton branches can be obtained using Eq. (11.13). In Fig. 11.3 the intensities are plotted against the

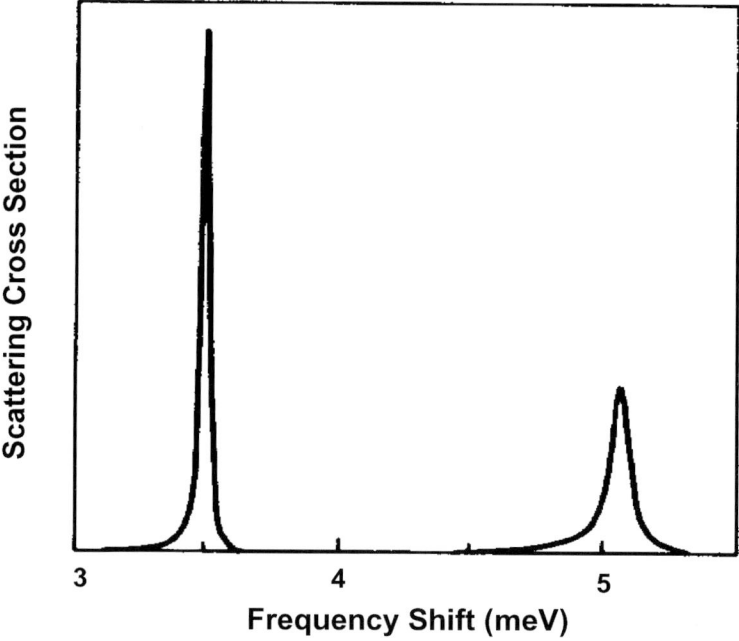

Fig. 11.2. Theoretical Raman spectra versus the frequency shift (in units of meV) for scattering from a two-component superlattice with a 2D electron-gas layers at the interface. See the text for parameter values (after Constantinou and Cottam [14]).

dimensionless factor $\delta = b/L$ for the transparent-medium limit (taking Im $\eta \to 0$) in which the scattering peaks become delta functions. We take the other physical parameters as in Fig. 11.2, and we also simplify by taking $\alpha \simeq k_x$, appropriate to the low-frequency region where retardation effects are small. It can be proved [14] that, for small δ, the ratio of the intensities associated with the lower (L) and the upper (U) branches is approximately

$$\frac{I_L}{I_U} = \left[\frac{n(\omega_L)+1}{n(\omega_U)+1}\right] \left[\frac{k_x L \sinh(k_x L)}{2[\cosh(k_x L) - \cos(QL)]}\right]^{1/2} \left(\frac{Q}{k_x}\right)^2 \delta^{3/2}. \qquad (11.30)$$

Thus a $\delta^{3/2}$ dependence is predicted for the intensity ratio, except for δ very small (less than about 0.05 in the case of Fig. 11.3), when the effect of the thermal factors leads to an overall proportionality to δ.

We turn our attention now to a *semi-infinite* superlattice, as shown in Fig. 11.4, where we have 2D charge layers at $z = 0$, $-a$, $-2a$, etc., separated by dielectric layers of thickness a and dielectric constant ϵ_B. At each interface we denote the areal carrier concentration by n, except at $z = 0$, where it may have a different value n_s. We consider also that the region $0 \leq z \leq d$ is filled by the same dielectric medium that fills the space between the charge layers, forming a capping layer of thickness d at the superlattice surface. Outside the superlattice,

11.1. RAMAN SCATTERING IN PERIODIC STRUCTURES

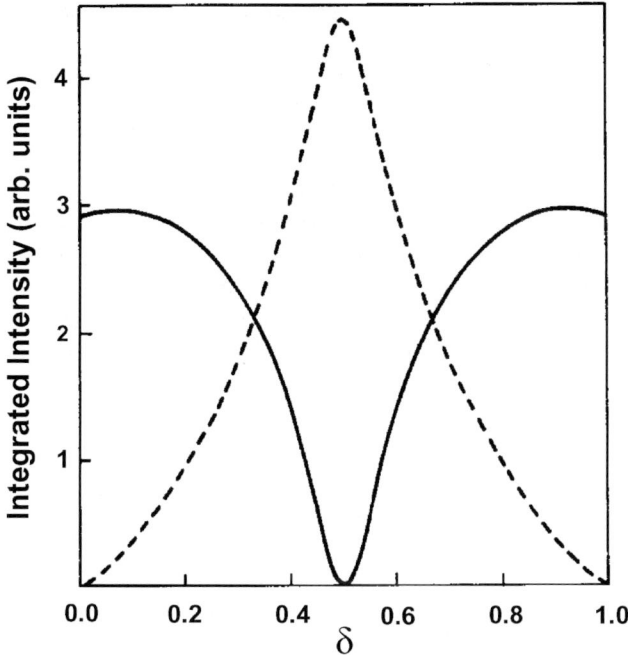

Fig. 11.3. Integrated Raman intensities as a function of $\delta = b/L$. The upper- and lower-frequency modes are represented by solid and broken lines, respectively. See the text for parameter values (after Constantinou and Cottam [14]).

there is a medium C with dielectric constant equal to ϵ_C. This model is suitable for a semiconductor superlattice like GaAs/Al$_x$Ga$_{1-x}$As, with appropriate layer thicknesses and doping levels. It allows us to deal with how the incident light enters and the scattered light leaves the sample.

Following the linear-response technique, as explained previously, we can generalize the preceding results that were obtained for an infinite superlattice (taking now $a = b$ for the semi-infinite structure). For simplicity, we consider just the low-frequency regime where retardation effects are unimportant. For the unretarded plasmon-polariton p-polarization mode, the Green functions are found to be [16]

$$G_{zz}(z, z') = (\pi k_x/\epsilon_0 \epsilon_B)\mathrm{sgn}(z - z') f_z^0$$

$$- (\pi k_x/2\epsilon_0 \epsilon_B \Delta)\Big(f_\theta \cosh\{k_x[z - (m-1)a]\} - \cosh[k_x(z - ma)]\Big)$$

$$\times \Big(\Xi \mathrm{sgn}(z') f_0^0 + f_{-\theta} \zeta k_x a [f_{(m-1)\theta} \mathrm{sgn}(a - z') f_a^0 - \mathrm{sgn}(ma - z') f_{ma}^0] \Big),$$

(11.31)

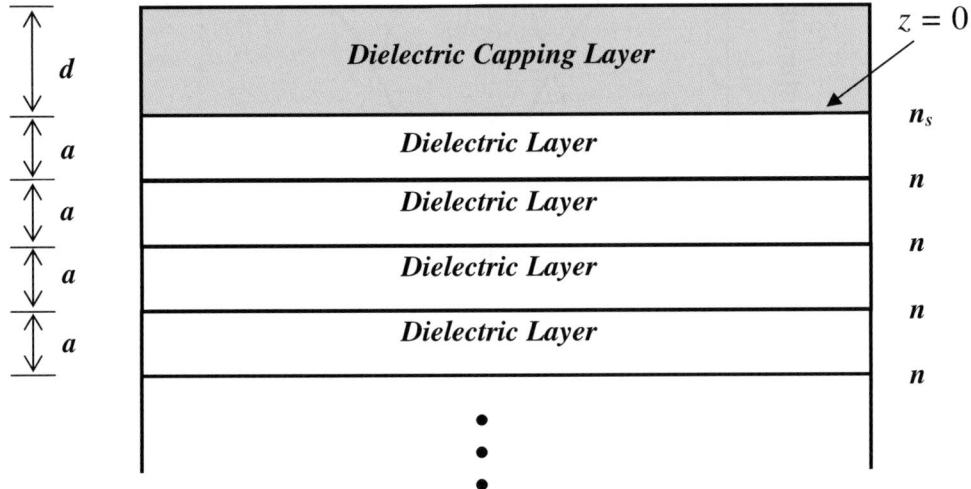

Fig. 11.4. Assumed geometry for a semi-infinite superlattice, with a dielectric capping layer of thickness d at the superlattice surface, and charge sheets at the interfaces.

with analogous expressions for G_{xx}, G_{xz}, and G_{zx}. Here m is zero or a positive integer, $\text{sgn}(z - z')$ is the sign function, and $\zeta = (\Omega/\omega)^2$, where Ω is the characteristic frequency defined after Eq. (11.27) with L replaced by a. The quantity $f_\theta = \exp(i\theta a)$ is a *transfer function*, where real values of θ correspond to the bulk modes of the superlattice, while complex values of θ (with Im $\theta > 0$) are related to surface modes. Also, $f_{n\theta} = (f_\theta)^n$ and $f_z^0 = \exp(-k_x|z - z'|)$, while

$$\Delta = \cosh(k_x a) - (\zeta k_x a/2)\sinh(k_x a) - (\mu^2 + 1)/2\mu, \qquad (11.32)$$

$$\mu = \cosh(k_x a) + [(n_s \sigma/n) + F]\sinh(k_x a), \qquad (11.33)$$

$$F = \Big[\epsilon_B \tanh(k_x d) + \epsilon_C\Big] \Big/ \Big[\epsilon_B + \epsilon_C \tanh(k_x d)\Big], \qquad (11.34)$$

$$\Xi = -f_{(m-1)\theta}\Big(\mu - f_{-\theta}\Big)\Big[\zeta k_x d + (1 + F)(\epsilon_B - \epsilon_C)(\epsilon_B + \epsilon_C)^{-1}\Big]. \qquad (11.35)$$

We recall that σ is related to the areal charge density defined by Eq. (4.23).

11.1. RAMAN SCATTERING IN PERIODIC STRUCTURES

These Green functions, as exemplified by Eq. (11.31), have a first term that depends on $|z - z'|$. This is characteristic of the spatial dependence of an infinite bulk sample, while the remaining terms with a spatial dependence on z and z' separately are due to the surface and to the charge layers. These latter terms have poles corresponding to $\Delta = 0$, which is just the condition giving the dispersion relation for the surface plasmon-polaritons (already discussed in Section 5.2).

The Green functions can also be used to determine the spectral intensities of the surface and bulk superlattice plasmon-polaritons in the superlattice. For example, the so-called power spectrum, $\langle |E_z(z)|^2 \rangle$, can be obtained from Eq. (11.31) through the fluctuation–dissipation theorem (see Section A.3 of Appendix A). The surface plasmon-polariton contribution consists of a delta-function spike, whose strength measures its integrated intensity. On the other hand, the bulk contribution consists of a continuous distribution over the range of bulk modes.

For a numerical example, we consider parameters appropriate to GaAs/Al$_x$Ga$_{1-x}$As, i.e. $a = 90$ nm, $n = 7.3 \times 10^{15}$ m^{-2}, $m = 6.37 \times 10^{-32}$ kg, and $\epsilon_B = 13.1$. These parameters give $\hbar\Omega = 11.1$ meV for the characteristic frequency Ω. Also we take medium C outside the superlattice to be vacuum so $\epsilon_C = 1$. We present results related only to the surface modes, since the bulk contributions can be estimated to be essentially independent of the ratio n_s/n and, for the cases treated here, are only weakly dependent on the capping layer thickness d.

Fig. 11.5 shows the integrated intensity of the surface plasmon-polariton mode versus n_s/n for a fixed in-plane wavevector corresponding to $k_x a = 1$. We have quoted only the contribution of the Green function $G_{zz}(z, z')$, since it is this function that has the dominant weight in the light-scattering cross section [11]. For the case without a capping layer (full line), the curve has a maximum for values of n_s/n between 0 and 0.5. In this region, the surface plasmon-polariton mode lies below the bulk continuum and emerges from it with a critical wavevector k_x^c [which is the solution of $|\mu| = 1$, with μ given by Eq. (11.33)] such that $k_x^c a < 1$. For $0.5 \leq n_s/n \leq 0.65$, the integrated intensity is zero, since $k_x^c a > 1$. Finally, for $n_s/n > 0.65$ the integrated intensity is non-zero with the surface plasmon-polariton lying above the bulk continuum and $k_x^c a < 1$. The behavior for the system with a capping layer (broken line) is similar. Fig. 11.6 illustrates the variation of the integrated intensities versus $k_x a$ for a fixed value of $n_s/n = 0.35$ (close to the maximum intensities in Fig. 11.5). The behavior is similar for cases where there is either a capping layer (broken line) or not (full curve).

The surface plasmon-polariton contribution for doped semiconductor superlattices now made up of 2D electron- and hole-gas layers in an alternating fashion separated by a dielectric medium of thickness a, can be calculated in a similar way. Allowing for effects due to a capping layer and charge depletion, we plot in Fig. 11.7 the Raman integrated intensity as a function of the dimensionless in-plane wavevector $k_x a$ [17]. Two different values are considered for the ratio r between the carrier concentration of the outermost n- and p-doped layers (denoted by n_s and p_s) and the carrier concentration of the bulk layers (denoted by n and p);

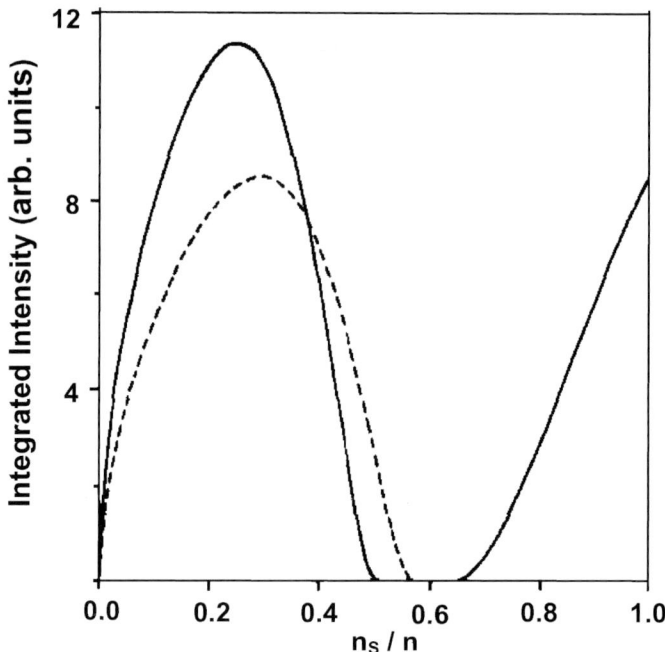

Fig. 11.5. Surface plasmon-polariton integrated intensity (from the imaginary part of the Green function G_{zz} with $z = z' = a/2$) plotted against the ratio n_s/n for the dimensionless in-plane wavevector $k_x a = 1.0$. The full curve represents the case of capping-layer thickness $d = 0$, while the dashed curve corresponds to $d/a = 0.44$. See the text for parameter values (after Constantinou and Cottam [16]).

also $z = z' = a/2$. The curves show, qualitatively speaking, essentially similar behavior, although they have different strengths, indicating their dependence on the choice of r.

11.1.2 *nipi* Superlattices

As another example of Raman scattering applied to plasmon-polaritons, we next consider the more complex *nipi* superlattice discussed in Section 5.3. Using the same procedure as in the previous subsection, the application of the electromagnetic boundary condition at any unit cell of the superlattice yields a matrix equation similar to Eq. (11.20), except that $\bar{M}$ now becomes an 8×8 matrix because the unit cell of the *nipi* superlattice has a unit cell with more layers. In a partitioned form, the matrix can be written as

$$\bar{M} = \begin{pmatrix} M_{11} & M_{12} \\ M_{21} & M_{22} \end{pmatrix} \qquad (11.36)$$

11.1. RAMAN SCATTERING IN PERIODIC STRUCTURES

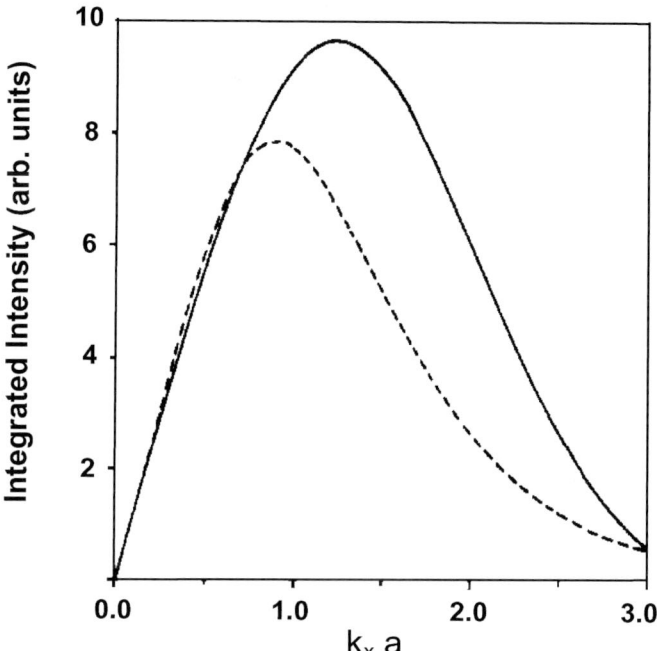

Fig. 11.6. Surface plasmon-polariton integrated intensity (from the imaginary part of the Green function G_{zz} with $z = z' = a/2$) plotted against the dimensionless in-plane wavevector $k_x a$ for the ratio $n_s/n = 0.35$. The full curve represents the case of capping-layer thickness $d = 0$, while the dashed curve corresponds to $d/a = 0.44$. See the text for parameter values (after Constantinou and Cottam [16]).

with the 4×4 blocks designated by

$$M_{11} = \begin{pmatrix} f_A & \bar{f}_A & -1 & -1 \\ -i\xi_A f_A & i\xi_A \bar{f}_A & i\xi_B - \sigma_h & -(i\xi_B + \sigma_h) \\ 0 & 0 & f_B & \bar{f}_B \\ 0 & 0 & -i\xi_B f_B & i\xi_B \bar{f}_B \end{pmatrix}, \quad (11.37)$$

$$M_{12} = \begin{pmatrix} 0 & 0 & 0 & 0 \\ 0 & 0 & 0 & 0 \\ -1 & -1 & 0 & 0 \\ i\xi_C - \sigma_e & -(i\xi_C + \sigma_e) & 0 & 0 \end{pmatrix}. \quad (11.38)$$

Also M_{21} is equal to gM_{12} [where g is the phase factor $\exp(iQL)$ defined earlier] provided we replace the subscripts C and e (for electron) by D and h (for hole),

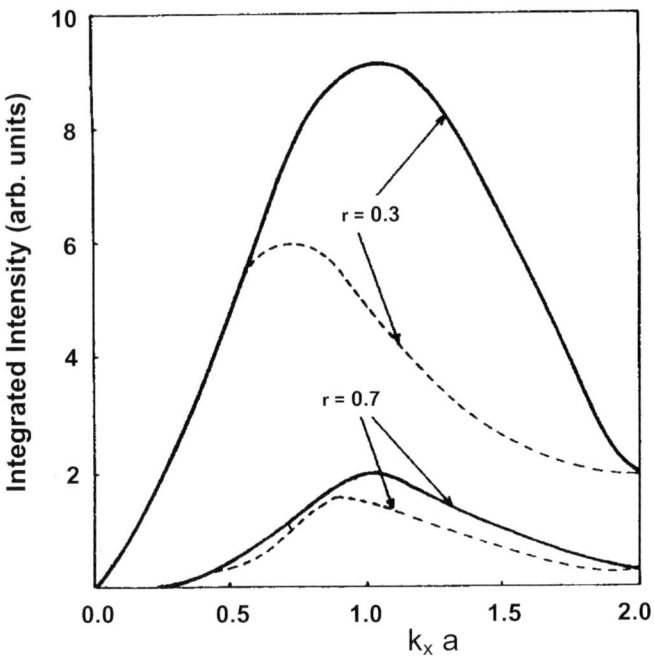

Fig. 11.7. Surface plasmon-polariton integrated intensity (from the imaginary part of the Green function G_{zz} with $z = z' = a/2$) plotted against the dimensionless in-plane wavevector $k_x a$ for the ratio $r = n_s/n = p_s/p$ equal to 0.3 and 0.7, as indicated. The full curves represent the case of capping-layer thickness $d = 0$, while the dashed curves correspond to $d = a$ (after Albuquerque [17]).

respectively. Likewise M_{22} is equal to M_{11} provided we replace the subscripts A, B, and h by C, D, and e, respectively. In this case $\bar{A}$ is an 8×1 column vector formed by the unknown coefficients A_{mJ} (with $m = 1, 2$, and $J = A, B, C, D$). Also the column vector $\bar{a}$ in Eq. (11.22) becomes

$$\bar{a} = \begin{pmatrix} h_A(V_{xB} - V_{xA}) \\ h_A[-(\varsigma_B - \sigma_h)V_{zB} + \varsigma_A V_{zA}] \\ h_A h_B (V_{xC} - V_{xB}) \\ h_A h_B [-(\varsigma_C - \sigma_e)V_{zC} + \varsigma_A V_{zB}] \\ h_A h_B h_C (V_{xD} - V_{xC}) \\ h_A h_B h_C [-(\varsigma_D - \sigma_e)V_{zD} + \varsigma_C V_{zC}] \\ g(V_{xA} - V_{xD}) \\ g[-(\varsigma_D - \sigma_h)V_{zA} + \varsigma_D V_{zD}] \end{pmatrix}, \qquad (11.39)$$

where h_J and V_{iJ} are defined in the previous subsection, and $\varsigma_J = \epsilon_J / k_x$.

11.2. RAMAN SCATTERING IN QUASIPERIODIC STRUCTURES

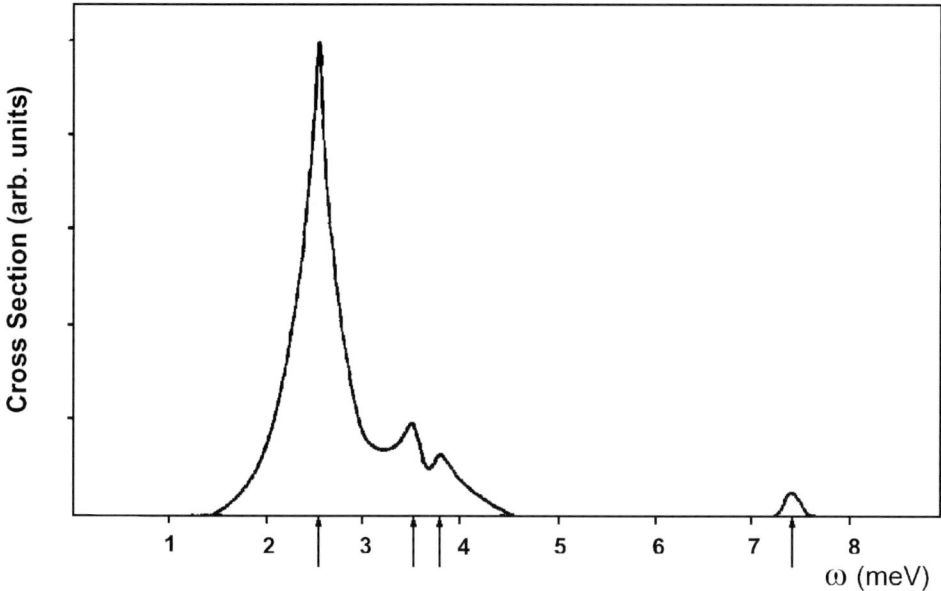

Fig. 11.8. Theoretical Raman spectra for scattering from bulk plasmon-polaritons in a *nipi* superlattice. The arrows below the horizontal axis indicate the peak positions. See the text for parameter values (after Albuquerque [15]).

The condition for the 8×8 matrix in Eq. (11.36) to have zero determinant is equivalent to the dispersion relation discussed in Section 5.3 for bulk plasmon-polariton in a periodic *nipi* superlattice. The relevant Green function can be derived following the same linear-response steps as before, and the light-scattering cross section can be deduced using Eq. (11.7). A numerical example is given in Fig. 11.8, where we plot the cross section versus the frequency shift ω (in meV) for a scattering geometry in Fig. 11.1, with angles $\theta = 20°$ and $\phi = 70°$ and wavelength $\lambda = 780$ nm. The other physical parameters are taken to be the same as employed in Section 5.3 for the *nipi* plasmon-polariton dispersion relation calculated there. The lowest-frequency peak at 2.6 meV is the analog of the peak that would be observed for a non-alternating sample. The more complex structure of the *nipi* superlattice gives rise to four branches of bulk polaritons, and the Raman spectrum correspondingly has additional higher-frequency peaks, as indicated by the arrows in Fig. 11.8.

11.2 Raman Scattering in Quasiperiodic Structures

Raman-scattering measurements of the spectra for acoustic phonons in GaAs/AlAs and Si/Ge$_x$Si$_{1-x}$ heterostructures, built up in a Fibonacci sequence, were reported by a number of authors (see e.g. Refs. [18–22]). They demonstrate that the spectra are sensitive to the quasiperiodic structure of the superlattice, as well as the phonon properties. The observed acoustic frequencies are shifted, depending on the size or stage number of the lattice. By contrast, periodic superlattices will typically exhibit peak broadening in similar-sized structures. Unlike the case of acoustic phonons, plasmons are strongly dependent on the geometry and generation of the Fibonacci sequence defining the superlattice. Thus they are good candidates for probing by Raman spectroscopy, and discrete plasmons in a finite-layered structure have indeed been observed in Raman experiments [23,24].

In this section we discuss the frequencies, intensities, and spectral lineshapes for Raman scattering by plasmon-polaritons in a quasiperiodic superlattice of the Fibonacci type. Our goal is to compare the theoretical results with the experimental Raman spectra found by Merlin et al. [25]. Again we use a method based on linear-response theory; it has been proved to be very effective for describing how plasmon spectral lineshapes and integrated intensities depend upon the polarization directions and scattering geometry [26]. We note that a different approach has been used by Das Sarma et al. [27,28] to compare the plasmon spectra in semiconductor superlattices with periodic, quasiperiodic (Fibonacci type) and random-layer structures.

The composite building blocks $\alpha = AB$ and $\beta = AC$ are first introduced. Here A is considered to be GaAs with thickness $a = 27$ nm, while B and C are $Al_xGa_{1-x}As$ layers with thicknesses $b = 80$ nm and $c = 43$ nm, respectively. Also $x = 0.3$ and the remaining physical parameters are given in Ref. [25]. The thickness of the α building block is $d_\alpha = a+b$, while the thickness of the β building block is $d_\beta = a+c$. The ratio d_α/d_β was chosen to approximate $(1+\sqrt{5})/2 \equiv \tau$, the gold mean number, which is the incommensurate ratio of the numbers of α and β building blocks in a Fibonacci sequence (see Subsection 2.3.2).

We consider the Fibonacci superlattice to be effectively infinite, and choose the Cartesian axes and light-scattering geometry as in Fig. 11.1. We consider, as in Section 11.1, that the incident light is p-polarized, with polarization vector $\hat{e}_I = (\cos\theta, 0, \sin\theta)$, while the scattered light is s-polarized, with polarization vector $\hat{e}_S = (0, 1, 0)$. Then the Raman-scattering cross section is described by Eqs. (11.7) and (11.8), where the terms $G_{\mu\nu}$ are wavevector-dependent, electric-field Green functions. Here we relate them to the corresponding position-dependent Green functions by

$$\int \langle\langle E_\mu(z); E_\nu^*(z')\rangle\rangle_{k_x,\omega} \exp(ik_z z) \exp(-ik_z' z') \, dz \, dz'$$

$$= [(2\pi)^3/\epsilon_0 V]\delta(k_z - k_z')G_{\mu\nu}(k_x, k_z). \tag{11.40}$$

The position-dependent form of the electric-field Green functions can be obtained first from the linear response of the electric field to an external applied polarization chosen, e.g., as in Eq. (11.14). The linear-response expression is [26]

$$\langle\langle E_\mu(z); E_\nu^*(z')\rangle\rangle_{k_x,\omega} = E_\mu(z)/P_\nu(z'), \tag{11.41}$$

which is an example of Eq. (A.31) in Appendix A. From Maxwell's equations the electric field inside each layer of the superlattice, which can be medium A (GaAs)

11.2. RAMAN SCATTERING IN QUASIPERIODIC STRUCTURES

with thickness a, or media B or C ($\text{Al}_x\text{Ga}_{1-x}\text{As}$) with thicknesses b or c, respectively, satisfies a differential equation in the presence of the external polarization. Thus we may again employ Eq. (11.15), whose solutions are given by Eqs. (11.16)–(11.18).

It then becomes a straightforward application of the method of induction to calculate the electric-field components, and hence the Green functions using Eq. (11.41), for the nth generation number of the Fibonacci superlattice. We first consider the infinite superlattice formed by the first two stages $S_0 = \beta$ (giving $\ldots /AC/AC/\ldots$) and $S_1 = \alpha$ (giving $\ldots /AB/AB/\ldots$). The standard electromagnetic boundary conditions are applied at the interfaces of the nth unit cell, i.e. at $z = nL + a$ and $z = (n+1)L$. Here the unit-cell size L is $d_\alpha = a + b$ (or $d_\beta = a + c$ as appropriate).

The result in the case of $S_1 = \alpha$, in matrix notation, is [26]

$$\bar{M}_1 \bar{A}_1 = \bar{a}_1, \tag{11.42}$$

where $\bar{M}_1$ is a 4×4 matrix that can be conveniently written in partitioned form as

$$\bar{M}_1 = \begin{pmatrix} M_{11} & M_{12} \\ gM_{12}(B \to A) & M_{11}(A \to B) \end{pmatrix}. \tag{11.43}$$

Here the 2×2 blocks for M_{11} and M_{12} are

$$\bar{M}_{11} = \begin{pmatrix} f_A & \bar{f}_A \\ -if_A\xi_A & i\bar{f}_A\xi_A \end{pmatrix}, \tag{11.44}$$

$$\bar{M}_{12} = \begin{pmatrix} -1 & -1 \\ i\xi_B & -i\xi_B \end{pmatrix}, \tag{11.45}$$

and the other two blocks in Eq. (11.43) are obtained as indicated, where $B \to A$ (or $A \to B$) means that we replace subscripts B by A (or vice versa). Similarly, the column matrices appearing in Eq. (11.42) are

$$\bar{A}_1 = \begin{pmatrix} A_{1A} \\ A_{2A} \\ --- \\ A_{1B} \\ A_{2B} \end{pmatrix} \equiv \begin{pmatrix} A \\ B \end{pmatrix}, \tag{11.46}$$

$$\bar{a}_1 = \begin{pmatrix} h_A\gamma_{BA} \\ h_A Y_{BA} \\ --- \\ g\gamma_{AB} \\ gY_{AB} \end{pmatrix} \equiv \begin{pmatrix} h_A\Gamma_{BA} \\ g\Gamma_{AB} \end{pmatrix}, \tag{11.47}$$

where the last expressions are these matrices written in a partitioned form. Γ_{AB} is the 2×1 column vector whose elements are γ_{AB} and Y_{AB}, respectively, and

Γ_{BA} can be obtained from Γ_{AB} by a replacement scheme as described above. In these equations $g = \exp(ik_z L)$ and other notations are (for $I, J = A$ or B) $f_J = \exp(-\alpha_J z)$, $h_J = \exp(ik_z J)$, and

$$\gamma_{IJ} = T_{xI} - T_{xJ}, \tag{11.48}$$

$$Y_{IJ} = (c/\omega)^2 \left[\gamma_{IJ} k_z - (T_{zI} - T_{zJ}) k_x \right]. \tag{11.49}$$

The case of $S_0 = \beta$ is identical to the above, provided we replace subscript B by subscript C.

For the next stage $S_2 = \alpha\beta$, giving the superlattice $\ldots/ABAC/ABAC/\ldots$, Eq. (11.42) becomes $\bar{M}_2 \bar{A}_2 = \bar{a}_2$ with

$$\bar{M}_2 = \begin{pmatrix} M_{11} & M_{12} & 0 & 0 \\ 0 & M_{11}(A \to B) & M_{12}(B \to A) & 0 \\ 0 & 0 & M_{11} & M_{12}(B \to C) \\ gM_{12}(B \to A) & 0 & 0 & M_{11}(A \to C) \end{pmatrix}, \tag{11.50}$$

$$\bar{A}_2 = \begin{pmatrix} A \\ B \\ A \\ C \end{pmatrix}, \quad \bar{a}_2 = \begin{pmatrix} h_A \Gamma_{BA} \\ h_A h_B \Gamma_{AB} \\ h_A h_B h_A \Gamma_{CA} \\ g \Gamma_{AC} \end{pmatrix}. \tag{11.51}$$

For the nth stage we have $\bar{M}_n \bar{A}_n = \bar{a}_n$, where it follows by induction that the matrices, in a partitioned form with 2×2 blocks, are

$$\bar{M}_n = \begin{pmatrix} M_{11} & M_{12} & 0 & \cdots & 0 \\ 0 & M_{11}(A \to B) & M_{12}(B \to A) & \cdots & 0 \\ \cdot & \cdot & \cdot & \cdots & \cdot \\ \cdot & \cdot & \cdot & \cdots & \cdot \\ gM_{12}(J \to A) & 0 & 0 & \cdots & M_{11}(A \to J) \end{pmatrix}, \tag{11.52}$$

$$\bar{A}_n = \begin{pmatrix} A \\ B \\ \cdot \\ \cdot \\ J' \end{pmatrix}, \quad \bar{a}_n = \begin{pmatrix} h_A \Gamma_{BA} \\ h_A h_B \Gamma_{AB} \\ \cdot \\ \cdot \\ g \Gamma_{AJ'} \end{pmatrix}. \tag{11.53}$$

For n even, we have $J = B$ (and $J' = C$), while $J = C$ (and $J' = B$) for the case of n odd. It may be shown that the condition for $\bar{M}_n$ to have a zero determinant is consistent with the dispersion relations of superlattice bulk plasmon-polaritons discussed in Section 6.1, provided the same choice of building blocks is made.

The relevant Green functions can now be derived from the linear-response theory given by Eq. (11.41), together with Eqs. (11.52) and (11.53). Although the calculations are straightforward, it is tedious to display the analytical expressions for these Green functions, even for relatively small n. Instead, we present some numerical computations and compare them with the Raman spectra measured by Merlin et al. [25]. From the Green functions for the nth stage of a Fibonacci superlattice, it is possible to evaluate numerically the Raman spectra using the scattering cross-section expression given by Eq. (11.7). In what follows, we consider the finite nth-stage Fibonacci superlattice to form an infinite periodic superlattice, where the repeat subunit is the finite multilayer structure given by the Fibonacci generation S_n. Although a fully realistic *sample-specific* model should consider the finite superlattice together with its substrate as well as a capping layer [17], the periodic model considered here should be a reasonable approximation. When the generation number n is small, deviations from the periodic model used here may be apparent.

As an example, we examine the case of $n = 6$, which has 8α and 5β building blocks. This generation number is chosen because, above it, in our periodic model, the calculated spectra are essentially independent of n. The physical parameters used are $k_x = 1.12$, 1.30, and 1.44 (in units of 10^5 cm^{-1}). We consider a scattering geometry with $\theta + \phi = \pi/2$, where the incident angle corresponds to $\sin\theta = k_x\lambda/4\pi$, with λ being equal to 800 nm in the experiment [25]. A useful simplification for the dielectric functions is to take them as constant with $\epsilon_A = \epsilon_B \equiv \epsilon$. The refractive index of the superlattice (which is complex due to the absorption of light) is considered to be $\eta = 3.6 + 0.07i = \epsilon^{1/2}$, which corresponds to the value of GaAs at this wavelength [25]. A damping factor $\gamma = 0.15$ meV can be included phenomenologically in the theory (as an imaginary part to the frequency).

Fig. 11.9 now shows the calculated Raman-scattering cross section as a function of the frequency (in meV) for the three values of k_x specified above. The Raman peaks occur at the following frequencies (all in units of meV): $\omega_1 = 2.53$, 3.05, and 3.32 (full lines); $\omega_2 = 4.25$, 4.91, and 5.22 (dashed lines); and $\omega_3 = 6.81$, 7.92, and 8.38 (chain-dotted lines). Apart from the lowest resonant frequency (ω_1), the other frequencies (ω_2 and ω_3) are close to the experimental data of Merlin et al. [25]. The overall spectra also show a very good agreement. The small discrepancies at the low-frequency end are probably due to the fact that this region is more sensitive to the *sample-specific* features. It is relevant to point out that the predicted resonant frequencies almost follow the *golden ratio* pattern, i.e. $\omega_3/\omega_2 \approx \omega_2/\omega_1 \approx \tau = (1+\sqrt{5})/2$. The small deviation is again probably due to inaccuracies in the low frequency ω_1.

The integrated light-scattering intensity for the Raman peak at any plasmon-polariton frequency ω_i can next be found from the Green functions using a formalism like that in Eqs. (11.7) and (11.8). We consider, as in the previous section, only the contribution of the z-component electric-field Green function, i.e. we calculate the integrated intensity corresponding to Im $\langle\langle E_z(z); E_z(z')^*\rangle\rangle_{k_x,\omega}$, for $z = z'$ and $\omega = \omega_i$ ($i = 1, 2, 3$). Thus in Fig. 11.10a we show the integrated intensity as a function of $\delta \equiv d_\alpha/L$, for $k_x = 1.12 \times 10^5$ cm^{-1}. We note that, in the two

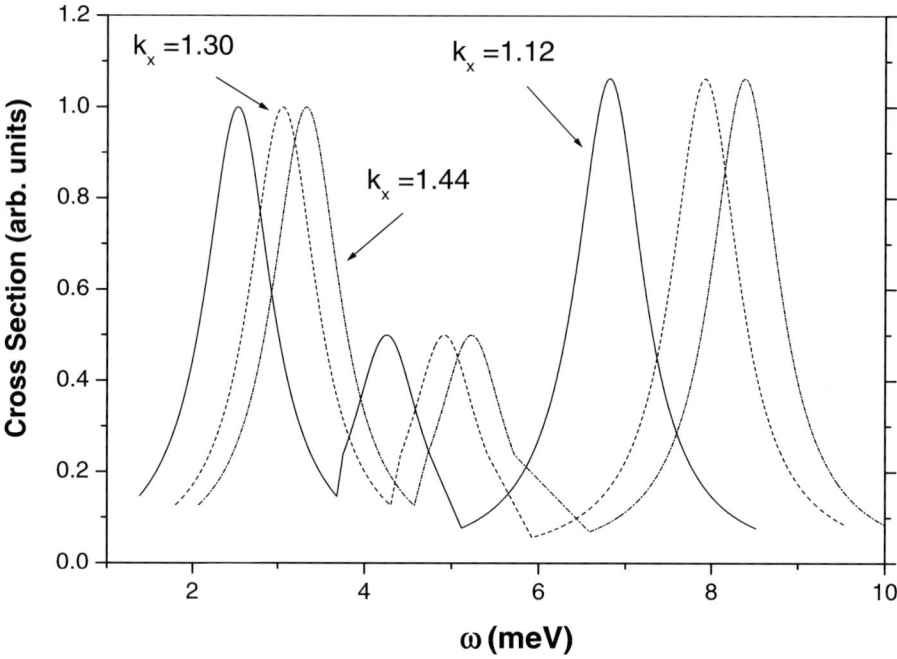

Fig. 11.9. Raman-scattering cross section calculated for three values of the in-plane wavevector k_x: 1.12×10^5 cm^{-1} (full line), 1.30×10^5 cm^{-1} (dashed line), and 1.44×10^5 cm^{-1} (chain-dotted line). The physical parameters are given in the text (after Albuquerque [26]).

limiting cases of $\delta = 0$ and 1, only the low-frequency mode ω_1 has an integrated intensity different from zero, since in these cases the superlattice is made up only of β (or α) building blocks. A similar behavior is also found in Figs. 11.10b and c for the other two values of k_x. The same qualitative behavior is found in all figures, which indicates that, provided a resonance frequency is obtained, the intensity is relatively independent of k_x. What matters are the relative values of the resonant frequencies in the dispersion curve for the excitations.

11.3 Brillouin Light Scattering

Brillouin light scattering (BLS) is the inelastic scattering of light from low-frequency magnetic and non-magnetic excitations. In the context of this book, it is mainly the ferromagnetic spin waves (and the magnetic polaritons formed from them) that are of interest. As in the case of Raman scattering, BLS provides a well-established, non-destructive optical technique. It can yield a great deal of information about magnetism in low-dimensional systems, such as thin films, multilayers, and patterned structures. Also it gives information about the dynamical properties of a spin system through the detection of long-wavelength spin waves. BLS can be implemented in situ to study thin magnetic films and nanostructures

11.3. BRILLOUIN LIGHT SCATTERING

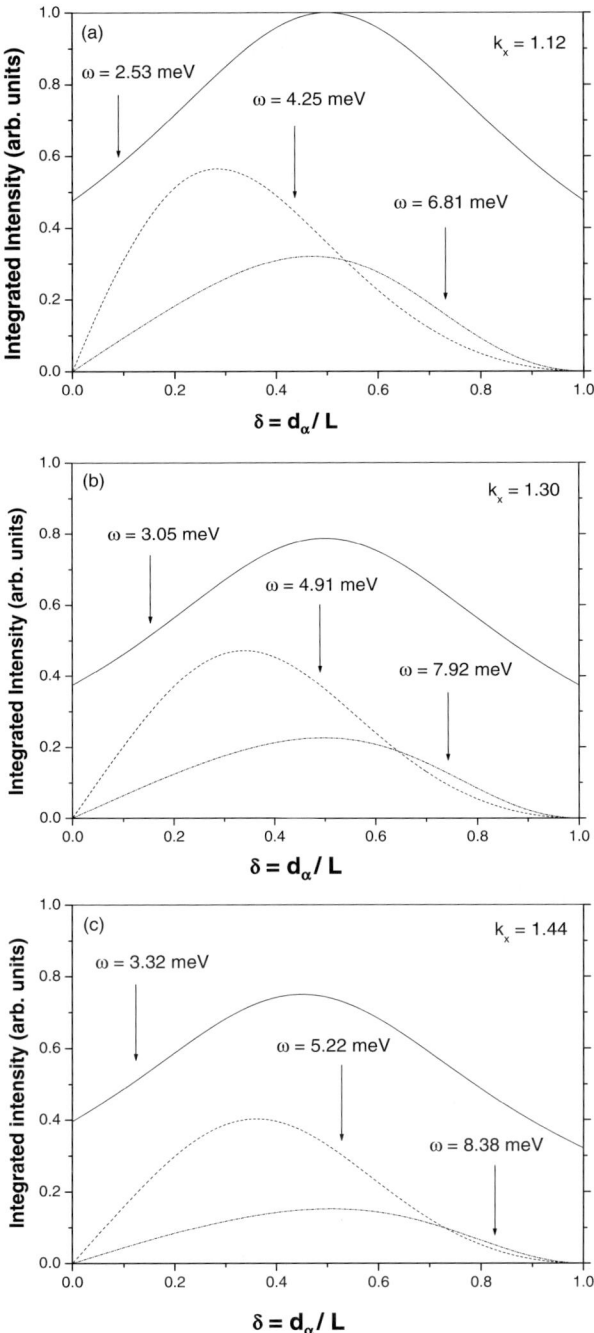

Fig. 11.10. Integrated intensities as a function of $\delta = d_\alpha/L$ for the dimensionless wavevector k_x (in units of 10^5 cm^{-1}) given by (a) 1.12, (b) 1.30, and (c) 1.44 (after Albuquerque [26]).

in ultrahigh-vacuum conditions [29]. Combined with the surface magneto-optical Kerr effect (SMOKE), which is discussed in Subsection 11.6.3, it can provide a detailed determination of the structural and magnetic parameters and of the anisotropy constants of ultrathin Fe films on Cu and Fe/Cu multilayers [30].

It can be a complementary experimental technique to ferromagnetic resonance or FMR (see Subsection 11.6.4) for magnetic materials. Like FMR, it can be used to measure magnetic anisotropies in ultrathin films, as well as the exchange coupling across spacer layers between films. It provides information on the saturation magnetization and on the magnetic ground state of the coupled layered system, i.e. the magnetization profile. Unlike FMR, however, in the BLS experiment the frequency is measured at a fixed magnetic field (in the FMR experiment the frequency is usually fixed and the magnetic field is varied until the frequency of the magnetic excitation matches the applied frequency). The resonant frequency can be determined in an FMR experiment with a precision of 0.01 GHz, although only one frequency can be measured (essentially the $\vec{k} \approx 0$ mode). By contrast, in BLS the $\vec{k}$ dependence of the modes can be studied, although only over a limited range of small $|\vec{k}|$ as in Raman scattering.

In a BLS experiment, frequencies can be measured over a range typically from 5 to 100 GHz. However, the precision is only of order 0.1 GHz (for a review see Ref. [31]). This is inferior to FMR, but is nevertheless typically several orders of magnitude better than the resolution in Raman light scattering. In BLS a multipass Fabry–Pérot interferometer (sometimes in a tandem arrangement) is used for the detection system replacing the grating spectrometer of the Raman case. BLS is often used to study linewidths, providing information about the damping, or spin relaxation. The origin of the damping of spin motions in ultrathin ferromagnetic films and multilayer structures, with focus on the linear-response regime probed by FMR or BLS, was recently reviewed [32].

Besides its magnetic applications, BLS is also an important tool to study acoustic phonons in periodic and quasiperiodic structures. As an example, high-resolution BLS can nowadays be a sensitive detector of acoustic phonons in complex structures like copolymer photonic crystals of styrene and isoprene, which possess a photonic stop band in the visible spectrum [33]. Based on the low, but finite, contrast in the mechanical properties between the styrene and isoprene components of the polymer, and taking into account the geometrical characteristics of the layered microstructure, acoustic band structures were determined showing acoustic-like and optical-like phonons [33]. Using BLS, the propagating and confined acoustic modes were also recently studied in crystalline polymer latex films [34], ferroelectric copolymer single-crystalline films [35], and nanosized carbon films [36].

Further, the vibrational modes in 3D ordered arrays of SiO_2 nanospheres were recently investigated by BLS [37]. Multiple, distinct Brillouin peaks were observed, and their frequencies were found to be inversely proportional to the diameter (approximately 200–340 nm) of the nanospheres, in agreement with Lamb's theory. This was the first Brillouin observation of acoustic-mode quantization in a nanoparticle arising from spatial confinement.

11.4 Resonant Brillouin Scattering

With a view to applications involving exciton-polaritons, we now provide an extension of the discussion for Brillouin scattering given in Section 11.3. RBS is perhaps the most powerful experimental technique to probe exciton-polaritons. It was a subject of considerable interest in the 1980s (for a review see Ref. [38]), being first proposed by Brenig et al. [39]. As we shall see, the cross section has denominator terms that may give a resonant enhancement if the incident light frequency ω_I is close to a phonon frequency in a spatially dispersive medium. The general theoretical predictions of this work were later confirmed experimentally by, among others, Ulbrich and Weisbuch [40], Winterling and Koteles [41], and Hermann and Yu [42]. On the theoretical side, a major step forward was made by Tilley [43] who used a Green-function method to calculate a backscattering Brillouin cross section for a normal-incidence geometry. Slightly different formulations were subsequently developed by So et al. [44], Fukui and Tada [45], and Albuquerque and Gonçalves da Silva [46].

As we emphasized previously (particularly in Section 10.1), one of the most important questions concerning exciton-polaritons is the choice of appropriate additional boundary conditions (ABCs) at interfaces of a medium with spatial dispersion (SD). As widely discussed (see Ref. [38]), reflectivity experiments are not sufficiently sensitive to allow for a choice among the many possibilities. This was disputed, however, by Halevi and Cocoletzi [47], who claim to have established the validity of Pekar's ABC for CdS. An alternative experimental approach employing picosecond induced absorption, e.g. as used by Segawa et al. [48] to study the ABCs in CuCl, was unable to clarify this matter. RBS provides, in principle, a far more sensitive test of the ABCs than reflectivity measurements. The challenges in RBS are with regard to the theory, requiring the evaluation of the complete RBS cross sections for realistic situations. Without this, a detailed interpretation of the experiments is not helpful regarding the choice of ABCs.

In this section we analyze the influence of the ABCs on the resonant RBS cross section. We do so in some detail because it emerges that the cross section is indeed sensitive to the choice.

11.4.1 Reflection and Transmission Spectra

It is convenient to start with simple reflection and transmission at an interface. We consider an incident s-polarized electromagnetic wave, with frequency ω_I and wavevector k_I, from the vacuum (taken in $z > 0$) into the SD medium ($z < 0$). For the vacuum the electric field is

$$\varepsilon_I^s = \exp(-ik_I^z) + R_s \exp(ik_I^z z). \tag{11.54}$$

Here, $k_I^z = k_I \cos\theta$, $k_I = \omega_I/c$, and R_s is the reflection coefficient. For the SD medium we have

$$E_I^s = T_1 \exp(-ik_{I1}^z z) + T_2 \exp(-ik_{I2}^z), \tag{11.55}$$

$$P = \epsilon_0\chi_{I1}T_1\exp(-ik_{I1}^z z) + \epsilon_0\chi_{I2}T_2\exp(-ik_{I2}^z z), \tag{11.56}$$

for the electric field and the polarization, with the electrical susceptibility χ_{Ij} ($j=1,2$) defined by Eq. (10.16). Also,

$$k_{I\lambda}^2 = \epsilon(k_{I\lambda},\omega)\omega_I^2/c^2, \tag{11.57}$$

$$k_I^x = k_{I\lambda}^x = \omega_I \sin\theta/c, \tag{11.58}$$

with $k_{I\lambda}^z = (k_{I\lambda}^2 - \omega_I^2\sin^2\theta/c^2)^{1/2}$ and λ can be 1 or 2.

Now, using the usual electromagnetic boundary conditions plus, as an example, ABC3 (namely, $\vec{P} + d\vec{P}/dz = 0$), we can find the reflection and transmission coefficients as

$$R_s = \frac{k_I^z(1-\xi) - (k_{I1}^z - \xi k_{I2}^z)}{k_I^z(1-\xi) + (k_{I1}^z - \xi k_{I2}^z)}, \tag{11.59}$$

$$T_1 = -T_2/\xi = \frac{2k_I^z}{k_I^z(1-\xi) + (k_{I1}^z - \xi k_{I2}^z)}, \tag{11.60}$$

where $\xi = \chi_{I1}(1-ik_{I1}^z)/\chi_{I2}(1-ik_{I2}^z)$. Figure 11.11 shows these coefficients, calculated using physical parameters appropriate to CdS and for several ABCs, as a function of the reduced frequency ω_I/ω_0, with ω_0 being the resonant frequency of the uncoupled exciton. Note the very close correspondence between the predictions of the different ABCs in this case.

On the other hand, for an incident p electromagnetic wave we have three additional light waves, two transverse and one longitudinal, with wavevectors $\vec{k}_{I3}$, $\vec{k}_{I4}$, and $\vec{k}_{I5}$, respectively. For $z > 0$ the expression for the incident electric field ε_I^p is the same as Eq. (11.54) but with a coefficient R_p instead of R_s. For $z < 0$ we have

$$E_I^p = T_3\exp(-i\vec{k}_{I3}\cdot\vec{r}) + T_4\exp(-i\vec{k}_{I4}\cdot\vec{r}) + T_5\exp(-i\vec{k}_{I5}\cdot\vec{r}). \tag{11.61}$$

As in the s-wave case the reflection and transmission coefficients can be obtained using the electromagnetic boundary conditions plus the ABCs. The results, taking ABC1, are

$$R_p = (1-r)/(1+r), \tag{11.62}$$

$$T_3 = T_4/A_1 = -T_5 f_I^x/A_2 f_I^z$$

$$= \frac{2k_I f_I^z}{(k_{I3}f_I^z + k_I f_{I3}^z) + (k_{I4}f_I^z + k_I f_{I4}^z)A_1 + k_I^2 f_I^z A_2/k_{I5}}, \tag{11.63}$$

where the ratios $f_{I\nu}^i = k_{I\nu}^i/k_{I\nu}$ and $f_I^i = k_I^i/k_I$ are introduced ($\nu = 3,4,5$ and $i = x,z$), and

$$r = [k_I(f_{I3}^z + f_{I4}^z A_1 + f_I^z A_2 k_I/k_{I5})]/[f_I^z(k_{I3} + k_{I4}A_1)], \tag{11.64}$$

$$A_1 = [\chi_{I3}(f_{I3}^z f_{I5}^x - f_{I3}^x f_{I5}^z)]/[\chi_{I4}(f_{I4}^z f_{I5}^x - f_{I4}^x f_{I5}^z)], \tag{11.65}$$

$$A_2 = (1/\epsilon_\infty k_I^z)(\chi_{I3}f_{I3}^z k_{I5} + \chi_{I4}f_{I4}^z k_{I5}A_1). \tag{11.66}$$

11.4. RESONANT BRILLOUIN SCATTERING

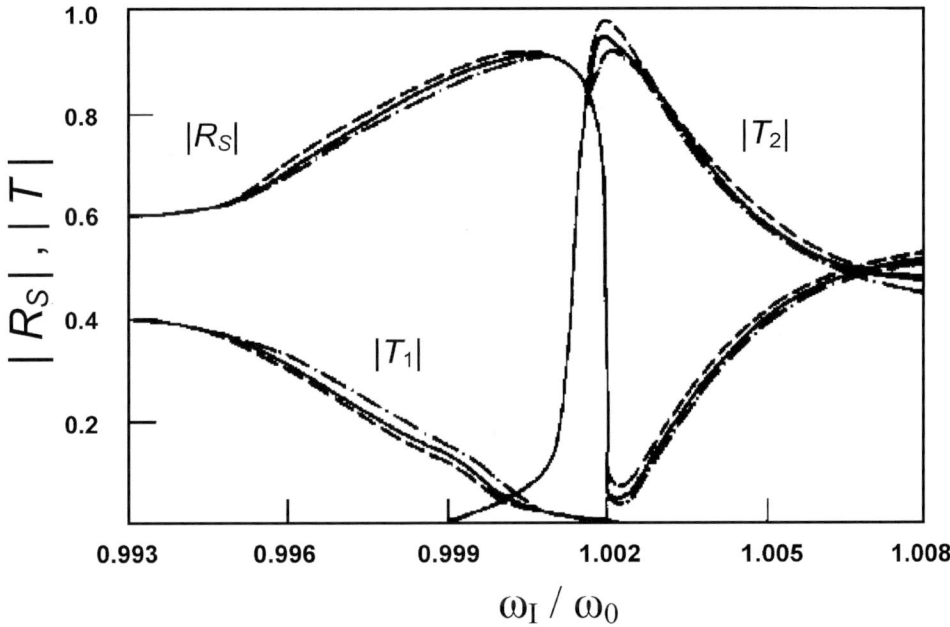

Fig. 11.11. Reflection (given by $|R_S|$) and transmission (given by $|T_1|$ and $|T_2|$) spectra in CdS as a function of the reduced frequency ω_I/ω_0, calculated using ABC1 (full curves), ABC2 (broken curves), and ABC3 (chain curves) at $\theta = 45°$ for s-polarized incident light (after Albuquerque and Gonçalves da Silva [46]).

Figs. 11.12a and b provide numerical examples of these results. While there is a good correspondence among the three ABCs, there are now some quantitative differences (e.g. in the transmission coefficient T_5, which corresponds to the longitudinal mode not excited in the s geometry). Thus we may conclude that the p geometry is more sensitive to the choice of the ABC, especially for the longitudinal mode [49].

11.4.2 Light-Scattering Formalism

The scattering process is conventionally considered in three stages. In the first stage, incident light of frequency ω_I and amplitude ε_I^s (s-polarized wave), or ε_I^p (p-polarized wave), is partially transmitted into the SD medium, where it couples to the exciton field to become

$$E_I^s = T_1 \varepsilon_I^s \exp(-i\vec{k}_{I1} \cdot \vec{r}) + T_2 \varepsilon_I^s \exp(-i\vec{k}_{I2} \cdot \vec{r}) \tag{11.67}$$

for the incident s-polarized wave, or

$$E_I^p = T_3 \varepsilon_I^p \exp(-i\vec{k}_{I3} \cdot \vec{r}) + T_4 \varepsilon_I^p \exp(-i\vec{k}_{I4} \cdot \vec{r}) + T_5 \varepsilon_I^p \exp(-i\vec{k}_{I5} \cdot \vec{r}) \tag{11.68}$$

for the incident p-polarized wave. The T_λ coefficients ($\lambda = 1\text{--}5$) are given in Subsection 11.4.1.

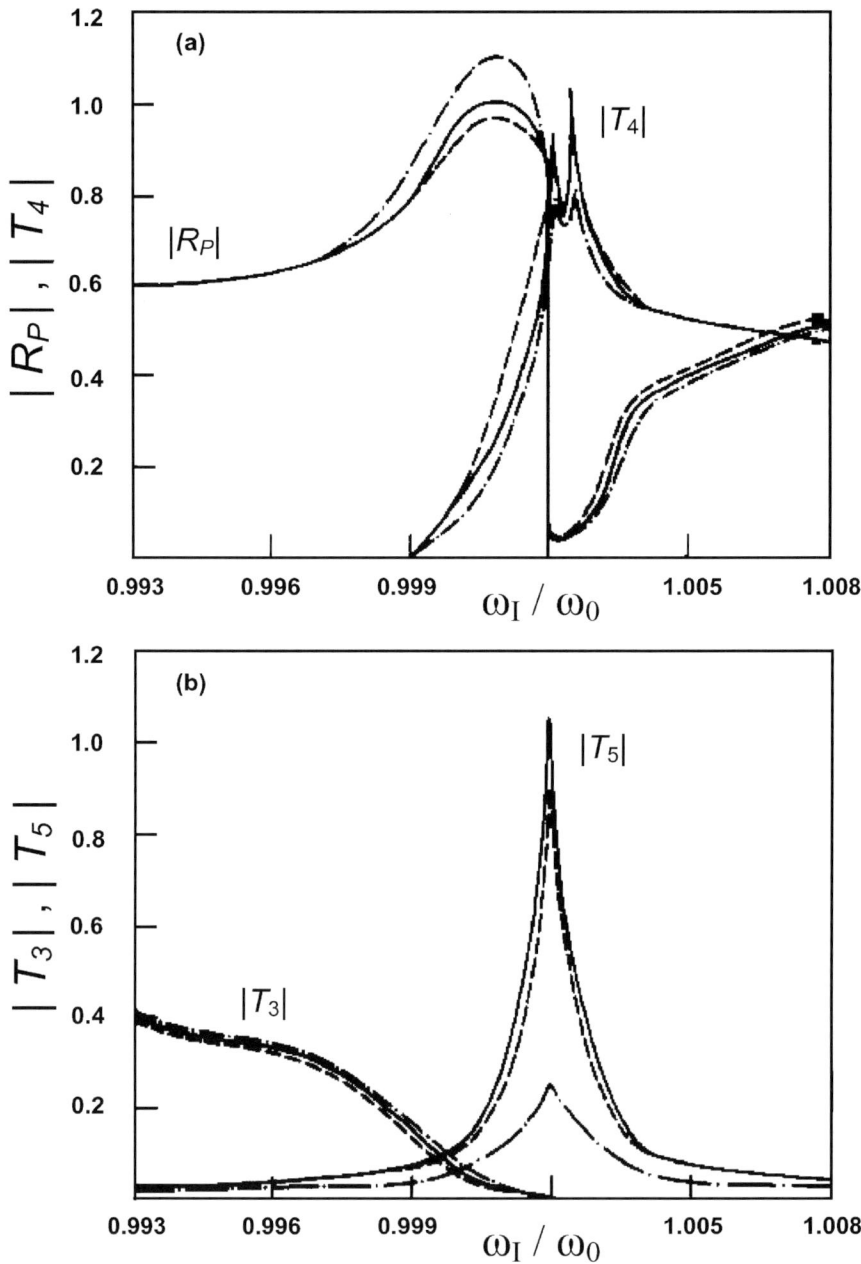

Fig. 11.12. (a) Reflection ($|R_P|$) and transmission ($|T_4|$) spectra in CdS as a function of the reduced frequency ω_I/ω_0, calculated using ABC1 (full curves), ABC2 (broken curves), and ABC3 (chain curves) at $\theta = 45°$ for p-polarized incident light. (b) The same as (a) for the other transmission spectra ($|T_3|$ and $|T_5|$) (after Albuquerque and Gonçalves da Silva [46]).

11.4. RESONANT BRILLOUIN SCATTERING

The second stage takes place inside the SD medium, where light interacts with an exciton-polariton of wavevector $\vec{Q} = (k_x, 0, Q^z)$ and frequency ω to produce a scattered wave of frequency $\omega_s = \omega_I - \omega$. This interaction occurs via a deformation potential, where an i-polarized acoustic-phonon mode (i can be x, y or z) with

$$u^i = u^i(Q^z)\exp(i\vec{Q}\cdot\vec{r} + i\omega t) \tag{11.69}$$

changes the band gap of the SD medium by an amount [44]

$$\Delta E_g = -C(Q^z)u^{ij}(Q^z)^*. \tag{11.70}$$

Here u^{ij} is a shorthand for $\partial u^i/\partial r^j$. This effect can be included in Eq. (3.19) by adding the term

$$\eta^{ijkl}P^j u^{kl}(Q^z)^*, \tag{11.71}$$

where i, j are labels related to the polarization of the incident and scattered light and k, l are labels related to the phonon polarization. Here, we are adopting the convention that repeated superscripts are summed over. If we consider an SD medium with cubic symmetry, the only non-vanishing components of the fourth-rank tensor η are [50]

$$\begin{aligned}\eta^{xxxx} &= \eta^{yyyy} = \eta^{zzzz} = \eta^{11},\\ \eta^{xxyy} &= \eta^{yyzz} = \eta^{xxzz} = \eta^{12},\\ \eta^{yzyz} &= \eta^{zxzx} = \eta^{xyxy} = \eta^{44}\end{aligned} \tag{11.72}$$

and the general property that

$$\eta^{ijkl} = \eta^{jikl} = \eta^{jilk} = \eta^{ijlk} \tag{11.73}$$

applies. These constraints on η imply some symmetry properties for the scattering process, i.e. for a given incident frequency there are four Stokes (or anti-Stokes) resonant lines in $s \to s$ scattering, six resonant lines in either $s \to p$ or $p \to s$ scattering, and nine resonant lines in $p \to p$ scattering [51].

The next steps are to incorporate Eq. (11.71) into Eq. (3.19) and to substitute

$$E^i = \sum_\lambda T^i_\lambda \exp(-i\vec{k}_{I\lambda}\cdot\vec{r})\exp(-i\omega_I t) + E^i_s \exp(-i\omega_s t) \tag{11.74}$$

and

$$P^i = \epsilon_0 \sum_\lambda \chi_{I\lambda} T^i_\lambda \exp(-i\vec{k}_{I\lambda}\cdot\vec{r})\exp(-i\omega_I t) + P^i_s \exp(-i\omega_s t) \tag{11.75}$$

into Eqs. (10.4) and (3.19), respectively, where $\lambda = 1, 2$ (incident s wave) or $3, 4, 5$ (incident p wave). One then finds to first order in η

$$\begin{aligned}D\nabla^2 P^i_s + (\omega_s^2 - \omega_0^2 + i\omega_s\Gamma)P^i_s + SE^i_s \\ = -\eta^{ijkl}u^{kl}(Q^z)^*\epsilon_0 \\ \times \sum_\lambda \chi_{I\lambda} T^j_\lambda \exp(-i\vec{k}_{I\lambda}\cdot\vec{r})\exp(-i\vec{Q}\cdot\vec{r}),\end{aligned} \tag{11.76}$$

$$\nabla^2 E_s^i + \epsilon_\infty (\omega_s^2/c^2) E_s^i + (\omega_s^2/c^2) P_s^i = 0. \tag{11.77}$$

The third and final stage of the process involves finding the scattered radiation produced in the medium $z > 0$ by the driving terms shown on the right-hand side of Eq. (11.76). For $z > 0$ we have for the scattered light

$$\varepsilon_s = \sum_\lambda A(\epsilon_0/S)^{1/2} \eta^{ijkl} u^{kl} (Q^z)^* \chi_{I\lambda} T_\lambda \exp(i\vec{k}_s \cdot \vec{r}), \tag{11.78}$$

while for $z < 0$ we have (denoting $\vec{k}_{0\lambda} = \vec{k}_{I\lambda} - \vec{Q}$)

$$E_s = \sum_\delta A_\delta \exp(-i\vec{k}_{s\delta} \cdot \vec{r}) + \sum_\lambda A_0 (\epsilon_0/S)^{1/2} \eta^{ijkl} u^{kl} (Q^z)^* \chi_{I\lambda} T_\lambda \exp(i\vec{k}_{0\lambda} \cdot \vec{r}), \tag{11.79}$$

$$P_s = \sum_\delta A_\delta \chi_{s\delta} \exp(-i\vec{k}_{s\delta} \cdot \vec{r})$$

$$+ \sum_\lambda A_0 \eta^{ijkl} u^{kl} (Q^z)^* (k_{0\lambda}^{z2} - \epsilon_\infty \omega_s^2/c^2) \chi_{I\lambda} T_\lambda \exp(i\vec{k}_{0\lambda} \cdot \vec{r}). \tag{11.80}$$

The combinations of polarization labels in Eqs. (11.79) and (11.80) give

$$\begin{aligned} \delta = \lambda = 1, 2 \quad &\text{implies } s \to s \text{ scattering,} \\ \delta = \lambda = 3, 4, 5 \quad &\text{implies } p \to p \text{ scattering,} \\ \delta = 1, 2; \ \lambda = 3, 4, 5 \quad &\text{implies } p \to s \text{ scattering,} \\ \delta = 3, 4, 5; \ \lambda = 1, 2 \quad &\text{implies } s \to p \text{ scattering.} \end{aligned} \tag{11.81}$$

The scattered field ε_s can now be calculated using the electromagnetic boundary conditions plus the ABCs. If the scattered light is s-polarized we have

$$\varepsilon_s^s = (\epsilon_0/S)^{1/2} \sum_\lambda \sum_{Q^z} \chi_{I\lambda} T_\lambda^j \eta^{ijkl} u^{kl} (Q^z)^*$$

$$\times \exp(i\vec{k}_s \cdot \vec{r}) B(k_{0\lambda}^z, \omega_s)(k_{0\lambda}^{z2} - k_{s1}^{z2})^{-1} (k_{0\lambda}^{z2} - k_{s2}^{z2})^{-1}. \tag{11.82}$$

Here we denote

$$B(k_{0\lambda}^z, \omega_s) = N(k_{0\lambda}^z, \omega_s)/D(\omega_s) \tag{11.83}$$

with

$$N(k_{0,\lambda}^z, \omega_s) = (\omega_s/c)^2 (S/\epsilon_0)^{1/2} \Big[(c/\omega_s)^2 (k_{0\lambda}^{z2} - \epsilon_\infty \omega_s^2/c^2)$$

$$\times (1 + ik_{0\lambda}^z)(k_{s2}^z - k_{s1}^z) + \chi_{s2}(1 - ik_{s2}^z)(k_{0\lambda}^z + k_{s1}^z)$$

$$- \chi_{s1}(1 - ik_{s1}^z)(k_{0\lambda}^z + k_{s2}^z) \Big], \tag{11.84}$$

11.4. RESONANT BRILLOUIN SCATTERING

$$D(\omega_s) = D\Big[\chi_{s2}(k_s^z + k_{s1}^z)(1 - ik_{s1}^z) - \chi_{s1}(k_s^z + k_{s2}^z)(1 - ik_{s2}^z)\Big]. \tag{11.85}$$

On the other hand, if the scattered light is p-polarized, we have

$$\varepsilon_s^p = \sum_\lambda \sum_{Q^z} \eta^{ijkl} u^{kl}(Q^z)^* \chi_{I\lambda} T_\lambda^j \exp(i\vec{k}_s \cdot \vec{r}) \frac{(\omega_s/c)^2 (\Delta_1 T_2 + \Delta_2 T_1)}{D(k_s \Delta_1 + f_s^z \Delta_2)}$$

$$\times \Big[(k_{0\lambda}^{z2} - k_{s1}^{z2})(k_{0\lambda}^{z2} - k_{s2}^{z2})(k_{0\lambda}^{z2} - k_{s3}^{z2})\Big]^{-1}, \tag{11.86}$$

where

$$\Delta_1 = f_{s3}^z - \frac{\chi_{s3} F_{35} a_{34} f_{s4}^z}{\chi_{s4} F_{45}} + \frac{\chi_{s3} a_{35}}{\chi_{s5} F_{45}} \Big(F_{45} f_{s3}^x - F_{35} f_{s4}^x\Big), \tag{11.87}$$

$$\Delta_2 = k_{s3} - \frac{\chi_{s3} F_{35} a_{34} k_{s4}}{\chi_{s4} F_{45}} + \frac{\chi_{s3} a_{35} k_{s5}}{\chi_{s5} f_{s5}^z F_{45}} \Big(F_{45} f_{s3}^x - F_{35} f_{s4}^x\Big), \tag{11.88}$$

$$T_1 = (k_{0\lambda}^z / k_{0\lambda}) - \Big[\gamma_1 (S/\epsilon_0)^{1/2} \gamma_2 / \chi_{s5} a_{50}\Big], \tag{11.89}$$

$$T_2 = k_{0\lambda} + \Big[k_{s5} \gamma_1 (S/\epsilon_0)^{1/2} \gamma_2 / \chi_{s5} f_{s5}^z a_{50}\Big], \tag{11.90}$$

$$\gamma_1 = (k_{0\lambda}^{z2} - \epsilon_\infty \omega_s^2 / c^2), \tag{11.91}$$

$$\gamma_2 = (k_{0\lambda}^x / k_{0\lambda}) + (F_{05} f_{s4}^x / F_{45}). \tag{11.92}$$

The notation is $f_{s\nu}^i = k_{s\nu}^i / k_{s\nu}$ ($\nu = 3, 4, 5$ and $i = x, z$), and $f_s^z = k_s^z / k_s$. Also

$$a_{ij} = (1 - ik_{si}^z)/(1 - ik_{sj}), \tag{11.93}$$

$$a_{50} = (1 - ik_{s5}^z)/(1 - ik_{0\lambda}^z), \tag{11.94}$$

$$F_{ij} = f_{si}^z f_{sj}^z + f_{si}^x f_{sj}^x. \tag{11.95}$$

11.4.3 RBS Cross Section

We start from the basic expression for the scattering cross section as given in Eq. (11.1). Using Eqs. (11.54) and (11.86) for the s-mode case, or the analogous equations for the p-mode case, we can obtain the RBS cross section for all types of scattering described in the previous subsection that are allowed by the symmetry selection rules [51]. To illustrate the results, we do this for a typical experimental scattering geometry. We shall eventually obtain a generalization of Eq. (11.2) with the resonant terms (due to SD) included.

Suppose we have an incident p-polarized light wave while the scattered wave is s-polarized. This implies that the polarization of the acoustic phonon is in the y direction. The relevant terms in the coupling tensor η are simply η^{yxyx} or η^{yzyz}, and therefore

$$\frac{\langle|\varepsilon_s|^2\rangle_\omega}{|\varepsilon_I|^2} = (\epsilon_0/S) \sum_{Q^z, Q'^z} |\eta^{ylyl}|^2 \langle u^{yl} u^{yl*}\rangle_\omega$$

$$\times \Big| \left[\chi_{I3} T_3 B(k_{03}^z, \omega_s)(k_{03}^{z2} - k_{s1}^{z2})^{-1}(k_{03}^{z2} - k_{s2}^{z2})^{-1}\right]$$

$$+ \left[\chi_{I4} T_4 B(k_{04}^z, \omega_s)(k_{04}^{z2} - k_{s1}^{z2})^{-1}(k_{04}^{z2} - k_{s2}^{z2})^{-1}\right]$$

$$+ \left[\chi_{I5} T_5 B(k_{05}^z, \omega_s)(k_{05}^{z2} - k_{s1}^{z2})^{-1}(k_{05}^{z2} - k_{s2}^{z2})^{-1}\right] \Big|^2, \quad (11.96)$$

where l is summed over x and z. Now using the fluctuation–dissipation theorem as in Eq. (A.39) of Appendix A, we have

$$\langle u^{yl} u^{yl*}\rangle_\omega = (k_B T / \pi \omega) \mathrm{Im} \langle\langle u^{yl}(Q^z); u^{yl}(Q'^z)^*\rangle\rangle_\omega. \quad (11.97)$$

We can express the above Green function, which was calculated by Loudon [52], as a double-Fourier transform:

$$\langle\langle u^{yl}(Q^z); u^{yl}(Q'^z)^*\rangle\rangle_\omega = \lim_{L \to \infty} \frac{1}{L^2} \int_{-L}^0 \int_{-L}^0 dz\, dz' \exp(-iQ^z z)$$

$$\times \exp(iQ'^z z') \langle\langle u^{yl}(z); u^{yl}(z')^*\rangle\rangle_\omega. \quad (11.98)$$

Next, rewriting the summations over Q^z and Q'^z in Eq. (11.96) as

$$\sum_{Q^z, Q'^z} \to \left(\frac{L}{2\pi}\right)^2 \int dQ^z\, dQ'^z \quad (11.99)$$

and carrying out the Q^z and Q'^z integrals directly, we obtain

$$\frac{\langle|\varepsilon_s|^2\rangle_\omega}{|\varepsilon_I|^2} = \frac{\epsilon_0 k_B T}{4\pi^3 S \omega} \mathrm{Im}\left(\frac{|\eta^{44}|^2}{|k_{1s}^{z2} - k_{2s}^{z2}|^2} \Delta\right), \quad (11.100)$$

where

$$\Delta = \int dz\, dz' \langle\langle u^{yl}(z); u^{yl}(z')^*\rangle\rangle_\omega Y^l \quad (11.101)$$

and

$$Y^l = \Big[H_{31}^l \exp(-ik_{31}^z z') + H_{32}^l \exp(-ik_{32}^z z') + H_{41}^l \exp(-ik_{41}^z z')$$

$$+ H_{42}^l \exp(-ik_{42}^z z') + H_{51}^l \exp(-ik_{51}^z z') + H_{52}^l \exp(-ik_{52}^z z')\Big]$$

$$\times \Big[H_{31}^{l*} \exp(ik_{31}^{z*} z) + H_{32}^{l*} \exp(ik_{32}^{z*} z) + H_{41}^{l*} \exp(ik_{41}^{z*} z)$$

$$+ H_{42}^{l*} \exp(ik_{42}^{z*} z) + H_{51}^{l*} \exp(ik_{51}^{z*} z) + H_{52}^{l*} \exp(ik_{52}^{z*} z)\Big]. \quad (11.102)$$

11.4. RESONANT BRILLOUIN SCATTERING

The additional notations are

$$k^z_{\lambda\delta} = k^z_{I\lambda} + k^z_{s\delta}, \quad (11.103)$$

$$H^l_{\lambda\delta} = T^l_\lambda \chi_{I\lambda} c_{s\delta}, \quad (11.104)$$

$$c_{s\delta} = 2k^z_{s\delta}(\omega_s^2/c^2)(S/\epsilon_0)^{1/2}(1 - ik^z_{s\delta})(\chi_{s2} - \chi_{s1})/D(\omega_s). \quad (11.105)$$

We have almost reached the final result; it remains for us to specify the branches involved in the scattering. Let us consider first the *intrabranch* scattering, i.e. scattering from branch 3 (*p*-mode) to branch 1 (*s*-mode) and from branch 4 (*p*-mode) to branch 2 (*s*-mode). The quantity Δ in Eq. (11.101) becomes

$$\Delta = \left(1/\rho \bar{S} v_T^2 q_T^z\right)\left(|H^l_{31}|^2 N^l_{31} + |H^l_{42}|^2 N^l_{42}\right), \quad (11.106)$$

where ρ is the density of the SD medium, q_T^z is the transverse acoustic wavevector defined by

$$q_T^z = (\omega^2/v_T^2 - k_x^2)^{1/2} \quad (11.107)$$

with v_T being the velocity of the transverse acoustic mode, and

$$N^l_{\lambda\delta} = \frac{(q_T^l)^2\left[(q_T^z) \text{ or } (q_T^z + 2i\,\mathrm{Im}\,k^z_{\lambda\delta})\right]}{(\mathrm{Im}\,k^z_{\lambda\delta})(k^{z*}_{\lambda\delta} - q_T^z)(k^z_{\lambda\delta} + q_T^z)}. \quad (11.108)$$

The alternatives in square brackets apply, respectively, for $l = z$ or x. The cross section is

$$\frac{d^2\sigma}{d\Omega\,d\omega_s} = \frac{\omega_I \omega_s \bar{S}\epsilon_0 k_B T \cos^2\theta_s}{16\pi^5 c^2 \rho v_T^2 S q_T^z \omega} \left|\frac{\eta^{44}}{k^{z2}_{1s} - k^{z2}_{2s}}\right|^2$$

$$\times \left[|H^l_{31}|^2 \mathrm{Im}(N^l_{31}) + |H^l_{42}|^2 \mathrm{Im}(N^l_{42})\right], \quad (11.109)$$

where l is summed over x or z.

The resonant terms in Eq. (11.108) are $(k^{z*}_{31} - q_T^z)$ and $(k^{z*}_{42} - q_T^z)$, and therefore there are two Stokes peaks at $q_T^z = k^{z*}_{31}$ and k^{z*}_{42}. Fig. 11.13 shows the form of the spectrum for CdS where the arrows indicate the positions of the resonant terms. The quantity plotted is the final pair of factors inside the brackets in Eq. (11.109) versus the reduced frequency $\omega/v_T k_x$. This theoretical spectrum is evaluated for an incident-light wavelength $\lambda = 504.7$ nm that corresponds to the ratio $\omega_I/\omega_0 = 1.003$. As expected there is good correspondence among the three ABCs since, for this case, there is no longitudinal-mode scattering, and only the transverse modes are involved. The lineshapes of the peaks are skewed, as expected in an absorptive medium [53].

Now, let us consider the *interbranch* scattering, which means four types of scattering, i.e.

(i) branch 3 to branch 2,

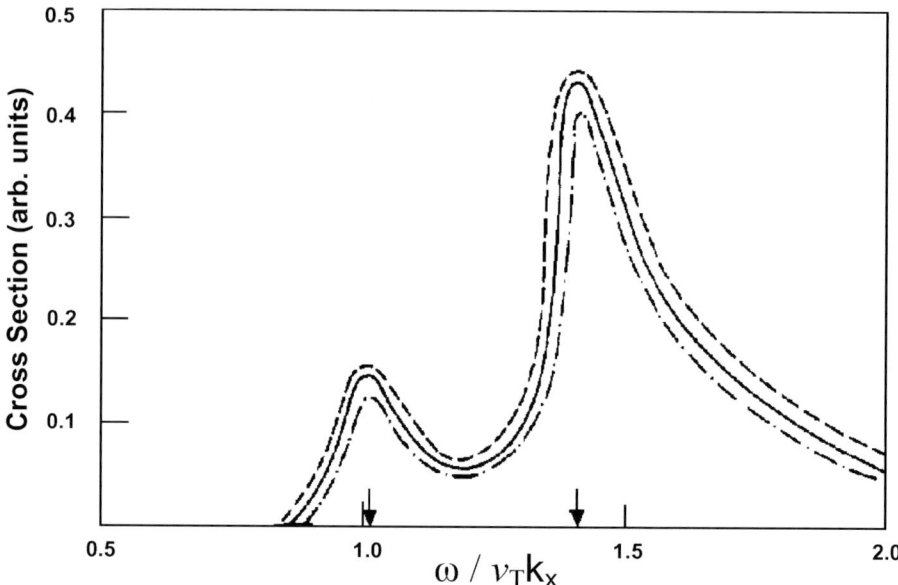

Fig. 11.13. Brillouin spectrum for intrabranch scattering as a function of the reduced frequency $\omega/v_T k_x$ calculated using ABC1 (full curves), ABC2 (broken curves), and ABC3 (chain curves) at $\theta_I = 45°$. The arrows denote the positions of the resonant peaks (after Albuquerque and Gonçalves da Silva [46]).

(ii) branch 4 to branch 1,

defining the p-transverse to s-transverse case, and

(iii) branch 5 to branch 1,

(iv) branch 5 to branch 2,

defining the p-longitudinal to s-transverse case. The RBS cross section can be calculated in a manner very similar to the intrabranch case and the result is [46]

$$\frac{d^2\sigma}{d\Omega\, d\omega_s} = \frac{\omega_I \omega_s S \epsilon_0 k_B T \cos^2\theta_s}{16\pi^5 c^2 \rho v_T^2 S q_T^z \omega} \left|\frac{\eta^{44}}{k_{1s}^{z2} - k_{2s}^{z2}}\right|^2 \Big[|H_{32}^l|^2 \mathrm{Im}(N_{32}^l) + |H_{41}^l|^2 \mathrm{Im}(N_{41}^l)$$

$$+ |H_{51}^l|^2 \mathrm{Im}(N_{51}^l) + |H_{52}^l|^2 \mathrm{Im}(N_{52}^l) \Big]. \tag{11.110}$$

There are now four resonant Stokes peaks at $q_T^z = k_{32}^{z*}$, k_{41}^{z*}, k_{51}^{z*}, and k_{52}^{z*}.

The shape of the predicted theoretical Brillouin spectrum is plotted, as a function of the dimensionless parameter $\omega/v_T k_x$, for CdS in Fig. 11.14, where we have evaluated the expression inside the square brackets in Eq. (11.110). The arrows again denote the positions of the resonant peaks. There are quantitative differences among the ABCs for scattering off the longitudinal mode, but almost no

11.4. RESONANT BRILLOUIN SCATTERING

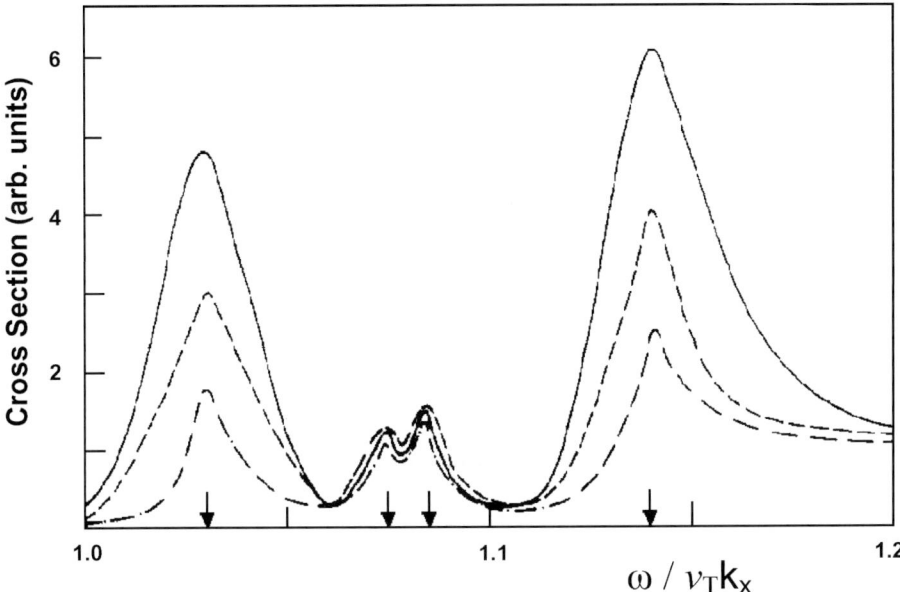

Fig. 11.14. The same as Fig. 11.13, but for interbranch scattering (after Albuquerque and Gonçalves da Silva [46]).

sensitivity to the ABCs for the transverse modes. It can be seen that the cross section varies by more than a factor of two with the use of different boundary conditions when the longitudinal exciton is excited. The sensitivity of RBS experiments and the strong dependence of the interbranch scattering cross section depicted in Fig. 11.14 should help in resolving the choice of ABCs. The outcome, however, is likely to depend on the material and its surface preparation, implying that RBS could be used as a surface-sensitive probe.

A related experimental technique is resonant Rayleigh scattering (RRS). It has possible applicability to the spectroscopy of other semiconductor nanostructures or to inorganic crystals (for a review see Ref. [54]). In semiconductors RRS occurs due to imperfections (including impurities, defects, or interfaces) breaking the translational symmetry of the crystal. The RRS observed from semiconductor quantum wells is mainly due to fluctuations in the lateral 2D potential confining the charge carriers and excitons [55,56]. When the lower exciton-polariton branch is optically excited, the strong dispersion results in a directional emission on a ring. The ring width converges with time to a finite value, a direct measure of an intrinsic momentum broadening of the exciton-polariton states localized by multiple-disorder scattering [57]. Further, a clear signature of enhanced backscattering of excitons is observed in the directional RRS from localized 2D excitons in disordered quantum wells, which is qualitatively different from backscattering phenomena in other branches of physics [58,59].

11.5 Far-Infrared Attenuated Total Reflection

An introduction to ATR was already given in Section 4.5. As mentioned, it was pioneered by Otto [1,2] and is very convenient for probing surface polariton modes, which are non-radiative in the sense that they cannot be excited by a light beam in the usual two-layer geometry. The basis of this method is a three-layer geometry depicted in Fig. 4.11 (the Otto configuration), where a glass prism with dielectric constant ϵ_p is placed above the sample and is separated from it by a gap of thickness d and dielectric constant ϵ_g (chosen such that $\epsilon_p > \epsilon_g$). The sample has a frequency-dependent dielectric function $\epsilon_s(\omega)$. When light is reflected internally off the base of the prism at an angle of incidence θ which exceeds the critical angle θ_C for total internal reflection, as given by Eq. (4.49), an evanescent wave with an in-plane wavevector k_x [see Eq. (4.50)] is set up in the region adjacent to the base of the prism. As explained in Section 4.5, the surface polariton is observed experimentally as a dip in the reflectivity. Some applications of ATR to the study of surface phonon- and plasmon-polaritons were also described.

The theory of ATR is well understood and is formally quite similar to the theory of polaritons in double-interface geometries, as described in Section 4.4. Since the surface polaritons are p-polarized modes, ATR occurs only with p-polarized light. Therefore we consider in medium 1 (the prism) the electric field in the form

$$\vec{E} = (E_0, 0, k_x E_0/k_{1z}) \exp(ik_x x - i\omega t) \exp(-ik_{1z} z)$$
$$+ r(E_0, 0, -k_x E_0/k_{1z}) \exp(ik_x x - i\omega t) \exp(ik_{1z} z), \quad (11.111)$$

where the first term describes the incident wave and E_0 is an (arbitrary) constant. The second term is the reflected term, and our objective is to calculate the complex reflection coefficient r. The electric fields in media 2 (the gap) and 3 (the sample) can be represented, as in Section 4.4, by Eqs. (4.28) and (4.29), respectively, with d now replacing the film thickness L. Using the electromagnetic boundary conditions at the two interfaces, we can solve the resulting system of four equations to find

$$r = \frac{[R_+ S_- + \exp(-2ik_{2z}d) R_- S_+]}{[R_+ S_+ + \exp(-2ik_{2z}d) R_- S_-]}, \quad (11.112)$$

$$R_\pm = \epsilon_s k_{2z} \pm \epsilon_g k_{3z}, \quad (11.113)$$

$$S_\pm = \epsilon_g k_{1z} \pm \epsilon_p k_{2z}. \quad (11.114)$$

As an example, we consider a *nipi* superlattice as the sample. In the long-wavelength limit we may define the effective dielectric function of the resulting (uniaxial) effective medium analogously to Eqs. (10.18) and (10.19), i.e.

$$\epsilon_{xx} = \epsilon_{yy} = f_a \epsilon_A + f_c \epsilon_C + 2 f_b \epsilon_B, \quad (11.115)$$

$$\epsilon_{zz}^{-1} = f_a \epsilon_A^{-1} + f_c \epsilon_C^{-1} + 2 f_b \epsilon_B^{-1}, \quad (11.116)$$

where the off-diagonal elements are zero. Here we have specifically considered a *nipi* structure in which medium B is identical to medium D (i.e. having the

11.5. FAR-INFRARED ATTENUATED TOTAL REFLECTION

same dielectric constant and same thickness). Also we define the fractions [as in Eq. (10.20)]

$$f_j = j/L, \quad j = a, b, c. \tag{11.117}$$

The unit-cell size is $L = a + c + 2b$, and a, b, and c are the layer thicknesses.

For the doped polar semiconductor layers of the *nipi* structure, the quasi-free electron (or hole) gas and the dipole-active long-wavelength transverse optical (TO) phonons are coupled over a wide frequency range in the infrared, forming a mixed mode. Generalizing Eq. (3.13) to include damping, we take the dielectric function for $J = A$ or C as

$$\epsilon_J = \epsilon_{\infty J}\left[1 + \frac{\omega^2 - \omega_{LJ}^2}{\omega(\omega + i\Gamma) - \omega_{TJ}^2} - \frac{\omega_{pJ}^2}{\omega(\omega + i\Gamma)}\right], \tag{11.118}$$

while ϵ_B is a constant. Here ω_L and ω_T are the longitudinal optical (LO) and TO phonon frequencies, respectively, ω_p is the plasma frequency, and Γ is a damping factor. In the effective-medium approximation for the semi-infinite superlattice, with vacuum outside, the surface polaritons follow the standard dispersion relation (see [60])

$$k_x^2 = (\omega^2/c^2)\epsilon_{zz}(\epsilon_{xx} - 1)/(\epsilon_{xx}\epsilon_{zz} - 1), \tag{11.119}$$

by analogy with results in Section 4.2 on anisotropic media. A necessary condition for the surface mode to exist is $\epsilon_{xx} < 0$. The surface-polariton dispersion curves are shown in Fig. 11.15, taking materials A and C to be n- and p-doped GaAs and AlAs, respectively, while media B and D are SiO_2. The physical parameters used are $\omega_{LA} = 54.96$ THz, $\omega_{TA} = 50.57$ THz, $\omega_{pA} = \omega_{pC} = 25.4$ THz, $\omega_{LC} = 69.79$ THz, $\omega_{TC} = 67.31$ THz, $\epsilon_{\infty A} = 10.9$, $\epsilon_{\infty C} = 10.22$, $\epsilon_B = \epsilon_D = 3.7$, $f_a = f_c = 2f_b = \frac{1}{3}$, and $\Gamma = 0$. There are three surface-polariton modes (dashed lines), labelled S_1, S_2, and S_3, all of them corresponding to both ϵ_{xx} and ϵ_{zz} negative. The ATR angular scan line (chain-dotted line) crosses the dispersion curves for the surface modes at the positions $k_x c/\omega_{TA} = 0.7$ (for S_1), 1.6 (for S_2), and 2.2 (for S_3). The term ATR *scan line* refers to the $\omega \propto k_x$ relation given by Eq. (4.50); as indicated, we are interested in where this line intersects the surface branches of the spectrum. The five bulk polariton modes are shown shaded, and the vacuum light line is shown.

The corresponding ATR spectra can be obtained using the reflectivity from Eq. (11.112). Here we quote the value of the ATR reflection coefficient r with the *nipi* structure as medium 3. The calculations make use of Maxwell's equation, together with the electromagnetic boundary conditions, and are straightforward but lengthy. The result is [61]

$$r = \frac{Q_{11} - (A_{11}Q_{11} + A_{12}Q_{21})}{Q_{12} - (A_{11}Q_{12} + A_{12}Q_{22})}, \tag{11.120}$$

where Q_{ij} and A_{ij} are elements of the following 2×2 matrices:

$$\bar{Q} = \bar{M}^{-1}(k_B, \epsilon_B, d)\bar{M}(k_2, \epsilon_g, d)\bar{M}^{-1}(k_2, \epsilon_g, 0)\bar{M}(k_1, \epsilon_p, 0), \tag{11.121}$$

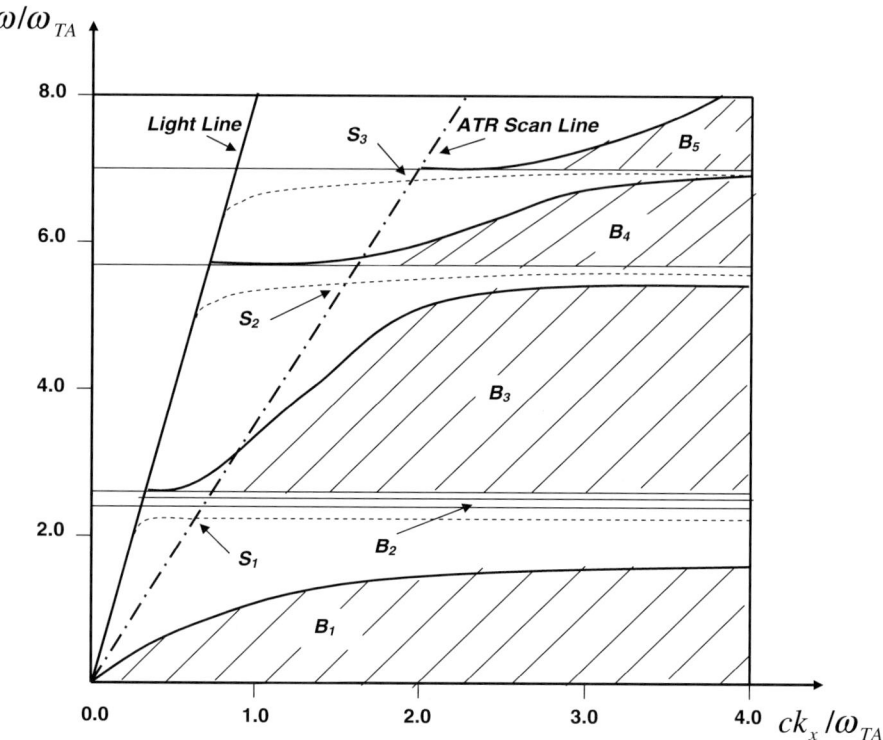

Fig. 11.15. Dispersion curves for surface polaritons (S_1, S_2, and S_3) in a *nipi* superlattice, as given by Eq. (11.119). The reduced frequency ω/ω_{TA} is plotted against the dimensionless in-plane wavevector ck_x/ω_{TA}. The light line in vacuum (full line) and the ATR scan line (chain-dotted line) are marked. The bulk modes (B_1, B_2, B_3, B_4, and B_5) are shown as shaded regions. See the text for parameter values (after Almeida et al. [61]).

$$\bar{A} = \bar{M}^{-1}(k_B, \epsilon_B, d+b)\bar{M}(k_C, \epsilon_C, d+b)\bar{M}^{-1}(k_C, \epsilon_C, d+b+c)$$
$$\times \bar{M}(k_B, \epsilon_B, d+b+c)\bar{M}^{-1}(k_B, \epsilon_B, d+c+2b)\bar{M}(k_A, \epsilon_A, d+c+2b)$$
$$\times \bar{M}^{-1}(k_A, \epsilon_A, d+L)\bar{M}(k_B, \epsilon_B, d+L). \quad (11.122)$$

Here the matrix $\bar{M}(k, \epsilon, z)$ is defined by

$$\bar{M}(k, \epsilon, z) = \begin{pmatrix} \exp(ikz) & \exp(-ikz) \\ (k/\epsilon)\exp(ikz) & -(k/\epsilon)\exp(-ikz) \end{pmatrix}. \quad (11.123)$$

In addition, k_1 and k_2 are the z-components of the wavevector $\vec{k}$ in media 1 (the prism) and 2 (the gap), respectively, and k_J (for $J = A, B, C$) are the z components of $\vec{k}$ inside the layers of the *nipi* superlattice. The other notation is as before.

The resulting theoretical ATR spectrum, in terms of $|r|^2$ versus a dimensionless in-plane wavevector ck_x/ω_{TA}, is drawn in Fig. 11.16 using Eq. (11.120) for r (full

11.5. FAR-INFRARED ATTENUATED TOTAL REFLECTION

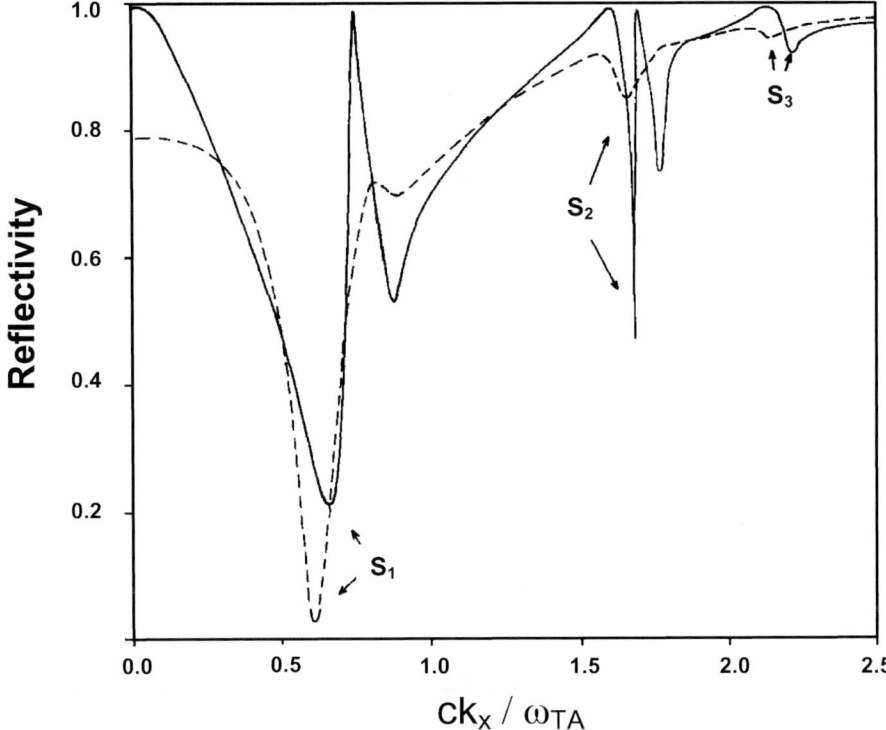

Fig. 11.16. Theoretical ATR curve (full line) for $|r|^2$, as a function of the dimensionless in-plane wavevector ck_x/ω_{TA}, for p-polarized light incident on a semi-infinite *nipi* superlattice, considered as an effective medium. The dips labelled S_1 to S_3 can be identified with the positions where the ATR angular scan line crosses the dispersion curves for the surface modes. For comparison, the broken line refers to the reflectivity from Eq. (11.112). See the text for parameter values (after Almeida et al. [61]).

line). For comparison, we have also shown the ATR spectrum with r described by Eq. (11.112) (broken line). We have taken $\epsilon_p = 10$ (Si prism), $\theta = 25°$, $d = 5$ μm, and $\Gamma/\omega_{pA} = 4 \times 10^{-4}$. One might expect a close correspondence between the two calculations in this long-wavelength regime (e.g. as was found in the case of normal-incidence-reflectivity spectroscopy [62]). However, as can be seen from Fig. 11.16, the two methods show only a fair agreement. The dips labelled S_1 to S_3 can be identified with the positions where the ATR angular scan line crosses the dispersion curves for the surface modes (see Fig. 11.15). In addition, there are other dips in Fig. 11.16 that can be identified as regions where the evanescent wave penetrates deeply inside the specimen and interacts with the bulk modes.

As a final example of an ATR spectrum, we consider the rare-earth superlattice considered in Section 7.4. Its theoretical ATR spectra, plotted here as a function of the reduced frequency $\omega/J_1 S$, are shown in Figs. 11.17a and b. We have considered $\epsilon_p = 10$ (suitable for an Si prism) and $\epsilon_g = 1$ (vacuum). The spectrum is sensitive

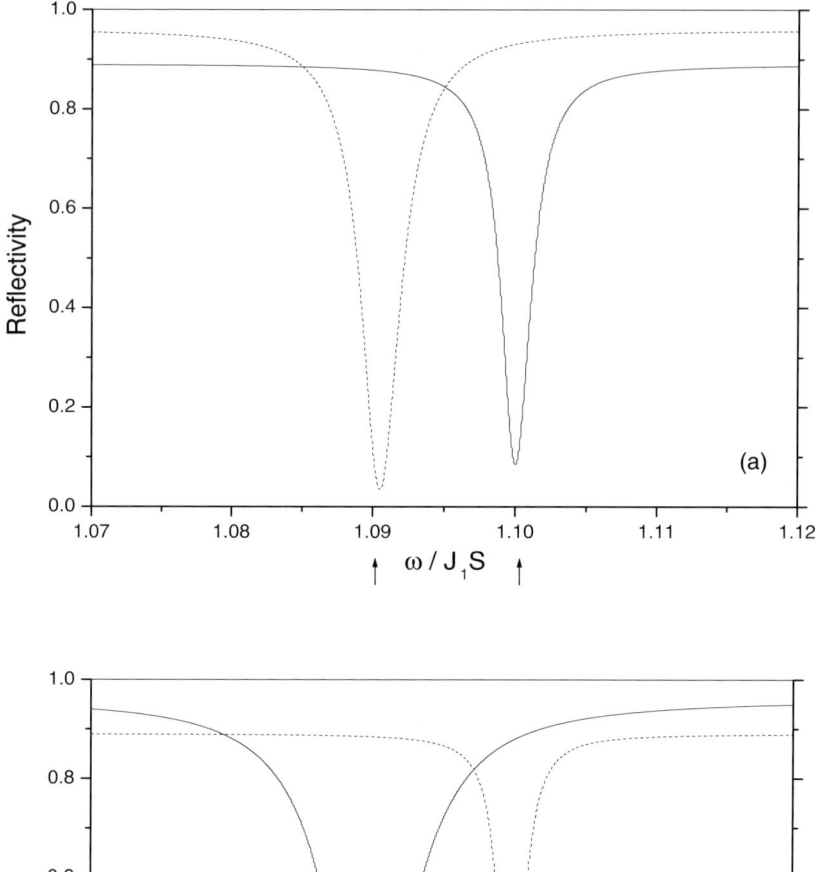

Fig. 11.17. Theoretical ATR curve ($|r|^2$ plotted versus a reduced frequency $\omega/J_1 S$) for p-polarized light incident on a semi-infinite rare-earth superlattice. The arrows are the points where the ATR angular scan line crosses the dispersion curve for the surface modes. (a) low-frequency surface polaritons; (b) high-frequency surface polaritons (after Albuquerque et al. [63]).

11.6. OTHER TECHNIQUES

to the air gap thickness d, which we take to be of the order of the wavelength of the surface wave (5 μm). The damping factor is 20 G in magnetic field units. The calculation refers to a scan of frequency for a fixed value 20° of the incident angle, which is slightly greater than the critical value 17.1° for total internal reflection at a silicon–air interface [63]. In Fig. 11.17a we show the ATR spectrum for the low-frequency surface modes of holmium in the cone state (see Fig. 7.1b) in the presence of a 200 G external magnetic field. Effects of the non-reciprocal propagation of the surface modes, identified by the sharp dips in the reflectivity spectrum, are evident. The arrows in the spectrum are the points where the ATR angular scan line crosses the dispersion curve for the surface modes. The full line is the ATR spectrum related to the +20° scan line (right-hand side of the polariton spectrum in Fig. 7.6b), while the dotted line describes the ATR spectrum for the −20° case (left-hand side of the polariton spectrum in Fig. 7.6b). Further evidence of the non-reciprocity is apparent in Fig. 11.17b, which depicts the ATR spectrum for the high-frequency surface modes.

11.6 Other Techniques

For completeness, we mention in this section some other techniques that may be useful to probe the excitations considered in this book.

11.6.1 Light-Emitting Tunnel Junction

In a light-emitting tunnel junction the tunnel current flowing between two metals is used to excite a surface plasmon. The plasmon then decays radiatively (because of roughness effects that are introduced at the junction) to produce a spectrum of emitted light whose intensity versus wavelength characteristics may be studied. The nature of the plasmon excited depends on the form of the junction(s) and on the value of the tunnelling current [64].

A simple example of a tunnel junction is an Al/Al$_2$O$_3$/Ag layered structure with the metal layer thicknesses being of the order of tens of nanometers and the Al$_2$O$_3$ tunnelling barrier (formed by oxidation of the Al film) being only 2 or 3 nm thick. The junction can be formed on a substrate material with a rough surface, thereby making all subsequent interfaces rough. If a d.c. voltage is applied, unpolarized light is emitted.

Another form of tunnel junction consists of a junction deposited on a diffraction grating, and this can be used to produce a p-polarized spectrum. Ushioda et al. [65] observed the direct light emission from the slow mode of surface plasmon-polaritons of metal–insulator–metal (MIM) tunnel junctions formed on an ultrafine grating. The grating, whose period was 100 nm, with groove depth 12 nm and area equal to 200 μm × 200 μm, was created on an Si(100) wafer, using a direct electron beam lithography technique and wet etching of SiO$_2$. The MIM junction was formed by evaporation of Al and Au films on this grating. Further details can be found in Ref. [65].

11.6.2 Far-Infrared Fourier-Transform Spectroscopy

Far-infrared (FIR) Fourier-transform spectroscopy has also been utilized to study the electronic [66,67], as well as phonon- and plasmon-related, properties of semiconductor superlattices [68,69]. For instance, FIR Fourier-transform spectroscopy and Raman analysis of phonons (and related interface modes) in GaN epilayers on GaAs and GaP substrates have enabled the determination of TO and LO phonon frequencies at the center of the Brillouin zone for propagation parallel and normal to the epilayers. Additionally, a number of interesting interface phonon modes have been studied. Comparison of the phonon properties of a selection of GaN epilayers deposited by molecular-beam epitaxy (MBE) on GaAs and GaP substrates indicates that their properties are strongly influenced by substrate material and orientation [70].

The various techniques now include both normal- and oblique-incidence dispersive Fourier-transform spectroscopy (DFTS), Fourier-transform spectroscopy (FTS), and ATR spectroscopy discussed in Section 11.5. In most experiments a Fourier-transform infrared spectrometer is used, consisting of a Michelson interferometer with the sample mounted within the output beam optics. Experiments to measure either the transmission through a sample or the reflection off its surface can be carried out directly. Additional information can be obtained through the use of prism couplers (as in ATR) or gratings. In the case of DFTS the sample is placed in one of the interferometer arms; this has the advantage that the phase of the radiation reflected off the sample can be measured as well as the amplitude.

The FIR spectroscopic techniques are complementary to Raman spectroscopy in the sense that Raman scattering provides a sensitive measure of the excitation frequencies whereas FIR experiments yield overall dielectric functions. So far, the Raman data on superlattice plasmons are more extensive, but it may be expected that the FIR spectroscopic techniques will yield further results in the near future. The existing experimental studies, which are mainly for long-period superlattices, are discussed in Refs. [68,69].

11.6.3 Magneto-Optical Kerr Effect

In recent years the magneto-optical Kerr effect (MOKE) has provided an important new means of probing a broad range of ultrathin magnetic films [71], magnetic trilayers [72,73], layered magnetic systems [74], and magnetic superlattices [75]. The possibility of an experimental observation of a magnetic Casimir-type interaction effect was investigated in ferromagnetic (FM) layers separated by vacuum (or a dielectric), in which perturbations of the zero-point electromagnetic energy are introduced by boundary conditions influenced by MOKE [76]. Furthermore, a representation of the Kerr effect in the complex rotation–ellipticity plane has been introduced to determine the magnetization state and hysteresis loop corresponding to one of the FM layers located at a given depth in a stack of FM/non-FM layers, also by means of MOKE [77].

Broadly speaking, magneto-optical effects in FM materials are produced by a combination of the net spin polarization that exists in the FM state and spin–orbit coupling. The spin–orbit interaction couples the spin components of the electronic wave functions to the spatial components which govern the electric-dipole matrix elements and optical selection rules. The general property that distinguishes MOKE from other magneto-optical effects in solids is that all manifestations of the Kerr effect are proportional to the magnetization $M(T)$, and so it vanishes at temperatures above the Curie temperature T_C. It provides an ideal method for studying thin-film magnetic anisotropy, as well as domain structure at surfaces (for a review see Ref. [78]). The surface-sensitive form of MOKE (known as SMOKE) was mentioned earlier in this chapter.

11.6.4 Ferromagnetic Resonance

Ferromagnetic resonance (FMR) is well established as a technique to measure basic physical quantities such as magnetic anisotropy energies or the interlayer exchange coupling J. Recently FMR measurement in coupled ultrathin Cu/Ni films showed that the temperature-dependent part of the effective exchange field follows a $T^{3/2}$ law over a wide temperature range, indicating that thermally excited spin waves at the interface of the ferromagnetic layers play a dominant role in the temperature dependence [79].

FMR is also a very sensitive technique that permits the study of excitations in ultrathin magnetic films (even down to one monolayer in thickness) [80,81]. It involves frequencies in the microwave range, and its signal is measured by monitoring the microwave losses in the studied film as a function of the applied d.c. external field H_0. Furthermore, in some ferromagnetic semiconductors (e.g. thin films of $Ga_{1-x}Mn_xAs$), it is possible to map out the dependence of the FMR position on temperature and on the angle between the applied magnetic field and the crystallographic axes of the sample. Besides, it allows us to obtain the values of the cubic and the uniaxial magnetic anisotropy fields, e.g. those which are associated with the natural (undistorted) zinc-blende structure and those arising from strain [82].

In FMR studies, the sample is inserted in a microwave cavity, and the external field is modulated with a low-frequency field component. The microwave cavity enhances the role of FMR losses and the external low-frequency modulation allows one to use lock-in amplifier detection. Modulation frequencies in the range 100–200 Hz are sufficient to improve significantly the signal-to-noise ratio. At the same time, spurious field-dependent signals associated with high modulation frequencies are avoided (for a review see Ref. [83]).

The choice of microwave cavities is crucial. The microwave cylindrical cavity is ideally suited for FMR studies for samples that exhibit a sufficient microwave reflection, such as bulk metallic substrates and the magnetic metallic superlattices that were discussed in Chapter 9.

References

[1] A. Otto, in: Advances in Solid State Physics, Festkörperprobleme XIV (1974) p. 1.
[2] A. Otto, in: Optical Properties of Solids: New Developments, Ed., B.O. Seraphim, North-Holland, Amsterdam, 1976.
[3] W. Hayes and R. Loudon, Scattering of Light by Crystals, Wiley, New York, 1978.
[4] M.G. Cottam and D.J. Lockwood, Light Scattering in Magnetic Solids, Wiley, New York, 1986.
[5] D. Olego, A. Pinczuk, A.C. Gossard and W. Wiegmann, Phys. Rev. B 25 (1982) 7867.
[6] R. Sooryakumar, A. Pinczuk, A.C. Gossard and W. Wiegmann, Phys. Rev. B 31 (1985) 2758.
[7] J.K. Jain and P.B. Allen, Phys. Rev. B 32 (1985) 997.
[8] S. Katayama and T. Ando, J. Phys. Soc. Jpn. 54 (1985) 1615.
[9] P. Hawrylak, J.W. Wu and J.J. Quinn, Phys. Rev. B 32 (1985) 5169.
[10] M. Babiker, N.C. Constantinou and M.G. Cottam, J. Phys. C 19 (1986) 5849; 20 (1987) 4581; 20 (1987) 4597.
[11] M. Babiker, N.C. Constantinou and M.G. Cottam, Solid State Commun. 57 (1986) 887; 59 (1986) 751; 68 (1988) 967.
[12] D.L. Mills, Y.C. Chen and E. Burstein, Phys. Rev. B 13 (1976) 4419.
[13] M. Babiker, N.C. Constantinou and M.G. Cottam, in: Properties of Impurity State in Superlattice Semiconductors, Eds. C.Y. Fong, I.P. Batra and S. Ciraci, Plenum, New York, 1988.
[14] N.C. Constantinou and M.G. Cottam, Solid State Commun. 69 (1989) 321.
[15] E.L. Albuquerque, Phys. Lett. A 150 (1990) 417.
[16] N.C. Constantinou and M.G. Cottam, Solid State Commun. 81 (1992) 321.
[17] E.L. Albuquerque, Solid State Commun. 91 (1994) 251.
[18] C. Wang and R.A. Barrio, Phys. Rev. Lett. 61 (1988) 191.
[19] M.W.C. Dharma-Wardana, A.H. MacDonald, D.J. Lockwood, J.-M. Baribeau and D.C. Houghton, Phys. Rev. Lett. 58 (1987) 1761.
[20] D.J. Lockwood, M.W.C. Dharma-Wardana, G.C. Aers and J.-M. Baribeau, Appl. Phys. Lett. 52 (1988) 2040.
[21] A.T. Macrander, G.P. Schwartz and J. Bevk, Phys. Rev. B 37 (1988) 8459.
[22] G.C. Aers, M.W.C. Dharma-Wardana, G.P. Schwartz and J. Bevk, Phys. Rev. B 39 (1989) 1092.
[23] A. Pinczuk, M.G. Lamont and A.C. Gossard, Phys. Rev. Lett. 56 (1986) 2092.
[24] G. Fasol, N. Mestres, H.P. Hughes, A. Fisher and K. Ploog, Phys. Rev. Lett. 56 (1986) 517.
[25] R. Merlin, J.P. Valladares, A. Pinczuk, A.C. Gossard and J.H. English, Solid State Commun. 84 (1992) 87.
[26] E.L. Albuquerque, Solid State Commun. 99 (1996) 311.
[27] S. Das Sarma, A. Kobayashi and R.E. Prange, Phys. Rev. Lett. 56 (1986) 1280.
[28] S. Das Sarma, A. Kobayashi and R.E. Prange, Phys. Rev. B 34 (1986) 5308.
[29] G. Carlotti and G. Gubbiotti, J. Phys.: Condens. Matter 14 (2002) 8199.
[30] S. Tacchi, L. Albini, G. Gubbiotti, M. Madami and G. Carlotti, Surf. Sci. 507 (2002) 535.
[31] M.H. Grimsditch, in: Light Scattering in Solids V, Eds., G. Güntherodt and M. Cardona, Springer-Verlag, Heidelberg, 1989.
[32] D.L. Mills and S.M. Rezende, Topics in Appl. Phys. 87 (2003) 27.

[33] A.M. Urbas, E.L. Thomas, H. Kriegs, G. Fytas, R.S. Penciu and L.N. Economou, Phys. Rev. Lett. 90 (2003) 108302.

[34] M. Pierno, C.S. Casari, R. Piazza and C.E. Bottani, Appl. Phys. Lett. 82 (2003) 1532.

[35] L. Otani, A. Yoshihara and H. Ohigashi, Phys. Rev. E 64 (2001) 051804.

[36] M.G. Beghi, C.S. Casari, A.L. Bassi, C.E. Bottani, A.C. Ferrari, J. Robertson and P. Milani, Thin Solid Films 420 (2002) 300.

[37] M.H. Kuok, H.S. Lim, S.C. Ng, N.N. Liu and Z.K. Wang, Phys. Rev. Lett. 90 (2003) 255502.

[38] C. Weisbuch and R.G. Ulbrich in: Light Scattering in Solids III: Recent Results, Eds. M. Cardona and G. Güntherodt, Springer-Verlag, Berlin, 1982.

[39] W. Brenig, R. Zeyher and J.L. Birman, Phys. Rev. B 6 (1972) 2482.

[40] R.G. Ulbrich and C. Weisbuch, Phys. Rev. Lett. 38 (1977) 865.

[41] G. Winterling and E. Koteles, Solid State Commun. 23 (1977) 95.

[42] C. Herman and P.Y. Yu, Solid State Commun. 28 (1978) 313.

[43] D.R. Tilley, J. Phys. C 13 (1980) 781.

[44] V.C.Y. So, J.E. Sipe, M. Fukui and G.I. Stegeman, J. Phys. C 14 (1981) 4487.

[45] M. Fukui and O. Tada, J. Phys. Soc. Japan 51 (1982) 172.

[46] E.L. Albuquerque and C.E.T. Gonçalves da Silva, J. Phys. C 18 (1985) 665.

[47] P. Halevi and G.H.-Cocoletzi, Phys. Rev. Lett. 48 (1982) 1500.

[48] Y. Segawa, Y. Aovagi, S. Komuro and S. Namba, Phys. Rev. Lett. 50 (1983) 436.

[49] J.S Nkoma, J. Phys. C 16 (1983) 3713.

[50] J.F. Nye, Physical Properties of Crystals, Oxford, Clarendon, 1957.

[51] E.L. Albuquerque and C.E.T. Gonçalves da Silva, J. Physique C5 (1984) 61.

[52] R. Loudon, J. Phys. C 11 (1978) 2623.

[53] A. Dervich and R. Loudon, J. Phys. C 9 (1976) L669.

[54] S. Haacke, Rep. Prog. Phys. 64 (2001) 737.

[55] G. Malpuech, A. Kavokin, W. Langbein and J.M. Hvam, Phys. Rev. Lett. 85 (2000) 650.

[56] A. Kavokin, G. Malpuech and W. Langbein, Solid State Commun. 120 (2001) 259.

[57] W. Langbein and J.M. Hvam, Phys. Rev. Lett. 88 (2002) 047401.

[58] W. Langbein, E. Runge, V. Savona and R. Zimmermann, Phys. Rev. Lett. 89 (2002) 157401.

[59] G. Kocherscheidt, W. Langbein, U. Woggon, V. Savona, R. Zimmermann, D. Reuter and A.D. Wieck, Phys. Rev. B 68 (2003) 085207.

[60] N. Raj and D.R. Tilley, in: The Dielectric Function of Condensed Systems, Eds., L.V. Keldysh, D.A. Kirzhnitz and A.A. Maradudin, North-Holland, Amsterdam, 1989.

[61] N.S. Almeida, E.L. Albuquerque and P. Fulco, Phys. Lett. A 160 (1991) 287.

[62] S. Perkowitz, D. Rajavel, I.K. Sou, J. Reno, J.P. Faurie, C.E. Jones, T. Casselman, K.A Harris, J.W. Cook Jr. and J.F. Schetzina, Appl. Phys. Lett. 49 (1986) 806.

[63] E.L. Albuquerque, J. Appl. Phys. 81 (1997) 4489.

[64] P. Dawson, D.G. Walmesley, H.A. Quinn and A.J.L. Ferguson, Phys. Rev. B 30 (1984) 3164.

[65] S. Ushioda, Y. Uehara, M. Takada, K. Otsubo and J. Murota, Jpn. J. Appl. Phys. 31 (1992) L870.

[66] A. Kuriyama, S. Takaoka, K. Ota, K. Murase, S. Shimomura, S. Hiyamizu, M. Chkr, T. Jungwirth and L. Smrcka, Solid State Commun. 111 (1999) 699.

[67] L. Hsu, M.D. McCluskey and E.E. Haller, Phys. Rev. B 67 (2003) 035209.

[68] T. Dumelow, A.A. Hamilton, K.A. Maslin, T.J. Parker, B. Samson, S.R.P. Smith, D.R. Tilley, R.B. Beall, C.T.B. Foxon, J.J. Harris, D. Hiltom and K.J. Moore, in: Light Scattering in Semiconductor Structures and Superlattices, Eds., D.J. Lockwood and J.F. Young, Plenum, New York, 1991.

[69] T. Dumelow, T.J. Parker, S.R.P. Smith and D.R. Tilley, Surf. Sci. Rep. 17 (1993) 151.

[70] G. Mirjalili, T.J. Parker, S.F. Shayesteh, M.M. Bulbul, S.R.P. Smith, T.S. Cheng and C.T. Foxon, Phys. Rev. B 57 (1998) 4656.

[71] G. Meyer, T. Crecelius, A. Bauer, I. Mauch and G. Kaindl, Appl. Phys. Lett. 83 (2003) 1394.

[72] J. Camarero, Y. Pennec, J. Vogel, M. Bonfim, S. Pizzini, F. Ernult, F. Fettar, F. Garcia, F. Lancon, L. Billard, B. Dieny, A. Tagliaferri and N.B. Brookes, Phys. Rev. Lett. 91 (2003) 027201.

[73] O. Zaharko, P.M. Oppeneer, H. Grimmer, M. Horisberger, H.C. Mertins, D. Abramsohn, F. Schafers, A. Bill and H.B. Braun, Phys. Rev. B 66 (2002) 134406.

[74] F. Matthes, A. Rzhevskii, L.N. Tong, L. Malkinski, Z. Celinski and C.M. Schneider, J. Appl. Phys. 93 (2003) 6504.

[75] V.I. Gavrilenko and R.Q. Wu, J. Mag. Magn. Mat. 260 (2003) 330.

[76] G. Metalidis and R. Bruno, Phys. Rev. A 66 (2002) 062102.

[77] J. Hamrle, J. Ferre, M. Nyvlt and S. Visnovsky, Phys. Rev. B 66 (2002) 224423.

[78] B. Heinrich, A.S. Arrott, J.F. Cochran, Z. Celinski and K. Myrtle in: Science and Technology of Nanostructured Magnetic Materials, Eds., G.C. Hadjipanayis and G.A. Prinz, Plenum, New York, 1991.

[79] J. Lindner and K. Baberschke, J. Phys.: Condens. Matter 15 (2003) S465.

[80] D.L. Mills, Phys. Rev. B 68 (2003) 014419.

[81] A. Cehovin, C.M. Canali and A.H. MacDonald, Phys. Rev. B 68 (2003) 014423.

[82] X. Liu, Y. Sasaki and J.K. Furdyna, Phys. Rev. B 67 (2003) 205204.

[83] B. Heinrich, in: Ultrathin Magnetic Structures, Vol. II, Eds., J.A.C. Bland and B. Heinrich, Springer-Verlag, Berlin, 1994.

Chapter 12

Concluding Topics

In the previous chapters we have assumed that the dielectric response of all the media is expressible in terms of a linear relationship between the electric displacement $\vec{D}$ and the macroscopic electric field $\vec{E}$, see e.g. Eqs. (3.1) and (3.2). However, all real materials are known to exhibit *non-linear effects* to a greater or lesser extent, and the previous linear assumption breaks down for sufficiently intense electric fields. As a consequence, in a non-linear medium the dielectric function $\epsilon(\omega)$, or more generally $\epsilon(\vec{k},\omega)$ if there is spatial dispersion (SD), has a dependence on the electric field. Likewise, in a non-linear magnetic material, there will be additional terms in the components of the magnetic permeability tensor.

The inclusion of non-linearities leads to a wide variety of new effects involving surface and bulk polaritons. This behavior is of considerable interest for the physics involved and for potential applications (e.g. for basic materials research, for the generation of new frequencies, and for all-optical signal processing). There are also propagating pulse-like excitations (non-linear wave packets), which include solitons as a case of particular interest and which have no counterpart in the linear regime.

The main emphasis in this final chapter will be on the non-linear effects in dielectric media. In Section 12.1 we describe the physical origin of the non-linear effects and the way in which the dielectric response will be modified. The books by Shen [1], Butcher and Cotter [2], and Mills [3] all provide a good introduction to the basics of non-linear optics, and they cover many of the applications also. Then in Sections 12.2–12.4 we discuss surface polaritons in several geometries, including the single-interface, double-interface, and multilayer cases. More details can be found in various articles by e.g. Stegeman et al. [4], Ponath and Stegeman [5], Chen and Maradudin [6], and Vassiliev and Cottam [7]. The emphasis will be on polaritons in non-magnetic materials.

The final part of this chapter, Section 12.5, consists of our overall conclusions in which we review the current situation and look ahead to some future directions and developments concerning the topics covered in this book.

12.1 Non-linear Dielectric Media

When the dielectric response is linear, the equations for the fields $\vec{E}$ or $\vec{H}$ in the presence of surfaces and interfaces are a set of linear differential equations [see e.g. Eqs. (5.3) and (5.4)] and boundary conditions. It follows that the principle of superposition holds, so that the sum of two solutions is itself a solution, and this property was used extensively in earlier chapters, e.g. in writing down the solutions for the field components in the nth unit cell of a superlattice in Eqs. (5.5)–(5.7).

In the case of non-linearity in the system, the above convenient properties no longer apply. Generally we express the polarization $\vec{P}$ in a non-linear medium as

$$\vec{P} = \vec{P}^L + \vec{P}^{NL}, \tag{12.1}$$

where the linear part can be written as $\vec{P}^L = \chi^{(1)}\vec{E}$, or more explicitly in component form as

$$P_i^L = \sum_j \chi_{ij}^{(1)} E_j. \tag{12.2}$$

Here $\chi^{(1)}$ denotes the second-rank (linear) susceptibility tensor, and i and j are Cartesian components. In an isotropic medium in the absence of spatial dispersion, $\chi^{(1)}$ is a scalar equal to $\epsilon_0[(\epsilon(\omega) - 1]$. The non-linear part $\vec{P}^{NL}$ takes different forms depending on the symmetry and other properties of the medium, and it will characteristically depend on the components of $\vec{E}$ to second or higher order. The inclusion of $\vec{P}^{NL}$ in Maxwell's equations can lead to new effects that were absent in the linear theory. It can result in a coupling (or "mixing") between the otherwise independent solutions obtainable in the linear approximation and, on the other hand, it can give rise to new types of surface polaritons that are essentially non-linear in nature. We shall discuss the occurrence of both phenomena.

The starting point is usually to expand $\vec{P}^{NL}$ as a Taylor series in powers of the components of the electric field(s) involved, leading to

$$P_i^{NL} = \sum_{jk} \chi_{ijk}^{(2)} E_j^a E_k^b + \sum_{jkl} \chi_{ijkl}^{(3)} E_j^a E_k^b E_l^c + \cdots, \tag{12.3}$$

where $\vec{E}^a$, $\vec{E}^b$, and $\vec{E}^c$ denote electric fields at frequencies ω_a, ω_b, and ω_c, respectively, and i, j, k, and l are Cartesian components. The lowest-order non-linear term involves $\chi^{(2)}$, which is a third-rank tensor whose non-zero elements can be determined from symmetry in the usual way (see e.g. Ref. [8]). It gives rise to wave-like solutions with the sum and difference frequencies $\omega_a \pm \omega_b$. These non-linear processes have been studied in the context of second harmonic generation, sum and difference frequency generation, optical rectification, etc. (see e.g. Refs. [2,5]).

The next term in Eq. (12.3), which involves the fourth-rank tensor $\chi^{(3)}$, is of particular interest for us because it generates wave-like solutions with the frequency combinations $\omega_a \pm \omega_b \pm \omega_c$. In particular, if $\vec{E}^a$, $\vec{E}^b$, and $\vec{E}^c$ are all at the same frequency ω, then one of the frequencies produced by the non-linear "mixing" is also equal to ω. This is in contrast to the situation with $\chi^{(2)}$. It follows that one of the effects is the appearance of an intensity-dependent dielectric function (for

12.1. NON-LINEAR DIELECTRIC MEDIA

frequency ω) in which the linear $\epsilon(\omega)$ becomes replaced by an effective non-linear dielectric function of the form

$$\epsilon^{NL} = \epsilon + \rho |\vec{E}|^2, \qquad (12.4)$$

where ρ is related to the appropriate non-vanishing tensor elements of $\chi^{(3)}$. It may, in principle, depend on frequency ω (as does the linear ϵ), although this is often neglected for simplicity. The form of the dielectric function in Eq. (12.4), with an intensity-dependent part proportional to $|\vec{E}|^2$, is often referred to as a *Kerr-type* non-linear dielectric function. The non-linear effects involving $\chi^{(3)}$ have been studied, for example, in the context of degenerate four-wave mixing, optical bistability, and coherent anti-Stokes Raman scattering (see e.g. Refs. [2,5]).

In the case of plasmon-polaritons, some of the earlier studies of non-linear effects concerned the second harmonic generation (SHG) of surface polaritons at a metal surface [9,10]. This is of interest because, for frequencies that are small compared with the plasma frequency ω_p, the surface plasmon-polariton dispersion curve is almost a straight line (see Fig. 4.2). Thus, corresponding to an incident surface polariton of frequency ω and in-plane wavevector k_x, the SHG output wave has frequency 2ω and wavevector $2k_x$. This means that its wavevector is very close to that for a freely propagating surface polariton in the metal, implying an enhancement of the efficiency for SHG as explained in Refs. [9,10]. This, and related work on SHG for plasmon-polaritons, has been discussed e.g. by Maradudin [11]. Further, it may be remarked that the non-linear processes corresponding to $\chi^{(2)}$ and $\chi^{(3)}$ can also contribute to the damping constant, e.g. as introduced phenomenologically in Eq. (4.43), and this is important for long-range surface plasmons (see Section 4.4).

However, the most interesting non-linear effects in the context of this book arise as a consequence of the effective dielectric function becoming dependent on the intensity, as in Eq. (12.4). It becomes possible, as we shall show in the next three sections for different geometries, to have new types of surface polaritons that are guided at the interface(s).

It is helpful first to make some further comments about Eq. (12.4). It can be re-expressed as an intensity-dependent refractive index n^{NL} defined by the relation $n^{NL} = (\epsilon^{NL})^{1/2}$. This means that to leading order in the term proportional to $|\vec{E}|^2$, assumed small, we have

$$n^{NL} = n_0 + n_2 |\vec{E}|^2 \qquad (12.5)$$

with $n_0 = \epsilon^{1/2}$ and $n_2 = \rho/2n_0$. The non-linear coefficient ρ, and hence n_2, can take either sign. However, the case of $\rho > 0$ corresponds to *self-focusing* of a laser beam, because the effective refractive index will then be largest at the beam center where the intensity is largest. This causes the beam to narrow, an effect opposed by the linear diffraction which gives a spreading effect [1]. Likewise, $\rho < 0$ corresponds to self-defocusing of a laser beam. We also remark that Eq. (12.4) eventually breaks down if the intensity becomes large enough, because the perturbation theory on which it is based is no longer applicable. Instead, a *saturation* effect is observed [2], whereby n^{NL} approaches a finite limit as $|\vec{E}|^2 \to \infty$.

In ferromagnetic and antiferromagnetic materials non-linear optics can, by analogy, be described in terms of an additional contribution to the magnetic permeability function that is proportional to $|\vec{H}|^2$ (see e.g. Ref. [12]). We mention this again in Section 12.4.

12.2 Non-linear Excitations in Single-Interface Geometries

In order to illustrate the general principles, we start in this section with a simple example where there is just one interface, and subsequently in Sections 12.3 and 12.4 we will generalize to thin films and multilayers (including superlattices), respectively.

We consider the case of an *s*-polarized (TE) mode at an interface between two semi-infinite isotropic media, where one is linear and the other is non-linear with an intensity-dependent dielectric function [13]. We use the same geometry and notation as in Fig. 4.1. As before, we deal with a single-frequency mode with in-plane propagation in the x-direction, so all amplitudes vary as $\exp(ik_x x - i\omega t)$, and the $\vec{E}$ field is in the y-direction for a TE mode.

Medium A, which occupies the half space $z > 0$, is taken to be non-linear and to have an intensity-dependent dielectric function $\epsilon_A + \rho_A |E_A|^2$, while medium B (occupying $z < 0$) is linear with dielectric constant ϵ_B. For simplicity, we ignore any frequency dependence of ϵ_A, ρ_A, and ϵ_B at this stage. The wave equation, obtained from Eq. (3.17), in the non-linear medium is

$$d^2 E_A/dz^2 - \left[k_x^2 - \left(\epsilon_A - \rho_A |E_A|^2\right)\omega^2/c^2\right] E_A = 0. \tag{12.6}$$

Taking E_A to be real, this equation can first be multiplied by dE_A/dz and then integrated to give an expression for dE_A/dz:

$$dE_A/dz = \left[2K + \left(k_x^2 - \omega^2 \epsilon_A/c^2\right)E_A^2 - \omega^2 \rho_A E_A^4/4c^2\right]^{1/2}, \tag{12.7}$$

where K is a constant of integration. However, as $z \to \infty$ in the non-linear medium, we require E_A and dE_A/dz to be zero for localization, and so $K = 0$. It is then straightforward to integrate once more, yielding

$$E_A = (2/\rho_A)^{1/2}(c\alpha_A/\omega)\operatorname{sech}[\alpha_A(z - z_0)], \quad z > 0, \tag{12.8}$$

where the real coefficient $\alpha_A = (k_x^2 - \epsilon_A \omega^2/c^2)^{1/2} > 0$ is defined as in Eq. (5.8), and z_0 is another constant of integration. Eq. (12.8) holds provided $\rho_A > 0$ (the self-focusing case). The corresponding solution for E_B in the linear medium B (in the half space $z < 0$) that corresponds to a guided wave with attenuation as $z \to -\infty$ is simply

$$E_B = A \exp(\alpha_B z), \quad z < 0, \tag{12.9}$$

12.2. NON-LINEAR EXCITATIONS IN SINGLE-INTERFACE GEOMETRIES

where A is a constant and α_B, which is real and positive, is defined in an analogous fashion to α_A. The usual electromagnetic boundary conditions at the interface $z = 0$ are equivalent to requiring that both E and dE/dz are continuous. After applying these and eliminating the constant A, we obtain the condition

$$\tanh(\alpha_A z_0) = \alpha_B/\alpha_A. \tag{12.10}$$

The above expression does not provide us with a dispersion relation for ω versus k_x because it contains the unknown constant z_0, and so another condition is required. Before considering this, we note that there are two necessary conditions for any solution. First, since the hyperbolic tangent function never exceeds unity, it follows from Eq. (12.10) that $\alpha_A > \alpha_B$ (from which we deduce $\epsilon_A < \epsilon_B$ and $z_0 > 0$). Second, since α_A and α_B must both be real, the dispersion curve lies to the right of the light line for both media.

The complete solution to this single-interface problem can now be found by recognizing that the field intensity $|\vec{E}_A|^2$ in the non-linear medium is a non-trivial parameter, and so we shall study it in terms of the time-averaged power flow. This is obtained starting from the Poynting vector $\vec{S} = \vec{E} \times \vec{H}$ in electromagnetism, which can have x- and z-components in our case. It is easy to conclude that the time average of the latter contribution is zero, whereas the time-averaged x-components are found to be

$$\bar{S}_x(z) = \frac{c^2 k_x \alpha_A^2}{\mu_0 \rho_A \omega^3} \operatorname{sech}^2[\alpha_A(z - z_0)], \quad z > 0, \tag{12.11}$$

$$\bar{S}_x(z) = \frac{k_x A^2}{2\mu_0 \omega} \exp(2\alpha_B z), \quad z < 0. \tag{12.12}$$

The time-averaged, x-directed power flow $\mathcal{P}$, per unit length in the y-direction, is therefore found from the integral

$$\mathcal{P} = \int_{-\infty}^{\infty} \bar{S}_x(z)\, dz. \tag{12.13}$$

On substituting Eqs. (12.11) and (12.12) into Eq. (12.13) and carrying out the integrations, we find after some algebra that

$$\mathcal{P} = \mathcal{P}_0 \left[1 + \frac{1}{2}\left(\frac{\alpha_A}{\alpha_B} + \frac{\alpha_B}{\alpha_A}\right)\right], \tag{12.14}$$

where $\mathcal{P}_0 = c^2 k_x \alpha_A / \mu_0 \rho_A \omega^3$. This result indicates that a physical solution exists only if the power $\mathcal{P}$ exceeds some threshold that is greater than $2\mathcal{P}_0$ (depending on the ratio α_A/α_B). The threshold power becomes large if the non-linearity term ρ_A is small or if α_A/α_B is large.

The solutions for the non-linear-mode frequency ω versus k_x can be obtained using graphical methods (see Ref. [14]). For example, Eq. (12.14) can easily be rewritten as an expression for the dimensionless quantity $\Omega = \mu_0 \rho_A \mathcal{P} \omega$ in terms of the dimensionless $\xi = c k_x / \omega$. For each value of ξ we may therefore calculate Ω

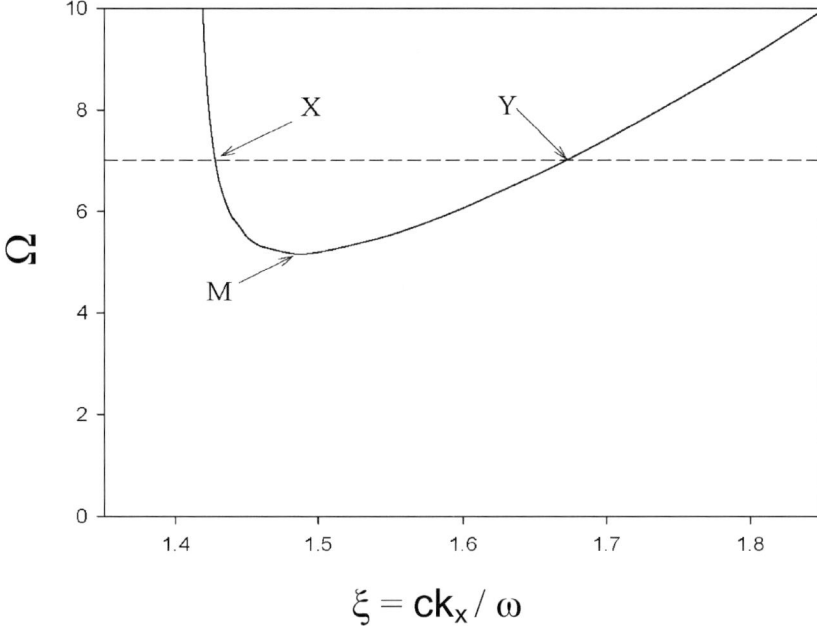

Fig. 12.1. Calculated dispersion curve for an *s*-polarized self-guided electromagnetic wave propagating at the interface between a non-linear dielectric medium (with $\epsilon_A = 1.5$) and a linear dielectric medium (with $\epsilon_B = 1.5$). We have plotted the dimensionless variables $\Omega = \mu_0 \rho_A \mathcal{P} \omega$ against $\xi = ck_x/\omega$.

(which is a scaled frequency). The scaled wavevector $\mu_0 \rho_A \mathcal{P} c k_x$ can be obtained from the product $\xi\Omega$. A numerical example, obtained taking $\epsilon_A = 1.5$ and $\epsilon_B = 2$, is shown in Fig. 12.1 as a plot of Ω against ξ. The requirement for α_B to be real implies that ξ must be greater than $\epsilon_B^{1/2}$. It is seen that the curve has a minimum near the point labelled M, where $\Omega \simeq 5.1$. For larger Ω the dispersion relation provides an example of *bistability*, i.e. in this case there are two modes with different k_x values that propagate with the same frequency. Thus, for $\Omega = 7$ (marked by the dashed line in Fig. 12.1), the two modes correspond to the points labelled X and Y).

The above calculation for a single interface has been generalized by Wendler [15] to include a thin polarizable dipole-active layer at the interface; this involves modifying the boundary condition for the H field (e.g. as was done for a 2D charge sheet in Section 4.3). He also treats the case of $\rho_A < 0$ as well as $\rho_A > 0$, and he allows for the possibility that the dielectric function of the linear medium is frequency dependent, taking the form in Eq. (3.11). Further discussion is provided by Stegeman et al. [4] and Maradudin [16], with the latter author discussing anisotropic materials.

The case of a *p*-polarized wave at an interface has also received attention, but it is more complicated because more field components are involved than in the case of s-polarization. However, Agranovich et al. [17] showed that non-linear p-polarized surface polaritons can exist at the interface between a linear and a non-linear medium when the dielectric constants of *both* media are positive. By contrast, for linear surface polaritons this is not possible (see Section 4.1). A more complete analysis has been developed by Leung [18,19]. In the case of weak non-linearity, meaning $\rho_A|\vec{E}|^2 \ll \epsilon_A$, he found

$$\frac{ck_x}{\omega} = \left(\frac{\epsilon_A \epsilon_B}{\epsilon_A + \epsilon_B}\right)^{1/2} \left(1 + \frac{\epsilon_B \rho_A |\vec{E}_1|^2}{4\epsilon_A(\epsilon_A + \epsilon_B)}\right), \quad (12.15)$$

where $\vec{E}_1$ is the electric field at the interface. This may be considered as a generalization of Eq. (4.4) to the non-linear case. In fact, an empirical expression of this form had previously been used by Chen and Carter [20] to analyze their experimental data on this mode.

Measurements of the non-linear susceptibility $\chi^{(3)}$, which is related to the magnitude of the non-linear coefficient ρ, by a technique using the intensity-dependent dispersion relation of p-polarized surface plasmon-polaritons were reported [20] for GaAs/Ag and Si/Ag systems (where GaAs and Si are the non-linear media). These results were subsequently applied to make ATR calculations for these same modes and materials [21]. Also it is known that some other semiconductors, such as InSb and HgCdTe, have enhanced third-order susceptibilities [15], making them suitable choices of materials for these studies.

12.3 Non-linear Excitations in Double-Interface Systems

There have been extensive studies for the two-interface geometry in which a thin film is sandwiched between two semi-infinite media. Various possibilities arise in these three-layer systems, depending on which of the layers or combinations of the layers are taken to be non-linear. In general, they are more useful systems than the single-interface case of Section 12.2 because the non-linear surface polaritons can have low-power limits. Moreover, self-focused fields are simultaneously possible in more than one medium, and so multiple guided-wave branches in the excitation spectrum can occur. Because the calculations are rather lengthy, we give only a brief, mainly qualitative treatment illustrated by a few numerical examples. Further details, including references to other studies, can be found in Refs. [4,6,7,15].

The single-interface calculation in Section 12.2 has been generalized by Wendler [15] to include the case of a surface-active film between two semi-infinite media, one of which has a Kerr-type non-linear dielectric function. However many studies have been made for the more general case of a thin film of a linear dielectric medium surrounded either symmetrically or asymmetrically by semi-infinite non-linear media as cladding (or bounding) layers. For example, Stegeman and Seaton [22] calculated the dispersion relations for p-polarized non-linear surface plasmons, making numerical applications to a linear InSb film with non-linear cladding layers. They found, depending on the conditions, that the usual two modes can be cut off and can undergo radical changes in their electric-field distributions, while new modes can exist above certain power thresholds.

For the case of s-polarized non-linear surface polaritons where the thin film is non-linear and the cladding media are linear, one may follow the same kind of calculation as in the previous section (see Refs. [23,6]). In particular, Eq. (12.7) still holds in the non-linear medium (extending now from $z = -L/2$ to $z = L/2$, say, for a film of thickness L), but the integration constant K need not be zero, because the non-linear medium is bounded in the z-direction. Instead of a sech-function dependence of the electric field, as in Eq. (12.8), the field in the film can be expressed in terms of Jacobian elliptic functions. The opposite case of s-polarized waves in a linear dielectric film bounded symmetrically on both sides by a non-linear dielectric medium was studied by Akhmediev [24] and Boardman and Egan [25].

Later, Vassiliev and Cottam [7] developed a theory of s-polarized non-linear electromagnetic waves in general multilayers with an arbitrary number of interfaces. As a special case, they were able to derive further results applicable to the two-interface geometries, including the case when all three media could be non-linear with dielectric functions having the form of Eq. (12.4) for $\rho > 0$. As well as reducing to the above-mentioned references in the appropriate limiting cases, their theory also led to the analytic results expressible in a more compact form convenient for numerical evaluation.

As a specific example, we shall consider the case of a non-linear film with ϵ^{NL} given by Eq. (12.4) with $\rho > 0$, bounded on each side by semi-infinite cladding layers composed of the same material with linear dielectric constant ϵ_c, where $\epsilon > \epsilon_c$. In s-polarization the only non-zero component of the electric field is the y-component and we express its dependence in the form $E(z)\exp(ik_x x - i\omega t)$ as in the previous section. It is found that the symmetric and antisymmetric modes (with respect to the mid-point of the film of thickness L) can be described by the electric-field distribution (see Ref. [7])

$$E(z)/E_1 = b_0\,\mathrm{cn}\Big([k_0 z + jK]|m\Big) \qquad (12.16)$$

for $-L/2 < z < L/2$, where j takes non-negative integer values. Here the electric field has been expressed relative to E_1, which denotes the field at the interface $z = -L/2$. It arises as an extra parameter, just as we obtained an extra parameter z_0 in the example given in Section 12.2; as before, it can be related to the power flow. In the linear media outside the film, the field $E(z)$ decays exponentially as $z \to \pm\infty$, just as in Chapter 4. The above expression involves a Jacobian elliptic function (see Ref. [26]) of the type $\mathrm{cn}(x|m)$. This has broadly similar features to $\cos(x)$, but a second argument m determines its shape and period. The constant K is defined as one quarter of the period, i.e. $K(m) \equiv \mathrm{cn}^{-1}(0|m)$. Hence, because of the periodicity of the cn function it is sufficient to take the set of j as 0, 1, 2, and 3. The other quantities appearing in Eq. (12.16) are

$$k_0(\omega) = (\omega/c)\Big[D(\omega)\Big]^{1/2}, \qquad (12.17)$$

12.3. NON-LINEAR EXCITATIONS IN DOUBLE-INTERFACE SYSTEMS

$$m(\omega) = p\big[b_0(\omega)\big]^2 \big/ D(\omega), \tag{12.18}$$

$$b_0(\omega) = \left\{ \big[(ck_x/\omega)^2 - \epsilon + D(\omega)\big] \big/ 2p \right\}^{1/2}, \tag{12.19}$$

where we define $p = \rho E_1^2/2$ as a scaled non-linearity coefficient and

$$D(\omega) = \left\{ \big[(ck_x/\omega)^2 - \epsilon\big]^2 + 4p(\epsilon - \epsilon_c + p) \right\}^{1/2}. \tag{12.20}$$

It is easily verified that $D(\omega)$ is a *real* function of frequency, and thus k_0, m, and b_0 are all real in this case. The quantity k_0 plays the role of an effective wavevector component (in the z-direction) for the guided polaritons in the film.

The implicit form of the polariton dispersion relation corresponding to Eq. (12.16) is found to be (see Ref. [7])

$$b_0 \operatorname{cn}\!\Big([k_0 L/2 - jK]\big|m\Big) = 1 \qquad (j = 0, 1, 2, 3). \tag{12.21}$$

This expression is analogous to a quantization condition for the wavevector component k_0, and the modes can then be classified according to the number of nodes, i, occurring in the cn function between $z = -L/2$ and $z = L/2$, where $i = 0, 1, 2, \ldots$. There is also a dependence in Eq. (12.21) on the parameter j entering the dispersion relation (and taking four distinct values). Therefore the modes can be labelled in terms of their (i, j) values. Eq. (12.21) can be solved numerically and some results are displayed in Fig. 12.2. These results are shown in terms of a plot of an effective refractive index $n = ck_x/\omega$, where k_x is the in-plane wavevector, versus $p^{1/2}$, where p is the scaled non-linearity coefficient defined earlier. Other parameters are chosen as $\epsilon_c = 1$, $\epsilon = 4$, and $L = 0.5\lambda$, where $\lambda = 2\pi c/\omega$ is the vacuum wavelength of light. Dispersion curves of this type were previously found by Chen and Maradudin [6].

The curves in Fig. 12.2 have the property that n is a double-valued function of $p^{1/2}$ for certain ranges of values of p. This is obvious for the first two curves labelled (0,0) and (0,1), but it is in fact a general property here. This is indicative of the possible existence of bistable states. A further comment about the modes represented by Eqs. (12.16) and (12.21), as well as in Fig. 12.2, is that they represent only symmetric and antisymmetric solutions. However, as discussed in Refs. [6,7], it is possible that spatially asymmetric modes might occur for some values of the parameters.

The case of a non-linear film bounded by non-linear cladding materials is, as might be expected, much more complicated [7], and so we deal with it more briefly. We assume now that the non-linear dielectric function of the cladding layers (on both sides) is $\epsilon_c + \rho_c |\vec{E}|^2$, while that of the film is denoted by Eq. (12.4) as before. It turns out that for symmetric and antisymmetric modes there are two main cases, depending on whether $\epsilon + p > \epsilon_c + p_c$ (case 1) or $\epsilon + p < \epsilon_c + p_c$ (case 2), where $p = \rho E_1^2/2$ and $p_c = \rho_c E_1^2/2$ are the scaled non-linearity coefficients. Case 1 turns out in some respects to be analogous to the previous example where only the film

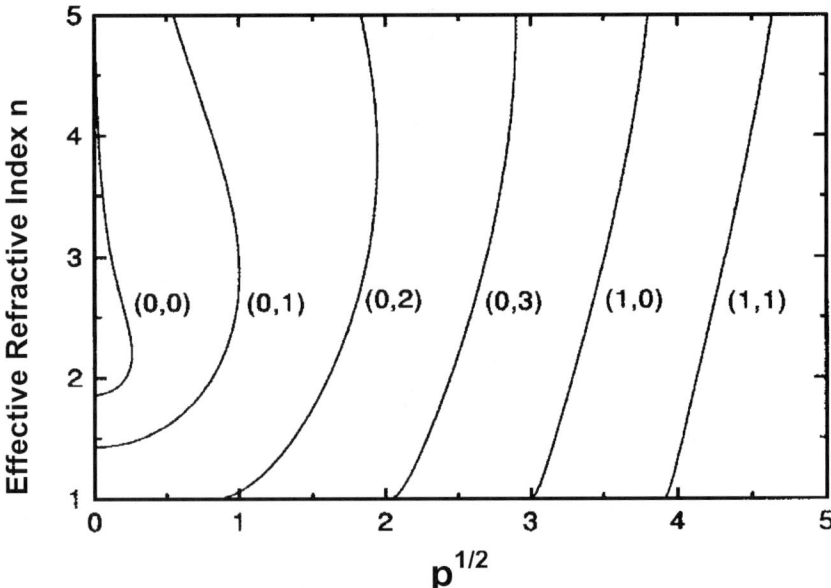

Fig. 12.2. Dispersion curves for s-polarized polaritons for a two-interface geometry consisting of a non-linear film bounded symmetrically by two linear media. The plots are shown as an effective refractive index n versus $p^{1/2}$, and the curves are labelled according to the integers (i, j). See the text for parameter values (after Vassiliev and Cottam [7]).

was non-linear, provided we redefine D in Eq. (12.20) as [7]

$$D(\omega) = \left\{ \left[(ck_x/\omega)^2 - \epsilon \right]^2 + 4p(\epsilon - \epsilon_c + p - p_c) \right\}^{1/2}. \quad (12.22)$$

The dispersion relation is then formally similar to Eq. (12.21), with four equations corresponding to $j = 0, 1, 2, 3$ and two extra equations with 0 (instead of 1) on the right-hand side with $j = 0, 1$. The two extra equations arise from the possibility that there can now be extrema of the electric-field distribution within the cladding layers (rather than just an exponential decay, as before). A numerical example is given for this case in Fig. 12.3, taking parameters $\epsilon = 2.45$, $\epsilon_c = 2.3$, $p/p_c = 2$, and $L = 0.5\lambda$. We notice that some curves, the ones labelled with the (i, j) notation, are analogous to those occurring in Fig. 12.2. The additional types of curves labelled A and B correspond to the extra modes with field extrema in the cladding layers. Another obvious difference compared with Fig. 12.2 is that all the physical modes lie above the line (shown dotted) corresponding to $n^2 = p_c + \epsilon_c$. The other case mentioned (case 2) can be treated in a similar manner [7]; it is found to involve other types of elliptic functions in the dispersion relation and electric-field distributions.

The calculations described in this section could be extended to study the power flow $\mathcal{P}$ in the layered system, just as was described for the single-interface

12.4. NON-LINEAR EXCITATIONS IN MULTILAYER SYSTEMS

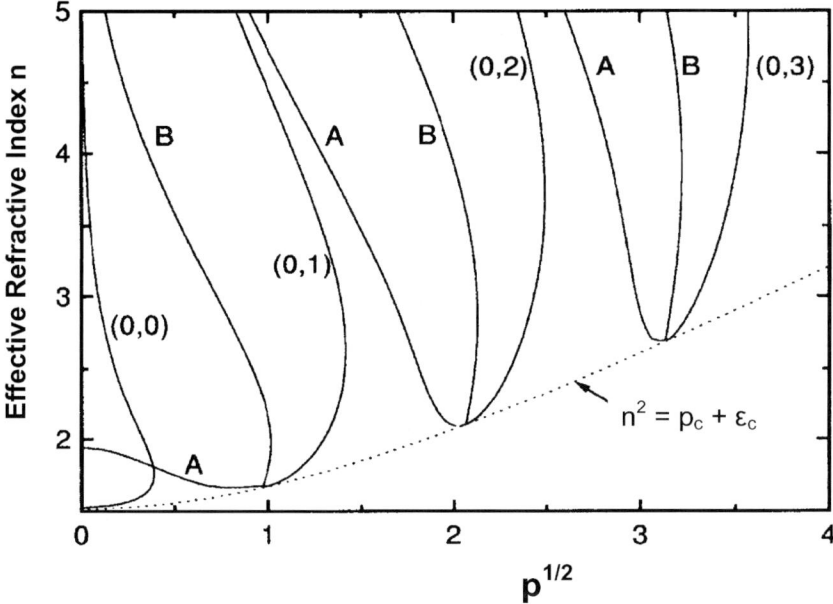

Fig. 12.3. Dispersion curves for s-polarized polaritons for a two-interface geometry consisting of a non-linear film bounded symmetrically by two non-linear cladding media, assuming $p/p_c = 2$ for the ratio of non-linearity coefficients. The plots are shown as effective refractive index n versus $p^{1/2}$. The labelling of the curves and the parameter values are explained in the text (after Vassiliev and Cottam [7]).

geometry in Section 12.2. This would enable the scaling term E_1 in the electric field to be related to $\mathcal{P}$ and would lead to an analysis of the threshold power levels required for some of the modes. The calculations can also be extended to include a characteristic frequency dependence of either ϵ for the film or ϵ_c for the cladding layers, or both quantities. Some preliminary studies using the plasma form of the dielectric function, Eq. (1.24), have been reported by Baher and Cottam [27].

12.4 Non-linear Excitations in Multilayer Systems

In contrast to the previous two sections, there has been much less work on polaritons in geometries with a large number of layers, including the superlattice case. However, the non-linear optical properties of such systems have received a lot of attention. The focus has been on studying the optical transmission and reflection characteristics for multilayers in which the individual films (or at least some of them) are described by a Kerr-type non-linear dielectric function. Such calculations for multilayers were reported, for example, by Chen and Mills [28] as an extension of their earlier work for a single film [29]. The calculations can be carried out using the so-called non-linear transfer-matrix method introduced by Dutta Gupta and Agarwal [30]; the exact boundary conditions at interfaces are used, but within each non-linear film an approximation known as the slowly varying amplitude approximation (SVA) is employed to make the analysis tractable. Essentially, SVA assumes that the amplitudes of the light beams in the sample vary over lengths that are long on the scale of the optical wavelength, and so it should be valid for weak non-linearity. Formally the optical reflection and transmission properties are found in terms of a matrix product of a large number of non-linear transfer matrices, so the numerical evaluation requires an elaborate calculational scheme; this is reviewed by Dutta Gupta [31].

There have been analogous studies to those of the preceding paragraph for finite superlattices composed of alternating layers of an antiferromagnet (such as FeF_2) and a non-magnetic dielectric material (such as ZnF_2); see Refs. [12,32,33]. The antiferromagnetic layer can be characterized by a permeability tensor that includes a non-linear part proportional to the square of the magnetic field, i.e. it is the magnetic-field analog of Kerr-type non-linearity for the dielectric function in Eq. (12.4). By calculating the transmission coefficient through a finite FeF_2/ZnF_2 superlattice (assuming ZnF_2 to be a linear medium), Kahn et al. [33] identified resonances that they associated with gap solitons. The same authors also studied finite quasiperiodic structures, generated using the Fibonacci sequence, for the same materials, but it is interesting that they found no evidence for solitons in the stop gap for this case.

Turning specifically now to non-linear polaritons, we mentioned in the previous section that the formalism developed by Vassiliev and Cottam [7] for s-polarized modes was developed for a general geometry consisting of a stack of N finite-thickness films bounded by two semi-infinite cladding media. The calculations are relatively straightforward for $N=1$ (i.e. the two-interface case already discussed), but become much more challenging numerically as N increases unless other assumptions are introduced. It will suffice just to show one numerical example here: the limiting case of an infinite two-component periodic superlattice in which both constituent media A and B (with thicknesses a and b, respectively) have Kerr-type non-linear dielectric functions $\epsilon_i + \rho_i |\vec{E}|^2$ (with $i = A, B$). While there is no rigorous analog of Bloch's theorem in a non-linear superlattice, we can still calculate the modes that are consistent with this symmetry (while modes with other symmetries are, of course, not excluded). Thus in Fig. 12.4 we show calculations corresponding to the Bloch factor, $\exp(iQL)$, in our usual notation where $L = a + b$ is the periodicity length, taking values equal to -1 for $QL = \pi$ (solid lines) and 1 for $QL = 0$ (dashed lines). The assumed parameters are $\epsilon_A = 2.3$, $\epsilon_B = 2.45$, $\rho_A/\rho_B = 0.5$, and $b/\lambda = 0.5$, while three different values of a/λ are considered. As before, the plots are in terms of the effective refractive index n versus $p^{1/2}$, where here $p = \rho_B E_1^2/2$. The two chosen values of the Bloch wavevector Q usually correspond to the band edges, and it can be seen that as the thickness a is increased (with b having a fixed value) the bands become narrower as expected, since the B layers are further apart, and they approximate to the (0,0) mode in Fig. 12.3.

We anticipate that non-linear excitations in superlattices, both periodic and quasiperiodic, are topics that should attract much more attention in the future.

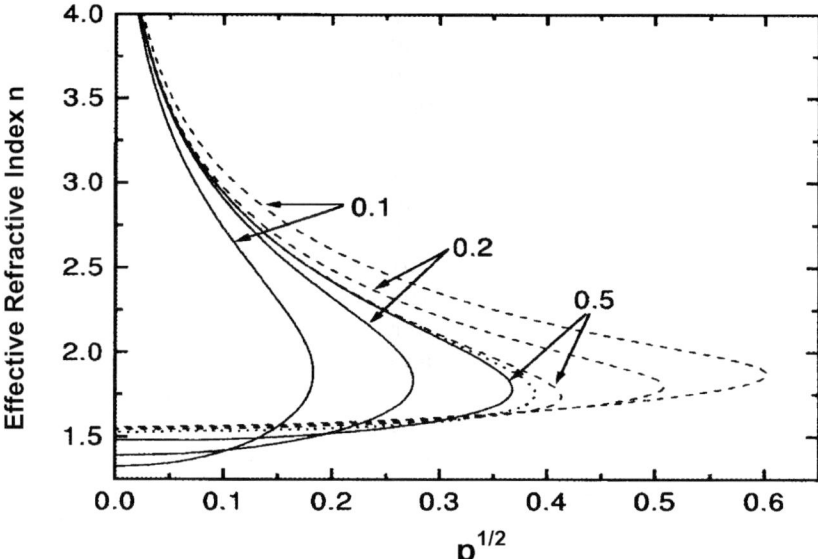

Fig. 12.4. Dispersion curves for s-polarized polaritons for a two-component superlattice where both constituents A and B have non-linear dielectric functions. Results are shown for three values of a/λ, as indicated, and the other parameter values are given in the text. Curves corresponding to the Bloch factor equal to -1 and 1 are indicated by solid and dashed lines, respectively. The plots are in terms of effective refractive index n versus $p^{1/2}$. For comparison the dotted line (unlabelled) shows a corresponding single-layer curve [mode (0,0) of Fig. 12.3] (after Vassiliev and Cottam [7]).

12.5 Conclusions and Future Directions

In this book, we have presented the current state of the art for various bulk and surface polariton modes (mainly those related to phonon, plasmon, exciton, and spin-wave excitations) that propagate in periodic and quasiperiodic structures of the substitutional type.

Periodic multilayered systems, such as the superlattice structures described here, opened up a new area of interdisciplinary investigations in the 1970s in the fields of materials science and device physics. A great variety of materials, including semiconductors, dielectrics, and magnetic crystals, have been exploited for their synthesis and novel properties. Ideas originating from these superlattices have found their applications in a number of important devices like quantum-well lasers, avalanche photodiodes, and spin valves, to name just a few. Superlattices even now constitute one of the most lively areas in solid-state physics, where new results are still being vigorously generated.

On the other hand, quasiperiodic systems are basically a novel class of materials in the sense that they have introduced ideas about condensed-matter structures with topological long-range order which are distinct from the usual periodicity. Their discovery and understanding have come about through a mixture of strong motivation from basic scientific interest and technological applications. Moreover, they give rise to many theoretical challenges and approaches in attempts to understand and predict the physical properties of long-range ordered structures, without the symmetry aspects and degeneracy associated with periodic invariance.

Since the defining rules of the quasiperiodic sequences impose long-range correlations on the excitations, it becomes plausible to search for *global* (or *universal*) consequences of these correlations. Indeed, the overall aspects of these sequences were found in structural features related to their bandwidths, in their localization and self-similarity behavior, and in fractal properties that can be described by power laws (scaling laws) not found in periodic crystals.

The field of magnetic ultrathin metallic structures is rapidly becoming one of the most active and exciting areas of current solid-state research. The magnetism in periodic and quasiperiodic superlattices, made up from the stacking arrangement of these materials, is nowadays an emerging important field of investigation. There are many novel physical properties, with potential technological applications, associated with these structures. Among them, the striking self-similar magnetoresistance profiles as a function of the biquadratic/bilinear exchange ratio found in quasiperiodic structures is a new magnetic phenomenon with no counterpart in the periodic arrangement, and undoubtedly deserves further theoretical and experimental investigations.

A quite interesting behavior is also found for the properties of the specific heat derived from the fractal spectra of the *real* excitations obtained in quasiperiodic structures. Theoretical results show that in the low-temperature limit, there is an unexpected oscillatory aspect to the specific-heat spectra. We believe that the physical origins of these properties are related to the nature of the excitation spectra in the quasiperiodic systems, since they have different scaling laws and, therefore, different fractal dimensions.

In closing, we would like to mention briefly some directions of the current research. The search for understanding the detailed nature of self-similar energy spectra is still attracting attention both from theoreticians and experimentalists. The self-similarity is responsible for almost all interesting physical properties described in this book. Another area of interest is the characterization of the transmission return maps, which behave like strange attractors defining the fingerprints of the quasiperiodic systems. They may be characterized by the singularities of the multifractal spectra and by Lyapunov exponents (see e.g. Ref. [34] for definitions of these terms). For instance, the oscillatory behavior of the specific-heat spectra at low temperatures may possibly be connected with the recently introduced non-extensive thermostatistics developed by Tsallis [35]. Indeed, the generalized specific heat $C_q(T)$ of the quantum one-dimensional harmonic oscillator does present oscillations if the entropic index q satisfies the inequality $q < 1$ [36]. Finally, as a consequence of the investigations related to the scale invariant properties in complex genomic sequences (like the first completely sequenced human chromosome 22 containing about 33.4×10^6 nucleotides), the nature of *long-range correlations* in DNA sequences has become the subject of intense debate [37,38]. The π-stacked array of DNA base pairs (made up from the nucleotides guanine

G, adenine A, cytosine C, thymine T) as well as a Fibonacci polyGC quasiperiodic sequence, constructed starting from a G nucleotide as seed and following the inflation rule $G \to GC$ and $C \to G$, can be used as prototypes of strongly correlated systems. These developments open the door for the understanding and engineering of biological processes.

In summary, we have presented an up-to-date account concerning the physical properties of a range of polaritons in periodic and quasiperiodic structures. We hope that this book can be useful for all those interested in this subject.

References

[1] Y.R. Shen, The Principles of Nonlinear Optics, Wiley, New York, 1984.
[2] P.N. Butcher and D. Cotter, The Elements of Nonlinear Optics, Cambridge University Press, Cambridge, 1990.
[3] D.L. Mills, Nonlinear Optics, 2nd. ed., Springer, Berlin, 1998.
[4] G.I. Stegeman, C.T. Seaton, W.M. Hetherington, A.D. Boardman and P. Egan, in: Electromagnetic Surface Excitations, Eds. R.F. Wallis and G.I. Stegeman, Springer, Berlin, 1986.
[5] H.E. Ponath and G.I. Stegeman, Eds., Nonlinear Surface Electromagnetic Phenomena, North-Holland, Amsterdam, 1991.
[6] W. Chen and A.A. Maradudin, J. Opt. Soc. Am. B 5 (1988) 529.
[7] O.N. Vassiliev and M.G. Cottam, Surf. Rev. Lett. 7 (2000) 89.
[8] J.F. Nye, Physical Properties of Crystals, Clarendon, Oxford, 1985.
[9] D.L. Mills, Solid State Commun. 24 (1977) 669.
[10] M. Fukui and G.I. Stegeman, Solid State Commun. 26 (1978) 239.
[11] A.A. Maradudin, in Festkorperprobleme XXI, Ed. J. Treusch, Vieweg, Braunschweig, 1981.
[12] W. Chen and D.L. Mills, Phys. Rev. Lett. 58 (1987) 160.
[13] A.E. Kaplan, Sov. Phys. JETP 45 (1977) 86.
[14] M.G. Cottam and D.R. Tilley, Introduction to Surface and Superlattice Excitations, 2nd. ed., Institute of Physics Publishing, Bristol, 2004.
[15] L. Wendler, Phys. Stat. Solidi (b) 135 (1986) 759.
[16] A.A. Maradudin, Z. Phys. B 41 (1981) 341.
[17] V.M. Agranovich, V.S. Babichenko and V.Ya. Chernyak, JETP Lett. 32 (1980) 512.
[18] K.M. Leung, Phys. Rev. A 31 (1985) 1189.
[19] K.M. Leung, Phys. Rev. B 32 (1985) 5093.
[20] Y.J. Chen and G.M. Carter, Appl. Phys. Lett. 41 (1982) 307.
[21] Y.J. Chen and G.M. Carter, Solid State Commun. 45 (1983) 277.
[22] G.I. Stegeman and C.T. Seaton, Opt. Lett. 9 (1984) 235.
[23] A.D. Boardman and P. Egan, IEEE J. Quantum Elect. 21 (1985) 1701.
[24] N.N. Akhmediev, Sov. Phys. JETP 56 (1982) 299.
[25] A.D. Boardman and P. Egan, IEEE J. Quantum Elect. 22 (1986) 319.
[26] M. Abramowitz and I.A. Stegun, Handbook of Mathematical Formulas, Dover, New York, 1968.
[27] S. Baher and M.G. Cottam, Surf. Rev. Lett. 10 (2003) 13.
[28] W. Chen and D.L. Mills, Phys. Rev. B 36 (1987) 6269.
[29] W. Chen and D.L. Mills, Phys. Rev. B 35 (1987) 524.

[30] S. Dutta Gupta and G.S. Agarwal, J. Opt. Soc. Am. B 4 (1987) 691.
[31] S. Dutta Gupta, Progress in Optics 38 (1998) 1.
[32] N.S. Almeida and D.L. Mills, Phys. Rev. B 36 (1987) 2015.
[33] L.M. Kahn, K. Huang and D.L. Mills, Phys. Rev B 39 (1989) 12449.
[34] S.H. Strogatz, Nonlinear Dynamics and Chaos, Addison-Wesley, Reading, 1994.
[35] C. Tsallis, J. Stat. Phys. 52 (1988) 479.
[36] G.R. Guerberoff, P.A. Pury and G.A. Raggio, J. Math. Phys. 37 (1996) 1790.
[37] P. Carpena, P.B.-Galván, P.Ch. Ivanov and H.E. Stanley, Nature 418 (2002) 955.
[38] S. Roche, D. Bicout, E. Maciá and E. Kats, Phys. Rev. Lett. 22 (2003) 2228101.

Appendix A

Some Theoretical Tools

For the most part in this book we have employed relatively standard theoretical methods. Thus, when deriving the dispersion relations corresponding to the various excitations in a multilayer system, our general approach has been to employ a basic equation of motion, often a differential equation or operator equation, together with boundary conditions to describe the interfaces and (where appropriate) transfer matrices to generate the layered structure. In some cases, however, we might need to calculate additional properties of an excitation, such as its contribution to a thermodynamic quantity (e.g. specific heat), a transport process (e.g. electrical conductivity), or a scattering intensity (e.g. in Raman or Brillouin scattering of light). While these are specialized topics, the purpose of this Appendix is to present a concise account of these methods in a form applicable to multilayered structures in order to supplement the main material of the earlier chapters.

The additional theoretical tools that are most useful involve using quantum-mechanical perturbation theory (for example, to calculate higher-order effects and transition probabilities). Often this will be done within the framework of many-body theory, where concepts such as second quantization and Green functions are important. Methods to calculate Green functions (which basically provide information about the correlations in an interacting system) include equation-of-motion methods, linear-response theory, and diagrammatic perturbation theory (Feynman-diagram methods). There are many textbooks giving comprehensive treatments and we mention, for example, those by Fetter and Walecka [1], Economou [2], Rickayzen [3], Negele and Orland [4], and Kittel [5]. Specific applications to surfaces (mainly in a thin-film geometry) have been reviewed by Cottam and Maradudin [6].

A.1 Perturbation Theory

We begin by quoting (mostly without proof) some relevant results from the standard formulation of quantum-mechanical perturbation theory. Details are to be found in textbooks (see e.g. Refs. [7,8]). Perturbation theory is useful in situations where the total Hamiltonian H of a system can be written as

$$H = H_0 + H_1, \tag{A.1}$$

and Schrödinger's equation, namely

$$i\frac{\partial|\Psi\rangle}{\partial t} = H|\Psi\rangle, \qquad (A.2)$$

can be solved *exactly* for the "unperturbed" part H_0. Here $|\Psi\rangle$ denotes the total wave function and we employ units such that $\hbar = 1$. The small perturbation is described by H_1, and this term may or may not (for any particular application) depend on time t.

Assuming H_0 to be independent of time, as is usually the case, we specify its stationary states using

$$H_0|\psi_n^0\rangle = E^0|\psi_n^0\rangle \quad \text{for } n = 1, 2, \ldots, \qquad (A.3)$$

where $|\psi_n^0\rangle$ are the time-independent Schrödinger eigenfunctions, with E_n^0 being the corresponding energy eigenvalues. The total unperturbed wave functions are thus

$$|\Psi_n^0\rangle = |\psi_n^0\rangle \exp(-iE_n^0 t). \qquad (A.4)$$

Time-independent perturbation theory (the case where the perturbation H_1 does not depend on t) then proceeds by writing down Schrödinger's equation for the full Hamiltonian $H = H_0 + H_1$ and collecting together terms up to a chosen order in the small parameter associated with H_1. Some standard results (in terms of matrix elements of the perturbation) are

$$|\psi_n\rangle = |\psi_n^0\rangle + \sum_{m \neq n} \frac{\langle \psi_m^0 | H_1 | \psi_n^0 \rangle}{E_n^0 - E_m^0} |\psi_m^0\rangle \qquad (A.5)$$

for the perturbed eigenfunction (up to first order in H_1), and

$$E_n = \langle \psi_n^0 | H_1 | \psi_n^0 \rangle + \sum_{m \neq n} \frac{|\langle \psi_m^0 | H_1 | \psi_n^0 \rangle|^2}{E_n^0 - E_m^0} \qquad (A.6)$$

for the energy eigenvalue (up to second order in H_1). It has been assumed that the quantum state $|\psi_n^0\rangle$ under consideration is non-degenerate, otherwise the above results must be generalized.

Suppose a quantum-mechanical system is initially (at time $t = 0$) in a particular stationary state $|\psi_n^0\rangle$ of H_0, and we want to describe its time evolution for $t > 0$. Of course, this would simply involve the factor $\exp(-iE_n^0 t)$ if the Hamiltonian comprised H_0 alone. The effect of H_1 will be to modify the wave function by "mixing" in other stationary states, i.e. there will be transitions between stationary states induced by the perturbation. When the perturbation analysis is generalized by starting from Schrödinger's time-independent equation, it is found that the transition probability per unit time (for non-degenerate states) is

$$W(n \to m) = 2\pi |\langle \psi_m^0 | H_1 | \psi_n^0 \rangle|^2 \delta(E_m - E_n), \qquad (A.7)$$

A.1. PERTURBATION THEORY

where the final term is a Dirac delta function. The above result is often called Fermi's golden rule.

We next outline a general time-dependent perturbation theory in a form that will be useful in the later sections regarding Green functions and diagrammatic expansions. The full Hamiltonian is again expressed as the sum of two terms, as in Eq. (A.1), and H_1 may be time dependent. We assume that the lowest eigenstate of H can be derived in terms of the lowest eigenstate of H_0 by adiabatically switching on the perturbation H_1 in the time interval $-\infty < t < 0$. Formally this involves replacing H_1 by

$$\lim_{s \to 0^+} \exp(-s|t|) H_1, \qquad s > 0. \tag{A.8}$$

Henceforth in this section we shall denote the unperturbed ground-state wave function by $|0\rangle$ and the exact ground-state wave function (of H) by $|0'\rangle$. It is easily proved from the respective eigenvalue equations that the shift ΔE in the ground-state energy produced by the perturbation is given exactly by

$$\Delta E = \frac{\langle 0 | H_1 | 0' \rangle}{\langle 0 | 0' \rangle}. \tag{A.9}$$

The task now is to calculate the exact ground-state wave function $|0'\rangle$, and for this we work with the perturbation in the interaction representation:

$$H_1(t) = \exp(iH_0 t) H_1 \exp(-iH_0 t) \exp(-s|t|), \tag{A.10}$$

which means that Schrödinger's time-dependent equation takes the form

$$i \frac{\partial |\Psi(t)\rangle}{\partial t} = H_1(t) |\Psi(t)\rangle \tag{A.11}$$

with the boundary condition that $|\Psi(-\infty)\rangle$ is equal to the unperturbed wave function $|\Psi^0\rangle$.

Next we make a formal definition of an operator $S(t, t')$ as

$$S(t, t') \equiv \hat{P} \left\{ \exp \left[-i \int_{t'}^{t} dt \, H_1(t) \right] \right\}, \tag{A.12}$$

where $\hat{P}$, the Dyson chronological operator, has the effect of ordering all quantities on its right in order of decreasing time. Operator S is an analog of the so-called S-matrix introduced in quantum field theory (see e.g. Ref. [9]). Eq. (A.12) is equivalent to the expanded form

$$S(t, t') = \sum_{n=0}^{\infty} \frac{(-1)^n}{n!} \int_{t'}^{t} \ldots \int_{t'}^{t} dt_1 \ldots dt_n \hat{P} \left[H_1(t_1) \ldots H_1(t_n) \right]. \tag{A.13}$$

By exploiting the properties of $S(t, t')$ we arrive at the important result that the exact ground state $|0'\rangle$ is given in terms of the unperturbed ground state $|0\rangle$ by (see e.g. Ref. [5])

$$|0'\rangle = \frac{S(t,-\infty)|0\rangle}{\langle 0|S(t,-\infty)|0\rangle}. \tag{A.14}$$

It can be demonstrated that the above result reduces in appropriate special cases to those of ordinary perturbation theory. For example, Eq. (A.6) can be obtained by keeping just the lowest few terms (up to second order in H_1) in the expansion of S. Its main advantage, however, is related to the development of a diagrammatic representation for perturbation theory (see Section A.4).

A.2 Second Quantization

The standard results of quantum-mechanical perturbation theory, such as those summarized by Eqs. (A.5)–(A.7), are often of limited use in many-body systems, partly because the unperturbed wave functions (with which the matrix elements would be evaluated) are not explicitly known in general, but also because of the complexity due to the large number of particles (10^{23} or more) that are typically involved. This is where the concept of *second quantization* for operators becomes useful.

We start by introducing the *occupation number representation*, defined with respect to the unperturbed Hamiltonian. For convenience we start with bosons and afterwards generalize to fermions. We suppose that, in specifying the total wave function of a many-particle system, there are n_1 particles in the single-particle state $|\phi_1\rangle$, n_2 particles in the single-particle state $|\phi_2\rangle$, and so on, where $\{|\phi_i\rangle\}$ denotes the complete set of single-particle states for H_0. In the occupation number representation, the total wave function is specified simply as

$$|n_1 n_2 \ldots n_i \ldots\rangle. \tag{A.15}$$

This wave function is assumed to be already properly orthogonalized and normalized, i.e.

$$\langle n'_1 n'_2 \ldots | n_1 n_2 \ldots\rangle = \delta_{n_1,n'_1}\delta_{n_2,n'_2}\ldots. \tag{A.16}$$

Next we define a *creation operator* and an *annihilation operator*, denoted by $a_i^\dagger$ and a_i, respectively, to have the properties that

$$a_i^\dagger |n_1 n_2 \ldots n_i \ldots\rangle = \sqrt{n_i+1}|n_1 n_2 \ldots n_i+1 \ldots\rangle, \tag{A.17}$$

$$a_i |n_1 n_2 \ldots n_i \ldots\rangle = \sqrt{n_i}|n_1 n_2 \ldots n_i-1 \ldots\rangle. \tag{A.18}$$

By definition, they increase or decrease the occupation number of single-particle state $|n_i\rangle$ by unity, leaving other states unaffected. The prefactors on the right-hand side are chosen to preserve the normalization.

The motivation for defining this pair of operators comes from the well-known operator treatment of the 1D simple-harmonic oscillator in quantum mechanics (see e.g. Ref. [8]). In that case the operators can be constructed explicitly as a linear combination of the position and momentum operators. In many cases the state label i is just the wavevector $\vec{k}$.

A.2. SECOND QUANTIZATION

From the definitions in Eqs. (A.17) and (A.18) it is easily proved that

$$a_i^\dagger a_i |n_1 n_2 \ldots n_i \ldots\rangle = n_i |n_1 n_2 \ldots n_i \ldots\rangle, \quad (A.19)$$

$$a_i a_i^\dagger |n_1 n_2 \ldots n_i \ldots\rangle = (n_i + 1) |n_1 n_2 \ldots n_i \ldots\rangle. \quad (A.20)$$

Two important conclusions follow. Firstly, $a_i^\dagger a_i$ behaves as a *number operator*, i.e. when it operates on the general wave function it gives the occupation number n_i of state $|n_i\rangle$. Secondly, by subtracting the above two equations we conclude that $a_i a_i^\dagger - a_i^\dagger a_i = 1$, or $[a_i, a_i^\dagger] = 1$ in commutator notation. The latter property is easily generalized to

$$[a_i, a_j^\dagger] = \delta_{i,j} \qquad \text{for bosons.} \quad (A.21)$$

The case of fermions is more complicated (see e.g. Refs. [2,3]) because the Pauli principle restricts the occupation numbers of each single-particle state to 0 or 1. Within these limitations we can again introduce $a_i^\dagger$ and a_i as creation and annihilation operators (e.g. $a_i^\dagger$ acts on the state $|n_i = 0\rangle$ to produce the state $|n_i = 1\rangle$, but it gives a null result when it acts on the state $|n_i = 1\rangle$). Hence the definitions in Eqs. (A.17) and (A.18) must be modified, but the important conclusions are:

(i) the combination $a_i^\dagger a_i$ still plays the role of a number operator;

(ii) the commutation relations of the boson case are replaced by anticommutation relations.

Thus for fermions, Eq. (A.21) is replaced by

$$\{a_i, a_j^\dagger\} = \delta_{i,j}, \quad (A.22)$$

where, by definition, $\{A, B\} \equiv AB + BA$ for any two operators A and B.

To conclude this section, we quote the form of a standard Hamiltonian for weakly interacting boson or fermion particles in the second-quantization notation (e.g. Ref. [5]):

$$H = \sum_{\vec{k}} \epsilon_{\vec{k}} a_{\vec{k}}^\dagger a_{\vec{k}} + \frac{1}{2} \sum_{\vec{k}_1, \vec{k}_2, \vec{q}} v(\vec{q}) a_{\vec{k}_1}^\dagger a_{\vec{k}_2}^\dagger a_{\vec{k}_2 + \vec{q}} a_{\vec{k}_1 - \vec{q}} \quad (A.23)$$

This is expressed in a wavevector representation. The first term on the right-hand side, which is quadratic in the operators, represents the total kinetic energy of the system, with $\epsilon_{\vec{k}} = k^2/2m$ being the kinetic energy of an individual particle with mass (or effective mass) m. The second term, which is quartic in the operators, represents the total potential energy due to a pairwise potential v between the particles. The potential (in real-space variables) is assumed to be a function of the distance between any pair of particles, and $v(\vec{q})$ is a Fourier component in the wavevector representation.

A.3 Basic Properties of Green Functions

Corresponding to any two operators A and B in quantum mechanics, we may now define the Green functions $\langle\langle A(t); B(t')\rangle\rangle$ by

$$\langle\langle A(t); B(t')\rangle\rangle = i\theta(t - t')\langle[A(t), B(t')]_\varepsilon\rangle, \tag{A.24}$$

where t and t' are time labels, and $\theta(t - t')$ is the unit step function defined as 1 for $t > t'$ and 0 for $t < t'$. The operators are in the Heisenberg representation, e.g.

$$A(t) = \exp(iHt) A \exp(-iHt) \tag{A.25}$$

with H denoting the *full* Hamiltonian of the system. The angular brackets $\langle \ldots \rangle$ denote a thermal average of the enclosed quantity, evaluated according to equilibrium statistical mechanics, and

$$[A(t), B(t')]_\varepsilon \equiv A(t)B(t') - \varepsilon B(t')A(t) \tag{A.26}$$

denotes either a commutator (if $\varepsilon = 1$) or an anticommutator (if $\varepsilon = -1$). The definition in Eq. (A.24) is equivalent to that introduced by Zubarev [10], the so-called retarded Green functions, except for the overall sign. It can be seen that the definition of the Green function contains thermal averages of products of operators; thus it can provide information about the correlation functions for the system.

Although it is not immediately obvious from the definition, $\langle\langle A(t); B(t')\rangle\rangle$ depends on t and t' only through the time difference $t - t'$. This convenient property then allows Fourier components $\langle\langle A; B\rangle\rangle_\omega$ of the time-dependent Green function to be defined by

$$\langle\langle A(t); B(t')\rangle\rangle = \int_{-\infty}^{\infty} \langle\langle A; B\rangle\rangle_\omega \exp\left[-i\omega(t - t')\right] d\omega. \tag{A.27}$$

Except in certain special cases, the Green function $\langle\langle A; B\rangle\rangle_\omega$ can be evaluated only by making some approximations. Various different approaches can be used, but one convenient method is based on writing down an equation of motion satisfied by the Green function. This may be obtained by differentiating Eq. (A.24) with respect to one of its time labels (say t). The result is

$$\frac{d}{dt}\langle\langle A(t); B(t')\rangle\rangle = i\delta(t - t')\langle[A, B]_\varepsilon\rangle - i\langle\langle[A(t), H]; B(t')\rangle\rangle. \tag{A.28}$$

Two properties used in obtaining the above expression are:

(i) the time derivative of the step function θ is a delta function;

A.3. BASIC PROPERTIES OF GREEN FUNCTIONS

(ii) the operator equation of motion [e.g. as proved by differentiating Eq. (A.25)] is

$$\frac{d}{dt}A(t) = -i[A(t), H]. \qquad (A.29)$$

Next, when Eq. (A.28) is re-expressed in terms of frequency labels, it yields the standard result (see e.g. Ref. [10])

$$\omega\langle\langle A; B\rangle\rangle_\omega = \frac{1}{2\pi}\langle[A, B]_\varepsilon\rangle + \langle\langle[A, H]; B\rangle\rangle_\omega. \qquad (A.30)$$

We notice that this produces (on the right-hand side) another Green function $\langle\langle[A, H]; B\rangle\rangle_\omega$, for which another equation of motion can be written down, and so on. This hierarchy of equations, generated from Eq. (A.30), can be terminated at any chosen stage by introducing a *decoupling approximation* so as to simplify the Green function in the final equation and yield a closed set of equations, which may be solved for the original Green function. An example of a decoupling approximation to simplify a product of operators is the random phase approximation (RPA), which was introduced for magnons in terms of the operator equation of motion in Section 1.7.

Another method to deduce approximate results for the Green functions is provided by *linear-response theory*. This is useful within a perturbation approach, where the Hamiltonian H can be expressed as the sum of two terms H_0 and H_1 as in Eq. (A.1). Again we will just quote the results that are relevant to this book; details are to be found in any of the general references listed at the beginning of this Appendix. Here we take H_0 to be independent of time, while H_1 is an external perturbation of the form

$$H_1(t) = -Bf(t). \qquad (A.31)$$

In this case $f(t)$ denotes an external field that couples linearly to a system variable represented by an operator B. For example, if B represents an atomic displacement then $f(t)$ is just the mechanical force. The basic idea in linear-response theory is that switching on the perturbation term H_1 will lead to a response in some other system variable represented, say, by operator A. This response can be quantified as the change $\bar{A}(t)$ produced in that variable, which in turn can be calculated by using properties of the density matrix [11]. Using the definition of a Green function in Eq. (A.24) for the commutator case ($\varepsilon = 1$) and neglecting second-order small quantities in the perturbation, the result for $\bar{A}(t)$ can eventually be expressed as

$$\bar{A}(t) = \int_{-\infty}^{\infty} \langle\langle A(t); B(t')\rangle\rangle f(t')\, dt'. \qquad (A.32)$$

This takes a simpler and more useful form in terms of frequency Fourier components, namely it becomes

$$\bar{A}_\omega = \langle\langle A; B\rangle\rangle_\omega f_\omega. \qquad (A.33)$$

Here we denote

$$f_\omega = \int_{-\infty}^{\infty} f(t) \exp(i\omega t)\, dt, \tag{A.34}$$

and there is a similar definition for $\bar{A}_\omega$ as the Fourier transform of $\bar{A}(t)$. From Eq. (A.33) we see that the Green function $\langle\langle A; B\rangle\rangle_\omega$ in the frequency representation plays the role of a linear-response function. In fact, this often provides a convenient and direct method for calculating the Green function (see examples given in Chapter 11).

In surface- and interface-related problems it is often useful to consider an external field that has the form corresponding to a delta-function stimulus $f(t)\delta(\vec{r}-\vec{r'})$ at position $\vec{r'}$ within the system. The interaction energy with a position-dependent operator $B(\vec{r})$ then becomes

$$H_1 = -\int B(\vec{r}) f(t) \delta(\vec{r}-\vec{r'}) d^3\vec{r} = -B(\vec{r'}) f(t). \tag{A.35}$$

Therefore, from the linear response produced in another operator $A(\vec{r})$ at position $\vec{r}$, we are able to deduce the position-dependent Green function $\langle\langle A(\vec{r}); B(\vec{r'})\rangle\rangle_\omega$ (see examples of this approach applied to electric-field Green functions in Chapter 11). Details may be found in Ref. [6].

It now remains for us to explain the connection between Green functions, as mathematical quantities, and what is actually measured in an experiment, which can usually be interpreted in terms of a correlation function. For example, in light scattering from a polariton the experiment consists of measuring a scattering cross section that can be expressed as a correlation function between two electric-field components (see e.g. Section 11.1 and Ref. [12]). Likewise, a thermodynamic property, such as the specific heat, can be related to a correlation function involving the dynamic variables.

At the beginning of this section we commented that the definition of the Green function in Eq. (A.24) involved thermal averages of time-dependent products of operators, i.e. the correlation functions $\langle A(t)B(t')\rangle$ and $\langle B(t')A(t)\rangle$. These two quantities are simply related, as might be expected because they involve the same two operators. The relationship can be expressed most concisely in terms of the frequency Fourier components $\langle AB\rangle_\omega$ and $\langle BA\rangle_\omega$, which are defined by analogy to Eq. (A.27); it can then be shown that [10]

$$\langle BA\rangle_\omega = \exp(\beta\omega)\langle AB\rangle_\omega, \tag{A.36}$$

where we define $\beta = (k_B T)^{-1}$ with k_B being Boltzmann's constant. Each of these correlation functions can be deduced from the corresponding Green function by means of the so-called *fluctuation–dissipation theorem* (see e.g. Ref. [10]), which states that

$$\langle BA\rangle_\omega = (1/\pi)\Big[n(\omega)+1\Big]\mathrm{Im}\langle\langle A; B\rangle\rangle_{\omega-i0^+}, \tag{A.37}$$

$$\langle AB\rangle_\omega = (1/\pi)n(\omega)\mathrm{Im}\langle\langle A; B\rangle\rangle_{\omega-i0^+}, \tag{A.38}$$

where $n(\omega) \equiv [\exp(\beta\omega) - 1]^{-1}$ is the Bose–Einstein thermal factor. At high temperatures (where $\beta\omega \ll 1$ or $k_B T \gg \hbar\omega$) both of the above expressions reduce to the classical limit

$$\langle AB \rangle_\omega = \langle BA \rangle_\omega = (k_B T/\pi\omega)\mathrm{Im}\langle\langle A; B \rangle\rangle_{\omega - i0^+}. \tag{A.39}$$

An identity that is often useful in taking the imaginary part of the Green function (at a frequency slightly displaced off the real axis of the variable) is

$$\frac{1}{\omega - \omega_0 - i0^+} = \mathrm{P}\left(\frac{1}{\omega - \omega_0}\right) + i\pi\delta(\omega - \omega_0), \tag{A.40}$$

where $\mathrm{P}(\ldots)$ is the principal value of the argument shown. Its presence here denotes that it should be taken in any integration over ω. The imaginary contribution from a term like that on the left-hand side of the above expression is therefore proportional to $\delta(\omega - \omega_0)$, i.e. there is a delta-function spike at the value ω_0 corresponding to a pole of the Green function.

A related property of the Green function $\langle\langle A; B \rangle\rangle_\omega$ is that it provides a description of the excitation spectrum, so its evaluation gives an alternative method to deduce the dispersion relations. Specifically the poles of the Green functions (or, in other words, the ω-values at which their denominator vanishes) correspond to the dispersion relations of the excitations. The use of the fluctuation–dissipation theorem provides us with additional information regarding the spectral intensities of the poles (i.e. the contributions of the frequency-dependent correlation functions).

A.4 Diagrammatic Perturbation Theory

So far, we have discussed equation-of-motion methods and linear-response theory as techniques for evaluating Green functions. Both methods are useful but they have their limitations when higher-order effects (such as damping terms in the dielectric response and non-linear processes in general) need to be considered. Another Green-function method, which is appropriate to these situations, is diagrammatic perturbation theory (or Feynman-diagram methods). We shall present just a few basic ideas here, while there are numerous textbooks giving detailed treatments (see e.g. Refs. [1–3,13,14]).

Conventionally, the Green functions of what will eventually become the diagrammatic technique are defined somewhat differently from Eq. (A.24). Instead, for any two operators A and B we take

$$\langle\langle A(\tau); B(\tau') \rangle\rangle = \langle \hat{T}_W A(\tau) B(\tau') \rangle, \tag{A.41}$$

where the operators are in a representation defined by

$$A(\tau) = \exp(H\tau) A \exp(-H\tau). \tag{A.42}$$

If we compare Eq. (A.42) with the Heisenberg representation defined in Eq. (A.25), we see that there is the replacement $it \to \tau$. Thus, although τ and τ' in the above

Green-function definition are real quantities, they can be thought of being like imaginary-time labels. The thermal averages are evaluated with respect to the full Hamiltonian H. Finally the operator $\hat{T}_W$, the Wick ordering operator, is defined in the same way as the Dyson operator $\hat{P}$ in Eq. (A.12), except that it has the added property of introducing a factor of -1 for every interchange of *fermion* operators involved in the ordering process.

Before proceeding further we make two comments. First, although the definition of the Green function is different from before, the Green functions still involve correlation functions for the operators A and B, and it can be shown that a fluctuation–dissipation theorem (see Section A.3) still applies. Second, we assume as before that the Hamiltonian H can be expressed as in Eq. (A.1), with the two terms specifically having the second-quantization form in Eq. (A.23) in terms of boson or fermion operators.

Next we note that in the non-interacting or unperturbed case (i.e. when H is everywhere replaced by H_0), the Green functions can be evaluated directly from their definition. For example, taking $A = a_{\vec{k}}$ and $B = a_{\vec{k}}^{\dagger}$ with H_0 as the first term in Eq. (A.23), we eventually find

$$\langle\langle a_{\vec{k}}(\tau); a_{\vec{k}}^{\dagger}(\tau')\rangle\rangle = \sum_{i\omega_m} G^0(\vec{k}, i\omega_m) \exp(-i\omega_m \tau), \tag{A.43}$$

where $G^0(\vec{k}, i\omega_m)$ is a Fourier-transformed Green function,

$$G^0(\vec{k}, i\omega_m) = \left(\frac{1}{\beta}\right) \frac{1}{\epsilon_{\vec{k}} - i\omega_m}. \tag{A.44}$$

The discrete (imaginary) frequencies appearing above are defined by $i\omega_m = 2mi\pi/\beta$ for bosons and $i\omega_m = (2m+1)i\pi/\beta$ for fermions, where $\beta = 1/k_B T$ and m takes all integer values from $-\infty$ to ∞. We see that the unperturbed Green function G^0 has a simple pole at the energy $\epsilon_{\vec{k}}$ of a free particle.

We might anticipate that, when the interaction term in Eq. (A.23) is included in a fuller calculation of the Green function, the resulting expression has a modified denominator. This does indeed turn out to be the case. The new Green function, denoted by $G(\vec{k}, i\omega_m)$ in the frequency representation, is usually found to have a denominator of the form

$$\epsilon_{\vec{k}} + \Sigma(\vec{k}, i\omega_m) - i\omega_m, \tag{A.45}$$

where the additional term denoted by Σ is referred to as the proper (or irreducible) *self-energy*. The new pole (or poles) of the Green function are obtained as the self-consistent solutions for the term in Eq. (A.45) vanish. If the perturbation is weak (so that Σ is small), the approximate solution for the frequency becomes $\epsilon_{\vec{k}} + \Sigma(\vec{k}, \epsilon_{\vec{k}})$. In general, however, Σ may be a complex function and we write $\Sigma = \Lambda + i\Gamma$. Thus the interpretation of the above-stated result for the Green function G is that free-particle energy $\epsilon_{\vec{k}}$ becomes shifted (or *renormalized*)

A.4. DIAGRAMMATIC PERTURBATION THEORY

to $\epsilon_{\vec{k}} + \Lambda(\vec{k}, \epsilon_{\vec{k}})$ due to the interactions, while the imaginary term $\Gamma(\vec{k}, \epsilon_{\vec{k}})$ represents the *damping* (or inverse lifetime) of the interacting particle.

The final step is to specify how the interaction terms are calculated. This can be developed in a rather analogous fashion to the time-dependent perturbation analysis described at the end of Section A.1. In particular, an operator S is introduced playing the same role as the quantity in Eq. (A.12); here it is defined formally by $\exp(-\beta H) = \exp(-\beta H_0) S(\beta)$. Using Eq. (A.1) it can be proved that this has the solution

$$S(\beta) = 1 + \sum_{n=1}^{\infty} \frac{(-1)^n}{n!} \int_0^\beta \ldots \int_0^\beta d\tau_1 \ldots d\tau_n \hat{T}_W \{H_1(\tau_1) \ldots H_1(\tau_n)\}, \quad (A.46)$$

which has a close resemblance to the expansion in Eq. (A.13) except that the τ representation and the $\hat{T}_W$ operator are used. The analog of Eq. (A.14) is that the partition function Z evaluated with the full Hamiltonian H is related to the corresponding Z_0 of the unperturbed system by

$$Z = Z_0 \langle S(\beta) \rangle_0, \quad (A.47)$$

where the thermal average is taken using H_0.

It can be seen by substituting Eq. (A.46) into Eq. (A.47) that the evaluation of Z (from which all other thermal equilibrium properties of the system can be deduced) requires the evaluation of unperturbed averages like $\langle \hat{T}_W \{H_1(\tau_1) \ldots H_1(\tau_n)\} \rangle_0$ in nth order of the perturbation scheme. This can be achieved through an important result known as *Wick's theorem*, which allows one to break down the $\hat{T}_W$-ordered product of $4n$ boson or fermion operators [assuming the Hamiltonian in Eq. (A.23)] into $\hat{T}_W$-ordered products involving the operators in *pairs* only. The latter quantities are just the unperturbed Green functions [see Eqs. (A.41) and (A.42) with H replaced by H_0]. It is a rather formidable task to carry this out algebraically, and so a diagrammatic representation (in terms of the so-called Feynman diagrams) is almost always used instead. We show the main elements of this in Fig. A.1: the Green functions G^0 are drawn as directed lines as in Fig. A.1a, and the interaction vertices (representing each H_1 term in the perturbation expansion) can then be drawn as in Fig. A.1b. The higher-order diagrammatic contributions to G then consist of one ingoing and one outgoing line with n interaction vertices in between, e.g. see Figs. A.1c and d for the simplest $n=1$ diagrams. A straightforward set of rules can be specified for the evaluation of these diagrammatic contributions. Details are given in the books cited earlier.

In summary, the diagrammatic method offers a rigorous approach (compared to the alternatives previously mentioned), because one may control the degree of approximation by going up to any chosen order n of perturbation. This is important for calculating some of the non-linear effects mentioned in Chapter 12 and studying the origin of the damping terms (in the dielectric functions) that we introduced phenomenologically into earlier discussions.

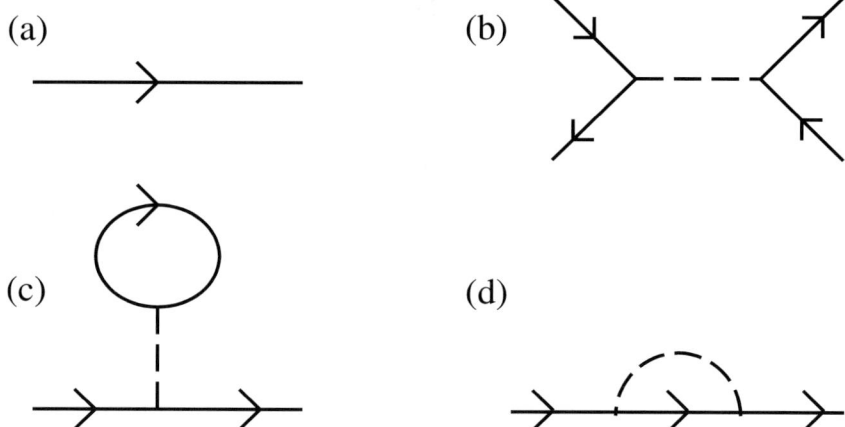

Fig. A.1. Diagrammatic representation: (a) unperturbed Green function G^0; (b) interaction vertex with the dashed line representing $v(\vec{q})$ in Eq. (A.23); (c, d) first-order diagrammatic contributions to the full Green function G.

References

[1] A.L. Fetter and J.D. Walecka, Quantum Theory of Many-Particle Systems, McGraw-Hill, New York, 1971.
[2] E.N. Economou, Green's Functions in Quantum Physics, Springer, Berlin, 1979.
[3] G. Rickayzen, Green's Functions and Condensed Matter, Academic, London, 1980.
[4] J.W. Negele and H. Orland, Quantum Many-Particle Systems, Addison-Wesley, Redwood City, 1988.
[5] C. Kittel, Quantum Theory of Solids, 2nd ed., Wiley, New York, 1987.
[6] M.G. Cottam and A.A. Maradudin, in: Surface Excitations, Eds., V.M. Agranovich and R. Loudon, North-Holland, Amsterdam, 1984.
[7] J.J. Sakurai, Modern Quantum Mechanics, Benjamin Cummings, Menlo Park, 1985.
[8] B.H. Bransden and C.J. Joachain, Quantum Mechanics, 2nd ed., Prentice-Hall, Englewood Cliffs, 2000.
[9] F. Mandl and G. Shaw, Quantum Field Theory, Wiley, New York, 1984.
[10] D.N. Zubarev, Sov. Phys. - Usp. 3 (1960) 320.
[11] L.D. Landau and E.M. Lifshitz, Statistical Physics, Pergamon, Oxford, 1980.
[12] W. Hayes and R. Loudon, Scattering of Light by Crystals, Wiley, New York, 1978.
[13] A.A. Abrikosov, L.P. Gorkov and I.E Dzyaloshinski, Methods of Quantum Field Theory in Statistical Physics, Dover, New York, 1975.
[14] R.D. Mattuck, A Guide to Feynman Diagrams in the Many-Body Problem, Dover, New York, 1992.

Subject Index

Additional boundary conditions, 51, 251, 289
   dead-layer model, 52
   see also exciton-polaritons
Anisotropic media, 42, 175
Antiferromagnets/antiferromagnetic, 22, 59, 61, 159, 162, 177, 179, 195, 197, 200, 202, 203, 205, 215, 231
   bilayer system, 197
   canted, 197, 204, 206, 209
   CoF_2, 197
   FeF_2, 203, 197, 207, 209
   films, 166, 209
   magnetic arrangement, 179, 195
   magnetic susceptibility, 54, 57
   MnF_2, 189, 198, 201, 208, 209
   non-reciprocal, 167
   non-linear effects, 313, 321
   resonance frequency, 56, 61, 159, 163, 189, 200, 205
   spin structure, 196
   superlattices, 171
   uniaxial, 55, 60, 162, 167, 187
   XF_2, 197, 204, 209
   see also spin waves
Attenuated total reflection, 83, 300
   critical angle, 83
   evanescent mode, 83
   nipi superlattice, 300
   Otto configuration, 84
   rare-earth superlattice, 303
   reflection coefficient, 300
   scan line, 301
   theory, 300

Bilinear exchange term, 168, 215, 216, 220, 226, 230, 237
   see also metallic magnetic multilayers
Biquadratic exchange term, 168, 216, 220, 226, 230, 234, 237
   see also metallic magnetic multilayers
Bloch wavevector, 93, 113
Bloch's theorem, 1, 7, 160, 271
Boundary conditions
   charge-sheet modes, 72
   elastic, 112, 114
   electromagnetic, 66, 74, 99, 106, 112, 114, 183, 187, 252, 261, 271, 278, 283, 290, 294, 315

electrostatic, 73, 112, 114
   see also additional boundary conditions
Bragg diffraction, 6
Bravais lattices, 2, 3, 6, 9, 10
   primitive cell, 2
   translation vector, 2, 8
   unit cell, 3, 12
Brillouin light scattering, 286
   Fabry–Pérot interferometer, 288
   frequencies, 288
   principles, 289
   see also resonant Brillouin scattering
Brillouin zones, 4, 7, 9, 26, 27, 54, 288
   mini-zones, 26, 90
   wavevector, 157, 197
Building blocks, 90
   see also superlattice

Cantor set, 30, 144
Charge-sheet, 72
   model, 72
   surface polariton dispersion relation, 73
Curie temperature, 20, 171
   see also Néel temperature
Cyclotron frequency, 117, 119, 263

Dielectric function, 13, 42
   doped polar semiconductor, 44
   electron gas, 13
   exciton-polaritons, 49
   ionic crystal, 43
Dielectric thin films, 73
   anisotropic modes, 75, 80
   guided-wave polaritons, 75
      p-polarized modes, 76
      s-polarized modes, 77
   surface polaritons, 75
Dipole–dipole interactions, 60, 157
Dipole-exchange region, 158
Dispersive Fourier-transform spectroscopy, 306

Elastic tensor, 111
Exchange region, 61, 158
Exchange term, 22, 157, 172, 177, 199
   antiferromagnetic interlayer, 181, 185
   ferromagnetic intralayer, 179
   inter-sublattice coupling, 57
   surface contribution, 159

339

Exciton, 1, 14, 48, 249
  A exciton mode, 250
  Frenkel exciton, 15, 16
  Wannier–Mott exciton, 15, 16, 256
  see also exciton-polaritons
Exciton-polaritons, 48, 249
  p-polarized mode, 254, 257
  s-polarized mode, 255
  CdS, 50
  dielectric approximation, 49
  dispersion relation, 257, 262
  effective dielectric tensor, 261
  effective mass, 49
  effective-medium theory, 257, 259
  electric field, 251
  electrical susceptibility, 49, 238
  equation of motion, 48, 251
  GaAs/Ga$_{1-x}$Al$_x$As superlattice, 267
  GaN, 250
  longitudinal mode, 50
  magnetoexciton-polariton dispersion relation, 263
  magnetoexciton-polariton modes, 259, 260
  non-reciprocal propagation, 265
  oscillator strength, 49
  polariton lasers, 256
  polarization field, 252
  resonant frequency, 49
  sapphire substrate, 253
  secular equation, 252
  spatial dispersion, 49
  spatial-dispersion term, 249
  superlattice, 252
  thin films, 250
  transverse modes, 49

Far-infrared Fourier-transform spectroscopy, 306
Ferromagnetic resonance curves, 231
  cubic anisotropy energy, 233
  dipolar energy, 233
  Fibonacci sequence, 237
    spin-wave acoustic modes, 237
    spin-wave optical modes, 238
  surface anisotropy energy, 233
  total magnetic energy, 233
  see also metallic magnetic multilayers
Ferromagnetic resonance spectroscopy, 307
  Cu/Ni films, 307
  magnetic anisotropy energies, 307
  microwave cavity, 307
Ferromagnets/ferromagnetic, 18, 22, 157, 159, 161, 162, 187
  dispersion relation, 20
  EuS, 188
  films, 166, 170
  interlayer exchange, 307
  magnetic arrangement, 179
  magnetic susceptibility, 58, 163
  non-linear effects, 313
  quantized bulk modes, 167
  resonance frequency, 20, 162
  semiconductor, 307
  superlattices, 168
  YIG, 197
  see also spin waves
Feynman diagram methods, 335
  Wick operator, 336
Fourier-transform infrared spectrometer, 306
Fractals, 34
  fractal dimension, 34

Green functions, 61, 296
  Bose–Einstein thermal factor, 269, 335
  correlation functions, 334
  definition, 332
  fluctuation-dissipation theorem, 296, 334
  Heisenberg representation, 332
  linear-response theory, 333
  thermal average, 334
  Zubarev notation, 332

Heisenberg Hamiltonian, 18, 19, 161, 198
  exchange term, 18
  Zeeman term, 18

Light-emitting tunnel junction, 305
  Al/Al$_2$O$_3$/Ag layered structure, 305
  diffraction grating, 305
Lyddane–Sach–Teller (LST) relation, 44

Magnetic superlattices, 171, 306
Magnetic thin films
  antiferromagnetic peak, 226, 237
  ferromagnetic–antiferromagnetic transition, 237
Magnetic-field induced surface spin wave, 195
Magnetic polaritons, 58, 157, 159
  s-polarized modes, 187
  bulk antiferromagnetic mode, 61
  bulk ferromagnetic mode, 60
  bulk modes, 59
  circularly polarized modes, 59
  Double-period sequence, 186

fractal spectra, 190
  power law, 190
 electromagnetic region, 158
 Fibonacci sequence, 186
  fractal spectra, 189
  power law, 189
 retarded thin-film surface mode, 183, 184
 transfer matrix, 188
 unretarded thin-film surface mode, 183, 184
 Voigt geometry, 187
 Voigt permeability, 183, 187, 205
Magnetization profiles
 Double-period sequence, 224
 Fibonacci sequence, 224
 gradient method, 225
 magnetic phases, 228
 self-similar pattern, 226, 228
 simulated annealing method, 225
 see also metallic magnetic multilayers
Magneto-optical Kerr effect, 306
Magnetoplasmon-polariton excitations, 116
 nipi superlattice, 120
 bulk mode dispersion relations, 118
 surface mode dispersion relations, 118
 see also plasmon-polaritons
Magnetoresistance, 215, 217, 219, 223
 Fibonacci sequence, 217, 224
 fractal properties, 224
 giant, 215
 phase transitions, 222
 self-similar pattern, 217, 221
 see also metallic magnetic multilayers
Magnetostatic modes, 161
 backward bulk modes, 165
 Damon–Eshbach surface mode, 163, 175, 195
 forward bulk modes, 167
 magnetic superlattices
  bulk modes, 170
  surface modes, 170
 magnetostatic potential, 164
 magnetostatic region, 61, 158
 non-reciprocal propagation, 165
 thin films, 161, 163, 164, 165
  bulk modes, 165
  surface modes, 165
 Voigt configuration, 164
Maxwell's equations, 45
 magnetostatic modes, 163
Maxwell–Boltzmann statistics, 145
Metallic magnetic multilayers, 215, 216, 224
 90° phase, 220
 anisotropy energy, 218

antiferromagnetic peak, 220
antiferromagnetic phase, 221, 222
antiferromagnetic–ferromagnetic transition, 220
bilinear energy, 218
biquadratic energy, 218
equilibrium orientations, 219
$Fe/Cr(100)$ structures, 217
Fibonacci sequence, 217
first-order phase transitions, 221
normalized magnetization, 219
saturated phase, 221
total magnetic energy, 219
total saturation magnetization, 219
ultrathin films, 217
Zeeman energy, 218
Metamagnetic materials, 177
 antiferromagnetic interlayer coupling, 179
 antiferromagnetic phase, 179, 185
 antiferromagnetic resonance frequency, 189
 effective field, 182
 $FeBr_2$, 179, 184
 $FeCl_2$, 179, 185
 ferromagnetic intralayer exchange, 179
 ferromagnetic phase, 179, 184
 ferromagnetic resonance frequency, 179, 184
 hexagonal structure, 179
 Lorentz term, 182
 magnetic susceptibility, 182
 spin Hamiltonian, 180
  anisotropy term, 181
  Zeeman term, 180
 spin waves, 179
 spin-wave antiferromagnetic phase, 181
 spin-wave ferromagnetic phase, 181
 thin films, 177
  guided modes, 179
  surface magnetic-polaritons, 179
 weak antiferromagnetic exchange, 177
Michelson interferometer, 306
Miller indices, 4

Néel temperature, 171, 199
Non-linear effects, 311
 bistability, 316
 dielectric function, 313
 dielectric media, 311
 double-interface systems, 317
 p-polarized wave, 316
 polarization vector, 312
 power flow, 315
 Poynting vector, 315

refractive index, 313
s-polarized wave, 314, 318
single-interface geometries, 314
solitons, 322
susceptibility tensor, 312
two-component superlattices, 322, 323

Perturbation theory, 327
  Dyson operator, 329
  eigenfunction, 328
  eigenvalue, 328
  Fermi's golden rule, 329
  Hamiltonian, 327
  interaction representation, 329
  time-dependent, 328
  time-independent, 328
Phonon, 1, 7, 9, 15, 18, 20, 21
  acoustic branch, 11, 12
  interface modes, 306
  optical branch, 12
  see also phonon-polaritons
Phonon-polaritons, 12
  bulk-modes spectra, 47
  single-interface mode, 71
Photon, 42
Piezoelectric materials, 107
  cubic symmetry, 109
    bulk solutions, 111
  equation of motion, 109
  hexagonal 6 mm symmetry, 109
    bulk solutions, 112
  piezoelectric tensor, 110
  superlattice
    surface plasmon-polariton modes, 114
    transfer matrix, 113
Plasma oscillation, 12
  frequency, 14
Plasmon, 1, 13, 89, 305
  see also plasmon-polaritons
Plasmon-polaritons, 14
  bulk mode spectrum, 46
Plasmon-polaritons in periodic superlattices, 89
  charge sheets, 101
    bulk and surface modes spectra, 102
    retarded modes dispersion relation, 103
    surface modes dispersion relation, 103
    unretarded modes dispersion relation, 103
  finite superlattices, 97
    dispersion relation, 100
    quantization of the bulk modes, 100
  nipi structures, 106
    bulk and surface modes spectra, 106
    transfer matrix, 107
  two-component
    bulk modes dispersion relation, 96
    p-polarized mode, 90
    surface modes dispersion relation, 95
    transfer matrix, 92
    unretarded modes dispersion relation, 93
Plasmon-polaritons in quasiperiodic structures, 125
  Cantor sequence, 135
    spectra, 133
    transfer matrix, 128
  Double-period sequence, 127
    fractal spectra, 132
    power law, 136
    transfer matrix, 127
  Fibonacci sequence, 126
    fractal spectra, 130
    power law, 134
    transfer matrix, 126
  multifractal analysis, 134
    $f(\alpha)$ function, 135
  nipi structures, 139
    fractal spectra, 141
    power law, 141
    transfer matrix, 140
  Thue–Morse sequence, 127
    fractal spectra, 131
    power law, 135
    transfer matrix, 127
Polariton, 1, 14, 41
  anisotropic media, 71
  see also exciton-polaritons
  see also magnetic-polaritons
  see also phonon-polaritons
  see also plasmon-polaritons
Polariton region, 60

Quasiperiodic structures, 30
  Cantor sequence, 33
  Double-period sequence, 37
  Fibonacci number, 35
  Fibonacci sequence, 34
  generalized Fibonacci structures, 35
  golden mean number, 35
  inflation rules, 33
  Pisot–Vijayraghavan (PV) irrational number, 36
  quasicrystals, 30
  substitutional sequences, 32
  Thue–Morse sequence, 36

SUBJECT INDEX

Raman light scattering, 267, 306
  differential cross section, 269
  electric-field Green functions, 269
  experimental geometry, 268
  integrated intensity, 271
  nipi superlattice, 281
  polarization vectors, 269
  power spectrum, 277
  quasiperiodic superlattice, 282
    cross section, 285
    integrated intensity, 285
  two-component superlattices, 274
Random-phase approximation, 19, 159, 199
Rare-earth superlattices, 171, 172
  cone state, 171
  dielectric function, 173
  dysprosium, 172
  effective dielectric tensor, 174
  effective magnetic permeability tensor, 174
  effective-medium theory, 174
  equilibrium condition, 174
  gyrotropic magnetic permeability tensor, 173
  holmium, 172
  magnetic susceptibility, 182
  magnetostatic bulk modes, 175
  magnetostatic surface modes, 175
  non-reciprocal propagation, 177
  resonance frequencies, 173
  retarded bulk modes, 175
  retarded surface modes, 175
  RKKY interaction, 172
  spiral state, 171
  Voigt geometry, 173
Reciprocal lattices, 5, 6
  primitive vectors, 6
  vector, 5
Resonant Brillouin scattering, 55, 289
  anti-Stokes process, 54, 293
  cross section, 295
  deformation potential, 293
  interbranch scattering, 299
  intrabranch scattering, 298
  kinematics, 54, 290, 291
  lineshapes, 297
  p-polarized mode, 290
  reflection spectra, 289, 291
  resonant peaks, 298
  s-polarized mode, 289, 294
  scattered radiation field, 294
  Stokes process, 54, 293
  transmission spectra, 289, 291
Resonant Rayleigh scattering, 299

Second quantization, 330
  annihilation operator, 330
  creation operator, 330
  number operator, 331
  occupation number representation, 330
Spin waves, 18, 201–203, 286, 307
  bulk ferromagnetic dispersion relation, 20
  dipole exchange, 165
  Fibonacci fractal spectra, 242
  inelastic neutron scattering, 21
  magnetic superlattices, 168
    bulk modes, 169
    surface modes, 169
  magnon, 1, 18, 21
  surface reconstruction, 168
  thin films, 159–161
    bulk frequency quantization, 160
    bulk modes, 160
    surface acoustic modes, 161
    surface modes, 161
    surface optical modes, 161
  transfer matrix, 169
Spin-canted, 195
  bulk spin-wave dispersion relation, 199
  canting angle, 199
  CoF_2, 198
    spin-wave bulk modes, 200
    spin-wave surface modes, 200
  effective field, 204
  $FeBO_3$, 195
  FeF_2, 197
    spin-wave bulk modes, 200
    spin-wave surface modes, 200
  ferrimagnetic iron oxides, 195
  geometry, 199
  magnetic permeability tensor, 204
  magnetic polaritons, 202
  magnetostatic bulk modes, 206
  magnetostatic surface modes, 206
  NiF_2, 195
  non-reciprocal propagation, 195
  resonance frequencies, 205
  retarded bulk modes, 206
  retarded surface modes, 206
  rutile structure, 204, 209
  spin dynamics, 197
  stability condition, 198, 199
  thin film FeF_2, 209
    bulk magnetic-polariton modes, 210
    non-reciprocal propagation, 211

surface/guided magnetic-polariton
  modes, 209
thin film MnF$_2$, 212
  bulk magnetic-polariton modes,
    212
  non-reciprocal propagation, 211
  surface/guided magnetic-polariton
    modes, 209
  thin films, 209
  total Hamiltonian, 198
Superlattice, 9, 26
  band offset, 28
  band gap, 28
  energy-band diagram, 27
  GaAs/Ga$_{1-x}$Al$_x$As, 27
  layer-growth methods, 26
  long-period mode, 306
  multiple quantum well, 26
  nipi, 29
  quantum well, 26
  superlattice excitation, 9
  two-component, 27
  unit cell, 93
Surface-active medium, 66

Thermodynamic properties
  plasmon-polaritons, 144

Fibonacci specific heat, 147
  even and odd parity, 149
  log-periodic behavior, 150
  spectra, 148
Thue–Morse specific heat, 150
  log-periodic behavior, 152
  spectra, 151
spin waves, 239
  Fibonacci specific heat, 243
    even and odd parity, 244, 245
    log-periodic behavior, 243
    spectra, 242
  specific heat, biquadratic influence,
    244, 245
surface plasmon-polaritons, 152
  Fibonacci specific heat spectra, 154
theoretical model, 145
Torque equation, 54
  effective magnetic field, 56
Transfer-matrix method, 92, 106, 112,
  126–128, 140, 188
Trilayer Fe/Cr/Fe structure, 215–217, 219,
  222, 231
  magnetostatic modes, 232
  spin-wave dispersion relation, 232

Wigner–Seitz primitive cell, 4